Palgrave Studies in Literature, Science and Medicine

Palgrave Studies in Literature, Science and Medicine is an exciting series that focuses on one of the most vibrant and interdisciplinary areas in literary studies: the intersection of literature, science and medicine. Comprised of academic monographs, essay collections, and Palgrave Pivot books, the series emphasizes a historical approach to its subjects, in conjunction with a range of other theoretical approaches. The series covers all aspects of this rich and varied field and is open to new and emerging topics as well as established ones.

More information about this series at
http://www.palgrave.com/gp/series/14613

David Fuller · Corinne Saunders ·
Jane Macnaughton
Editors

The Life of Breath in Literature, Culture and Medicine

Classical to Contemporary

palgrave
macmillan

Editors
David Fuller
University of Durham
Durham, UK

Corinne Saunders
Department of English Studies
University of Durham
Durham, UK

Jane Macnaughton
University of Durham
Durham, UK

ISSN 2634-6435 ISSN 2634-6443 (electronic)
Palgrave Studies in Literature, Science and Medicine
ISBN 978-3-030-74442-7 ISBN 978-3-030-74443-4 (eBook)
https://doi.org/10.1007/978-3-030-74443-4

Cover credit: The British Library, MS Royal 1 E VII, f.1v. Bible, England (Canterbury, Christ Church), 3rd quarter of 11th century. Detail of a drawing of God over the circle of the world with two long trumpets issuing from his mouth (the breath of God), holding a scale, and the Holy Spirit over the waters, enclosed within a hemisphere, at the beginning of Genesis.

This Palgrave Macmillan imprint is published by the registered company Springer Nature Switzerland AG
The registered company address is: Gewerbestrasse 11, 6330 Cham, Switzerland

Praise for *The Life of Breath in Literature, Culture and Medicine*

"*The Life of Breath* was born of a brilliantly varied years-long project in the critical medical humanities which brought together artists, humanists, medical practitioners, scientists and patients to study and perform arts and acts of breathing. This essay collection breaks new ground in establishing the foundational role of respiration in the (inter)subjective workings of desire, the interdependence of interior and exterior environments and 'conspiration'—the often-hidden commonality of breathing. Because we all breathe the same air, 'breath, intimately connected with life, connecting mind and body, opens onto profound—and timeless—ethical questions,' write the editors. Those questions are richly and magisterially addressed in essays that trace histories of living and thinking the breath, and articulate what the editors call 'the potential of the arts to help people live well with breathlessness.' This volume is required reading for anyone, in any discipline, devoted to any of the many arts of living, who recognizes the urgency, today, of returning 'to every living thing—human, plant, and animal—the space and conditions required for its breathing.'"

—Aranye Fradenburg Joy, *Psychoanalyst and Professor Emerita, University of California, Santa Barbara, USA*

"Across centuries and countries, we have one thing in common: we all breathe. But after reading this volume it is not possible to draw breath without a nuanced and awed awareness of what that breath means, how attitudes to it differ and what it means to lack or be denied breath. Ranging from Homeric epic to the twenty-first-century clinic, this mesmerizing collection investigates highly diverse topics, but even more impressive than the variety of essays is their seamless intersection. The ways in which they relate to each other is testament to the ways in which breath and breathing affect every aspect of our body, our environment and our politics. It is easy to say that this is an 'inspiring' collection but such a pun does disservice to the complexity of the topics addressed. Just as it is rare to find a collection of essays of this span so integrated, it is unusual to find a medical humanities topic that is so symbiotically valuable to both medical and humanities communities. From start to finish, this book is a remarkable achievement."

—Laurie Maguire, *Professor of English Language and Literature, University of Oxford, UK*

"An inspired collection of essays on the life, origin, meaning and metaphor of breath over the centuries in the sacred and secular worlds, a wonderful fruition of an interdisciplinary Wellcome research project *The Life of Breath*. It is timely, as breathing and breathlessness have been brought to the centre of the world's attention in the last year by the devastation and human suffering wrought by the coronavirus. It is rich, bringing to readers' attention the many interpretations of the significance of breath and breathing over millennia. It is definitive—there is no other volume that brings together such depth and breadth on this subject."

—Sara Booth, *Lecturer, University of Cambridge, and Honorary Consultant, Cambridge University Hospitals, UK*

"*The Life of Breath* is a truly breathtaking panorama of the newly-formed respiratory paradigm of our time, taking in a vast range of philosophical, psychological, religious, medical, artistic and political articulations of the quality of air. Its own rhythm orchestrates the anxious or oppressive constrictions of breath with the many positive ways of producing, unloosing and augmenting it. The volume is truly voluminous in every sense—in the reach of its themes, occasions and instances, and the giant span of its historical focus, from the classical world to the contemporary agonistics of respiration. Breath gives life, but this unbated volume reciprocally imparts new kinds of life to breath."

—Steven Connor, *Director of CRASSH, Grace 2 Professor of English, University of Cambridge, and Fellow of Peterhouse, Cambridge, UK*

Contents

List of Illustrations

Acknowledgements

Our primary acknowledgement is to the Wellcome Trust. The *Life of Breath* project (2014–20), of which this collection of essays is one principal outcome, was generously funded by Wellcome (103339/Z/13/Z), which also funds Durham University's Institute for Medical Humanities, in which the project was based. The support of the Wellcome Trust is fundamental to the Institute and all its activities.

Among individuals to be thanked, we would like first to acknowledge the *Life of Breath* project manager, Dr Sarah McLusky, who is also Manager of the Institute. Sarah's vision, initiative, and unfailingly scrupulous attention to detail were enabling and essential contributions both to the conference from which this collection originated and to the project which provided its context.

The conference, 'The Life of Breath: History, Texts, Contexts', planning for which began in late 2016, was held in St John's College Durham in July 2018: https://lifeofbreath.org/event/history-texts-contexts/. The conference was structured and run by Mary Robson, facilitator for interdisciplinary research projects carried out by the Institute. Working with a team of assistants, Mary organised a variety of modes of interactive working in which exchange within and between panels, and across the whole conference membership, enabled creative dialogue between periods and between disciplines. This conduct of the conference was fundamental to the ways in which participants considered the focus of their individual contributions in the context of the subject overall, and to the ways in which the papers were developed into essays for this collection.

Almost all the contributors to this collection were members of the 2018 conference. There was one other background to the collection—a series of public lectures and workshops, 'The Arts of Breath' (autumn 2017 to spring 2018), on breath in dance, singing, visual art, fiction, and poetry. We are grateful to contributors to that series whose lectures and workshops on the practice of performance arts were not amenable to representation in a collection such as this, but all of whom distinctly contributed to our thinking about the subject: Professor Susan Jones (Oxford), on dance; Julian Wright, Francesca Massey, and the Durham Singers, on choral singing; and Miranda Wright, with David Murray (piano), and her students, Zoë Jackson, Rebecca Roberts, and Patrick Owston, on solo singing. Podcasts from the series can be found on the project website: https://lifeofbreath.org/.

Dr. Michael Baker has assisted, in a huge variety of ways and with magnificently comprehensive care for detail, in preparing the typescript for publication. His unfailing total competence has transformed the potential torments inherent in editing into pleasant diversions.

Aspects of publication were generously supported by funding from the *Life of Breath* project and Durham University's Department of English Studies.

We have incurred many other debts in the course of publication. Allie Troyanos, Rachel Jacobe, and other editorial and production staff at Palgrave Macmillan have been unfailingly helpful, and especially in accommodating delays caused by the COVID-19 pandemic which disrupted so much of the world during the final stages of the book's preparation. It is also a pleasure to thank our original editor at Palgrave, Ben Doyle, for his initial encouragement with the project. Finally, we are grateful to our production editor at Springer Nature, Brain Halm, to Punitha Balasubramaniam and her colleagues at Scientific Publishing Services (P) Ltd., especially Shukkanthy Siva, who was unfailingly helpful in accommodating a variety of complex issues in the preparation of the text for printing.

Our greatest thanks goes to our contributors, for their original commitment to the conference, and in some cases the public lecture series, on which the book is based, and for their contributions, not only to the book as it finally appears, but also to the extensive processes of its evolution.

Notes on Contributors

Peter Adey is Professor in Human Geography at Royal Holloway University of London, and works at the intersections of space, security, and mobility. He is former Chair of the Social and Cultural Geography research group of the Royal Geographical Society with the Institute of British Geographers, has published widely in academic journals and edited collections, and is co-editor of the journal *Mobilities*. His research interests have focused primarily on air travel and the substance of air itself, as set out in his books *Aerial Life: Mobilities, Subjects, Affects* (2010), *Air: Nature and Culture* (2014), and *Levitation: the Science, Myth and Magic of Suspension* (2017). He is currently finishing the monograph *The Way We Evacuate: Emergency and the Aesthetics of Mobility*, forthcoming with Duke University Press.

Marco Bernini is Assistant Professor in Cognitive Literary Studies in the Department of English Studies at the University of Durham, UK. His research focuses on narrative theory, modernist fiction, and cognitive science. He works on the relationship between mind and narrative, and chiefly on how literary narratives explore and model cognitive processes. He has also worked on the extended mind theory and authorial agency, on empirical studies on readers, and on narrative and cognitive theories of complexity and emergence. His *Beckett and the Cognitive Method: Mind, Models and Exploratory Narratives* (2021) has recently been published by Oxford University Press.

Stephen J. Chapman is Consultant and Senior Lecturer in Respiratory Medicine at Oxford University Hospitals, UK. He has subspecialty clinical interests in adult cystic fibrosis, pneumonia, and complex lung infection and has published extensively on the subject of lung infection. He was named an 'Outstanding Young Investigator' by the American Thoracic Society and awarded the Linacre Medal by the Royal College of Physicians, London. He is co-author of the *Oxford Handbook of Respiratory Medicine* (2014) and editor-in-chief of the international respiratory medicine journal *BMJ Open Respiratory Research*.

Christine L. Corton graduated from the University of London and worked in publishing before gaining a doctorate in 2010 from the University of Kent. A revised and extended version of her thesis was published as *London Fog: The Biography* in 2015 by Harvard University Press to great acclaim. She has spoken at literary festivals all over the world including in Jaipur, Lahore, Althorp, and Hay, and has given lectures in New Zealand and Canada. She is now working on a literary history of divorce. She is a founding member of the Dickens Fellowship in Cambridge and a member of the advisory board of 'Women without Borders'. She is a Senior Member of Wolfson College, Cambridge.

Katharine A. Craik is Reader in Early Modern English Literature at Oxford Brookes University, UK. She is the author of *Reading Sensations in Early Modern England* (2007), co-editor of *Shakespearean Sensations: Experiencing Literature in Early Modern England* (2013), and editor of *Shakespeare and Emotion* (2020). She has recently published essays in *Renaissance Studies*,*Shakespeare Studies*, and *Shakespeare Survey*. Katharine led the Wellcome Trust-funded project *Watching*, a creative exploration of early modern sleep, and is a founding editor of *Beyond Criticism*, a new series of books exploring the radical forms literary criticism might take in the twenty-first century.

David Fuller is Emeritus Professor of English in the University of Durham, UK. From 2002 to 2007 he was the University's Public Orator. He is the author of monographs on Blake (1988), Joyce (1992), Shakespeare (2011), and (co-authored) literary treatments of the sacraments (1995), editor of texts by Marlowe (Clarendon), Blake (Longman), and the Pearl-poet (Enitharmon), and of two co-edited collections of essays (Oxford, Palgrave). He has written on a range of poetry, drama, and novels from Medieval to contemporary, on editorial theory, on opera, and on dance. His *Shakespeare and the Romantics* (2021) recently appeared in the series 'Oxford Shakespeare Topics'.

Abbie Garrington is Associate Professor of Modern and Contemporary Literature at the University of Durham, UK, where she is currently completing the Leverhulme-funded monograph *High Modernism: A Literary History of Mountaineering, 1890–1945*. Primarily a modernist, her research interests also include the representation of the body and of sense experience, particularly touch and the history of the human hand, in literary contexts and the wider culture in the early twentieth century. She is the author of *Haptic Modernism: Touch and the Tactile in Modernist Writing* (2013).

Patrick Gray is Associate Professor of English Studies at the University of Durham, UK. He is the author of*Shakespeare and the Fall of the Roman Republic: Selfhood, Stoicism, and Civil War* (2019), co-editor of *Shakespeare and Montaigne* (2021), editor of *Shakespeare and the Ethics of War* (2019), and co-editor of *Shakespeare and Renaissance Ethics* (2014). His essays have appeared in *Textual Practice, Comparative Drama, Shakespeare Survey, Shakespeare Jahrbuch,* and *The Journal of Medieval and Early Modern Studies*, as well as collections including *Shakespeare and the Soliloquy in Early Modern English Drama* (2018)and *Shakespeare au risque de la philosophie* (2017).

Stefanie Heine was until recently a senior researcher and Lecturer in the Department of Comparative Literature at the University of Zürich, Switzerland, and is currently Assistant Professor in the Department of Arts and Cultural Studies at the University of Copenhagen, Denmark. She studied English, Philosophy, and Comparative Literature; and completed her Ph.D. in 2012 on Virginia Woolf, Maurice Blanchot, and impressionist painting. From 2016 to 2018, she was a postdoctoral fellow at the University of Toronto (Centre for Comparative Literature), where she worked on a book project recently published as *The Poetics of Breathing: Modern Literature's Syncope* (SUNY, 2021). She is also the author of *Visible Words and Chromatic Pulse: Virginia Woolf's Writing, Impressionist Painting, Maurice Blanchot's Image* (2014) and a joint author of *Reading Breath in Literature* (2019).

Phillip Sidney Horky is Associate Professor of Ancient Philosophy and Co-Director of the Durham Centre for Ancient and Medieval Philosophy (DCAMP) at the University of Durham, UK. He is author of *Plato and*

Pythagoreanism (Oxford, 2013) and editor of *Cosmos in the Ancient World* (Cambridge, 2019), which features the fruits of other work undertaken in the *Life of Breath* project. He is now finishing a source book, *Pythagorean Philosophy: 250 BCE–200 CE* (Cambridge), and continues to investigate the intersection of philosophy, natural science, and theology in the ancient world.

Thomas E. Hunt is Senior Lecturer in Theology at Newman University, UK. He is interested in the cultural and religious history of late antiquity. His other publications on breath include 'Breathy Shame and the Place of Hebrew in the Work of Jerome of Stridon', *Religion and Theology* 26 (2019). His most recent publication is *Jerome of Stridon and the Ethics of Literary Production in Late Antiquity* (Brill, 2020). His current project traces the development of the academic study of late antiquity in France and the wider French empire between 1945 and 1962.

Rina Knoeff is Associate Professor at the Faculty of Arts and the Aletta Jacobs School of Public Health at the University of Groningen, the Netherlands. She has worked on the history of medicine, health and the body in the Enlightenment. Her recent work is on histories of healthy ageing. Taking a medical humanities perspective, she has brought these histories into conversation with today's concerns. With James Kenn-away she has edited *Lifestyle and Medicine in the Enlightenment. The Six Non-Naturals in the Long Eighteenth-Century* (Routledge, 2020).

Clark Lawlor is Professor of Eighteenth Century and Romantic Literature at Northumbria University, UK. He is Principal Investigator for the Leverhulme Trust Major Projects *Fashionable Diseases: Medicine, Literature and Culture, ca. 1660–1832,* and *Writing Doctors: Representation and Medical Personality ca. 1660–1832.* His monographs include *Consumption and Literature: The Making of the Romantic Disease* (2006) and *From Melancholia to Prozac: a History of Depression* (2012). He has recently edited (with Anita O'Connell) *Fashion and Illness in Eighteenth-Century and Romantic Literature and Culture*, a special issue of *The Journal of Eighteenth-Century Studies* 40.4 (2017), and (with Jonathan Andrews) *'An Exclusive Privilege ... to Complain': Framing Fashionable Diseases in the Long Eighteenth Century*, a special issue of *Literature and Medicine* 35:2 (2017). Two volumes on Literature and Medicine co-edited with Andrew Mangham were recently published by Cambridge University Press: Vol 1, *The Eighteenth Century*; Vol. 2, *The Nineteenth Century* (2021).

A. A. Long is Professor Emeritus of Classics and Affiliated Professor of Rhetoric and Philosophy at the University of California, Berkeley. He is a graduate and former faculty member of University College London. From 1973 to 1983, he served as Gladstone Professor of Greek at the University of Liverpool. He is the author of many books on ancient philosophy, including most recently *Greek Models of Mind and Self* (2015) and *Epictetus. How to be Free* (2018). Long is a Fellow of the British Academy and of the American Academy of Arts and Sciences, and he holds an honorary doctorate from the University of Crete.

Jane Macnaughton is Professor of Medical Humanities at the University of Durham, UK, and Director of the University's Institute for Medical Humanities. She was Dean of Undergraduate Medicine at Durham until the medical programme transferred to Newcastle University in 2017, and is now Professor in the Department of Anthropology. She continues to be clinically active as Honorary Consultant in Obstetrics and Gynaecology at the University Hospital of North Durham. She is Principal Investigator on the Wellcome-funded *Life of Breath* project, which won the Inspiration Medal in the 2018 Health Humanities Awards. Her interest in breathlessness stems from her research focus on the idea of the 'symptom': its initial appearance, development, and evolution in connection with medical contexts, habits, and technologies. With Corinne Saunders she has edited three books on connections between medicine and the arts, and, with Angela Woods et al., has co-edited *The Edinburgh Companion to Critical Medical Humanities* (2016).

Francis O'Gorman was educated as C. S. Deneke Organ Scholar of Lady Margaret Hall, Oxford, where he read English. The author or editor of twenty-four books, mostly on English and Irish literature, 1780–1920, he also writes widely on music. He is one of the organists of the Priory Church of the Holy Trinity, York. Francis O'Gorman is Saintsbury Professor of English Literature at the University of Edinburgh.

Carole Rawcliffe is Professor Emeritus of Medieval History at the University of East Anglia, Norwich, UK, where she has taught since 1992. She has published widely on the history of hospitals, disease, and the pursuit of health in the premodern period. Her books include *Medicine in Later Medieval England* (1995), *Medicine for the Soul* (1999), *Leprosy in Medieval England* (2006), and *Urban Bodies: Communal Health in Medieval English Towns and Cities* (2013), as well as co-edited collections of essays on *Society in an Age of Plague* (2011) and *Policing the Urban Environment in Premodern Europe* (2019).

Denis Renevey is Professor of Medieval English Language and Literature at the University of Lausanne, Switzerland. His work on devotional and mystical literature includes two recent collections of essays: *Revisiting the Medieval North of England: Interdisciplinary Approaches,* edited with Anita Auer, Camille Marshall, and Tino Oudesluijs, Religion and Culture in the Middle Ages (Cardiff: University of Wales Press, 2019); and *Late Medieval Devotional Compilations in England*, edited with Marleen Cré and Diana Denissen, Medieval Church Studies 41 (Turnhout: Brepols, 2020). His monograph *Name above Names: Devotions to the Name of Jesus in Medieval English Literature c.1100–c.1530* is forthcoming from Oxford University Press.

Fraser Riddell is Assistant Professor of English and Medical Humanities at the University of Durham, UK. He has recently completed *Music and the Queer Body in English Literature at the Fin de Siècle*, which is forthcoming from Cambridge University Press. His work on John Addington Symonds and the Victorian chorister was recently published in *Victorian Literature and Culture*. Other publications include an article on queer musical geographies in Decadent literature in *Journal of Victorian Culture*. His current project explores the significance of tactile sensory perception and embodied cognition in Victorian and early twentieth-century literature.

Michael Symmons Roberts has published eight collections of poetry—most recently *Ransom* (2021)—and two novels with Cape (Random House in the US), and (in collaboration with Paul Farley) two books of non-fiction. He has won the Forward Prize, the Costa Poetry Prize, and the Whitbread Poetry Award; been shortlisted for the Griffin International Poetry Prize and the T. S. Eliot Prize; and received major awards from the Arts Council and the Society of Authors. He is a Fellow of the Royal Society of Literature and of the English Association. In collaboration with composer Sir James MacMillan he has written libretti for choral works, song cycles, music theatre, and operas, which have been performed internationally. His broadcast work includes the BBC4 verse film *Men Who Sleep in Cars*, and radio dramas, features, and documentaries. He is Professor of Poetry at Manchester Metropolitan University, UK.

Arthur Rose was until recently a Vice-Chancellor's Fellow in English at the University of Bristol, and is currently a Senior Research Fellow in the Wellcome Centre for Cultures and Environments of Health and the Department of English at the University of Exeter, UK. He is the author of *Literary Cynics: Borges, Beckett, Coetzee* (2017), co-editor of *Theories of History: History Read across the Humanities* (2018), and a joint author of *Reading Breath in Literature* (2019). He is currently writing *Asbestos: The Last Modernist Object* for Edinburgh University Press.

Andrew Russell is Professor of Medical Anthropology at the University of Durham, UK, and was Collaborator on the Wellcome Trust funded *Life of Breath* project. He has recently co-edited a special issue, 'Interdisciplinary Perspectives on Breath, Body and World', for the journal *Body and Society*. As a member of Durham's Anthropology of Health Research Group and degree tutor for the M.Sc. in Medical Anthropology, he is interested in tobacco as a global phenomenon and public health risk. He works collaboratively with 'Fresh', the UK's first tobacco control office, and with the Framework Convention Alliance, the civil society consortium supporting the WHO's Framework Convention on Tobacco Control. His publications include *The Anthropology of Tobacco: Ethnographic Adventures in Non-Human Worlds* (2019), and (co-edited with Elizabeth Rahman), *The Master Plant: Tobacco in Lowland South America* (2015).

Corinne Saunders is Professor of English and Co-Director of the Institute for Medical Humanities at the University of Durham, UK. She specialises in medieval literature and history of ideas, with an emphasis on medicine, emotions, gender, and the body. She is Co-Investigator on the *Hearing the Voice* project and was Collaborator on the *Life of Breath* project, both funded by the Wellcome Trust. Her third monograph, *Magic and the Supernatural in Medieval English Romance*, was published in 2010. She is a joint author of *Reading Breath in Literature* (2019). Recent co-edited books include *The Recovery of Beauty: Arts, Culture, Medicine* (2015), *Emotions in Medieval Arthurian Literature: Body, Mind, Voice* (2015), *Romance Re-Written* (2018), and *Visions and Voice-Hearing in Medieval and Early Modern Contexts* (2020). She is Editor for English Language and Literature of the journal *Medium Ævum*.

Gillian Skinner is Associate Professor at the University of Durham, UK. She is the author of *Sensibility and Economics in the Novel, 1740–1800: The Price of a Tear* (Macmillan, 1999) and of scholarly articles on various aspects of eighteenth-century literature, including women writers, the novel, drama, and life writing.

Naya Tsentourou is Lecturer in Early Modern Literature at the University of Exeter, Penryn, UK. She is the author of *Milton and the Early Modern Culture of Devotion: Bodies at Prayer* (2017) and co-editor, with Lucia Nigri, of *Forms of Hypocrisy in Early Modern England* (2017). Her current project, 'The Breathless Renaissance', investigates the relationship between breath and emotions in sixteenth- and seventeenth-century literature and medicine. She is currently working on the feminist politics of breath in the writings of Lucy Hutchinson.

Oriana Walker is an historian of medicine and the human body. She is currently a postdoctoral fellow at the Berlin Centre for the History of Knowledge and Humboldt University, having formerly held postdoctoral fellowships on the Wellcome project *Life of Breath* (Bristol), and at the Max Planck Institute for the History of Science in Berlin.

Jayne Wilton is a visual artist who explores the breath as a unit of exchange between people and their environments. Her practice uses darkroom processes with drawing, photography, video, and sound to capture the usually invisible trace of breath as it moves across a surface. Since graduating with an M.A. from The Slade School of Fine Art in 2010 she has collaborated with physicists, hospitals, a hospice, and most recently with the *Life of Breath* Project to explore further technologies and processes to make visible the breath and enable audiences to engage with this invisible animating force.

CHAPTER 1

The Life of Breath: Contexts and Approaches

David Fuller, Jane Macnaughton, and Corinne Saunders

Contexts and Approaches: 'The Age of the Breath'?

'The Age of the Breath': in the view of the philosopher Luce Irigaray, this characterises the late twentieth century and beyond.[1] Irigaray's idea is a variation of the threefold scheme of Christian history of the medieval theologian Joachim da Fiore: the Age of the Father (the Old Testament, the Law), the Age of the Son (the New Testament, freedom from the Law), and the Age of the Spirit (a utopian age of universal love). Like the Age of the Spirit, Irigaray's 'Age of the Breath' potentially transcends major limitations of history, specifically on issues of gender and all that follows from differently conceived relations between men and women. Breath is central to this in her reworking of a major philosophical predecessor, Martin Heidegger.[2] Heidegger is admired: he thought radically, working not only from what had already been thought but attempting to see nakedly from the bases of thinking. Irigaray's critique is not of Heidegger himself, but of Heidegger as representative of even the best in the tradition of Western philosophy, limited by its unrecognised assumption of the thinking subject as male. For Heidegger the primary element is earth—solid, and for Irigaray, masculine. For Irigaray more primary,

D. Fuller (✉) · J. Macnaughton · C. Saunders
University of Durham, Durham, UK

D. Fuller et al. (eds.), *The Life of Breath in Literature, Culture and Medicine*, Palgrave Studies in Literature, Science and Medicine,
https://doi.org/10.1007/978-3-030-74443-4_1

more utterly essential to Being, is the element of air—fluid, and feminine; the basis of life, the substance of the breath.

One need not accept Irigaray's arguments about gender to see the interest of her claim and the fecundity of its ramifications.[3] Breathing can be recognised (often elicited in retrospective analysis) as a major issue in many areas of twentieth- and twenty-first-century thinking: in philosophy, in feminism, in the arts, in psychoanalysis, in education, in religion, in politics, and in cultural geography, especially ecological issues including the contemporary global problems of air pollution and climate change. As with other conditions of life so axiomatic that attention has often scarcely been paid to them, recent writing in many areas, by bringing breath more into view, opens up a wide variety of new perspectives. In the current context of a global breathing-illness pandemic, COVID-19, with all that has exposed about global health issues and national and international relations, and with myriad implications as yet far from fully recognised, it can hardly be contested that Irigaray's characterisation of the late twentieth and early twenty-first centuries has been prescient.

Many areas of thought elicit a related sense of the period, sometimes from quite different starting points. In philosophy, Peter Sloterdijk's *Terror from the Air* argues that the use of poison gas in World War I was a fundamental reorientation of warfare, attacking not the enemy's body but the enemy's breathing environment. Extended in other terrors of atmospheric violence, from the gas chambers of the extermination camps of World War II to recent attempts to control the weather for military purposes, this has brought the relation of human beings to the atmosphere on which life depends, formerly taken for granted, newly into consciousness. It has also generated a new counter-awareness of the need for atmospheric hygiene and techniques for monitoring and maintaining air quality.[4] The issue of poison gas is only one element in Sloterdijk's argument. He presents the twentieth century as an 'age of explication', meaning that many aspects of existence formerly tacit are brought to more conscious notice and newly explored. As with Irigaray, a central focus is air and breathing, though brought to attention by a quite different route.[5]

Sloterdijk's extension of his argument to include the arts, with Surrealism seen as initiating modes of art as 'atmo-terrorism' designed to attack audiences, has proved less persuasive than his central thesis, and can be detached from his fundamental claim about an age of explication. With the arts Sloterdijk's thesis might more convincingly be extended to the theatre, to the twentieth century's revival of theatrical traditions less verbal, less cerebral, most obviously in the misleadingly named 'theatre of cruelty' of Antonin Artaud, in which the issue is not cruelty in any ordinary sense but the visceral nature of fully theatrical experience. It is a mode of theatre in which text is recognised as only one element, with movement, dance, costume, setting, lighting, but above all the body of the actor in all the viscerality of its emotional experience: the blood, the breath. The total art work with its address to the whole mind–body; but activated not with the familiar defamiliarisation of Brecht, addressing the detached intellect, but with what is permanently unfamiliar to the composed social being: myth, by which, 'using breathing's hieroglyphics', the audience is assaulted, disconcerted, disturbed, as by anxiety, fear, the erotic.[6] Artaud's ideal is more truly an aesthetic 'atmo-terrorism' than Sloterdijk's Surrealism.[7]

Sloterdijk's 'age of explication' thesis might more comprehensively be extended not to what the arts are in the twentieth century and beyond but how they are understood, with explication—an ever-increasing sophistication of consciously-applied critical techniques—replacing education through practice (the teaching of rhetoric, drawing, musical performance), with its concomitant address to cultivated intuition. Nevertheless, twentieth-century art has thematised breathing, most famously in Samuel Beckett's textless playlet-cum-happening, *Breath*, written originally (with Beckettian humour) for Kenneth Tynan's erotic review, *Oh! Calcutta!* (1969). A recent study has shown the resonance of Beckett's birth-cry to death-rattle encapsulation in a range of creative work before and after Beckett, including ways in which breathing can be presented in new modes of visual art (often with associated new problems for art criticism about the very nature of art), from happenings, performance-art, and anti-art to conceptual art and work in more traditional modes.[8]

Breathing can also be seen as foregrounded in relation to Lacanian psychoanalysis, drawing on Lacan's seminars on anxiety of 1962–1963.[9] The fundamental idea of this series is that the object of anxiety is not known: anxiety is fear without focus. Lacan calls the hidden provocation '*object (petit) a*' (*autre*/other). While the syndrome comes into being through post-Freudian Lacanian norms (primary deprivations of desire) which to the non-believer may sound unpersuasive if not fantastical, the syndrome itself—fear without focus, dependent on desire concealed by displacement—may be recognised outside specifically Lacanian frameworks of origin. Given the obvious relation of anxiety to breathing (constricted breathing, suffocation, as symptom or effect) it is surprising that Lacan himself has so little to say about breath—little more than a brief serio-comic episode on the conception of Christ by the entry of the breath of the Holy Spirit (*spiritus* of the *Sanctus Spiritus*) through the Virgin's ear, in which he relates the (divine) mouth to other orifices and the (divine) breath to other excretions.[10]

In *Staying Alive* Aranye Fradenburg includes an extended discussion of Lacan on anxiety which suggests what he might have said about breath.[11] A passionately-written anti-utilitarian defence of the arts as fundamental to the possibilities of humane living, the book is also a deeply well-informed critique of the contemporary university as semi-automated learning-factory, in which the aim is not knowledge and understanding but certification as a passport to employment. Exchange between intellectuals (albeit a great deal of what passes for this in universities is a parody of the real thing), real exchange, models a humane community. This is reflected in the organisation of *Staying Alive*, in which Fradenburg is in dialogue with an interactive counterpoint of related views ('fugues'). It is a mode that conveys a human presence relating felt thought to the subtle, mysterious, even bizarre—those aspects of human experience to which the arts are addressed, which are antithetical to the antiseptic of institutional bureaucracy.

In the chapter specifically concerned with Lacan and breathing, Fradenburg takes the highly interpretable *objet a* to be a 'conceptualization of the embodied mind's experience of change' (*Staying Alive*, 164), an index therefore of the crucial presence of the body in intellectual activities, which are often (wrongly) understood as not shaped by their basis in corporeality. As the primary experience of change, respiration, suddenly independent at the trauma of birth (thrust from a protected to a vulnerable condition—to anxiety), helps us to think, she argues, about the psychosomatic nature of rhetorical structures—the not unusual modern argument that writing is from the body, which when (as here) performed as well as affirmed requires an active and sympathetic reader. In Ruth Evans' response *objet a*, as reconceived by Fradenburg, becomes breath as a catalyst that sets off love: respiratory shapes in literary language (Frank O'Hara, Margery Kempe) brought off the page by real interaction with a responsive reader; breath heard and seen in the work of performance-artists (the duo, Smith/Stewart). Evans exemplifies the claim made from various perspectives by the whole book: the arts (and not the arts alone, but the arts understood in relation to contemporary conceptions of the life sciences as comparably interpretable), the arts are as necessary to living as breath.

One aspect of the fundamental issue Fradenburg addresses—'staying alive' in an academic context—involves escaping norms of academic 'professionalism' which encourage people to act as semi-automata, minds without bodies, not as human beings emotionally as well as intellectually responsive to interpretive complexity. In the liberal arts some traditions of criticism connecting literary study with life values and experience continue to recognise interactions between culturally situated and individually idiosyncratic readers from whom art requires active, engaged responses. In academic contexts, however, these have often been displaced by a pseudo-science of scholarship designed to demonstrate supposed objective presence (in a text, in a context) analogous, as Fradenburg argues, to a superseded notion of 'hard' science. Fradenburg aims to reverse this dehumanising process, which funding difficulties created by the COVID-19 pandemic have now newly intensified in universities worldwide. As institutions seek to fund their activities by moves to online learning that do away with people meeting together physically, with all the interpretable signals of actual life such meetings entail, and replace these with virtual meetings significantly evacuated of human presence—the living, breathing, emotionally-signalling body—the breathing illness potentially contributes several turns of the screw to intellectual-emotional suffocation. But, like properly holistic medicine, teaching in the Arts and Humanities must engage, Fradenburg argues, with the embodied mind.

Philosophy, feminism, the arts, psychoanalysis, structures and practices of higher education: and the editors of a recent collection on air and breathing offer re-orientations in yet more directions.[12] Like Irigaray, they begin from a critique of Western philosophy (tendentiously exemplified by a reading of a famous passage of Descartes), though they also acknowledge predecessors in Western tradition: apart from Irigaray, Gaston Bachelard, Elias Canetti, Merleau-Ponty and Levinas. The collection draws positively too on the place of disciplines of breathing in ancient Chinese philosophy (the Daoist philosopher traditionally known as Chuang Tzu, now Zhuang Zhou), modern Sufism (the Sufi master, Inayat Khan), and (again like Irigaray) in Eastern religious and spiritual traditions. From this they propose a 'respiratory philosophy' based in more conscious attention to and practice of breathing. If their critique of the Western philosophical tradition seems less substantial than Irigaray's this is in part because it is simpler. Whereas Irigaray grapples—tentatively, speculatively—with a profound problem of ontological consciousness in Heidegger, they convict Descartes of straightforward error: his supposed new beginning ignores his breathing. Descartes, one imagines, would have thought this critique readily answered: the observation is correct, but has no bearing on his reorientation of thinking. Descartes' writings began a new phase of Western philosophy not because he, and the most powerful minds of succeeding centuries, failed to identify this supposed omission, but because of his work's genuine and powerful originality. (Irigaray's new beginning is quite different: fundamentally she agrees with Heidegger about the aims and methods of philosophy, but plausibly—like Heidegger himself—looks to a new place for a first ground. The questioning, tentative and often syntactically inconclusive openness of her critique also recognises—as did Heidegger—that a new mode of thinking requires new modes of expression.)

That the critique of Descartes is tendentious does not, however, impugn the renewed and new attention to breathing drawn from it, the validity of which can be best considered in terms of its results. Much of the new reading in the collection is concerned with issues and figures in Western culture examined afresh in relation to theories and practices that variously foreground breathing from a range of Eastern traditions. The stress on practice is recurrent. The new philosophy is a new way of being: not on our lips only but in our lives also. This is exemplified by a discussion of Derrida's exploration of breath in Artaud, background to a new way of *performing* philosophy: in their different spheres both Artaud and Derrida match new ideas with new modes of expression.[13] Similarly with a phenomenology of breathing illnesses: it requires a new imagination of mind–body integration; a holistic understanding of the subjective experience of illness complementary to objective clinical analysis, treating the whole person in his or her context.[14]

The collection as a whole, like Sloterdijk's *Terror from the Air*, and much other recent writing on breath, breathing and breathlessness, recognises ramifications that are international and urgently relevant, as indexed by the relatively new focus of the World Health Organisation on air quality (http://www.who.int/airpollution/en/). As well as involving elements of individual choice—the choice to evolve and exercise a 'respiratory philosophy'—there is a larger sense in which 'atmospheres of breathing' affect health issues with the widest social ramifications. Choice can be exercised about these only by international political co-operation, and through negotiations in which the principal sufferers are often those with least political and economic power. Nevertheless, understanding that air pollution is a major cause of poor health—not only of breathing illnesses directly but also of heart disease and strokes—underlies efforts to clean up the polluted cities of developing industrial economies. Along with its consequences in climate change it also underlies the drive for clean air legislation in many parts of the world, with its potentially radical consequences for how we all live.

A comparably international perspective on the twentieth century and beyond as an 'age of the breath' in religion has two prominent and very different strands: Christian Pentecostalism, emphasising direct personal experience of God through baptism by the descent of the Holy Spirit, the breath of God; and the discovery in the West of ancient practices derived from the *Sūtras of Patañjali* (*ca.* 400 BCE to 400 CE), texts connected with Hinduism, Buddhism, and Jainism, concerned with disciplines of breathing in relation to spiritual awareness.[15] In the second of these areas Irigaray is again a significant figure, though interest in the Sūtras in European culture dates from earlier, one of the most prominent early translators, Charles Johnston, being a member of the late nineteenth-century literary and theosophical circle which included the poet W. B. Yeats.

Irigaray's *Between East and West* presents the issue of learning new practices of breathing as fundamental to a positive redirection of consciousness.[16] Only through this new practice, bringing into being a new relation of the mind to the body, Irigaray argues, is it possible to move beyond what she presents as the destructive elements of Western metaphysics and the spiritually empty (male) struggle to dominate nature: its aim is a new relation between the sexes, new possibilities of community, and the coexistence of diverse communities required by contemporary societies. Irigaray's argument has the strength of her training in Western intellectual traditions and their modes of critique complemented by profound reorientation through her own lived and evolved knowledge of Indian spiritual practices.[17] She writes, that is, from a complementary basis of intellectual analysis and whole-person experience. While Irigaray's specific arguments about gender and community are her own, her fundamental approach is congruent with a general tendency in contemporary Western societies, more than ever conscious of religious diversity and the cultural bases of belief, to value religion less in terms of faith and doctrine and more in terms of spiritual knowledge and practice.

Pentecostalism is quite different—a version of an antithetical strand in contemporary religions, the return to renewed fundamentalisms, Christian, Muslim, and Hindu. The understanding of the Christian God as a triune figure, Father-Son-Spirit, means that breath, and the crucial multivalent terms—Hebrew *ruach*, Greek *pneuma*, Latin *spiritus*—have always been central to Christianity.[18] Disciplines of breathing, though more prominent in Eastern religions, have also been present in many periods of the history of Christian prayer—in the prayers of medieval mystics,[19] in the practices of prayer proposed in the widely-distributed *Spiritual Exercises* (1541) of St Ignatius of Loyola,[20] and in the methods recommended for saying the 'Jesus prayer' (sixth-century) from the quite different background of Eastern Orthodoxy. This became widely known in Western Christendom through the nineteenth-century compilation, *The Way of the Pilgrim*,[21] which, after its translation in the 1930s, became one of the most widely-circulated books about Christian practices of prayer, conveying something of its background in an eighteenth-century collection of fourth- to fifteenth-century texts, the *Philokalia*, the most significant and authoritative compilation in Orthodoxy after the Bible. Here too psychosomatic techniques of prayer, based on a view of the body as 'an essential aspect of total personhood', emphasise the importance of disciplines of breathing.[22]

Nevertheless, breathing disciplines are less evident in Christian practices of prayer and meditation than in those of Eastern religions. Even the visitation of the Holy Spirit, the breath of God, though important in the scriptural account of the accomplishment of Christ's mission of salvation (Resurrection, Ascension, Descent of the Spirit), became suddenly prominent in a new way in the early twentieth century. Now a worldwide church with an estimated 280 million adherents in Africa, India, the Americas, Europe, and Scandinavia, Pentecostalism takes its origin and much of its character from African American charismatics meeting in Los Angeles in 1906. Central to Pentecostal belief is baptism by the Holy Spirit, the breath of God, characteristic manifestations of which are the mutually-inspired in-and-out-breathings of congregational communities in whooping, shouting, laughing, singing, and speaking in tongues. It is a religion with political implications: unstructured forms of worship in which any member of a congregation can take the lead model non-hierarchical forms of society in the world beyond the church. One recent account of black Pentecostalism sees it as rejecting the whole of Western civilisation as fundamentally white supremacist and hetero-patriarchal, in its intellectual frameworks (philosophy, theology) as well as its social and political structures; its positive aim as to imagine 'otherwise' modalities and epistemologies, which connect the movement with whatever is broadly non-heteronormative and liberationist.[23]

Stressing embodied thinking, this account, though ostensibly a radical rejection of Western traditions, can, nevertheless, be seen as congruent with other contemporary critiques of dualist mind–body modes of thought. How difficult it is to think embodied experience, however, is strikingly demonstrated by the account's contrast between an experiential narrative of breathing and emotion in episodes from two Pentecostal sermons, incorporating community responses, and a 'scientific' version of the relationship between breathing and emotion.[24] The two approaches are so radically different that the language of the scientific account positively excludes the mode of what is to be conveyed in the Pentecostal experience. The associated critique of major European and Scandinavian theologians of Pneumatology and the Pentecostal-Charismatic (Jürgen Moltmann, Veli-Matti Kärkäinen)—that they exclude black Pentecostalism because they are unable to deal with the physicality of its breathing—may be weakened by overworked invocations of 'otherwise' possibilities; but the book certainly succeeds in avoiding the all too usual intellectual's embarrassment about emotion and in illustrating the difficulties of intellectualising about bodily experience. As with Irigaray on Heidegger, as with Fradenburg on Lacan, as with Derrida's interaction with Artaud, it indicates the need for new modes of intellectual performance to match and engage with more embodied conceptions of intellectual experience.

As with breathing and air pollution, the issue is not only how to think but also how to live. Pentecostal practices of worship imply practices of social organisation, particularly because in its early twentieth-century beginnings the co-breathing of brothers and sisters in sharing the spirit recognised no racial limitations: black and white members of a congregation prayed together, contravening then current segregation laws in the southern states of the USA. Pentecostalism was therefore an early example of political movements in which breathing becomes a metaphor for freedom, constriction of breathing a metaphor for oppression.

'I can't breathe': the last words of Eric Garner, an African American who died as the result of a prohibited chokehold applied during arrest by police in New York in July 2014. 'I can't breathe': also the last words of George Floyd, an African American who died as the result of violent arrest by police in Minneapolis in May 2020. As a result of the death of George Floyd there were demonstrations against police brutality, and more generally against racial oppression, in every state of the USA, and internationally. 'I can't breathe': the words originally associated with protests against the killing of Eric Garner, after the death of George Floyd became the slogan-symbol of an international movement against racism and racial oppression, Black Lives Matter.[25]

Even before the death of George Floyd the wider political implications of the words had been developed by the political theorist, Achille Mbembe.[26] 'Caught in the stranglehold of injustice and inequality, much of humanity is threatened by a great chokehold' (§2): in terms that refer back to the death of Eric Garner but are also prescient of the events that gave rise to Black Lives Matter, Mbembe interprets the international inequalities exposed by the COVID-19 pandemic as quasi-apocalyptic signs. Dehumanising digital communication; oppressive exploitation of postcolonial and developing communities; destructive exploitation of nature, all on an international scale: the political metaphors of oppression and the literal consequences of polluted environments point in the same direction. They are signs of the need to return to every living thing—human, plant, and animal—the space and conditions required for its breathing.

The *Life of Breath* Project

This book was inspired by the *Life of Breath* project (https://lifeofbreath.org/), a collaborative interdisciplinary study based at the Universities of Durham and Bristol, UK, and funded by the Wellcome Trust. The project has brought together researchers in arts and humanities, social science and clinical science, healthcare professionals, activists and also 'experts by experience'—those with personal experience of breathing-related diseases. It takes up the complex, mysterious yet crucial aspects of experience connected with breath and breathlessness. It speaks to the present as an 'age of breath'. At the same time it addresses the relative invisibility of breath within the medical community, the silences surrounding breathlessness and breathing illnesses, and the continuing need for language and ways of expressing breath and its lack—needs so powerfully evoked by thinkers and writers from the late twentieth century onwards.

The project was founded on the proposition that breathing and breathlessness can only be understood fully through the insights of cultural, historical, and phenomenological sources, and through incorporating perspectives derived from the arts and humanities into the clinical understanding of the physical symptom of breathlessness. By transforming understandings of breath and breathlessness the research aimed to reduce stigma and empower those who live with breathing illnesses. It also aimed to offer new possibilities for therapy and the management of diseases for which breathlessness is the primary symptom.

The integration of cultural and clinical understandings has been central to the *Life of Breath* as a 'critical medical humanities' project. While medical humanities was for many years associated with broadening the education of clinicians, in particular medical students, it is now recognised that the humanities also have a key role to play in influencing the evidence base for clinical interventions that goes far beyond enhancing the empathy of practitioners. 'Critical' medical humanities is characterised by interdisciplinarity, ensuring that knowledge and methods from arts and humanities as well as social sciences are mobilised to understand and address health problems in ways that are integrated with clinical knowledge. The force of 'critical' asserts the value of the humanities in calling attention to the richness and importance of the contexts in which lives are lived and illness experienced.[27] The arts serve both to reflect and to illuminate experience: they play crucial roles in offering language and other forms of expression to articulate experience and frameworks for understanding.

The *Life of Breath* project took up the approach and methods of critical medical humanities to explore breath and breathlessness from an interdisciplinary perspective alongside the insights of those who live with breathlessness.[28] One aim was to use the outcomes of research drawing on medical humanities perspectives to inform and improve clinical practice, expanding the evidence base, addressing the lack of knowledge surrounding the embodied experience of breathing and breathlessness, and exploring how this connects with cultural attitudes and assumptions concerning breath. Research strands spanned varying cultural conceptions of breath, the medical history of breathlessness, the development of a phenomenology of breathing, including through work with trained and aware or 'interesting' breathers (diving, exercising, singing, playing a wind instrument, and even sleeping), and the experience of clinical breathlessness, with a focus on the ways in which the clinical encounter shapes notions of breathlessness and the problems of 'symptom discordance', the mismatch between objective measurements of lung function and patients' experience of breathlessness.

Though common to many diseases, chronic breathlessness is most often caused by the condition known as chronic obstructive pulmonary disease (COPD), the third most common cause of death globally.[29] In western countries, COPD is frequently caused by smoking, which is highly stigmatised. Physical constraints which are consequences of COPD are often compounded by shame, with the result that breathlessness sufferers hide away from society and may feel undeserving of help.[30] Conditions causing chronic breathlessness are also typically associated with older age groups and with low socio-economic status.[31] Partly because of this, research into respiratory diseases has had few energetic champions to redress the pressing need for improved funding and political action.

The COVID-19 pandemic has, however, brought breathlessness into sharp relief with peculiar global urgency. The *Life of Breath* project thus seems eerily prescient. In 2015 when the project began it was on the basis that breathing illnesses were relatively unnoticed and research into them was underfunded. In 2017 when the essays in this volume were commissioned, in 2018 when the contributors met together to hear and engage with each other's work, in 2019 when the essays were completed, that situation had not substantially changed. In 2020, as the editors worked on putting the volume finally together, and in 2021, as the volume goes to press, the pandemic has transformed the world. A virus that literally takes away our breath has caused many thousands of deaths—at the time of writing (late February 2021) in the UK over 120,000 people, in the USA over 510,000, and globally over 2,520,000, with over 110 million cases confirmed worldwide.[32] The illness has devastated the lives of individuals, families, and communities. Its as yet incalculable but evidently huge repercussions are a primary preoccupation of medical research, national governments, and international relations and organisations. Attempts to limit the spread of the virus have driven economies and businesses to ruin, and radically changed how we live, work, and relate to each other. Breath—and its lack—now seem more than ever to signal the ways in which human beings are united: every act of care taken not to spread the virus contributes to the common good, every act of carelessness to the common suffering. It has also shown how we are different. 'We're all in this together' has been a common statement of solidarity by politicians, but COVID-19 has emphasised that some are more vulnerable than others, in Europe and the USA especially Black, Asian, and minority ethnic communities. COVID-19 has underlined the key role of social deprivation in making people more vulnerable to ill health and shown that the poorest communities are often among those that have suffered the most.[33] While 'conspiration', 'breathing together', is a unifying theme of this book, so too is its converse: as the *Life of Breath* project has also shown, at the individual level obstructed or difficult breath is a profoundly isolating and lonely experience.

When the *Life of Breath* project group welcomed contributors to this book to Durham in 2018 to reflect on and initiate a critical medical humanities account of breath in Western culture, a world so engaged with breath was unthinkable. In our transformed world, it might seem that the rationale for this book is less urgent. If everyone is now talking about breath, can we still assert its absence in our culture and the need to raise awareness about it? In fact people are *still* not really talking about breath or breathlessness, except in a specific context. Major restrictions on the lives of people around the world may be in place to prevent the spread of COVID-19, yet the focus of attention is not on the symptom of breathlessness itself but rather on ways of preventing the spread of the virus and ways in which lives have changed. Paradoxically, COVID-19 seems to be deflecting attention from the articulation of what breath means, and how we cope with its lack in less extreme, more everyday contexts. Breath and breathlessness are subjects with powerful contemporary resonance, literal and metaphorical. Nevertheless, we do not have adequate terms and conceptions with which to discuss the actual experience of illnesses of the breath.

This theme of invisibility was a key stimulus to the development of the *Life of Breath* project.[34] A major aim was to fill a gap in critical writing and awareness of breath and breathlessness in Western thought because the lack of articulation of the meanings and significance of breath can adversely affect those who suffer from breathlessness. Research undertaken by the project emphasised the negative consequences of the difficulty of explaining or expressing breathlessness for those who suffer from it, for societal understanding of it, and for political investment in addressing the growing incidence of lung disease. Findings highlighted the lack of language to express breathlessness, its 'incommunicable' quality,[35] its isolating effects on sufferers, who experience an absence of social connection and a 'shrinking lifeworld'.[36] They revealed too the neglect and underfunding of the condition from a research perspective, with the result that the mortality rate across the last decade has remained static, by contrast to that for heart disease, which has reduced by 15%, reflecting significant material and scientific investment.[37] Research benefited from bringing together analysis of lived experience from empirical field work with philosophical phenomenology and literary insights to show that breathlessness is invisible in a complex range of interconnected ways that need to be addressed not just by clinicians but by changes of attitude in politics and society. This change of attitudes requires enhanced awareness and understanding of deep-rooted, culturally driven ideas and assumptions about breath.

The invisibility of breath, then, may be considered in relation to the self, society, and the sphere of policy, where evidence and political will are needed to make real changes for people whose breath is problematic. Individual experience of breathing, like many important bodily functions, occurs largely in the background and is not usually the object of conscious awareness. The philosopher-physician Drew Leder describes these functions as aspects of the 'recessive body', that is, the body outside our conscious influence.[38] However, unlike other 'recessive' functions, such as the cardiac or gastrointestinal systems, we have some control over breathing. And breathing becomes more conscious as it is taxed by high levels of physical exertion. This provides some experience of breathlessness but, as Havi Carel emphasises, it does not replicate the existential fear of pathological breathlessness: 'it is not like running for a bus; it is not like hiking in high altitude; it's more what I imagine dying is like'. Carel, who herself suffers from chronic breathlessness, speaks of it as expressible only through comparisons such as 'like dying' or 'like drowning'. Breathlessness is 'an overwhelming sensation, to which we are deeply sensitive, but it is also behaviourally subtle, and so often invisible to others'.[39]

Breath and breathlessness were brought into focus by the *Life of Breath* project through co-produced and engaged research activities, externally focussed communications, an exhibition and public events, and hence, the creation of a diverse and unprecedented community, including experts-by-experience, healthcare professionals, artists, and academics from a range of disciplines dedicated to exploring breathing and breathlessness in their own right. A research partnership with British Lung Foundation 'Breathe Easy' support groups for people with respiratory illness both informed the research and led directly into the development of creative writing, singing, and dance programmes which explored the potential of the arts to help people live well with breathlessness. These initiatives led in turn to the creation of online resources for breathlessness sufferers, made freely available on the project website and recommended in national health guidance.[40] Further work addressed the culture of pulmonary rehabilitation, and the barriers presented by clinical settings and language, while collaboration with clinicians built on insights into the cultural formation of experience and the deep connections between breath and embodiment.[41] This generated new hypotheses concerning the sensation of breathlessness, the cultural contexts that shape the experience of breathlessness, and the problem of symptom discordance.

Central to the project's aim of transforming public understanding of breath and breathlessness was the curation of the public exhibition *Catch Your Breath*.[42] The first exhibition ever to focus on breathing and breathlessness, *Catch Your Breath* drew on the project's research both to raise public awareness and challenge individuals to think differently about a bodily activity often taken for granted. The exhibition (running from November 2018 to February 2020) was hosted by venues academic, medical, and public: Palace Green University Library, Durham, the Royal College of Physicians, London, Southmead Hospital, Bristol, and Bristol Central Library. A smaller version toured libraries and scientific and medical conferences. Each venue attracted different communities and was accompanied by public events ranging from lectures and poetry readings to interactive activities, writing workshops, and mindfulness breathing classes. The exhibition included literary and cultural artefacts from medieval manuscripts to contemporary glass sculptures and short films, and newly commissioned interactive displays exploring the embodied experience of breathing. Through the themes of visibility and invisibility, the exhibition traced historical and cultural connections between breath, body, mind, creativity, and spiritual inspiration. Cultural, religious, and literary conceptions of breath and breathlessness from the classical period to the present were set alongside the medical history of breathlessness, its diagnosis and treatments, the histories of tobacco and air pollution, and the narratives of breathers themselves.

Another focal point of the project's exploration of invisibility was a 'Breath Lab', which brought together those with lived experience of breathlessness, their families and carers, clinicians, and policy-makers to explore the 'language of breathlessness'.[43] Discussion revealed the difficulty of describing breathlessness. Whereas a wide range of words existed to convey the 'character' of pain, there were few words to characterise breathlessness.[44] The language of breathlessness seemed to have been usurped by the clinical context: the three 'characters' employed by clinicians, 'air hunger', 'the work of breathing' and 'tightness',[45] left those suffering from breathing illnesses dissatisfied and struggling to find more accurate ways to express their experience. *Life of Breath* researchers also found that similar linguistic issues render clinical questionnaire tools for assessing the sensory experience of breathlessness confusing and difficult for patients.[46] Clinical language also inhibited those suffering from breathlessness from taking up opportunities for pulmonary rehabilitation. For the participants in the 'Breath Lab', the inability to find words to describe their experience was not only frustrating but also compounded the invisibility to others that defined their experience. Just as the ability to breathe easily facilitates ordinary social life, so breathing illness inhibits normal social interaction. Those with lived experience of breathlessness described stratagems they adopt to avoid being seen to struggle for breath in public, actions also prompted by the stigma they perceive as associated with their condition on account of its negative connections with smoking, age, and social deprivation.[47]

Language and its lack, the *Life of Breath* project suggested, are at the heart of the problem of the invisibility of those experiencing breathlessness. The lack of language to express what breathing means, how it feels, and especially what it is like not to be able to breathe, renders understanding opaque for people with breathlessness and for those around them. The inadequacies of the abstract, detached language of the clinic removes agency from those who struggle to breathe and be in the world. Jean-Paul Sartre's characterisation of the nexus of language, body, and the Other captures such disengagement:

> Language by revealing to us abstractly the principal structures of our body-for-others ... impels us to place our alleged mission wholly in the hands of the Other. We resign ourselves to seeing ourselves through the Other's eyes; this means that we attempt to learn our being through the revelations of language. Thus there appears a whole system of verbal correspondence by which we cause our body to be designated for us as it is for the Other by utilizing these designations to denote our body as it is for us.[48]

Those experiencing breathlessness found it frustrating and shaming, as well as inaccurate, that their experience was articulated only through clinical terms, or the real or imagined disparagement of others. They were also seeking ways of articulating this experience that made sense to themselves. For one support group, working with a writer-in-residence at the *Catch Your Breath* exhibition to produce poetry expressing their experience was revelatory.[49] They described their 'gratitude' at being offered, through the skill of the writer, metaphors that enabled them to find ways of explaining what breathlessness meant for them:

> We have the thoughts.
> Mostly hidden.
> But words?
> Denied, or rather not asked for
> Over the millennia.
> (From 'A Chance', by Jill Gladstone).

This book explores the language and conceptions that have been used in relation to breath and breathlessness 'over the millennia' from the classical period to the present, and the richness and power of ideas associated with them. It is one step in the larger project of rendering the invisible visible.

The Life of Breath: From Classical to Contemporary

The imaginative worlds of literature from the classical period onwards demonstrate the complex significance and symbolic power of breath and breathlessness across time, illustrating both cultural shifts and continuities. Breath and breathlessness are flashpoints in a range of discourses, complex terms linked to ideas of health and life and to their converse, illness and death. Breath can signal the most fundamental aspects of human existence—and the most ephemeral.

While breath and breathing have never been such resonant and urgent subjects as they are now, they have not been the subject of systematic cultural or literary study. Studies have focused on particular topics related to breathing—allergy, asthma, the air and pollution, smoking.[50] This collection, the first of its kind, adopts a wider perspective, tracing the origins and development of ideas concerning breath and breathlessness to explore their imaginative power and to demonstrate how literary texts and the cultural discourses that shape them reflect and reflect upon current ideas, understandings, assumptions, and preconceptions.

The collection was developed through an invited international conference, which brought together contributors to tease out cultural attitudes and understandings, and to probe the imaginative and affective power of ideas connected with breath across time. Discussion and dialogue were informed by the clinical, sociological, and empirical work of other members of the project team as well as by researchers across a range of literary and cultural disciplines. The volume also draws on *The Arts of Breath*, a series of public workshop-performances, lectures, and interviews exploring breath in creative arts contexts and forms—poetry, fiction, drama, solo and choral singing, and dance. These events laid the foundations for a volume reflecting the full trajectory of historical ideas of breath and breathlessness, their cultural and creative significance, and their resonances for contemporary understanding and experience.

The book is structured chronologically to present a wide range of cultural reflections within a framework of historical development—classical, medieval, and early modern to the present, with the period from the late sixteenth century to the present represented by some exemplification from each half-century, and over a range of literary, scientific, and cultural discourses, because, with the development of science in every area (but particularly with the chemistry of air, the physiology of breathing, and the more general comprehension and treatment of illness and disease), understandings shifted more quickly in this period. Within this structure contributors trace connections, contrasts, and continuities, with a view also to speaking to current experience of breathing, normal and pathological. The history of breath is not linear: rather, it circles and loops around essential, recurring, often difficult concepts. It is written deeply into religious belief—and into concepts of nature and being. It is inextricable from notions of spirit, inspiration, voice, and movement. It underpins the performing arts—poetry, music, drama, and dance. Its lack—breathlessness—can signal profound emotion but also illness and death. Breath may liberate, but also poison, infect, and contaminate. Breath is longed for, its purity guarded, and its loss feared: synonymous with life and being, it connects body, psyche, and world. While the volume focuses on writing in English and the western cultural discourses that underpin it, individual essays look beyond—to earlier and other literatures and discourses, to other nations and continents, to different epochs and modes of thinking.

The presence and meanings of breath are elicited in a variety of ways. One need not be a Nietzschean to accept in some form the famous proposition of Nietzsche that 'Against positivism, which goes no further than the phenomenon and says, "there are only facts", I would say: no, facts are precisely what there are not, only interpretations. We can establish no fact "in itself"'.[51] The issue with critical medical humanities is where and how to look for interpretations.[52] Its typical terms for method have anti-methodological implications of unpredictability: entanglement, entwining, imbrication (where the usage has left behind the word's origin in geometric patterning [tiling] to imply interactions of a more free-wheeling kind). Its 'weaving' voices may be on different wavelengths. Its 'dialogic' voices may be speaking at a tangent to each other. In its 'polyphony' dissonances can be passing or unresolved (Palestrina or Ligeti). In its 'heteroglossia' multiple languages may understand each other and communicate, or speak in terms that profoundly complicate if not defy translation and harmonisation (Pentecost or Babel). Binary oppositions are more than superseded: they are extended to a dissolution of boundaries: interdisciplinary becomes 'post-disciplinary'.[53] The interaction of a range of disciplines—arts and humanities and social science with biomedical science and medical practice—often involves experiments in interpretation, taking the view that nothing has meaning in and of itself but only within some context or mode of understanding which more or less inflects its meanings. While in some modern philosophies of science this is seen as apparent within science itself (Michael Polanyi, Paul Feyerabend, Thomas Kuhn),[54] the kinds of contextual and cultural interpretations offered by arts and humanities and social science disciplines are characteristically of a different kind. In this collection they are seen, for example, in complexities of history and usage of multivalent words, the complexities of how contexts that inflect interpretation may be assumed, or implied, or change over time, and how meanings arise not only from the reconstructed past and the actual present but also from the accreted history of meanings. As the Romantic-period polymath Friedrich Schlegel puts it, 'every great work, of whatever kind, knows more than it says':[55] in new contexts works may acquire new meanings, ideas may acquire new applications that were not visible to their author, originator, or earlier interpretive communities.

The volume takes as its starting point classical literature, philosophy, and medical theory from Homer to Galen, which lay the foundations for much later thought, through the Middle Ages and beyond.[56] Anthony Long demonstrates the long roots of the connections between breath, mind, and body and the startling contemporaneity of ancient ideas concerning breath. Breath and breathing are essential concepts in Platonic and Aristotelian philosophy: breath is a fundamental principle of both individual life and the universe. Stoic philosophers took up the notion of *pneuma* or vital spirit, air and fire, the active generative principle of the universe, which was connected by Galen with the individual physiology of breath and breathing. Late classical philosophy also developed radical ideas of 'conspiration', the subject of Phillip Horky's essay. This notion of the reciprocal breathing of human and divine came to be central not only to classical cosmology but also to early Christian thought. As Thomas Hunt shows, Christian theology drew on both classical concepts of *pneuma*, the life-giving spirit, and Hebrew notions of *ruach*, the breath of God, to develop and debate conceptions of the Holy Spirit—conceptions that had political resonances, relating to ideas of order and mastery. From the start, concepts of air, life, spirit, psyche and soul, external and internal *pneuma*, interweave: blurring into each other, they provoke dynamic responses embedded in changing notions of vitality, consciousness, and power, while the idea of co-breathing resonates with later notions of the interdependence of human and natural worlds.

Medieval medical theory was deeply rooted in classical thought and its Arabic reworkings, invoking the notion of the vital spirits to explain the physiology of the emotions and the intersection of thought with feeling. Corinne Saunders explores how the interrelated ideas of breath and vital spirits underpin and shape representations of affective experience in medieval imaginative texts, from popular romances to the medically alert fictions of Chaucer, arguing that representations that may seem to modern readers purely conventional are rooted in the medieval physiology of the breath. A similar understanding of the apparently conventional as physiological can also be seen, she argues, in devotional and visionary works, in which the concepts of vital spirits and the Holy Spirit intersect, giving breath a peculiar force. *The Book of Margery Kempe* offers an extended narrative of these intersections of physical and spiritual in Kempe's deeply embodied piety. Medieval physiological models provide a context and framework for Kempe's experience allowing the reader to place it not simply as performative or conventional but as rooted in learned ideas that were passing into general currency. The play of breath in tears, sighs, and swoons writes feeling on the body, creating a living tapestry of emotional experience from romantic love to mystical vision. Denis Renevey explores the possibility that medieval mystical experience was in part rooted in 'volitional breathing' resulting in changes to consciousness, drawing, in the absence of evidence in mystical texts themselves, on the insights of professional brass-players, eastern spiritual practices and the breathing techniques of 'new age' therapies. The repetition of prayers such as that on the Name of Jesus, he suggests, may have allowed for the conscious manipulation of the deep connection between breath and consciousness. At the same time, as Carole Rawcliffe shows, consciousness of the dangers of breathing infected air manifested itself in actions designed to improve air quality in medieval cities in a period repeatedly threatened by plague and epidemic. Being in the world depended not only on the movement of the bodily spirits but also on the purity of the air breathed in to form the vital spirits and influence the health of mind and body.

In turning from the medieval period to the early modern, the collection explores how, over the following centuries, these notions endured but took on new forms as understandings of physiology, disease, and the spiritual changed and developed. Katherine Craik and Stephen Chapman offer a novel perspective on present-day breathlessness by considering this within the unfamiliar context of early modern literature and culture. They argue that cross-disciplinary study can work not only by applying disciplines of interpretation from the arts and humanities to medicine but also in the other direction: medical science can shed new light on Shakespeare. As they demonstrate in relation to *King Lear*, early modern writing took up traditions of thought in which there was no simple separation between physiology and metaphysics. Breath—and its loss—define individual identity, but also human relationality, presaging the ways that breathlessness shapes not only its sufferers but also those who 'con-spire' with and care for them. Breath, fundamental to life, connecting mind and body, opens onto profound—and timeless—ethical questions. The early modern period retained the connection of breath with devotion and spiritual inspiration, enacted and sought after in highly physical affective encounters; it also extended physiological and psychological theories concerning the emotions. Naya Tsentourou revisits treatises on the passions to elucidate the place of breath within the early modern history of emotion, with a particular focus on the sigh, a response signalling overwhelming emotion that deregulates and disrupts. Science and art, thinking and feeling, form and meaning, intersect and clash as writers engage with the disruptive emotional valences of breath. Intertextual references to emotional breathing blur the distinction between patient and physician: bodies and texts become spaces where the detached witness conspires with the lovesick subject, and in turn, with the reader. Early Christian debate concerning the relations between breath, soul, and the Holy Spirit was reanimated in the political context of questions of the divine right of kings, exploited, as Patrick Gray shows, by John Donne in his sermon on a text from the Lamentations of Jeremiah, 'the breath of our nostrils'. Here, through Donne's complex play with Hebrew terms signifying life, spirit, and soul, breath becomes deeply ambiguous, its valences dependent on both political and religious interpretation.

Eighteenth-century medical and scientific discoveries complicated long-standing ideas of the connection between interior and exterior, individual and environment through new understandings of the nature of air and debates surrounding its potential role in disease. Rina Knoeff probes the medical, cultural, and imaginative effects of these, which were taken up in theories of pathology and environment, leading to new emphases on the importance of exercise, clean air, and landscape, both interior and exterior, and shaping artistic consumption. Ideas connecting breath with well-being were closely allied to understandings of embodied emotional experience. Gillian Skinner explores the formative role of breath and breathlessness in eighteenth-century notions of sensibility, in particular feminine sensibility, and in the literary genre of sentimental fiction these inspired. Again, images of sudden loss of breath—fainting and swooning—recur, brought out with peculiar force in Frances Burney's epistolary novel *Evelina* where the writing of emotion on the body and on the page intersect. Attention to the breath reveals a proto-feminist heroine, actively involved in scenarios that both challenge her capacity for moral conduct and demonstrate her power to act. At the same time, there were threats related to breathlessness, including, as Andrew Russell shows, through the introduction of a new and powerful material agent, tobacco, a primary cause of breathing illnesses worldwide, with profound and enduring effects on health and medicine. Adding smoke to breath instituted a culture fuelled by the perceived intellectual and creative possibilities of tobacco. Russell argues that the arrival in Europe of tobacco, with its ability to change processes of cognition, influenced literary developments, including the 'poetry of attention', with its interest in minute detail, and the distinctive 'it'-narratives of the period, which emphasise the division between self and other. In tobacco, with its apparent offer of inspiration from heightened experience, promise and danger combined.

Cultural and literary conceptions of breath were similarly dualistic—on the one hand, opening onto the sublime, and on the other, signalling human frailty. The concept of divine, life-giving breath retained its connections with Christianity, as in the familiar Victorian hymn, 'Breathe on me, breath of God', but was also extended. Romantic writers developed a theory of inspired composition rooted in nature rather than the supernatural, with the metaphor of a 'correspondent breeze', a quasi-divine breath whose power operates through the poet. As Clark Lawlor shows, such notions of inspiration were both shadowed and enhanced by the threat of the loss of breath and the fading out of vitality through illness, in particular, the Romantic disease of tuberculosis, more commonly known as 'consumption' owing to its effects on the body. Consumption was 'fashionable', a disease that in the popular imagination illuminated the spirit as the body wasted, and which became a powerful artistic and literary topos, while in reality mortality was marked all too acutely on the consumptive breath. In Romantic constructions of consumption, ancient connections of breath with death, life, spirit and genius take on new force, heightened by the experience of breathlessness.

Nineteenth-century writers sustained such images of consumption, with their complex interweaving of respiratory difficulty with intensity of life. The industrial revolution also brought a new interest in the possibility that disease could be carried by air and inhaled, and in new subjects connected with breath and illness: emphysema and other diseases caused by, for example, cotton processing and mining. Victorian engagement with breath in relation to industrial shifts was marked by duality. Progress could seem to signal movement towards immortality, as Francis O'Gorman demonstrates in relation to the invention of the mechanically powered pipe organ—an instrument with seemingly endless breath, which inspired new literary engagements with the eternal. Yet the contrast with limited human breath also signalled the frailty of human life, the limits of possibility and the inevitability of death. A similar duality characterised the ways that the nineteenth century engaged with the effects of industrialisation on the natural environment, which had severe negative consequences for breathing, most marked in the phenomenon of London fog. Christine Corton explores how, in literature and culture, this densely polluted air came to be represented as food, a soup that was inhaled and ingested—a metaphor that paradoxically appeared to celebrate this aspect of London, perhaps delaying legislation for clean air, even as fog's breath-damaging qualities were acutely recognised by Victorian medicine, as contemporary records and reports connecting high mortality rates with dense fogs demonstrate. The intersection of ideas of poison and nutrition offered rich and enduring creative possibilities for both writers and artists. Alongside this emphasis on the relation between air and illness, at the *fin de siècle*, as Fraser Riddell shows, new sexological discourses placed the breathing body centre stage. In both aesthetic theory and poetry, the ideas of consumptive wasting, lung disease, and air that kills came together to shape queer notions of embodiment that highlighted forbidden but all-consuming and inspiring experiences of materiality, loss, and desire. Breath and breathlessness animated treatments of the homosexual subject.

Modernism acted as a crucible for ideas of breath and breathlessness. Within a context of dramatic scientific, intellectual, cultural, and aesthetic shifts, new forms of writing emerged to which breath was central in radically different ways. The notion of divine, life-giving breath was questioned and complicated in a world where religious faith was profoundly challenged. New developments in medicine and psychoanalysis extended and altered understandings of body, mind, and affect, and their connections. As Arthur Rose and Oriana Walker argue, breath played a complex role in psychoanalysis from the late nineteenth century to the 1930s, as a potential signifier of psychic experience, and as a psycho-physical variable in its own right for the theorists who followed and challenged the founders of the discipline. Breath becomes an 'uncanny object' and a key to the unconscious: it also continues as a focus for debates concerning vitality, materiality, spirit, and consciousness. Breath and its politics illuminate in new ways the histories of psychoanalysis and psychotherapy. At the same time, changes in relation to the lived environment led to new ideas of breathing in the world. Abbie Garrington explores the encounter of science, culture, and art in modernist attitudes to mountaineering, an activity that tested the limits and possibilities of breath, engaging with deeply ingrained ideals of masculinity and bodily endurance. New developments in the oxygen rig, like the Victorian invention of the mechanical organ, promised more enduring breath, yet also pointed to man's limits on the mountains, challenging ideas of heroism. Such possibilities and paradoxes contributed to the new poetic breath of modernism, as did new questions concerning the relation of mind and body. Taking up William James's idea of 'consciousness-as-breath', Marco Bernini probes Samuel Beckett's interweaving of physiology and psychology in representations across his oeuvre of breath as intimately bound up with mind, revealing the workings and textures of consciousness and prefiguring more recent ideas of embodied cognition.

The long-standing connection between breath and inspiration was newly rewritten not only in European thought but also by American poets from Walt Whitman to the Black Mountain School, who located the formal and rhetorical structures of poetry in the rhythms of breathing. David Fuller takes up the subject of breath in poetry and aesthetic theory in America with a particular focus on Charles Olson's concept of the breath-line and the related experiments with poetic form of William Carlos Williams, setting these experiments alongside those of the German poet Paul Celan, whose poetry and poetics of difficulty enact a 'breath-turn', a profound change of direction, a new beginning, in response to the chilling questions for art raised by the cataclysm of World War II. The poetry of breath is necessarily embodied, intimately connected with the voice, a topic Fuller also addresses through consideration of the implications of and need for reading poetry aloud. Twenty-first-century poetics, as Stefanie Heine shows, has returned to the 'pneumatic turn' of the 1950s and 1960s. The contemporary African American writer M. NourbeSe Philip develops from Olson her own ideas of respirational poetics, beginning from the ethically positive idea of pre-natal 'conspiration', a mother breathing for her unborn child. Heine elaborates Philip's exemplification of a poetics of fragmentation in her anti-narrative narrative poem, *Zong!*, which engages with and radically remakes Olson's interests in the syllable and in word-materials taken over from a problematic 'mother-text', a case report related to a massacre of African captives thrown overboard from a slave-trading ship in the late eighteenth century in order to collect insurance on the 'cargo'. Philip's 'conspiration' is also a deconstruction of language, speaking to the violent silencing of the voices which *Zong!* evokes and to conspiration/conspiracy against black lives.

Modern technology has yielded new possibilities of realising breath in art. The contemporary visual artist Jayne Wilton has drawn on the possibilities offered by modern technology to examine the relationship between breather and spectator, individual and environment. Her work translates breathing into art, rendering the unseen visible—an extraordinary moment in the long history of the art of breath and breathlessness, which reaches back to primitive art and is refigured across cultural epochs. Wilton's work engages directly with sufferers from breathlessness, a refiguring of conspiration that challenges comfortable assumptions but also shows the potential of the arts to articulate embodied experience. The potential unease and violence of conspiration are presented in a radical new light by the contemporary English poet Michael Symmons Roberts, whose novel *Breath* (2006) explores a lung transplant that also becomes a deeply political act within a context of civil war. Breath opens onto questions of the intersection of mind and body, identity and consciousness, spirit and voice. These issues emerge across Symmons Roberts' writing, which is also concerned with how the poet, poem, and reader may realise structures of breathing implied by the printed page. In the arts of breath, frailty and resilience meet.

As Peter Adey shows in his Afterword, across the book themes and issues interweave: the body and pathology, vitality and emotion, soul and spirit, inspiration and creation, conspiration and community, politics and environment, nature and technology, voice and silence, and, above all, identity and consciousness. He also returns to the contexts in which the book was completed of an international breathing illness pandemic and the Black Lives Matter movement. As he suggests, the movements, meanings, vocalisations, and violences of breath are among the defining moments of our time. Lives, beings, and imaginings are inextricably bound up—for better or worse—with breathing in the world. From the ancient world to the present, philosophy, literature, and the cultures they reflect and shape write the life of breath.

Notes

1. 'The Age of the Breath', *Luce Irigaray: Key Writings*, ed. by Luce Irigaray (London: Continuum, 2004); Part IV: Spirituality and Religion, 165–70.
2. Luce Irigaray, *The Forgetting of Air in Martin Heidegger* (1983), trans. by Mary Beth Mader (Austin: University of Texas Press, 1999). For the variety of areas into which Irigaray's fundamental argument can be extended, including religion, politics, and animal and environmental ethics, see Lenart Škof and Emily A. Holmes (eds), *Breathing with Luce Irigaray* (London: Bloomsbury, 2013), which includes a new essay by Irigaray herself, 'To begin with breathing anew'. For a further extension see Anne Emmanuelle Berger, 'Irigaray's Breath, or Poetry after Poetics', in *Philosophy and Poetry: Continental Perspectives*, ed. by Ranjan Ghosh (New York: Columbia University Press, 2019), 209–34. For recent developments of Irigaray's views on the centrality of air and breathing, the crucial importance of human relations with the natural environment, and implications of this for animal and environmental ethics, see Luce Irigaray and Michael Marder, *Through Vegetal Being: Two Philosophical Perspectives* (New York: Columbia University Press, 2017).

3. On the contrary, even within the area of their primary application, to feminism, Irigaray's claims have been radically contested as 'essentialist', mistaking cultural constructions of gender for conditions of nature. See, for example, Judith Butler, *Gender Trouble: Feminism and the Subversion of Identity* (London: Routledge, 1990; 2nd edn, 2006).
4. Peter Sloterdijk, *Terror from the Air*, trans. by Amy Patton and Steve Corcoran (Los Angeles, CA: Semiotext(e), 2009). For a compressed but very well documented extension and complication of Sloterdijk's argument about 'atmo-terrorism', showing how knowledge of air's relationship to breathing has informed not only twenty- and twenty-first-century military technology but also processes of governance ('atmospheric policing'); see Marijn Nieuwenhuis, 'The Politics of Breathing: Knowledge on Air and Respiration', in *Atmospheres of Breathing*, ed. by Lenart Škof and Petri Berndtson (New York: SUNY, 2018), 199–218.
5. That Sloterdijk implicitly dissents from Irigaray's view of Heidegger as 'forgetting air' (*Terror*, 60–63) hardly affects their confluence of focus. (Sloterdijk interprets Heidegger's later discourse of *Heimatlosigkeit* [homelessness] as related to the profound change in the environment of air that is Sloterdijk's subject: air transformed from life-giving to death-dealing.).
6. Antonin Artaud, *The Theatre and its Double* (1964; essays *ca.* 1932–1938), trans. by Victor Corti (Richmond: Oneworld Classics, 2010); quotation from 'An Affective Athleticism', 93–99 (99). For a reading of Artaud that foregrounds the issue of breath see Jacques Derrida, 'La Parole Soufflée', *Writing and Difference* (Chicago: University of Chicago Press, 1978), and the discussion of this by Jones Irwin, 'Theater of Breath', in *Atmospheres of Breathing*, ed. by Škof and Berndtson, 167–78.
7. On the presence of breath in modern and contemporary poetry, see the essays in this collection by Fuller and Heine, and the interview with Michael Symmons Roberts.
8. Sozita Goudouna, *Beckett's 'Breath': Anti-theatricality and the Visual Arts* (Edinburgh: Edinburgh University Press, 2018). Central figures of Goudouna's account include happening, performance, and conceptual artists, Marcel Duchamp (1887–1968), Joseph Beuys (1921–1986), and Piero Manzoni (1933–1963), but also artists working in more traditional modes, Bram van Velde (1895–1981), and the video-artist Bill Viola (b. 1950). On breathing in visual art, ancient to contemporary, see the essay in this collection by Jayne Wilton.
9. Jacques Lacan, *Anxiety: The Seminar of Jacques Lacan. Book X*, ed. by Jacques-Alain Miller, trans. by A. R. Price (Cambridge: Polity, 2015).
10. *Anxiety*, 298–99. For the background treated allusively by Lacan (speculations of the Church Fathers on the fecundating divine breath) see

Leo Steinberg, '"How shall this be?" Reflections on Filippo Lippi's *Annunciation*', *Artibus et Historiae*, 1987 (8), 25–44.

11. L. O. Aranye Fradenburg, *Staying Alive: A Survival Manual for the Liberal Arts*, ed. by Eileen O. Joy (New York: Punctum, 2013); especially ch. 3, 'Breathing with Lacan's Seminar X: Expression and Emergence', and the response of Ruth Evans, 'Fugue 3: The Object Breath'.
12. Škof and Berndtson (eds), *Atmospheres of Breathing*: see Note 4.
13. See Note 6 (Irwin, 'Theater of Breath').
14. Havi Carel, 'Invisible Suffering: The Experience of Breathlessness', in *Atmospheres of Breathing*, ed. by Škof and Berndtson, 233–45. (Professor Carel was the leader of the Bristol team for the *Life of Breath* project.).
15. These have been more or less adapted or transformed in various contemporary Western practices. For recent examples, see Robin Rothenburg, *Restoring Prana: A Therapeutic Guide to Pranayama and Healing through the Breath for Yoga Therapists, Yoga Teachers, and Healthcare Practitioners* (London: Jessica Kingsley, 2019); and James Nestor, *Breath: The New Science of a Lost Art* (London: Penguin, 2020).
16. Luce Irigaray, *Between East and West: From Singularity to Community*, trans. by Stephen Pluhácek (New York: Columbia University Press, 2003).
17. As with Irigaray's arguments about gender seen as 'essentialist' (see Note 3), her views on Eastern religions have been subject to critique as 'orientalist' (based in uncritically Western views of Eastern cultures): see Morny Joy, *Divine Love: Luce Irigaray, Women, Gender and Religion* (Manchester: Manchester University Press, 2006), especially ch. 6, 'Irigaray's Eastern Excursion'.
18. See the essays in this collection by Long, Horky, Hunt, and Gray.
19. See the essay in this collection by Renevey.
20. See George G. Ganss, SJ, *The Spiritual Exercises of Saint Ignatius: A Translation and Commentary* (St Louis, MI: Institute of Jesuit Sources, 1992), §258. Ignatius began the *Spiritual Exercises* in 1521 and completed the final substantive version in 1541. A plausible estimate (Ganss, 8) suggests the work has been through over 4,500 editions worldwide.
21. *'The Way of a Pilgrim'; and 'The Pilgrim Continues his Way'*, trans. by R. M. French (1931; 1943; London: SPCK, 1954), 54. The method of breathing recommended here (an in-breath for the first half of the prayer, an out-breath for the second half) is just one of various suggested methods: see Per-Olof Sjögren, *The Jesus Prayer* (1966), trans. by Sydney Linton (1975; 3rd edn, London: SPCK, 1996), 41–45. See also the interview in this collection with Michael Symmons Roberts.
22. *The Philokalia: The Complete Text*, ed. and trans. by G. E. H. Palmer, et al., 4 vols (London: Faber, 1979–95); quotation, vol. 4, 66; and see,

for example, St Gregory Palamas, 'In Defence of Those Who Devoutly Practice a Life of Stillness', vol. 4, §7.

23. Ashon T. Crawley, *Blackpentecostal Breath: The Aesthetics of Possibility* (New York: Fordham University Press, 2017).
24. Crawley, *Blackpentecostal Breath*, 44–47. The 'scientific' account to which Crawley refers is taken from Yutaka Haruki, et al. (eds), *Respiration and Emotion* (New York: Springer, 2001).
25. For further discussion of the Black Lives Matter movement in this collection, see Peter Adey's extended 'Afterword'. See also the essay by Stefanie Heine, which discusses M. NourbeSe Philip's *Zong!*, based on a slave ship massacre and its legal consequences central to the eventual abolition of the slave trade in Britain and the British Empire.
26. Achille Mbembe, 'The Universal Right to Breathe', trans. by Carolyn Shread, *Critical Inquiry*, Posts from the Pandemic, 13 April 2020: https://critinq.wordpress.com/2020/04/13/the-universal-right-to-breathe/.
27. For elaboration and development of these arguments, see Jane Macnaughton, 'Medical Humanities' Challenge to Medicine', *Journal of Evaluation in Clinical Practice* 17 (2011), 927–32, and William Viney, Felicity Callard, and Angela Woods, 'Critical Medical Humanities: Embracing Entanglement, Taking Risks', *Medical Humanities* 41 (2015), 2–7.
28. For a particular example of this approach, see Alice Malpass, et al., 'Disrupted Breath, Songlines of Breathlessness: An Interdisciplinary Response', *Medical Humanities* 45/3 (2019), 294–304.
29. World Health Organisation, 'The Top 10 Causes of Death', 2016, https://www.who.int/news-room/fact-sheets/detail/the-top-10-causes-of-death.
30. British Lung Foundation (BLF), *Invisible Lives: Chronic Obstructive Pulmonary Disease (COPD)—Finding the Missing Millions* (London: BLF, 2007).
31. British Lung Foundation, *The Battle for Breath: The Impact of Lung Disease in the UK* (London: BLF, 2016), 5: https://www.blf.org.uk/policy/the-battle-for-breath-2016.
32. Figures from national public health agencies as collected by Johns Hopkins University: https://coronavirus.jhu.edu/map.html [last consulted: 27 February 2021].
33. Michael Marmot, et al., *Build Back Fairer: The COVID-19 Marmot Review. The Pandemic, Socioeconomic and Health Inequalities in England* (London: Institute of Health Equity, 2020).
34. Jane Macnaughton, 'Making Breath Visible: Reflections on Relations between Bodies, Breath and World in the Critical Medical Humanities', *Body and Society* 26/2 (2020), 30–54.

35. Havi Carel, *Phenomenology of Illness* (Oxford: Oxford University Press, 2016), 121.
36. Rebecca Oxley and Jane Macnaughton, 'Inspiring Change: Humanities and Social Science Insights into the Experience and Management of Breathlessness', *Current Opinion in Supportive Palliative Care* 10/1 (2016), 256–61 (259).
37. BLF, *The Battle for Breath*, 4.
38. Drew Leder, *The Absent Body* (Chicago: University of Chicago Press, 1990), 53.
39. Carel, *Phenomenology of Illness*, 110, 109.
40. Resources include downloadable breathing exercises, dance class videos, a podcast series, and multiple blog posts, all of which can be found via the project website: https://lifeofbreath.org/. They are recommended in lists of patient resources created by (among others) the UK National Institute for Health Research, the British Thoracic Society, and Lung Foundation Australia. Cf. the initiative recently launched by the singer Renée Fleming, *Healing Breath: breathing exercises for health and wellbeing*: https://artsandculture.google.com/project/healing-breath.
41. Jane Macnaughton, et al., 'Chronic Breathlessness: Rethinking the Symptom', *European Respiratory Journal* 51 (2018), 1702331.
42. On the exhibition and associated events, with an extensive series of related blogs and an online version of the exhibition by its curator, David Wright, see: https://catchyourbreath.org/. Two clinical fellows of the Royal College of Physicians, Nigel Cook and Noel Snell, created an accompanying booklet, *Breath and Breathing: An In-depth Look at the Processes Involved in Breathing, and the Difficulties Faced by Sufferers of Breathlessness*, available on the exhibition website.
43. Rebecca Oxley, et al., 'The Meaning of the Name of "Pulmonary Rehabilitation" and Its Influence on Engagement with Individuals with Chronic Lung Disease', *Chronic Respiratory Disease* 16 (2019), 1–9.
44. Macnaughton, 'Making Breath Visible', 40.
45. Robert W. Lansing, Richard H. Gracely, and Robert B. Banzett, 'The Multiple Dimensions of Dyspnea: Review and Hypothesis', *Respiratory Physiology and Neurobiology* 167/1 (2009), 53–60.
46. Alice Malpass, Coreen McGuire, and Jane Macnaughton, '"The Body Says It": The Difficulty of Measuring and Communicating Sensations of Breathlessness', *Medical Humanities* (Epub ahead of print: 28 January 2021), 0:1–13. doi:10.1136/medhum-2019-011816.
47. Macnaughton, 'Making Breath Visible', 36.
48. Jean-Paul Sartre, *Being and Nothingness: An Essay on Phenomenological Ontology*, trans. by Hazel E. Barnes (1958; London: Routledge, 2003), 377.

49. See https://catchyourbreath.org/the-secret-to-breathe/ which shows and describes the work of the project's writer-in-residence, Christy Ducker; on her workshops see https://catchyourbreath.org/sharing-the-unshareable/.
50. Gregg Mitman, *Breathing Space: How Allergies Shape our Lives and Landscapes* (New Haven, CT: Yale University Press, 2008); Mark Jackson, *Asthma: The Biography* (Oxford: Oxford University Press, 2009); Steven Connor, *The Matter of Air: Science and the Art of the Ethereal* (London: Reaktion Books, 2010); Virginia Berridge, *Demons: Our Changing Attitudes to Alcohol, Tobacco, and Drugs* (Oxford: Oxford University Press, 2013).
51. *The Will to Power* (1901; rev. edn 1906) [a compilation from Nietzsche's Notebooks overseen by his sister], trans. by R. Kevin Hill and Michael A. Scarpitti (London: Penguin, 2017), 287. For the German text in its original notebook form see Nietzsche, *Werke*, VIII.1, *Nachgelassene Fragmente. Herbst 1885 bis Herbst 1887* (Berlin: de Gruyter, 1974), 7 [60], 323.
52. For the history and current state of this developing discipline, see the introduction to *The Edinburgh Companion to the Critical Medical Humanities*, ed. by Anne Whitehead, Angela Woods, et al. (Edinburgh: Edinburgh University Press, 2016), 1–31.
53. On interdisciplinary studies and the 'post-disciplinary' implications of their methodological variety, see David Fuller, '"There is no method …"? Contact and Conflict in Interdisciplinary Studies', *Contact and Conflict in English Studies*, ed. by Sabine Coelsch-Foisner and Herbert Schendl (Frankfurt: Peter Lang, 2015), 133–49.
54. Michael Polanyi, *Personal Knowledge: Towards a Post-critical Philosophy* (London: Routledge, 1958); Paul Feyerabend, *Against Method: Outline of an Anarchistic Theory of Knowledge* (1975), 3rd edn (London: Verso, 1993); Thomas S. Kuhn, *The Structure of Scientific Revolutions* (1962), 3rd edn (Chicago: Chicago University Press, 1996).
55. 'On Goethe's *Meister*', in *Classic and Romantic German Aesthetics*, ed. by J. M. Bernstein, Cambridge Texts in the History of Philosophy (Cambridge: Cambridge University Press, 2003), 269–86 (281). For the German text see Friedrich Schlegel, *Kritische Neuausgabe*, ed. by Ernst Behler, et al., 35 vols (Munich: Schöningh, 1958–2006), II. *Charakteristiken und Kritiken I (1796–1801)*, ed. by Hans Eichner, 1967, 126–46 (140).
56. For historical understandings of the physiology of breathing, see the essays on the classical period by Long, on the medieval period by Saunders, and by Knoeff on the eighteenth century before Lavoisier, whose experiments on human and animal breathing initiated a revolution in the understanding of respiratory physiology. The ideas are set in historical

conspectus, with an idiosyncratic emphasis on the work of John Mayow (1641–79), in Donald F. Proctor (ed.), *A History of Breathing Physiology* (New York: Marcel Dekker, 1995). For an alternative account of Mayow and other mid-seventheenth-century developments, see Robert G. Frank, *Harvey and the Oxford Physiologists: Scientific Ideas and Social Interactions* (Berkeley: University of California Press, 1980).

Selected Bibliography

Boston, Jane, and Rena Cook, eds 2009. *Breath in Action: The Art of Breath in Vocal and Holistic Practice*. London: Jessica Kingsley.

Connor, Steven. 2010. *The Matter of Air: Science and Art of the Ethereal*. London: Reaktion.

Crawley, Ashon T. 2017. *Blackpentecostal Breath: The Aesthetics of Possibility*. New York: Fordham University Press.

Fradenburg, L. O. Aranye. 2013. *Staying Alive: A Survival Manual for the Liberal Arts*. Ed. Eileen O. Joy. New York: Punctum.

Goudouna, Sozita. 2018. *Beckett's 'Breath': Anti-theatricality and the Visual Arts*. Edinburgh: Edinburgh University Press.

Irigaray, Luce. 1999. *The Forgetting of Air in Martin Heidegger* (1983). Trans. Mary Beth Mader. Austin: University of Texas Press.

———, ed. 2004. *Luce Irigaray: Key Writings*. London: Continuum.

Irigaray, Luce, and Michael Marder. 2016. *Through Vegetal Being: Two Philosophical Perspectives*. Chicago: Chicago University Press.

Shusterman, Richard. 2008. *Body Consciousness: A Philosophy of Mindfulness and Somaesthetics*. Cambridge: Cambridge University Press.

Škof, Lenart, and Petri Berndtson, eds 2018. *Atmospheres of Breathing*. New York: SUNY.

Škof, Lenart, and Emily A. Holmes, eds 2013. *Breathing with Luce Irigaray*. London: Bloomsbury.

Sloterdijk, Peter. 2009. *Terror from the Air*. Trans. Amy Patton and Steve Corcoran. Los Angeles, CA: Semiotext(e).

Whitehead, Anne, Angela Woods, et al., eds 2016. *The Edinburgh Companion to the Critical Medical Humanities*. Edinburgh: Edinburgh University Press.

PART I

The Classical Period

CHAPTER 2

Pneumatic Episodes from Homer to Galen

A. A. Long

Introduction

Ancient Greece is an apt time and place for the start of this book's explorations of the life of breath. Greek breath (*pneuma*) was all around, in the guise of wind or air; it was what Hellenic people inhaled and exhaled every moment; it was the vital spirit (*psyche*) that coursed through their blood vessels; and it was the 'holy spirit' (*hagion pneuma*) of the New Testament. Our chemistry and physiology have moved on exponentially, but we have inherited the Greeks' *pneuma* in numerous ways, including language (think of 'pneumonia'), metaphor ('breath of life'), or even giving up the *ghost*. The 'pneumatic episodes' of this essay start with the earliest texts and contexts of Homeric epic. They then range through salient passages of philosophy and medicine, including Hippocrates' extraordinary essay on epilepsy and a Stoic account of

This essay is largely built around a selection of texts. I am responsible for the translations unless otherwise indicated. For access to the Greek originals I include a selection of editions in the notes and bibliography.

A. A. Long (✉)
Classics Department, University of California, Berkeley, CA, USA

D. Fuller et al. (eds.), *The Life of Breath in Literature, Culture and Medicine*, Palgrave Studies in Literature, Science and Medicine,
https://doi.org/10.1007/978-3-030-74443-4_2

embryology. The episodes conclude with Galen's ideas about psychic *pneuma*, the use of breathing, and the pathologies of breathlessness. These ancient authors explored vital connections between breath, heart, mind (*psyche*), and brain. They also considered how breath functions to control the body's temperature, how animal life begins with breath, and why breath and breathing are symptomatic for health and sickness. Their ancient thoughts resonate with this book's generic themes, especially breath's significance and importance for our inhabiting and understanding the world.

Greek breath is most familiar to English speakers in our words pneumonia and pneumatic. Respiratory illness and air pressure, conveyed by these words respectively, have breath or wind in common. Unsurprisingly, then, *pneuma* was a key term for Greek medical writers and also for philosophers engaged with 'physics' or study of nature (*physis*). The Latin equivalent of *pneuma* is *spiritus*, giving us vital spirits and spirituality. The obsolete word 'pneumatology' was coined in the seventeenth century to convey the study of spiritual or divine beings. And in the next century so-called pneumatologists took on the secular subject of the human mind or soul, prefiguring the field that later came to be called psychology. In extending the range of *pneuma* from respiration and air to topics of theology and psychology our British forebears were as close to ancient Greek usage as when they were speaking in the contexts of medicine or physics. As for the subject of psychology, whose word derives from *psyche* (commonly rendered by soul or mind), breath or *pneuma* plays a central role in many Greek speculations concerning the source and make-up of our mental life and faculties.

Greek breath, then, or *pneuma*, is a notion of enormous scope. We can study its usage in a huge range of different texts and contexts—in poetic and prose authors who describe everyday conditions of body and mind; in technical treatments by medical writers and philosophers; in meteorology and cosmology; and in the word's extension from the strictly physical notions of air, respiratory breath, and vaporous substance transmitted in the blood vessels, to the domain of immaterial (non-physical) reality or spirituality. To cover all of this in a single essay is obviously out of the question. What I can do, to open our volume's vast topic, is present some key episodes and texts in the ancient life of breath as recorded by a selection of Greek authors, starting with Homer and concluding with Galen, a period of almost a millennium.

We should remember from the outset that Greek authors, like everyone else until two hundred and fifty years ago, knew nothing of oxygen and carbon dioxide. There was much debate and disagreement, therefore, about the nature of breath and the biological function of breathing. Greek philosophers and physicians lacked all chemistry in our sense of the term. Their so-called 'physical' elements were just the hoary quartet, earth, air, fire, and water, with dry, cold, hot, and moist as the corresponding elementary qualities. They had no instruments to measure temperature and air pressure. Their ideas were more imaginative than accurate concerning the cardiovascular system, the pumping function of the heart, and the relationship between blood flow and breathing. They treated cold and hot, especially what they called 'innate heat', as quasi-substances. Yet, in spite of these scientific vagaries, we can still be impressed by the empirical soundness of many Greek findings, their grasp of basic anatomy, and the amazing range of their omnium gatherum concept of *pneuma*.

Breath and Life in the Earliest Greek Literature

T1 Of all that breathe (*pneiei*) and crawl on the earth, none is more piteous than man.—Homer, *Iliad* 17, 446–47.

T2 His life (*psyche*) left Sarpedon, and mist poured over his eyes; but he inhaled again (*empnunthe*), and a gust (*pnoie*) of the North Wind blowing around him restored (*zogrei*) his flagging spirit (*thumos*).—*Iliad* 5, 696–98.

T3 Zeus saw Hektor sprawled on the battlefield, his comrades kneeling around him; and he had difficulty breathing (*argaleo asthmati*), losing consciousness in his heart (*ker apinusson*) and vomiting blood (*haima*).—*Iliad* 15, 9–11.

T4 Odysseus digs a trench, to evoke the shades (*psychai*) of the dead from the underworld. After drinking sacrificial blood, the shades recover their wits and converse with him.—Homer, *Odyssey* 11, summary of lines 24–50.

T5 [The shade of Antikleia speaks about death to her son Odysseus:] Our sinews no longer grip our flesh and bones, but they are destroyed by the fierce fire as soon as the *thumos* departs from our white bones, and the *psyche* flies away like a dream.—*Odyssey* 11, 219–22.

The Greek noun *pneuma* in its earliest instances signifies wind or breeze, not inhaled or exhaled breath. But the corresponding verbs *pneo* and *empneo*, meaning breathe and inhale respectively, occur in two of the Homeric passages cited above (T1, T2). For the extraordinary range of *pneuma* that I have already outlined the word's primary sense of wind is of paramount importance. What we inhale is something that is ubiquitous on and around the earth, invisible, mobile, varying in temperature and speed, and capable of extreme force. These properties of wind have obvious bearing on the connotations of breath as a (or rather *the*) vital power in our bodies irrespective of breath's precise role and properties in biology and anatomy. Our life's dependence on breath and breathing puts us into a direct and intrinsic relationship with one of the planet's great powers along with water, solar heat, and light.

In Homer, probably our earliest Greek literary text, a person's sentient life is frequently expressed by the word *thumos* (which is etymologically related to our word 'fume', as in 'fuming with anger' (T2 and T5).[1] Within the human body the breathy *thumos*, vaguely situated in the region of the lungs or diaphragm (*phren* or *phrenes*) not the head, is the location and instrument of thoughts and feelings. Homer typically refers to what we call the mind with a hendiadys, saying that some person had such and such a thought or feeling in his or her *thumos* **and** *phrenes*. Loss of *thumos* (*i.e.* life breath) signifies loss of consciousness, which results in death if *thumos* permanently leaves the body. To express that ultimate loss of life breath, Homer has recourse to the word *psyche* (T4 and T5), meaning vital spirit. For Greek philosophy and medicine, *psyche* became the standard term not only for the human mind but also for the animating principle of all animals. But in Homer, whose text predates the earliest scientific literature by about two and a half centuries, *psyche* largely occurs in contexts of dying and of death, much as we used to talk about 'giving up the ghost'. The word's etymology, as the Greeks supposed, draws on two notions, breathe (the verb *psycho*) and cold (the adjective *psychros*). Like *thumos* Homer's *psyche* constitutes life breath, which we inhale as cold air. Even in thinkers as brilliant and erudite as Aristotle and Galen, this etymology underlay their curious conviction that the primary anatomical purpose of breathing is refrigeration—cooling and regulating the vital heat that flows out from our heart and moves around in our blood vessels.

When the Homeric *psyche* leaves the body permanently at death, it survives as a mere ghost or shade, a wisp of the spirit that had previously animated the body. To recover any semblance of life sacrificial blood is necessary, as we find Odysseus administering to the vampire-like shades in the book of the dead (T4). While this spooky episode makes for a wonderful fairy-tale, the breathy make-up of *thumos* and *psyche* together with the draughts of revivifying blood foreshadow medical theories about the vital relationship between respiration and blood. In grasping that crucial connection, Homeric Greeks were already onto one of biology's great truths.

I turn now from breath in Homeric poetry to its role and uses in the earliest Greek scientific literature.

Air/Breath/Wind as Life Force in Early Greek Philosophy and Medicine

T6 As our spirit (*psyche*) being air (*aer*) controls us, so do wind (*pneuma*) and air encompass the entire world. —Anaximenes.[2]

T7 The Pythagoreans said that there is a void, and that it enters the universe from the infinite breath (*pneuma*) as the universe also inhales the void.—Aristotle, *Physics* 4, 213b22.

T8 Pythagoras and Empedocles ... say that we not only have a social relationship (*koinonia*) with one another and with the gods but also with the non-rational animals. For there is a single breath (*pneuma*) that permeates the entire universe like a spirit (*psyche*), and it unifies us with them.—Sextus Empiricus, *Against the Mathematicians* Bk 9, §127.

T9 Heraclitus (according to the Stoic Zeno) called the vital principle (*psyche*) a percipient vaporization (*aisthetike anathumiasis*).—Heraclitus.[3]

T10 All creatures including human beings live by inhaling air. It is this that is their spirit (*psyche*) and intelligence (*noesis*), and if this is taken away they die and their intelligence fails.—Diogenes of Apollonia.[4]

T11 In my opinion the source of intelligence is what people call air (*aer*), and everyone is directed by this and it is omnipotent. It is this that I take to be divinity, that reaches to everything and disposes everything and is in everything. ... It is of many kinds, hotter and colder, drier and moister, more stationary and more swiftly mobile. ... In all living creatures the spirit (*psyche*) is the same, air that is warmer than the air outside in which we exist, but much cooler than the air close to the sun. In none of them is this warmth identical, but different enough for them to be similar. —Diogenes.[5]

T12 Pleasure and pain come about in this way: whenever a lot of air mixes with the blood and reduces its density, in the properly natural way, penetrating right through the body, pleasure comes about. But when the air is present contrary to nature and does not mix, the blood coagulates becoming weaker and denser, and pain is the result... Sentience is due to air that is clear and dry; for moisture impedes intelligence. Hence sentience is diminished in sleep and intoxication and over-eating... Plants are completely insentient because they are not hollow and do not take in air...—Diogenes, in Theophrastus, *De sensu* 39–44.[6]

T13 The bodies of human beings and other animals are nourished by a trio of nutriments; their names are food, drink, and wind (*pneuma*). Wind in bodies is called breath (*phusa*) and outside bodies it is called air (*aer*). This wind is the greatest power of all [as evident in storms on land and sea]. ... For mortals air is the cause of life, and the cause of disease in the sick. So great is the body's need for wind that a person can be deprived of food and drink for several days and survive, but if the windpipe is cut off death will rapidly ensue. Furthermore, all other activities are intermittent, but breathing alone is continuous, by inhalation and exhalation.—Hippocrates, *On Breaths* 3–4.[7]

The Greeks knew 'air' in the spelling *aer*. In its earliest uses (from Homer again) *aer* signifies mist or haze rather than the stuff we inhale. In the 500–400s BCE, when Greek speculation into nature got going in a big way, *aer* has become atmospheric air, but it retains its early associations with mist and haze from its susceptibility to condensation and rarefaction. Thus the Presocratic philosophers Anaximenes of Miletus and Diogenes of Apollonia pressed air into service as the unitary substance that constitutes everything and causes natural processes by its inherent capacity to change in density and temperature.[8] Most germane to our theme (as we see in T6, T9, T10, and T11) was the recognition of atmospheric air and breath as the explicit source of life or *psyche* and the extraordinary promotion of air to the role of world-controlling divinity. The Athenian comic poet Aristophanes mocked this air god in his play *Clouds*, where the chorus is composed of ethereal goddesses. His principal target was the character he called Socrates whose comic adoration of these unlikely divinities was tantamount to the virtual atheism that soon brought the real Socrates a death sentence.

Diogenes' theories were well known in late fifth-century Athens when Socrates was in his maturity. In Diogenes' philosophy air took on the role of global force and animistic power. When appropriately embodied it endowed creatures with sentience, intelligence, feelings, and emotions (T11, T12), thus constituting the main attributes of a mind. Some hundred years later, Stoic philosophers started to exploit the pantheistic and spiritual potentialities of divine *pneuma* (retrojected anachronistically onto early Greek philosophers in T8).[9] As we see in T7, breath and breathing could be extended from the microcosm or individual person to the macrocosm or the world as such. In this way breath and the process of breathing offered themselves as models for understanding cosmology while also serving as the basis for biology and sentient life.

Diogenes' interests in air and breath led him to engage in anatomical investigations of the blood vessels and their function in transmitting breath along with blood (T12). Interplay between breath and blood, and their respective contributions to animal life, soon became a constant theme in Greek medicine and philosophy of biology.[10] In the early medical text T13 we find breath treated as a *nutriment* along with food and drink—indeed as the prime nutriment because of the immediate mortality consequent on suffocation and also because of the unceasing continuity of respiration throughout life.

I turn now from early Greek speculations about life, breath, and air to ideas concerning the actual process of breathing, the functions of breath within the body, and the relation of breathing to blood flow.

Breath, Blood, and Respiration in Early Greek Thought

T14 This is how all things inhale (*anapnei*) and exhale (*ekpnei*). There are bloodless tubes in their flesh, stretched all over their bodies, and at the mouths of these tubes the outermost surface of skin is pierced right though with many pores, so that the blood is kept in but an easy path is cut for the air (*aither*) to pass through. Then, when the thin blood rushes away from there, the bubbling air rushes in with violent surge, and when the blood leaps back, the air is breathed out again.—Empedocles.[11]

T15 Diseases are caused by the condition of the body in the following way. When … the whole body breathes well and the breath passes through without hindrance, health ensues. For breathing takes place not only through the mouth and nostrils, but also throughout the whole body. When the body does not breathe well, diseases ensue in different ways.—Philistion in *Anonymus Londinensis* 20.[12]

T16 When a person takes in breath (*pneuma*) by the mouth or the nostrils, it first goes to the brain, then most of it goes to the belly, though some goes to the lungs and some to the veins. From these parts it spreads into the rest by means of the veins. The amount that goes into the belly has merely a cooling effect, but the air that goes into the lungs and the veins, when it enters the cavities and the brain, causes sentience (*phronesis*) and motion of the limbs. So if phlegm completely cuts off the veins from the air, the patient becomes speechless and loses consciousness. ... When the air is shut off in the limbs, and cannot pass through to the outside because of phlegm, the patient kicks; the air, rushing back and forth through the blood, causes convulsion and pain; hence the kicking. These sufferings occur when the phlegm flows cold into the blood which is warm, causing it to chill and stop moving. If the flow of phlegm is extensive and dense, it kills the patient at once; for it takes hold of the blood by its chill and congeals it. If it is not so great, it takes hold at first and blocks respiration; but when it disperses through the veins and mixes with the copious and warm blood, if it loses power in this way, the veins let the air in and consciousness returns.—Hippocrates, *On the Sacred Disease* 10.[13]

T17 As the breath is being discharged, the area of the chest and the lungs fills up again with the air that surrounds the body, air that goes through the cycle of displacement and penetrates the porous flesh. And again, when the air is turned back and passes outward through the body, it comes round to push respiration inward by way of the mouth and nostrils. ... As the oscillation goes on, the heat of the body pumps digested food from the belly and packs it into the veins. This is the mechanism by which the streams of nourishment continue to flow throughout the bodies of all living beings.—Plato, *Timaeus* 79c, 80d (trans. by Donald Zeyl, slightly modified).[14]

Veins and arteries (which were not systematically distinguished from one another at this date) were regarded as channels for the motion of breath as well as blood. That notion persisted right down to the time of Galen for whom the vital *pneuma* manufactured by the heart was transmitted throughout the body via the blood vessels. Rather than regarding the heart as a pump, Empedocles (T14) and Hippocrates (T16) assigned the cause of blood flow to breath and breathing. According to the physician Philistion (T15) breathing takes place not only through the nose and mouth but also through the pores of the skin.[15] Skin-breathing seems to be implicit in the passages from Empedocles and Hippocrates as well, and to be central to their claims about how breath activates the blood. The main subject of T16 is epilepsy, which Hippocrates, the attributed author, held to be no more 'sacred' than any other illness. Among many notable features of this passage, I draw particular attention to four points:

1. The cooling function of breath and breathing in general (which will be strongly defended, by both Aristotle and Galen).
2. The brain as the special destination of inhaled breath.
3. The causal connection between brain, breath, and sentience or consciousness (*phronesis*).
4. The necessity to normal health of regular respiration and blood flow, which are interrupted by epileptic attacks.

In opting for the brain rather than the heart as the centre of consciousness, the Hippocratic author was way ahead of the time, approximately 400 BCE.[16] Until the nervous system was discovered in the Hellenistic era, the heart held sway over the brain as the location of mind and emotion.[17] The Hippocratic author of T16 was also distinctive in making no mention of *psyche* in this account of breath and the contribution of breath to vital processes. In modern terms he or she looks like a brain/mind identity theorist, who sees no need to specify a special substance or soul, something other than the brain or additional to the brain, as the vital core of conscious and intelligent living.

This silence concerning *psyche* is most noteworthy. In earlier texts, as we have seen, *psyche* is embodied in the life breath that we inhale, making it something physical and tangible. The word thus became the standard name for spirit or 'vital principle' (T9–11). In this usage *psyche* signified a special kind of thing within the body constitutive of both life and mind. By the time of the authors presented in T14–17, *psyche* could also signify a self or person or soul—a human being viewed from the perspective of character and ethical identity as distinct from body and biological process. *Psyche* as vital breath—the stuff of respiration—looks straightforwardly physical and empirical. Yet if, as Homeric Greeks had supposed, we are survived after death by a disembodied *psyche* that is a wraith of our lived self, can vital breath be all that there is to living human beings? Perhaps breath and breathing, though necessary to our embodied life, are merely contingent to our true identity as souls or as mental and moral beings?

Plato was the Greek philosopher who most sharply distinguished *psyche* as the life of the mind from biological processes. In his late work *Timaeus* (excerpted in T17) Plato made a significant contribution to the contemporary interests in respiration, blood flow, nutrition, and vital heat. But even there he took little interest in the life of breath as such. What chiefly interested him throughout his copious writings were moral philosophy and philosophy of mind, not physiology and biology. Thus he elaborated the notion that what we are, essentially, is an incorporeal and immortal *psyche*, capable of inhabiting a sequence of bodies, both human and non-human.[18] In this conception, breathing is simply one of many physical processes that incarnate souls activate and experience during their temporary attachment to a particular body. The *psyche* itself is not a physical being and therefore has no more to do with breath than it has to any other material substance.

Plato's extraordinary dualism became influential only in late antiquity. I mention it here because even hardline materialists such as Plato's near contemporaries, Democritus and Epicurus, found it appropriate to draw a distinction between 'body' and 'soul' in thinking about our biological make-up. Epicurus composed the *psyche* from atoms (microscopic bodies) compounded in 'a blend of breath (*pneuma*), heat and a third still more refined constituent'; these especially mobile atoms were proposed to penetrate all parts of the body, much in the way that our nerves actually do, and with comparable functions of transmitting sensations and activating movement.[19]

Does the *pneuma* of the Hippocratic author (T16) foreshadow the role that Aristotle, Stoic philosophers, and Galen will attribute to 'congenital breath' (*symphyton pneuma*) as the all-encompassing life force manufactured in the body's blood vessels?[20] Like the ether once imagined to fill outer space, 'congenital breath' had few scientific credentials from a strictly empirical perspective; but there is nothing mysterious or magical or spiritual about the *pneuma* that 'causes awareness and motion of the limbs' according to the Hippocratic author of T16. This *pneuma* is simply the breath of respiration, and the thrust of the passage is entirely empirical. It would be interesting to know how the author would have modified these theories if he or she had known about the nervous system and the circulation of the blood.

Thus far I have illustrated roles assigned to breath and breathing in early Greek treatments of life's vital processes, including health and sickness, thought and feeling, loss of consciousness, and death.[21] We may also wonder how breath pertains, if it does pertain, to the beginning of life. Death as expiration is self-evident. Pregnancy, on the other hand, is not an outcome of inhalation. How does life get transmitted, and was breath credited with a function in the process of sexual reproduction?

The earliest Greeks in our record, as we have seen, intuited the connection between breathing, continuity of life, and finally death or ultimate loss of breath. As Greek medicine developed, breath (in the form of the warm vaporous substance also called *pneuma*) acquired the further vital function of fostering the onset of life and collaborating with blood and innate heat, in the body's vital processes.

Breath/*Pneuma* as Life-Generating Substance

T18 Animals and plants are formed in the earth and in the water because in earth water is present, and in water *pneuma* is present, and in all *pneuma* soul-heat (*thermotes psychiche*) is present, so that in a way all things are full of psyche.—Aristotle, *Generation of Animals* 3.11.[22]

T19 The semen contains within itself that which causes it to be fertile—the so-called hot substance, which is not fire nor any similar substance, but the *pneuma* enclosed within the semen or foam-like stuff, and the nature in the *pneuma*, which is comparable to the element of the stars. ... As to the principle of *psyche*, one portion is separable from body, and belongs to those creatures that include something divine (the sort of thing called intellect), the other portion is inseparable, fluid and watery, dissolves and evaporates.—Ibid., 2.3.

T20 If the semen falls into the womb at the suitable moment, and if at the same time it is gripped (i.e., conceived) by the receptacle when that is in good health, it no longer stays still as it did hitherto, but is set in motion and begins its own specific activities. Drawing material from the pregnant body, it fashions the embryo in accordance with certain fixed patterns up to the time when it reaches its end and makes its product ready to be born. Yet throughout all this time—I mean the time from conception to birth—it persists as merely a 'growth' (*physis*), i.e., as a *pneuma* that has changed from semen and is developing systematically from its beginning to its end. In the first stages the 'growth' is breath of a rather compact kind and considerably distant from *psyche*. But later, and when it has almost reached the moment of birth, the *pneuma* becomes refined, since it is fanned by the continuous [labour] exertions], and (so far as quantity is concerned) it is *psyche*. So when it passes outside, it is adequate for the atmosphere, with the result that it is tempered, as it were, by the external air and changes into *psyche*. For, just as the *pneuma* in stones is quickly kindled by a blow on account of its readiness for this change, in the same way the 'growth' of the foetus, once it has become ripe, is ready to change into *psyche* when it encounters the atmosphere. Thus whatever issues forth from the womb is at once an animal.—Stoic philosopher of about 150 CE, *Elements of Ethics* 1.[23]

Both Aristotle and Stoic philosophers had recourse to breath or *pneuma* as life-generating substance—not the breath of directly inhaled air that has largely occupied us so far, but *pneuma* as the special, rarefied, warm, and watery substance of which I was writing just now, 'congenital breath', as Aristotle terms it. Passages T18–20 could prompt discussion for an entire book. I include them here not so much for their own fascinating sake, as to illustrate the conceptual connection supposed to obtain between breath or *pneuma* as a material substance, qualified by attributes like hot and moist, and *psyche* as the principle of life.

In T18 Aristotle is treating supposedly spontaneous generation of plants and simple animals. In T19 his focus is on sexual reproduction especially of human beings. What chiefly distinguishes our species from other animal kinds is *nous* or our capacity to engage in rational thinking. Unlike the five senses, which are functions of bodily organs, Aristotelian *nous* has no physiological correlate. Hence, as he states in T19, the human *psyche* requires a 'portion' that 'includes something divine', which is not transmitted directly in the *pneuma*. This non-material, separable and ethereal entity is a clear gesture in the direction of Platonic dualism. Aristotle's 'divine portion' is not literally pneumatic because it is immaterial, but its metaphysical status may be compared to the Judaeo-Christian idea of Holy Spirit (*hagion pneuma*).

In the text of Hierocles (T20) the Stoic philosopher gives a remarkable account of animal embryology, from the moment of conception to the moment of birth. *Pneuma* provides the foetus with physiological continuity and consistency from the first moment of its formation to eventual emergence at birth from the womb. This pneumatic or breathy substance, generated by the interaction of semen and female 'matter', gives the foetus initially the status of a mere 'growth' (*physis*) or plant-like entity. During the embryological process the foetal *pneuma* develops from this vegetative stage into a fully formed animal.[24] Transition from the intra-uterine condition to self-sustaining life is marked and caused by the creature taking its first literal breath. We are to presume that this intake of cold air tempers the warm *pneuma* of the pre-natal creature. That inhalation transforms the embryological *pneuma* from being purely nutritive and vegetative into sentient and automotive life.

Galen on the Use of Breathing and the Causes of Breathlessness

To conclude this survey, I turn to Galen, classical antiquity's greatest physician, whose prolific writings include several works on respiration. In the book entitled *On the use of breathing* Galen reviews five possibilities: (1) to provide the source of the vital principle (*psyche*); (2) for nourishment; (3) to cool the innate heat; (4) to nourish *and* to cool; and (5) to replenish the arteries with vital/psychic *pneuma*.[25] The first four possibilities had all been canvassed by earlier figures I have reviewed. For the fifth possibility, regarding psychic *pneuma,* Galen drew on and contested the theories of the great Alexandrian physician, Erasistratus.[26] Galen concludes that what we require from inhaled air is not its substance (air as such) but the fourth of his five possibilities, a combination of refrigeration and nutriment:

> T21 We breathe for regulation of heat, and secondly to nourish the psychic *pneuma*. The first function is brought about by both parts of breathing, both inhalation and exhalation. To inhalation belong cooling and fanning, and to exhalation, evacuation of the smoky vapour. The second function—nourishment—is brought about by inhalation only.—Galen, *On the use of breathing* 5.8.[27]

At the beginning of this essay I called attention to the ancient Greeks' ignorance of oxygen and carbon dioxide. Yet Galen actually posited an analogue to oxygen in the nutritive function he ascribed to respiration and an analogue to carbon dioxide in the evacuative function of exhalation. As to the mysterious psychic *pneuma,* Galen showed admirable restraint:

> T22 Let us consider ... whether it is possible for the psychic *pneuma* to draw sustenance from respiration. Let us first say how we give something the name 'psychic *pneuma*,' ignorant as we confessedly are of the substance of the *psyche*. Since the emptying of the *pneuma* from the ventricles in the brain, when it is wounded, at once makes people both motionless and insentient, it must surely be that this *pneuma* is either the very substance of the *psyche* or its primary instrument. ... This *pneuma* must be nourished. From what other source, therefore, will it get nourishment unless from that which is drawn in through inhalation? But it is not improbable that it may be nourished also from the vapour arising from the blood.—Ibid., 5.5.

This cautious statement, keeping options open, is characteristic of Galen's expository style. Like his contemporary intellectuals and notable predecessors, he continued to invoke a notion of *psyche* and of psychic *pneuma* in his accounts of physiology and respiration. But in T22, as elsewhere, he confessed complete uncertainty about the 'substance' of the *psyche*.[28] He showed himself prepared to reduce the *psyche* to its pneumatic accompaniments whose existence he took to be confirmed by dissection of the brain's ventricles and by the effect of pressure on the unfortunate dissected animal's behaviour.[29]

Because the problem of breathlessness is one of this volume's main themes, I attach an excerpt from Galen's lengthy work entitled *On difficulty in breathing*.[30] Its notable features include Galen's focus on the physiological conditions that constrict breathing and increase its speed and 'thickness', and his references to pregnancy, asthma, and cancer. I do not have the medical knowledge to comment on the text in detail, but it may provide a fitting passage to conclude this survey of Greek pneumatic episodes.

> T23 Breathing has to increase in speed and density under the following conditions—when the airways contain lumps of undigested food, or are blocked by gluey and thick juices (humours), or some hard excrescence has developed in them, or anything in the adjacent bodies has become excessively elevated for any reason, the breath is inhaled less than it should be on account of the constriction. There are many types of this difficulty in breathing when different parts of the airways are constricted. Irregular growths in the stomach and liver, and also in the spleen and colon and neighbouring parts, by constricting the diaphragm, make breathing weak and rapid and thick. This type of inhalation occurs similarly in people with dropsy or who are pregnant or have eaten too much, or who have a distended belly. In asthmatics (so-called vertical breathers) gluey and thick juices (humours) coagulate in the windpipe. So then, if something hardens there or a lump develops that compresses and narrows the pulmonary passages (arteries), which are the recipients of the incoming air, the breath will be weak and thick and rapid. In the same way, if a lump develops in the chest, as often happens to people who are hunchbacked or bent or crooked in the spine, the breathing will become weak and thick and rapid because the irregular lump clearly blocks the place of the lung into which it has expanded and extended itself. Often too such conditions constrict the entire chest if women go suddenly into labour or even experience labour subsequently.—Galen, *On difficulty in breathing*.[31]

* * *

Conclusion

Ancient Greek *pneuma* provided philosophers and physicians with a notion that was central to their understanding of life and vital processes. This 'breath' came to comprise not only the material of respiration but also the vaporous substance supposed to be responsible for the generation of life as such and for vital processes within the body. Because of its refined texture, fluidity, and inherent vitalism, *pneuma* was also amenable to becoming spirit as distinct from matter and to becoming soul as distinct from body. With the demise of alchemy and the development of empirical science spirit and spirits dropped out of intellectual discourse. But since we continue to wonder how organic life could originate from inorganic matter, may there still be a role for the Greeks' 'congenital breath' if we could gain access to the primordial ooze?

Acknowledgements I am most grateful to Professors Corinne Saunders and Jane Macnaughton for inviting me to participate in 'The Life of Breath' colloquium. The occasion was exceptionally stimulating, and I benefited greatly from all the discussions. I also take this opportunity to thank my classical period co-panelists, Dr Phil Horky and Dr Tom Hunt, and our panel's chair, Professor George Boys-Stones.

Notes

1. For full discussion see R. B. Onians, *The Origins of European Thought about the Body, the Mind, the Soul, the World, Time, and Fate* (Cambridge: Cambridge University Press, 1951), David B. Claus, *Toward the Soul. An Inquiry into the Meaning of Psyche Before Plato* (New Haven: Yale University Press, 1981), Michael Clarke, *Flesh and Spirit in the Songs of Homer: A Study of Words and Myths* (Oxford: Clarendon, 1999), and A. A. Long, *Greek Models of Mind and Self* (Cambridge, MA: Harvard University Press, 2015).
2. Hermann Diels and Walther Kranz, *Die Fragmente der Vorsokratiker* (Zürich: Weidmannsche, 1964), 13 B2; G. S. Kirk, J. E. Raven, and M. Schofield, *The Presocratic Philosophers* (Cambridge: Cambridge University Press, 1983), 160.
3. Diels and Kranz, *Fragmente*, 22 B12.
4. Diels and Kranz, *Fragmente*, 64 B4; Kirk, et al., *Presocratic Philosophers*, 602.
5. Diels and Kranz, *Fragmente*, 64 B5; Kirk, et al., *Presocratic Philosophers*, 603.

6. Diels and Kranz, *Fragmente*, 64 A19; Kirk, et al., *Presocratic Philosophers*, 612.
7. W. H. S. Jones, *Hippocrates, with an English Translation*, vol. 2 (Cambridge, MA: Harvard University Press, 1923).
8. For details see Kirk, et al., *Presocratic Philosophers*, and A. A. Long, ed., *The Cambridge Companion to Early Greek Philosophy* (Cambridge: Cambridge University Press, 1999).
9. See Horky in this volume.
10. See James Longrigg, *Greek Rational Medicine: Philosophy and Medicine from Alcmaeon to the Alexandrians* (London: Routledge, 1993).
11. Diels and Kranz, *Fragmente*, 31 B100, 1–8.
12. W. H. S. Jones, *The Medical Writings of Anonymus Londinensis* (Cambridge: Cambridge University Press, 1947).
13. Jones, *Hippocrates.*
14. Plato, *Timaeus*, trans. by Donald J. Zeyl (Indianapolis: Hackett, 2000).
15. See D. J. Furley and J. S. Wilkie, *Galen. On Respiration and the Arteries* (Princeton: Princeton University Press, 1984).
16. Empedocles (the T14 author) identified intelligence (*noema*) with the blood around the heart.
17. See Friedrich Solmsen, 'Greek Philosophy and the Discovery of the Nerves', *Museum Helveticum* 18 (1961), 150–97.
18. See Long, *Greek Models.*
19. See Julia Annas, *Hellenistic Philosophy of Mind* (Berkeley: University of California Press, 1992).
20. On *symphyton pneuma* in Aristotle see A. L. Peck, ed., *Aristotle: Generation of Animals* (Cambridge, MA: Harvard University Press, 1943), Martha C. Nussbaum, *Aristotle's* De Motu Animalium (Princeton: Princeton University Press, 1978), and Gad Freudenthal, *Aristotle's Theory of Material Substance: Heat and Pneuma, Form and Soul* (Oxford: Clarendon, 1995); on *pneuma* in Stoicism, see Annas, *Hellenistic Philosophy*, and in Galen, see *The Cambridge Companion to Galen*, ed. by R. J. Hankinson (Cambridge: Cambridge University Press, 2008).
21. For reasons of space I omit discussion of Aristotle's account of respiration and its contribution to vital processes: see his work *On Respiration* in *Aristotle on the Soul, Parva Naturalia, On Breath*, ed. by W. S. Hett (Cambridge, MA: Harvard University Press, 1936).
22. A. L. Peck, ed., *Aristotle: Generation of Animals* (Cambridge, MA: Harvard University Press, 1943).
23. This is my translation of the papyrus text of Hierocles which I edited along with Guido Bastianini in *Corpus dei Papiri Filosofici Greci e Latini* 1.1 (Florence: Olschki, 1992). Hierocles' text was completely unknown before 1900. This particular passage has been barely discussed in print so far and not yet incorporated in standard works on ancient physiology. For

an accessible version of Hierocles' work, see *Hierocles the Stoic: Elements of Ethics*, ed. by Ilaria Ramelli (Leiden: Brill, 2011).

24. Stoicism envisioned a four-rank *scala naturae* in the *pneuma* that constitutes the world's 'cohesive principle'. All substances, ranging from inanimate bodies to plants, animals (ensouled creatures), and human beings (creatures with rational souls), are held together and activated by the *pneuma* that constitutes the kind of being that they are.
25. For discussion see Furley and Wilkie, *Galen*, and Armelle Debru, 'Physiology', in *The Cambridge Companion to Galen*, 275–78.
26. See Solmsen 'Greek Philosophy'.
27. Furley and Wilkie, *Galen*.
28. See Pierluigi Donini, 'Psychology', in *The Cambridge Companion to Galen*, 185.
29. See Julius Rocca, 'Anatomy', in *The Cambridge Companion to Galen*, 247.
30. I only discovered the existence of this work while I was working on this essay. What put me on to it was a reference by Ben Morison in *The Cambridge Companion to Galen*, 149, in a discussion of Galen's remarks about language. As far as I have been able to discover, the most recent edition of the work and the only accessible translation of it (into Latin) is that of C. G. Kühn, *Galeni Opera Omnia*, vol. 7 (Leipzig: Knobloch, 1824). During the Middle Ages the work was translated into Arabic and Syriac, and there are three Latin translations before Kühn's by Niccolo da Reggio, Johannes Vassaeus, and Janus Cornarius. I owe this information to the bibliography of the *Corpus Galenicum*, ed. by Gerhard Fichtner and Roland Wittwer (Berlin: Berlin-Brandenburgische Akademie der Wissenschaften, 2018).
31. *Galeni Opera Omnia*, vol. 7, 781–82.

Bibliography

Annas, Julia. 1992. *Hellenistic Philosophy of Mind*. Berkeley: University of California Press.

Beare, John I. 1906. *Greek Theories of Elementary Cognition from Alcmaeon to Aristotle*. Oxford: Clarendon.

Clarke, Michael. 1999. *Flesh and Spirit in the Songs of Homer: A Study of Words and Myths*. Oxford: Clarendon.

Claus, David B. 1981. *Toward the Soul. An Inquiry into the Meaning of Psyche Before Plato*. New Haven: Yale University Press.

Hankinson, R. J., ed. 2008. *The Cambridge Companion to Galen*. Cambridge: Cambridge University Press.

Long, A. A., ed. 1999. *The Cambridge Companion to Early Greek Philosophy*. Cambridge: Cambridge University Press.

———. 2015. *Greek Models of Mind and Self*. Cambridge, MA: Harvard University Press.
Longrigg, James. 1993. *Greek Rational Medicine: Philosophy and Medicine from Alcmaeon to the Alexandrians*. London: Routledge.
Nussbaum, Martha C. 1978. *Aristotle's* De Motu Animalium. Princeton: Princeton University Press.
Onians, R. B. 1951. *The Origins of European Thought About the Body, the Mind, the Soul, the World, Time, and Fate*. Cambridge: Cambridge University Press.
Solmsen, Friedrich. 1961. Greek Philosophy and the Discovery of the Nerves. *Museum Helveticum* 18: 150–97

CHAPTER 3

Our Common Breath: 'Conspiration' from the Stoics to the Church Fathers

Phillip Sidney Horky

This essay tracks the metaphorical expansion of the concept of cosmic breathing by locating it in ancient Greco-Roman discourses on cosmology and religion.[1] Its chronological scope runs from the mid-third century BCE, with the philosophy of the great Stoic Chrysippus, and advances as far as the third century CE, with the Church Father Clement of Alexandria. Over these six centuries, which witnessed extensive transformations of the philosophical and religious worlds of the Mediterranean and Near East, we see an expansion of the concept of breathing through the introduction of a notion of 'conspiration' (the Ancient Greek abstract noun is *sumpnoia* [συμπνοία], and its correlative verb is *sumpnein* [συμπνεῖν]). 'Conspiration' is my own Anglicization of a concept that refers to the reciprocal breathing that obtains between animate entities, and functions to explain how it is that divine and mortal come together in the cosmos. The concept of 'conspiration' (*sumpnoia*) is sometimes thought to arise for the first time in the fifth to fourth centuries BCE, being ascribed to

P. S. Horky (✉)
Department of Classics and Ancient History,
University of Durham, Durham, UK

D. Fuller et al. (eds.), *The Life of Breath in Literature, Culture and Medicine*, Palgrave Studies in Literature, Science and Medicine,
https://doi.org/10.1007/978-3-030-74443-4_3

the early Pythagorean Ecphantus of Syracuse and the legendary physician Hippocrates of Chios. I will present an alternative account in which this concept of 'conspiration' actually arises in the third century BCE, in the writings of the Stoics, and in particular Chrysippus of Soli (279–206 BCE). In order to do so, I will examine a range of evidence from Greek and Latin didactic, philosophical, and theological texts that helps both to contextualize the concept of 'conspiration' in time and place and understand its peculiar aspects.

The Latin version of this notion, *conspiratio*, which is formed of the verb 'to breathe' (*spirare*) with the prefix 'with/together' (*con-*), is first attested in the middle of the first century BCE, in the writings of Cicero, and it had two standard meanings among the Romans. The first meaning is well-known, that is, the notion of 'conspiracy', or an agreement between individuals for the purposes of hostile or illegal activity;[2] this is a notion that I will not discuss here, since, while it does represent a true cognate in Latin, it would appear to be a *derivation* that does not represent the original meaning of the term *conspiratio*; moreover, at least for our purposes, it is notable that this meaning does not get translated back into Greek συμπνοία.

The second standard meaning of *conspiratio*, which appears to be the older and more original one, is also more valuable to our collaborations in this volume: it refers to a kind of agreement *in voice*—a *reciprocal breathing*, a concord, a harmony of sorts.[3] This sort of *conspiratio* or συμπνοία as 'concord' is used in domestic environments to refer to the agreement that, for example, might obtain between a husband and wife: this appears to be a non-marked usage among philosophical schools from the first century BCE onward.[4] It also appears in the context of civic communities uniting under terms of agreement, such as a league or treaty.[5] Interestingly, however, 'conspiration' as 'reciprocal breathing' is also used metaphorically to describe how the many parts of the harmonious universe—especially the stars and the planets—relate to the cosmic universe as a whole. Two roughly contemporaneous examples—one in Latin, and one in Greek—demonstrate this sort of expression:

T1: This fabric which forms the body of the boundless universe, together with its members composed of nature's diverse elements, air and fire, earth and level sea, is ruled by the divine force of Soul; **by sacred dispensation the god draws things together [*conspirat*] and governs with hidden reasoning, and he apportions mutual bonds among all parts, so that each may furnish and receive another's strength**

and that the whole may stand fast in kinship despite its variety of forms.—Manilius, *Astronomica* 1.247–54 (trans. by G. P. Goold, modified) (early first century CE).[6]

T2: It seems to me that the fact that the nature of every animal is adapted (*harmoktai*) to the cosmos, and to the things in it, is evidenced through many indications. **For [each animal's nature], by conspiring (*sumpneiousa*) with it [sc. the cosmos] and by being bound to a path that is simultaneously best and necessary, follows the stream (*rhumai*) of the universe that encompasses it**, in relation to both the universal good-order (*eukosmia*) and the peculiar permanence of each [thing?]. Hence, this very thing is called a cosmos, and it is the most perfect of animals that exist. In its parts, which are many and diverse in nature, this animate [sc. cosmos] rules according to its innate properties, and because it has a greater share of what is divine. And, in the nature of the divine continually running (*aei theontos theiou*),[7] the things that follow the first and greatest path follow it…[lacuna]…and the wandering planets. —pseudo-Ecphantus of Syracuse, *On Kingship*, 79.9–14 (mid-first century BCE-mid-first century CE?).[8]

It is important to contextualize these wonderful passages. Manilius' *Astronomica*, which we can date around 15 CE, was written as a cosmic celebration of two Caesars—Augustus Caesar and Tiberius Caesar, the first emperors of Rome. As a whole, the poem is often considered representative of the didactic genre, borrowing much from other didactic poems of the Greek and Roman traditions, including Hesiod's *Works and Days*, Aratus' *Phaenomena*, Lucretius' *On the Nature of Things*, and Virgil's *Georgics*. It is the earliest surviving explicitly astrological work, and it enjoyed some popularity both in antiquity and in the early twentieth century, having been a particular favourite of A. E. Housman, who devoted thirty years of his life to working on the text.

At punctuated moments, Manilius refers to the organization of the universe—and indeed the organization of the poem itself—as subject to types of primal breathing; this takes the form of, respectively, *conspiration* and *inspiration*. To take the latter, a programmatic passage in Book 2 refers to the 'divine exhalation' (*divino flatu*) that literally inspires his poem to reach the stars themselves (*ad sidera*).[9] This is of course a twist on the familiar epic/didactic trope of the poet being inspired by the divine in order to produce a poem that is sufficiently lofty for the genre; but that is not all. As scholars have noted, the poem is infused—no pun intended—with demonstrably Stoic concepts and themes.[10] The Stoics had developed a sophisticated materialist cosmological system in which breath functioned as the primary mechanism for the articulation and sustenance of the parts of the universe. Many passages could be cited and discussed, but Diogenes Laertius presents one of the clearest examples:

> T3: The cosmos, they [sc. the Stoics] say, is one and this is limited, having a spherical shape; for a shape of this sort is most suitable for motion, according to Posidonius in the fifth book of his *Natural Reason* and the followers of Antipater in their works *On the Cosmos*. Outside this [sc. the cosmos] is diffused the unlimited void, which is incorporeal. By 'incorporeal' they mean that which, although it is capable of being occupied by corporeal things, is not occupied by them. There is no void in the cosmos, but it is wholly unified. **For this [sc. unification] necessarily results from conspiration (*sumpnoia*) and tension (*suntonia*) of the heavenly things towards the terrestrial things.**—Diogenes Laertius, *Lives of the Eminent Philosophers*, 7.140 (text: early third century CE, looking backwards to the third century BCE).[11]

The entire description is remarkable,[12] but for the purposes of this essay I want to focus on the last sentence. We can identify here three attributes of the Stoic cosmos: first, the unity of the cosmos, which results from conspiration, is somehow 'necessary'; second, that this conspiration is related to the notion of the 'tension' that obtains between heavenly and terrestrial objects; and third, that the tension and conspiration proceed from heavenly objects down to earthly ones (*i.e.* are downwardly directed from higher to lower objects ontologically). Given the notion of 'inspiration' by the divine breath we noted previously in Manilius—and which is well attested throughout the poetic tradition from Homer onwards—the downward motion of conspiration and tension is not especially surprising. But we should not take it for granted that Manilius' view that cosmic conspiration is caused by God, or maintained by God, reflects a canonical Stoic view.[13]

Let us now turn from the Stoic notion of cosmic conspiration and focus on how the Post-Hellenistic Pythagoreans developed these ideas. To do this, we can return to the description of cosmogony according to pseudo-Ecphantus of Syracuse in *On Kingship*. The text of pseudo-Ecphantus belongs to a series of Pythagorean pseudepigrapha—texts that were ascribed to the ancient Pythagoreans (who lived at the beginning of the fifth century BCE but probably wrote little if anything), but that were composed starting from, at the earliest, around 350 years later, in the mid/late second century BCE.[14] This particular text appears to have been composed around the turn of the millennium, although we cannot be sure exactly when.[15] Pseudo-Ecphantus hints at the notion of cosmic harmony in the first line, where the Greek term *harmoktai*, which means 'adjusted to' or 'fitted to', refers to the Pythagorean concept of the 'harmony (*harmonia*) of the spheres'—well attested both in early Pythagorean philosophy, and in the pseudepigraphical works of this period. Pseudo-Ecphantus aims to demonstrate, by appeal to evidence (the 'indications'), that each 'animal' is properly fitted to, or harmonized with, the cosmos as a whole. These 'animals' could include not only mundane objects, like human beings or dogs, but also, and more importantly, stars and planets, which, by virtue of being capable of *moving themselves*, were considered animate by the Pythagoreans of this period.[16] But they are all subject to the greatest and most perfect animal of all—the cosmos, which rules over everything inside it. Of special importance to our subject is the idea that each animate being that is ruled over by the cosmos 'conspires' with it, and this act of conspiration is emblematic of each animal's engagement with the 'stream' or flux of the universe. In another fragment of this work, we find that all humans have a divinely portioned inspiration (*empnoiesis*), and, at least among terrestrial creatures, they appear to share to the greatest extent in the divine breath.[17]

That the ascription of a theory of cosmic conspiration is being associated with a great Pythagorean cosmologist is itself unsurprising, since the Pythagoreans were among the very first Greek philosophers to conceptualize the articulation of the cosmos as an act of breathing.[18] However, what differentiates pseudo-Ecphantus' *On Kingship* from what we can trace back to early Pythagoreanism 350 years prior, is that here each ensouled animal itself breathes in unison, or 'co-breathes', *along with* the cosmos, whereas among the early Pythagoreans it was the cosmos as a whole that did the breathing and not its parts (at least so far as the surviving evidence suggests).[19] That is to say, for pseudo-Ecphantus, in the act of conspiring with the universe, each animal plays its own part in the cosmic system and carves its own path in accordance with its peculiar capacities. Such an act of 'conspiration' might seem to indicate agency on the part of the lower animate beings. But how much agency do these Post-Hellenistic Pythagoreans really accord to lower animate beings vis-à-vis conspiration? Other roughly contemporary evidence shows that some figures, like Moderatus of Gades (a self-proclaimed Pythagorean[20] who probably lived at the end of the first century CE), understood *the one*[21] to be the ultimate cause of such a cosmic conspiration from, just as we saw for the Stoics, the top down. Moderatus believed that the number one is the 'cause of the conspiration and the sympathy of wholes':

> T4: Geometricians too, since they do not have the ability to express incorporeal things in speech, have recourse to figural diagrams when they say 'this is a triangle': they do not mean that what falls to the eye *is* the triangle, but that it [sc. the triangle] is *of this sort*, and it is through this that they express the concept of the triangle. And the Pythagoreans did the same thing in the case of primary reasons and types: since they did not have the ability to express the incorporeal types and first principles in speech, they had recourse to numerical demonstration. And in this way they called the one the reason of unity, and of identity and equality; and they called it **the cause of the conspiration (*sumpnoia*) and the sympathy (*sumpatheia*)** of wholes, and of the preservation of what is absolute and as such: for, indeed, when it is among things [divided] part by part, the one subsists by unifying with the parts, and **by conspiring (*sumpnoun*) with them according to participation in the primary cause.**[22]—Porphyry of Tyre, citing Moderatus of Gades, *Life of Pythagoras* 49 (text: late third century CE, looking back at end of first century CE).[23]

Here we see a remarkable account of how cosmic conspiration works. Moderatus explains that the early Pythagoreans needed to be able to explain how incorporeal things, such as the ultimate causes of everything in the universe or the ultimate species of existence, could be formulated in speech, when they could not be seen by the eye. They appealed to *numbers*, which belong to both the world of the concept and the world of sense-perception: for example, we can see the twoness of two pints of beer with our eyes, but at the same time we can think 'twoness' in our heads without having to associate it with the two pints of beer we are seeing. For these Pythagoreans, numbers were substantial rather than (as Aristotle might hold) quantitative predicates, which indicated something fundamental about the identity of the object under examination. The number one, which is solicited here, for example, is understood to be the explanation for unity, equality, and identity of enumerated objects. It is equally understood to be the cause of 'conspiration and sympathy' among those objects, that which allows them to participate in the ultimate cause of all, namely, the One (not the number one, but the One that is above all Being and Existence).[24] As with the Stoics, conspiration is understood to be a top-down mechanism that explains the unity of the diversity of the cosmos; higher beings—whether we call them 'God', or 'the divine', or 'the number one', or 'the One'—are responsible for sustaining the act of cosmic breathing, and lower animate beings simply reciprocate the action by necessity. Among the Stoics and the Hellenistic/Post-Hellenistic Pythagoreans, then, conspiration is similarly asymmetrical and top down.

In the final part of this essay, I want to turn to the origins of the notion of cosmic conspiration. If the concept was not really developed by the Early Pythagoreans, where, and when, we might ask, did it ultimately arise? The ostensible answer reveals as many questions about the problem as it does solutions. It comes in the writings of the famous physician Galen (*ca.* 130 CE–*ca.* 210 CE):

> T5: If, on the other hand, we were to bring forward philosophers from the Stoa to the council chamber and entrust the vote to them, on the basis of the doctrines they themselves have put forward, they will crown Hippocrates. For Hippocrates was the first to propose [the elemental qualities of] hot, cold, dry and moist and, after him, Aristotle proved them.
>
> And the followers of Chrysippus, when they accepted these things already to hand, were not embroiled in contention. Rather, they say that all things are compounded from these [four elemental qualities], and that these things are affected by and act on each other, and that nature is proficient. And they approve all the other doctrines of Hippocrates regarding nature, apart from one minor point which is a difference between them and Aristotle. For when Hippocrates correctly says that **'the whole body conspires (*sumpnoun*) and flows together (*surroun*)'**, and that **'all the members of animate things are [in] sympathy (*sumpathea*)'**, both accept this very point. However, they differ in this: Aristotle held that the qualities alone interpenetrate each other and are mixed together completely, whereas those from the Stoa suppose that not these qualities only but also the substances themselves [are mixed completely].—Galen, *Method of Medicine* 1.2,[25] attacking Thessalus and defending Hippocrates (trans. by Ian Johnston and G. H. R. Horsley, modified) (text: mid-second century CE).

Thankfully, we actually have the passage Galen is summarizing, from the treatise *On Nutriment*, ascribed to Hippocrates of Chios (*ca.* 460–*ca.* 375 BCE), who was a rough contemporary of the early Pythagoreans:

T6: XXIII. **Conflux (*surroia*) one, conspiration (*sumpnoia*) one, [in] sympathy (*sumpathea*) all things**; in wholeness of limbs, all things, but severally, the parts in each part relative to the function.

XXIV. The great beginning comes to the furthest part; from the furthest part it travels to the great beginning: one nature, to be and not to be. —Pseudo-Hippocrates, *On Nutriment* 23–24, p. 143.1 Joly, 9.106,6 (before mid-second century CE).[26]

Now when we examine these texts, we face a similar problem to that faced with the text of pseudo-Ecphantus. It is not certain that *On Nutriment* reflects Hippocrates' own words: rather we are dealing here with a textual tradition in which Hippocratic texts were produced over a long period of time. Scholars have attempted to date this text to a period as early as the beginning of the fourth century BCE;[27] at any rate, we need to date it prior to Galen, who knew it in the mid-second century CE. Does the style help us to adjudicate the matter? Someone might describe it as archaic, or as archaizing: as Craik notes, the style is aphoristic,[28] and we might think it could reflect anything from Heraclitean riddle to a Hippocratic physician's crib notes. But here Galen is instructive: he places the document within a 'first discoverers' *topos*, using (his own interpretation of) the Hippocratic text to demonstrate that Hippocrates was the first to advance a theory of the elemental qualities. He is at pains to demonstrate that Hippocrates anticipated the elemental theories of the Stoics (especially Chrysippus) and Aristotle, and that they agreed with him in most ways. This is a worrying sign, because it is not obvious that Aristotle agreed with Hippocrates at all on this point; and while the Stoics would appear to have agreed with him (as well as the Post-Hellenistic Pythagorean pseudo-Ecphantus, and the Neo-Pythagorean Moderatus), this would point to a Stoic origin for the aphoristic statement 'Conflux one, conspiration one, [in] sympathy all things'; after all, the evidence for Stoic conspiration as sympathy is robust, whereas this is the only evidence for a similar idea among Hippocratics. It seems likely, then, that the Hippocratic *On Nutriment* is a text influenced by Stoicism, rather than the other way around, and moreover that similarities in the formulation found among the Post-Hellenistic Pythagoreans also point to a Stoic origin.

What are we left with at the end of this selective discussion of notions of cosmic conspiration? After examining texts written by a Roman didactic poet (Manilius), some Post-Hellenistic Pythagoreans (pseudo-Ecphantus and Moderatus), a materialist philosophical school (the Stoics), a famous doctor (Galen), and an unknown author of an enigmatic medical text (the Hippocratic author of *On Nutriment*), we can conclude that cosmic conspiration as described here is ultimately the brainchild of the Stoics, and probably the transformative Chrysippus of Soli, whose contributions to philosophy are so wide and varied it becomes almost impossible to determine them. In this way he resembles the legendary Pythagoras and Hippocrates—with the exception that it is less common to see ancient philosophers and doxographers crediting Chrysippus with major contributions to the history of thought.

By way of a coda, to give some indication of where things go from Galen, it is worth considering a text written by Clement (*ca.* 150 CE–*ca.* 215 CE), a Christian theologian and convert who lived and taught in Alexandria, a centre for philosophical activity and culture since the time of Aristotle. As teacher of the immensely influential Origen of Alexandria, Clement is to be counted as especially important in the transformation of Greek philosophy to Christian doctrine. In the following passage of his *Miscellanies* Clement both *appropriates and rejects* the philosophical and medical models of 'conspiration', in favour of something apparently different:

> T7: What, then, is their [sc. the worshippers'] idea as to the breathing of God? Is it by means of transpiration as in the demons? Or by inspiration only, as in fishes through the dilation of their gills? Or by circumspiration, as in insects through the pressure of the membranes on the waist? No, they would not liken God to any of these, if they were in their senses. But as for creatures that live by respiration, they draw in the air by rhythmic beats corresponding to the counter-dilatation of the lungs against the chest. Then if they assign viscera and arteries and veins and sinews and members to God, they will exhibit Him as in no respect differing from man.
>
> The word 'conspiration' (*sumpnoia*) is that which is properly used of the Church. For the Church's sacrifice is indeed speech rising, like incense, from holy souls, while every thought of the heart is laid open to God along with the sacrifice. They are fond of talking about the purity of the most ancient altar at Delos, that altar which, we are told, was the only one approached by Pythagoras, because it was unpolluted by slaughter and death: will they then refuse credence to us when we say that the truly hallowed altar is the righteous soul, and the incense which ascends from it, the prayer of holiness? Sacrifices, I believe, are an invention of mankind to excuse the eating of flesh, though, even apart

from such idolatry, it was always possible for one who wished it to partake of flesh.—Clement of Alexandria, *Miscellanies* VII 6.32 (trans. by Hort and Mayor) (late second century CE).[29]

It may be that Clement found in his doxographical sources the rather typical claim that ascribed a high value to the notion of breath and breathing to Pythagoras; and, indeed, this association is as old as Aristotle in the mid-fourth century BCE.[30] Clement's rejection of anthropomorphizing God rests on the implication that He cannot 'breathe' in any way similar to the animals He created; and it is a point well taken, if not very original.[31] However, this rejection also places certain special demands on Christian readers of, for example, Genesis 2:7, where famously God breathes into the nostrils of Adam in order to give him life: how are we to explain this obvious analogy drawn between God and Man, both of whom breathe?[32] Clement's solution to the problem, to find in *the Church* the proper space for deployment of the powerful metaphor of conspiration, is inflected with Pythagorean and Stoic ideas that trace back, at the very latest (and possibly earlier), to the first century BCE, whereby conspiration is a collaborative effort that somehow connects mere mortals to the highest reaches of the cosmos—even to God. In this way, Clement offers a resonant example of how the Christian writers of the second and third centuries CE sought to reject pagan philosophical doctrines as absurd and contrary to Christian wisdom, while still carefully manipulating and repurposing their content in almost imperceptible ways.

Notes

1. Also see Phillip Sidney Horky, 'Cosmic Spiritualism among the Pythagoreans, Stoics, Jews, and Early Christians', in *Cosmos in the Ancient World*, ed. by Phillip Sidney Horky (Cambridge: Cambridge University Press, 2019), 270–94. For a fuller account of ancient cosmic breathing, readers are encouraged to consult that piece, which was also produced from work undertaken for the Life of Breath project. For a more scientific approach to the history of breath and its meaning in Greco-Roman culture, see Long's contribution in Chapter 2 of this volume. Unless otherwise noted, translations of Greek and Latin texts are my own.
2. The earliest evidence of this usage is mid-first century BCE: Caesar, *Commentarii. I. Bellum Gallicum*, ed. by R. L. A. Du Pontet (Oxford: Oxford University Press, 1963), 3.10.3; T. Livius, *Ab Urbe Condita*, vol.

1, bks 1–5, ed. by R. M. Ogilvie (Oxford: Oxford University Press, 1974), 3.64.1. etc.

3. The earliest reference of this usage in Latin occurs around 140 BCE: Accius, in *Remains of Old Latin, Volume II: Livius Andronicus. Naevius. Pacuvius. Accius.*, ed. by E. H. Warmington (Cambridge, MA: Harvard University Press, 1936), Frs. 648–49: '*O suavem linguae sonitum! o dulcitas conspirantum animae!*' (O pleasant sound of the tongue! O sweetness of the soul of those who breathe as one!).
4. Epicurean: T. Lucretius, *De Rerum Natura: Libri Sex*, 2nd edn, ed. by C. Bailey (Oxford: Oxford University Press, 1923), 4.1212–17; Stoic: G. Musonius Rufus, *Reliquiae*, ed. by O. Hense (Leipzig, Teubner, 1905), 13a.
5. At Plutarch, *Life of Aratus* 24.5, in Plutarch, *Plutarch's Lives*, trans. by B. Perrin (Cambridge, MA: Harvard University Press, 1926), Aratus considers how the weak Greek *poleis* could be preserved by mutual support through a common agreement, and that 'just as the members of the body have a common life and breath because they cleave together in a common growth, but when they are drawn apart and become separate they wither away and decay, in a like manner the several states are ruined by those who dissever their common bonds, but are augmented by mutual support, when they become parts of a great whole and enjoy a common foresight' (trans. by Perrin).
6. Manilius, *Astronomica*, ed. and trans. by G. P. Goold (Cambridge, MA: Harvard University Press, 1977).
7. This is an apparent etymologization of 'aether' as 'always running' (*aei theontos*). The 'upper' aether constituted the region associated with the sphere of the sun, stars, and planets in Hellenistic/Post-Hellenistic Pythagoreanism. See, e.g., Alexander Polyhistor at Diogenes Laertius, *Lives of the Eminent Philosophers*, ed. by T. Dorandi (Cambridge: Cambridge University Press, 2013), 8.26–27.
8. The edition of the Pythagorean Pseudepigrapha and related testimonies used throughout is Holger Thesleff, *The Pythagorean Texts of the Hellenistic Period* (Åbo: Åbo Akademi, 1965).
9. Manilius, *Astronomica*, 2.136–37.
10. See Katharina Volk, *Poetics of Latin Didactic: Lucretius, Vergil, Ovid, Manilius* (Oxford: Oxford University Press, 2002), ch. 6.
11. Cf. Hans von Arnim, ed., *Stoicorum Veterum Fragmenta*, 3 vols (Berlin: de Gruyter, 1978), 3.43 (Antipater) (part) and 2.543 (part).
12. See, *e.g.*, Horky, 'Cosmic Spiritualism', 275–77.
13. There is even some room for debate on this point among the Stoics, if we are to believe pseudo-Plutarch's account of Stoic cosmology: 'this cosmos, being conspirative (*sumponoun*) and co-affected (*sumpathe*) itself

with itself, is governed by nature' (von Arnim, ed., *Stoicorum Veterum Fragmenta*, 2.192).

14. For general accounts, see Giulia De Cesaris and Phillip Sidney Horky, 'Hellenistic Pythagorean Epistemology', *Lexicon Philosophicum* Special Issue: *Hellenistic Theories of Knowledge* (2018), 221–23; and Bruno Centrone, 'The pseudo-Pythagorean Writings', in *A History of Pythagoreanism*, ed. by Carl A. Huffman (Cambridge: Cambridge University Press, 2014), 321–26.
15. For a review of the various positions advanced, see Francesca Calabi, *God's Acting, Man's Acting: Tradition and Philosophy in Philo of Alexandria* (Leiden: Brill, 2007), 186–88.
16. The origins of this notion are to be found ultimately in the activity of the World-Soul, identified as a living being in Plato, *Timaeus* 30b6-c1, in Plato, *Opera, Volume IV: Clitopho, Respublica, Timaeus, Critias*, ed. by J. Burnet (Oxford: Oxford University Press, 1922).
17. Cf. pseudo-Ecphantus, *On Kingship*, Thesleff, *Pythagorean Texts*, 79.2–7-14. On this passage and its relations to Philo, see Calabi, *God's Acting*, 200–03. It is surprising that Calabi does not note that all other animate beings also 'conspire' with the cosmos in particular. Hence, her interpretation of the passage, which claims that 'it seems ... far more likely that the breath [of man] derives from god', does not address the point that it is the cosmos that engages in conspiration, and not god (as in Philo, who is influenced by Genesis).
18. See Horky, 'Cosmic Spiritualism', 272–75.
19. Ibid.
20. He is often referred to as a 'Neo-Pythagorean' (cf. John Dillon, *The Middle Platonists: 80 B.C. to A.D. 220*, rev. edn (London: Duckworth, 1996), 341).
21. Typically, we would be inclined to capitalize as 'the One'; but it is clear, as we see in the passage, that Moderatus does not mean to refer to the divine One, or the god One, to which the Middle Platonists deferred.
22. That this passage ultimately derives from Moderatus is clear from Porphyry, *Life of Pythagoras* 48, where Porphyry tells us that Moderatus treated Pythagorean numerology and compared their approach to that of the geometricians.
23. Porphyre, *Vie de Pythagore, Lettre à Marcella*, ed. by Edouard des Places (Paris: Les Belles Lettres, 1982).
24. On the three Ones in Moderatus' ontology, see Dillon, *Middle Platonists*, 347, and the cautionary remarks of Svetla Slaveva-Griffin, *Plotinus on Number* (Oxford: Oxford University Press, 2009), 44–46.
25. Galen, *Method of Medicine*, vol. 1, bks 1–4, ed. and trans. by I. Johnston and G. H. R. Horsley (Cambridge, MA: Harvard University Press, 2011), 1.27–28. Cf. von Arnim, ed., *Stoicorum Veterum Fragmenta*, 2.411.

26. Hippocrate, *Du régime des maladies aiguës, De l'aliment, De l'usage des liquides*, vol. 6.2, ed. by R. Joly (Paris: Les Belles Lettres, 1972), 9.106,6.
27. *E.g.*, Elizabeth M. Craik, *The 'Hippocratic' Corpus: Content and Context* (London: Routledge, 2015), 26 (who follows Joly).
28. Ibid.
29. Fenton John Anthony Hort, and Joseph B. Mayor, *Clement of Alexandria: Miscellanies Book VII* (London: Macmillan, 1902), 5.5.
30. See Horky, 'Cosmic Spiritualism', 272–75.
31. The rejection of anthropomorphic gods is as old as Xenophanes of Colophon in the late-sixth-century BCE (see Hermann Diels and Walther Kranz, *Die Fragmente der Vorsokratiker*, 6th edn, vol. 2 (Berlin: Weidmann, 1952), 21 B 11–12, 14, and 23).
32. The Hellenic Jew Philo of Alexandria (*ca.* first century CE) attempts to explain it in different ways. See Horky, 'Cosmic Spiritualism', 281–86.

Bibliography

Calabi, Francesca. 2007. *God's Acting, Man's Acting: Tradition and Philosophy in Philo of Alexandria*. Leiden: Brill.

Centrone, Bruno. 2014. "The Pseudo-Pythagorean Writings." In *A History of Pythagoreanism*, ed. Carl A. Huffman, 315–34. Cambridge: Cambridge University Press.

Craik, Elizabeth M. 2015. *The "Hippocratic" Corpus: Content and Context*. London: Routledge.

De Cesaris, Giulia, and Phillip Sidney Horky. 2018. "Hellenistic Pythagorean Epistemology." *Lexicon Philosophicum*. Special Issue: *Hellenistic Theories of Knowledge*: 221–62.

Dillon, John. 1996. *The Middle Platonists: 80 B.C. to A.D. 220*, Rev. edn. London: Duckworth.

Horky, Phillip Sidney. 2019. "Cosmic Spiritualism among the Pythagoreans, Stoics, Jews, and Early Christians." In *Cosmos in the Ancient World*, ed. Phillip Sidney Horky, 270–94. Cambridge: Cambridge University Press.

Hort, Fenton John Anthony, and Joseph B. Mayor, eds 1902. *Clement of Alexandria: Miscellanies Book VII*. London: Macmillan.

Slaveva-Griffin, Svetla. 2009. *Plotinus on Number*. Oxford: Oxford University Press.

Thesleff, Holger. 1965. *The Pythagorean Texts of the Hellenistic Period*. Åbo: Åbo Akademi.

Volk, Katharina. 2002. *Poetics of Latin Didactic: Lucretius, Vergil, Ovid, Manilius*. Oxford: Oxford University Press.

CHAPTER 4

Late Antique Cultures of Breath: Politics and the Holy Spirit

Thomas E. Hunt

Around the turn of the fifth century CE the Roman grammarian Servius wrote a commentary on the *Georgics*, a poem written by Vergil in the first century BCE.[1] Near the beginning of the first book of the *Georgics*, Vergil talks about the processes by which farmers begin to till the land. Servius weaves his comments into the poem's text (in italics), clarifying philological points and connecting these lines to other works by Vergil and the poet Lucretius:

> *Before we split the unknown plane with iron*, that is, before it has been ploughed ... Moreover, we should understand 'plane' to refer to the earth, saying it's flat, like 'in fury he forces Dares across the whole plane' [Vergil, *Aeneid* 5, l. 456]. From this it is said that the sea is a plane.
>
> [*let it be our concern to learn in advance*] *the winds* because different winds are common in different places. This is meant: you ought to know to which wind your land is subject.

T. E. Hunt (✉)
Theology Department, Newman University, Birmingham, UK

D. Fuller et al. (eds.), *The Life of Breath in Literature, Culture and Medicine*, Palgrave Studies in Literature, Science and Medicine,
https://doi.org/10.1007/978-3-030-74443-4_4

> *and the changeable custom of the heavens*, that is, air: Lucretius: 'this heaven which is called air' [Lucretius, *De rerum natura* l. 132]. So this is said: Moreover, you should know your sky, whether it rejoices in rain, whether it's warm or cold.[2]

The first book of *Georgics* is a complicated and allusive work. Focussed on farming and the development of culture that agricultural settlement brings, the poem also positions human communities as radically contingent, inevitably threatened by political collapse and civil war.[3] In the early lines quoted here, the poem looks ahead to the moment when plough breaches earth for the first time. In the poem this ploughing leads to crops, livestock, a hearth, a household, all the goods of culture that follow, and then, later, civil war and dissolution. In her analysis of this section of *Georgics*, Ika Willis argues that this ploughing is a mark on the ground that makes the earth political space. After the metal strikes soil, there is a difference between the untilled—unknown—earth and the cultivated terrain of farming and human settlement. In the passage above, however, all this is mere potential. Before he can cut a furrow that will realise this future, the farmer must know the air and earth.[4]

Servius understood this. His comments emphasise the plough's capacity to make space political and situate this future-making as an event of cosmic significance. He tells us that this soil is called a 'plane' (*aequor*) because it is flat and undifferentiated (*ab aequalitate dictam*). The comparison is to the sea, another place unmarked, unknown, and unowned.[5] Servius draws out the cosmological import of the first furrow with a reference to Lucretius: the air the farmer observes is merely one part of a much larger thing reaching up into the sky, including the stars and what lies beyond them.[6] Air was therefore part of the wider cycling of time and space which determined how humans lived in the world.[7] In this way, Servius draws out the poem's claims about nature and human communities. The dip-and-ridge of ploughing makes a place for people to live together, a moment in time that should be understood in the wide, cosmic context of earth, air, and human life.

In late antiquity, commentaries like this served to order people's knowledge of the world.[8] Certain works, like Vergil's, were held to be reservoirs of divine knowledge that might be used to learn about far more than poetic form and grammar. For men like Servius, to know the world was to approach it through the stores of divine knowledge available in this textual patrimony to those who knew how to cultivate it.[9] This constant working and re-working of texts was 'circular', in the words of Raffaella Cribiore.[10] One text could be laboured over again and again, like a tilled field, producing new knowledge about the world. Commentators like Servius linked these texts to other works by the 'ancients' (*ueteres*).[11] As they did so, they laid out the subtle connections that revealed a deeper cosmic order subtending these inscribed words. Education in late antiquity was participation in this textual patrimony. Vergil told them that the beginning of all this was the plough marking soil, a moment around which space and bodies were ordered as the commentator ordered verse. But the beginning of ploughing is wind and 'this heaven which is called air'. It was this cosmic, primordial thing that was breathed in late antiquity.

This essay argues that late antique discussions of breath worked in two registers. Breath was part of material human life and so was drawn into wider models of community and politics. In effect, to breathe placed one in the human world that Vergil said began with the plough's furrow. The manner of one's breathing was a means by which humans could be differentiated. At the same time, breath also took on cosmic significance as a site of primordial becoming. The beginning of things and how they were sustained could be described using the language of breath and breathing. These two registers overlapped with and determined each other. In late antiquity, political accounts of breath had metaphysical overtones, and cosmic accounts of breath also said something about the everyday life of embodied humans. This blending of the cosmic and the political is the principal character of late antique cultures of breath.

Breath and Human Difference in Late Antique Rome

One place where breath becomes visible in late antiquity is in the work of grammarians. Grammarians taught boys and girls the structures of language, how to read, and how to write. As Servius' commentary indicates, instruction in grammar was also an induction into the wider social and cultural world of late antiquity. As the grammarian guided students through the canonical texts of late antique society, lessons covered areas as diverse as the climate, the social structures and alimentary habits of animals, the precise ways in which bird flight presaged disaster, and all the other information that constituted the knowledges of late antique people. These topics were not introduced haphazardly but were part of a careful induction into the structures by which people orientated themselves in the world. Reading texts, then, was an ethical practice because it showed boys and girls the principles and standards by which they should behave. Grammatical instruction carried young people into the world of adulthood. It made them visible as users of language and hence as political actors. It taught them to live in the world.

One of the things that these students learned was theories of language. Aelius Donatus was a grammarian who lived and worked in Rome in the third quarter of the fourth century CE. His curriculum began with a focus on the nature of sound and its relationship to human language:

> A sound is air that has been struck (*uox est aer ictus*) and which, barring other factors, can be heard. Every sound is either articulated or confused. An articulated sound is one that can be captured by the letters of the alphabet; a confused sound is one that cannot be written down. The letter is the smallest element of articulated sound. Some letters are vowels, some are consonants.[12]

Noise (*uox*) is made by striking air, a definition found in other grammars from late antique Rome.[13] The distinction in Donatus' text is between noise that can be differentiated, ordered, articulated, and noise that cannot be differentiated and is mere confusion. The former—*uox articulata*—is human language. This *uox* is held by the letters inscribed on the page, tablet, or stone. Donatus is referring particularly here to the human voice, as it sounds human language that can be inscribed. This discussion of the vocal character of language opens the formal study of grammar and begins the student's movement into the social world of late antiquity. The struck air in this context is breath.

These remarks are particularly important given the position of spoken language in late antique society. In the later Roman Empire, relationships between people were often understood in terms of mastery. Mastery over others began with a surveillance and mastery of the self.[14] Control of the emotions and body denoted fitness to hold mastery over others and in this way models of mastery and slavery determined how people imagined themselves in relation to others.[15] This discourse was also gendered. To be a man meant being sovereign over the self—not overwhelmed and unmanned by emotions—and over others. Spoken language was a key part of this model of relationality. Words said, and the timbre of the saying, were images of an interior state. To master the tongue was to speak well and, consequently, to hold off the violence that inflected life in the late ancient world. One's position in this complex web of mastery and violence therefore became socially legible through the way that one spoke. The manipulation of language was one of the principal means by which humans were differentiated from each other. The beginning of language was breath.

Donatus addresses the breath's role in the production of spoken language. Given the importance of speech to late antique relationality, the passage implicitly addresses the role of breath in the constitution, reproduction, and maintenance of difference. In this sense, breath is part of the wider play of mastery and power in late antique society. Breath is struck, as a master might strike a child, slave, or animal.[16] In late antique society, such a blow indicated the striker's mastery over the struck. In Donatus's example, the striking of the breath denotes the speaker's mastery over the breath, and hence, over speech. Breath is therefore positioned within the wider discourse of mastery and slavery that determined relationships between people. However, the importance of speech to late antique understanding of human relationality actually means that, for Donatus, breath is the grounds of this discourse. Without breath, there is no language, not even writing, because all writing is breath being held (*conprehendi*). The striking of breath splits the air moving out of the mouth, producing a *uox* made of parts joined together (*articulatus*). This incision begins grammar and makes it possible to pass on the patrimony to the next generation of students. Without this patrimony, there can be no society. Like the cleft in the soil that heralded the possibility of cultivation, culture, and imperial civilisation, this breathy cut opens out the space of the political in later Roman society.

To speak of breathing in late antiquity, then, is to refer to the movement of air around bodies rendered master and mastered. The way this air is moved, struck, and inscribed has a direct isomorphic relationship to the way that the bodies of slaves, children, women, and animals might be moved, struck, and inscribed. In this way, breath is subsumed within the discourses of mastery and property ownership by which social and political relations were understood. Breath is an index of the properties held by a body: its masculinity, its mastery, the material things over which it holds sovereignty.[17] It concerns the capture and ownership of things and the disposition of people in space. This emphasis recurs in other late antique texts.

The poet Prudentius wrote a generation after Donatus. In the first book of the poem written in reply to Symmachus he presents the Emperor

Theodosius, victorious in battle against usurpers in 394, encouraging Roman senators to give up their paganism.[18] At the opening of this speech, Theodosius addresses the city itself. Under his rivals Rome had been 'besieged by the black smoke' of pagan sacrifices and covered by 'hovering clouds' that were the habitat of demons.[19] Theodosius encourages the city to 'raise your exalted face above the air' and remember that God has decreed that all of the earth should be subject to Rome.[20] In this section of the poem, Rome is restored to her divinely appointed mastery of the world. In Prudentius's account, air becomes a way of marking the political transition from pagan usurper to rightful, Christian emperor. The dirty air—a billowy substance that sustains the demonically non-human—is swept away and Rome is restored.

Embedded in Theodosius's speech is an account of how the city of Rome was conquered by the emperor Constantine and first made Christian. In this passage Constantine triumphed over Maxentius, the usurper who held the city. This usurper had disrupted the natural order of things, but Constantine's victory makes Rome the greatest city in the world, placing the Christian God at Rome's head. Natural order is therefore restored, a process presented as mirroring Theodosius's own achievements. In his description of Rome under the usurpers, Prudentius states that women were rounded up and taken away from their husbands and fathers:

> The cruel emperor's prisons were full of the fathers of girls. If a father murmured and complained too bitterly when his daughter was taken away, he could not betray his anger or heave too frank a sigh without punishment.[21]

Drawing on the earlier example of Constantine, Theodosius's speech argues that the triumph of an orthodox Christian emperor restores a natural order which has been disrupted by pagan usurpers. In this context, the sighing and murmuring of fathers indicates this wider disruption: children are taken from their rightful guardians. Reading the fathers' sighs as part of a late antique culture of breath, however, should draw our attention to the particular ways that mastery works in this natural order. In late Roman familial law fathers exercised absolute rights of life and death over their children, at least theoretically.[22] In this wider context it is clear that Prudentius's primary concern in this passage is not the women's welfare, but the property rights of their fathers. Abducting female members of the family is an attack on the father's mastery over his children—an assault on his masculinity and status.[23] The proscription against sighing sits alongside references to smoke and clouds. The way that air moves in weather, pollution, and breath signals a breakdown in human relations and the proper cosmic order of Roman power.

When we think about breath in late antiquity, we translate words like *pneuma*, *aer*, or *spiritus* into English, but we should also think about the kind of ideas that cluster around respiration. In Donatus and Prudentius, the physical exhalation of air is the thing that determines the organisation of community. Donatus's account of speech as struck air opens his grammar; it is the beginning of rightly ordered speech and the induction into appropriate human relationships. For Prudentius the pollution of air betokens a wider disruption of the natural order: the arrival of a usurper or the praise of false gods. For both authors, the circulation of air in breath is formally related to social and political structures. When fathers' natural rights over their children are disrupted, this disorder manifests itself in failures of respiration and muttering speech. Breathing is always an embodied action and—by virtue of its embodiment—it indicates the structures of interrelationship in the human community.

The Holy Spirit and Breath

Donatus and Prudentius both wrote in the later fourth century CE. This was also an important period in the development of Christian accounts of the triune God. In 325 the Council of Nicaea had established a formula for describing the relationship of the Son to the Father, but disagreements about the Son's status and the nature of God continued to rumble on. These were not simply abstract arguments. Rather, they were debates over the parameters of the material world, the nature of human life, and the role this humanity played in the wider cosmos. From the 360s until the end of the fourth century a key focus of this theological wrangling was the nature and work of the Holy Spirit. While there was a common agreement that the Holy Spirit did the work of God—sanctifying the world and its creatures—it was also clear that the biblical traditions that described the Spirit and its work were ambiguous. This established the field within which Christians argued about the Spirit. As they did so, their language for describing God and his *pneuma/spiritus* ('breath') became increasingly precise. Particular areas of debate included the divinity of the Spirit itself, the role of the Spirit in the Trinity, and the Spirit's relationship with human beings. In each of these cases there was engagement with the wider cultures of breath and breathing already identified above.

Sometime before 360 some Christian groups in Egypt began to argue that the Spirit was an angel and therefore a creature rather than fully divine. Those who rejected this argument responded by resuscitating the venerable idea that the Spirit was Creator. For example, *The Letter to Serapion* (360), written by Athanasius of Alexandria (296–373), describes the Spirit as uncreated.[24] This coequivalence of Creator and Spirit made it possible to argue that the Spirit was truly divine and that it sanctified ('made holy') other things. As Didymus of Alexandria put it, 'the Holy Spirit, as all acknowledge, is the immutable sanctifier, the bestower of divine knowledge and all goods'.[25] Didymus took ancient arguments about God's immutability and applied them to the Spirit: God is not diminished or changed when God acts or gives. Those who receive gifts from God do so not according to God's capacity to give, but according to their own capacity to receive. So, when faced with arguments that the Spirit was of lesser status than the Father, Didymus drew on the ancient argument for divine immutability, applying it to the Spirit to show how the ongoing work of the Spirit in the world was the ongoing work of God. To argue that the Holy Breath drifted over the waters at Creation and now drifted across the bodies of humans resurrected an old theological position.[26] Its restoration in the 360s allowed fourth-century theologians to show with greater precision how the Holy Spirit might be considered to be God.

With the Spirit's divinity more precisely described, theologians could also define the relationship between the Spirit and the world which it sanctified. To do this, they often drew on pre-existing ways of imagining relationality and community. For example, in Basil of Caesarea's work *Against Eunomius*, God and Creation are distinguished and their relationship described:

> Now it is said that there are two realities: the divinity and the created order, sovereignty and servitude, the power that sanctifies and the one who is sanctified, the one who has virtue by nature and the other that achieved virtue by free will.[27]

Basil argues that the Spirit should be placed in the first, divine, class. The Holy Spirit is the thing that sanctifies. It does not receive its holiness from God, but rather is holy by its divine nature. Basil's concern here is to show

how the divine reality of Father, Son, and Spirit is related to the visible and tangible reality around us. The relationship between the two realms is described as that of sovereignty and servitude. The term ‘sovereignty’ (*despoteia*) is often applied to the legal holding of property, as when a rich person owns a house. It is a political description of ownership. Servitude (*douleia*) here refers to the nature of a particular thing that has the property of being a slave. At the heart of this passage is Basil’s attempt to situate divine *pneuma* as part of the reality that is the Godhead. As he does so, however, political notions like property ownership intertwine with ontological claims about the difference between slave and master. Basil tries to show how the Spirit is distinct from this created world; to speak of the Spirit is to speak of the structure and pattern of their relationship. As he describes this relationship, he draws on a wider culturally recognised model of human community, one that understood relationality in terms of dominance and subjugation.

Another example of relationality in the developing theologies of the Holy Spirit occurs in the attempts of theologians to define the interrelationship of the three persons of the Trinity (Father, Son, and Spirit). At the start of the 360s, Eunomius, a bishop of Cyzicus in Asia Minor, had argued that the Spirit was of lesser status than the Father and the Son. In response, Didymus of Alexandria wrote *On the Holy Spirit*, arguing that the Spirit was of equal status with the Father and the Son, and positing a mutual indwelling of all the persons of the Trinity. In making this argument, he includes a long discussion of biblical texts concerning the Spirit, beginning with John’s Gospel. When Didymus comes to John 16:13 (‘[the Spirit of Truth] will not speak on his own, but will speak whatever he hears, and he will declare to you the things that are to come’) he deploys an image already familiar to us:

> When human beings speak to one another about something, we first conceive what we want to say in our mind without speech. Then when we want to convey it into the mind of another, we set the tongue in motion as an instrument, and by striking it like a kind of plectrum on the strings of the teeth, we emit an articulate sound (*uocalem sonum emittimus*). So … we control how we knock our tongue on the palate and the teeth and combine the struck air into various utterances (*et ictum aerem in diuersa temperamus eloquia*) in order to communicate to others what we have in mind.[28]

Didymus refers to speech as consisting of *aer ictus*, the same description already deployed in the work of grammarians like Aelius Donatus.[29] Didymus's argument is that John 16 is not referring to physical speaking of the kind that humans use to communicate with each other. Rather, he says that the biblical text is deploying an analogy to describe the common will shared by the three persons of the Trinity. Humans speak and by this speaking reach accord. The 'speaking' of John 16 denotes the accord and equality between the three persons of the Trinity. Didymus does not think that the three persons communicate as humans do, nor does he position the Holy Spirit as the medium of communication within the Trinity. He does, however, integrate Trinitarian theology into an analogy of human speech and therefore positions the Spirit in the same late antique culture of breath evident in Donatus and Prudentius. When theologians imagined the interrelationship of the three persons of the Trinity, the notions of breath and the mechanics of speaking could provide useful and culturally legible tools to work with.

The development of Christian theology of the Holy Spirit has some important consequences for our models of late antique breath. In the work of Donatus and Prudentius, human breathing bore a formal relationship to the wider social and political structures of human communities. Manners of breathing differentiated humans, binding them into social and legal relations. As theology of the Holy Spirit developed in the fourth century, differences of holiness within the Christian community were explained through reference to the Spirit. For example, Basil's work on the Holy Spirit says:

> [The Spirit] fills all things with his power, but only those who are worthy may share it. He distributes his energy in proportion to the faith of the recipient, not confining it to a single share. He is simple in being, his powers are manifold: they are wholly present everywhere and in everything.[30]

This passage answers an apparent dilemma. The Holy Spirit sanctifies the world and is immutable and perfect. Despite this, our experience of human communities suggests that not all humans are equally holy.[31] In response, Basil says that humans only engage with the Spirit as far as they are able. The better one is able to participate in the Spirit, the more God-like one becomes, a doctrine called theosis.[32] Didymus introduced the idea that the breath of God, the *pneuma*, was limitless and immutable. The following generation developed this to logical conclusions, arguing that this breath was everywhere but that not everybody was able to partake of it to the same extent. Those who could, were sanctified. This is a fundamentally spatial position, for the Spirit is extended through the cosmos. It is also a political position, giving a clear account of why there are differences between humans and how those differences should be valued.

From the 360s to the end of the fourth century, Christian theologians became more careful in their statements about the nature and work of the Holy Spirit. They patterned their descriptions on wider accounts of how relationships worked, emphasising the sovereignty of the divine and the subjugation of the world. This was, therefore, inherently a political discourse, for it understood the ontological structure of reality through the prism of contemporary human community. This is clearly visible in Didymus's accounts of the interrelationship of the persons of the Trinity, in which he deploys accounts of breath and speech taken from the classrooms of grammarians like Donatus. This grammatical education was always political in character, for it positioned breath and speech as the beginning of human community. Didymus uses the same terms to describe the 'community' of the Trinity. The readiness of Christian writers to draw on accounts of human community as they fashioned analogies for the divine life had important consequences for how they came to picture the social life of people. Differences between humans—someone's relative sanctity and their superior holiness—could be presented as evidence of their capacity to participate in the divine community of the Godhead. Neither Basil nor Didymus would accept that the divine *pneuma* is a material thing; it is not of this reality. Nevertheless, their accounts of the Holy Spirit implicate it in wider models of community and human difference, and in so doing, they place the Holy Spirit within the wider cultural politics of breath in late antiquity.

Conclusion

By the end of late antiquity, *pneuma* and *spiritus* referred to the Holy Spirit, carefully distinguished from the material and immaterial reality of creation. As the articles by Long and Horky in this volume demonstrate so well, accounts of divine breath were already extant in Hellenistic philosophy and it might be argued that these accounts, combined with the ambiguities of the biblical text, provide sufficient cause for the development of Christian theologies of the Holy Spirit. Similarly, the accounts of breath and language outlined by the grammarians are indebted to a Stoic tradition that distinguished rational from irrational speech. Based on these facts we might say that the late antique culture of breath laid out here is simply a particular manifestation of a wider ancient physics of air.

It would be wrong to assert that Hellenistic accounts of divine *pneuma* did not play a role in the development of Holy Spirit theology, but something else is going on here as well. As C. Michael Chin observes, late antique grammars are cultural products.[33] Just so for late antique Christian theology. Cultural products are determined by the material conditions of their production: models of property ownership and the protection of these by legal, social, and political structures. Few people in late antiquity knew or understood Hellenistic philosophy, but everyone breathed. The sources laid out above are explicit about the materiality of breath as it moves through and around bodies. In these accounts, human respiration is an index of social relations, whether in the city of Rome or, by analogy, in the Godhead. The cultures of breath in late antiquity shaped the language and theology of the Holy Spirit as it developed through the latter half of the fourth century.

While not a sufficient cause, then, late antique cultures of breath determined the forms taken by theology of the Holy Spirit. The consequences of this would seem to be far-reaching. The political and social differentiation of humans carried great ontological and soteriological significance. Breath, and one's capacity to breathe, was placed in the same register as one's capacity to engage in divine life and be sanctified. Human respiration was intertwined with the imaginative structures of Christianity and, consequently, notions of good, evil, blame, election, salvation, freedom, servitude, and damnation could be interwoven with the character and deportment of one's breathing body. Understanding how late antique people thought about respiration and the movement of *aer* around bodies requires working simultaneously in a cosmic and in a political register.

Notes

1. For Servius, see Anne Uhl, *Servius als Sprachlehrer: zur Sprachrichtigkeit in der exegetischen Praxis des spätantiken Grammatikerunterrichts* (Göttingen: Vandenhoeck & Ruprecht, 1988). My translation.

2. Servius, *Servii Grammatici qui feruntur in Vergilii carmina commentarii*, ed. by Georg Thilo, vol. 3, fasc. 1 (Leipzig: Teubner, 1887), 145–46. He is commenting on *Georgics*, bk 1, ll. 50–51.
3. Virgil, *Eclogues. Georgics. Aeneid: Book 1–6*. trans. by H. R. Fairclough (1916), rev. by G. P. Goold (Cambridge, MA: Harvard University Press, 1999), 98–135. For a commentary, see R. A. B. Mynors, *Virgil, Georgics* (Oxford: Clarendon, 1990). For an introduction to the poem and secondary literature, including the debate about whether it is pessimistic: William Batstone, 'Virgilian diaxis: value and meaning in the *Georgics*', in *The Cambridge Companion to Virgil*, ed. by C. Martindale (Cambridge: Cambridge University Press, 1997), 193–215.
4. Ika Willis, *Now and Rome: Lucan and Vergil as Theorists of Politics and Space* (London: Continuum, 2010), 21–22.
5. Carl Schmitt, *The Nomos of the Earth in the International Law of the Jus Publicum Europaeum* (New York: Telos Press, 2006), 172.
6. Gregory Smith, 'Physics and Metaphysics', in *The Oxford Handbook of Late Antiquity*, ed. by Scott Fitzgerald Johnson (Oxford: Oxford University Press, 2012), 531–33.
7. David O. Ross, *Virgil's Elements: Physics and Poetry in the Georgics* (Princeton: Princeton University Press, 1987), 38–43.
8. Jason König and Tim Whitmarsh, 'Ordering Knowledge', in *Ordering Knowledge in the Roman Empire*, ed. by Jason König and Tim Whitmarsh (Cambridge: Cambridge University Press, 2007), 29.
9. C. M. Chin, *Grammar and Christianity in the Late Roman World* (Philadelphia: University of Pennsylvania Press, 2007), 18.
10. Raffaella Cribiore, *Gymnastics of the Mind: Greek Education in Hellenistic and Roman Egypt* (Princeton: Princeton University Press, 2001), 129 and 187.
11. Chin, *Grammar and Christianity*, 21–24.
12. Louis Holtz, *Donat et la tradition de l'enseignement grammatical: étude sur l'Ars Donati et sa diffusion (IVe-IXe siècle) et édition critique* (Paris: Centre national de la recherche scientifique, 1981), 603. The translation of this passage, with changes: Shane Butler, *The Ancient Phonograph* (New York: Zone, 2015), 112–13.
13. For a near contemporary of Donatus, see Marius Victorinus, *Ars Grammatica*, ed. by Italo Mariotti (Firenze: Felice Le Monnier, 1967): '*Vox est aer ictus auditu percipibilis …*' (Sound is struck air that can be heard …).
14. Pierre Hadot, *Philosophy as a Way of Life: Spiritual Exercises from Socrates to Foucault* (Oxford: Blackwell, 1995), 130–36.
15. Peter Brown, *Power and Persuasion in Late Antiquity* (Madison, WI: University of Wisconsin Press, 1992), 50–51.
16. Geoffrey S. Nathan, *The Family in Late Antiquity: The Rise of Christianity and the Endurance of Tradition* (London: Routledge, 2000), 33–34.

17. Breath's importance as a site from which to observe and critique juridical and epistemic violence is clear in some recent studies on aesthetics and ethics. See, *e.g.*, Ashon T. Crawley, *Blackpentecostal Breath: The Aesthetics of Possibility* (New York: Fordham University Press, 2016), 32–86; Christina Sharpe, *In the Wake: On Blackness and Being* (Durham, NC: Duke University Press, 2016), 108–11.
18. Dated to between 394 and 395 by Jill Harries, 'Prudentius and Theodosius', *Latomus* 43 (1984), 79.
19. Black smoke: Prudentius, *Reply to Symmachus*, bk 1, l. 414. The demons flitting through clouds: *Reply to Symmachus*, bk 1, ll. 419–24. Critical edn: *Aurelii Prudentii Clementis Carmina*, ed. by Maurice P. Cunningham (Turnhout: Brepols, 1966). Eng. trans.: *Prudentius*, trans. by H. J. Thomson, 2 vols (Cambridge, MA: Harvard University Press, 1949–53).
20. Prudentius, *Reply to Symmachus*, bk 1, ll. 425–29.
21. Prudentius, *Reply to Symmachus*, bk 1, ll. 477–80.
22. Richard P. Saller, *Patriarchy, Property and Death in the Roman Family* (Cambridge: Cambridge University Press, 1994), 115.
23. Kate Cooper, *The Virgin and the Bride: Idealized Womanhood in Late Antiquity* (Cambridge, MA: Harvard University Press, 1996), 12–13.
24. Lewis Ayres, 'Innovation and *Ressourcement* in Pro-Nicene Pneumatology', *Augustinian Studies* 38 (2008), 187–205, at 194.
25. Lewis Ayres, 'The Holy Spirit as the "Undiminished Giver": Didymus the Blind's *De Spiritu Sancto* and the Development of Nicene Pneumatology,' in *The Holy Spirit in the Fathers of the Church: The Proceedings of the Seventh International Patristic Conference, Maynooth, 2008*, ed. by Vincent D. Twomey and Janet Rutherford (Dublin: Four Courts, 2010), 58.
26. For example, see Clement of Alexandria, *Exhortation to the Greeks*, ch. 1, para. 5. Edn and trans.: Clement of Alexandria, *Clement of Alexandria*, trans. by G. W. Butterworth (Cambridge, MA: Harvard University Press, 1919), 11–13. See also Miguel Herrero de Jáuregui, *Orphism and Christianity in Late Antiquity* (Berlin: De Gruyter, 2010), 292.
27. Basil of Caesarea, *Against Eunomius*, bk 3, § 2. Edn: Basil of Caesarea, *Contre Eunome*, ed. and trans. by B. Sesboüé, G. M. de Durand, and L. Doutreleau, 2 vols (Paris: Éditions du Cerf, 1982–83). Eng. trans.: St. Basil of Caesarea, *Against Eunomius*, trans. by Mark DelCogliano and Andrew Radde-Gallwitz (Washington, DC: Catholic University of America Press, 2011).
28. Athanasius the Great and Didymus the Blind, *Works on the Spirit: Athanasius's Letters to Serapion on the Holy Spirit and Didymus's On the Holy Spirit*, trans. by Mark DelCogliano, Andrew Radde-Gallwitz, and Lewis Ayres (Yonkers, NY: St Vladimir's Seminary Press, 2011), 191–92. Critical edn and French trans.: Didymus the Blind, *Traité du Saint-Esprit*, ed. and trans. by Louis Doutreleau (Paris: Éditions du Cerf, 1992).

29. Didymus wrote in Greek but the text is extant in a Latin translation by Jerome of Stridon, who was a pupil of Aelius Donatus.
30. Ayres, 'The Holy Spirit as the "Undiminished Giver"', 67. Basil, *De spiritu sancto* 9.22. Edn and French trans.: Basil of Caesarea, *Sur le Saint-Esprit*, ed. and trans. by Benoît Pruche (Paris: Éditions du Cerf, 1968).
31. A comparable and near contemporary position might be found in the work of Jovinian (*ca.* 390) whose positions on baptism and the differences within the Christian community were informed by his account of the actions of the Spirit. See David Hunter, *Marriage, Celibacy, and Heresy in Ancient Christianity: The Jovinianist Controversy* (Oxford: Oxford University Press, 2007), 35–39.
32. Ayres, 'The Holy Spirit as the "Undiminished Giver"', 69.
33. Chin, *Grammar and Christianity*, 5–6.

Bibliography

Ayres, Lewis. 2008. Innovation and *Ressourcement* in Pro-Nicene Pneumatology. *Augustinian Studies* 38: 187–205.

Chin, C. M. 2007. *Grammar and Christianity in the Late Roman World*. Philadelphia: University of Pennsylvania Press.

Crawley, Ashon T. 2016. *Blackpentecostal Breath: The Aesthetics of Possibility*. New York: Fordham University Press.

Cribiore, Raffaella. 2001. *Gymnastics of the Mind: Greek Education in Hellenistic and Roman Egypt*. Princeton: Princeton University Press.

DelCogliano, Mark, Andrew Radde-Gallwitz, and Lewis Ayres. 2011. General Introduction. In *Athanasius the Great and Didymus the Blind, Works on the Spirit: Athanasius's Letters to Serapion on the Holy Spirit and Didymus's On the Holy Spirit*, trans. Mark DelCogliano, Andrew Radde-Gallwitz, and Lewis Ayres, 11–51. Yonkers, NY: St Vladimir's Seminary Press.

Hunt, Thomas E. 2020. Breathy Shame and the Place of Hebrew in the Work of Jerome of Stridon. *Religion and Theology* 26: 85–111.

Kaster, Robert A. 1988. *Guardians of Language: The Grammarian and Society*. Berkeley, CA: University of California Press.

MacKendrick, Karmen. 2016. *The Matter of Voice: Sensual Soundings*. New York: Fordham University Press.

PART II

The Medieval Period

CHAPTER 5

From Romance to Vision: The Life of Breath in Medieval Literary Texts

Corinne Saunders

> Therwith the sorwe so his herte shette
> That from his eyen fil there nought a tere,
> And every spirit his vigour in knette,
> So they astoned or oppressed were.
> The felyng of his sorwe, or of his fere,
> Or of aught elles, fled was out of towne;
> And down he fel al sodeynly a-swowne.
> (*Troilus and Criseyde*, III, 1086–92)

> And sumtyme sche myth not stondyn but fel downe amonge the pepil and cryid ful lowde, that many man on hir wonderyd and merveylyd what hir eyled, for the fervowr of the spiryt was so meche that the body fayld and myth not endur it.
>
> (*The Book of Margery Kempe*, 6682–85)

Two images of extreme emotion: Troilus, overcome by the sight of his beloved lady Criseyde's tears, falls into a swoon; Margery Kempe, ravished

C. Saunders (✉)
Department of English Studies, University of Durham, Durham, UK

D. Fuller et al. (eds.), *The Life of Breath in Literature, Culture and Medicine*, Palgrave Studies in Literature, Science and Medicine,
https://doi.org/10.1007/978-3-030-74443-4_5

by visions of Jesus and Mary, falls down to the ground, crying out loudly. Such descriptions seem part and parcel of the conventions and excesses of medieval writing: love-sickness, madness, longing sighs, and visionary swoons. Yet they speak to medieval understandings of breath and breathlessness, and the physiological models of emotion that underpin these, in ways that go far beyond convention. Consciousness and breath are intimately interwoven. Thinking about medieval breath requires frameworks that are not our own, yet have their own complex logic. Breath forms an essential element in medieval theories of the bodily spirits which govern life, feeling, and thought. In romance, the imaginative fiction of the Middle Ages, the play of emotions is dependent on the movements of the spirits, which engender the states of deep feeling, in particular, love and grief, that distinguish the genre. Such feeling is dramatically written on the body through the flight of breath in and out of the heart. In religious writing, these affective movements are the catalysts for moments of revelation and vision. Breath and breathlessness animate the individual relationship with the divine and the enactment of piety, often in extreme ways. Middle English romances, the works of Chaucer and Gower, the visionary poem *Pearl*, the *Revelations of Divine Love* of Julian of Norwich, and the *Book of Margery Kempe* all offer compelling examples of the centrality of breath to medieval conceptions of affect and the shaping of experience from romantic love to spiritual epiphany. The breathless body becomes the focus for explorations of the boundary between death and life, body and soul. The realisation of embodied being in medieval writing may be foreign in its workings, yet the meanings it conveys concerning the profound connections between breath, feeling, and consciousness remain powerfully and vividly relevant.

Breath, *Pneuma*, and Embodied Being

Breath and breathlessness become central to medieval understandings of feeling and emotion through the concept of the 'vital spirits' (*pneuma/spiritus*), to which breath is integral. These ideas take up Galenic models and Arabic extensions of them, to envisage the heart itself as 'breathing', as the vital spirits move in and out.[1] Galen (129–*ca.* 216 CE) followed the emphasis of Alexandrian medicine on the brain as the site of cognition and sensory perception. He also reworked Hippocratic and Aristotelian physiology to develop the theory of the bodily spirits.[2] Aristotle had privileged the heart, conceiving of a mysterious, innate cardiac 'vapor' or breath, *pneuma*, as enabling perception, movement, and generation; Galen's model, by contrast, was rooted in his theories of pulsation.[3] He did, however, sustain Aristotle's notion of a tripartite soul, with vegetable, animal, and rational elements, following Platonic identifications of these with the liver, heart, and brain.[4] All these required the 'instruments' (*organa*) of *pneuma* and heat, which originated in the heart; the liver provided blood. Galen's *pneuma* was 'the vaporous substance ... formed ... in part by the inspired air and in part by the vaporization of the arterial blood',[5] responsible 'for the transmission of sensation and other psychological and physiological faculties'.[6] It 'exhale[d]' a purer, 'psychic *pneuma*' contained in the ventricles of the brain, which governed sensation and movement.[7] Breathing, effected through the pores, supplied and replenished the air required to mix with blood in order to create *pneuma* and to cool the heart.

Later Galenism took up Galen's ideas, but also modified them. The concept of *pneuma* was elaborated, in particular, by the tenth-century Persian philosopher-physician, Avicenna (Ibn Sīnā). While Galen's main focus was the psychic *pneuma* (with some reference to that originating in the heart), Avicenna's *Canon* (1037), a comprehensive compendium rooted in Galenic medicine but also drawing extensively on Aristotle, set out a tripartite structure according to which the *pneuma*/breath was envisaged as modified into three kinds in the three principal organs.[8] *Pneuma* (Latin, *spiritus*) formed 'the link between the body, soul and spirit'.[9] Avicenna identified three 'faculties' or 'drives', corresponding with the three-part soul: natural or vegetative, located in the liver; vital, connected with the breath; and animal or sensitive, situated in the brain. A vital force, which Avicenna also terms 'breath' ('formed of the light and vapory part of the humours' (6.6.477) and produced in the heart), gave to all members of the body their 'virtues' or powers. Avicenna presented this vital force enabling the drives in mystical terms as 'a divine emanation' comparable to light (7.1.488). His model, then, was firmly cardiocentric: while he adopted Galen's view of the brain as essential to physiology, he followed Aristotle in placing the heart as origin of the life force, 'the one single breath which accounts for the origin of the others'. This was differentiated in the heart into the vital 'breath' or drive, and in turn enabled the natural and animal 'breaths' (7.1.489). From the vital breath arose the emotions; from the animal breath the senses and 'common sense', imagination, cognition, memory, and understanding.

Early twelfth-century Latin translations of Arabic and Greek medical texts, many originating from the schools of Salerno and Toledo, disseminated Galenic and Arabic physiology to the Christian West, and provided the basis of Western medical theory. Particularly influential were the translations of Constantine of Africa, who practised at Salerno. Constantine's *Pantegni theorica*, which translated parts of the tenth-century Arabic medical encyclopaedia of 'Haly Abbas' ('Alī Ibn al-'Abbas al-Mağūsī), in turn based on Galenic works, and his translation of the *Isagoge Johannitii in Tegni Galeni*, a treatise on the Galenic theory of humours and spirits by the ninth-century scholar and physician 'Johannitius' (Ḥunayn ibn Ishāq), became foundation texts for the *Articella* or *Ars medicine*, the collection of six medical works that entered the university curriculum in the thirteenth century. Translation and circulation of other medical texts was rapid and extensive. A translation of Avicenna's *Canon*, attributed, along with a range of Galenic treatises, to Gerard of Cremona (d. 1187, translator of many scientific works), probably made in the thirteenth century by Gerard de Sabbioneta (also often known as/confused with Gerard of Cremona), was taken up in Toledo and introduced in the curricula of Montpellier, Paris, and Bologna at the end of the thirteenth century.[10] Avicenna's influence is evident through references to his work in the later thirteenth century, including in the scholastic writings of Roger Bacon, Albertus Magnus, and Bartholomaeus Anglicus. Vincent of Beauvais' widely circulated *Speculum naturale* (*ca.* 1235–64) draws on the *Canon*, and copies were owned in England by the late thirteenth century, including by the Benedictine foundations Christchurch Priory and St Augustine's Abbey, Canterbury. Ideas concerning the *pneuma/spiritus* were of interest to theologians on account of their bearing on debates concerning the relationship between material and spiritual and the unity of the soul.[11] The *pneuma* was most often seen as the 'instrument of the soul' but sometimes, in its 'animal' form in the brain, as the corporeal soul itself.[12] Breath, then, played a key part in physical, intellective, and spiritual being.

By the later fourteenth century, this physiological model was widely circulated, rendered accessible in lay circles through vernacular works such as John Trevisa's translation of Bartholomaeus Anglicus' encyclopaedic *De proprietatibus rerum*. Trevisa's description closely parallels Avicenna's and draws on Constantine's *Pantegni*:

> Out of þe herte comeþ lif to al þe limes. [...] Þis vertu of lif openiþ þe herte by worchinge of þe longen and draweþ in aier to the hert and sendiþ forþ from þe herte to oþir limes by smale weyes. And by help of þe vertu þat closith and riueþ and openith þe herte þis vertu worchiþ and makeþ breþinge in a beest. And by breþinge þe brest meueþ continualliche, but sinewis and brawnes beþ first imeued. Þis blast, breþ, and onde [wind] is nedeful to slake þe kindeliche hete, and to foode of þe spirit of lif, and also to þe gendringe of þe spirit þat hatte *animalis* þat ȝeueþ felinge and meuynge.[13]

Breath cools the heart where the 'virtue' or power of life begins. Through breath are generated the vital spirits—here seemingly synonymous with the 'vertue vitall'—which in turn create the animal spirits. The lungs take in air and convey it to the heart, cooling it and providing the air that will be transformed, mixed with blood, to the vital spirits:

> Þe kepinge of þe kinde hete is a temperat indrawinge of coold aier and þe kepinge of þe spirit þat hatte spiritus vitalis 'of lif'. Of þe temperament of þis spirit is þe spirit gendrid þat hatte animalis þat ȝeueþ felinge. Þerefore noþing is more nedeful to kepe and to saue þe lif þan breþ, wel disposid and ordeyned in alle pointis. Al þis seiþ Constantinus in Pantegni. (III.15, 105)

As Carole Rawcliffe shows in this volume, these ideas, originating from Constantine of Africa's translations and Avicenna's *Canon*, also found their way into the regimens of health that circulated widely across Western Europe.[14]

The framework of the bodily spirits was crucial to medieval understandings of the emotions and psychology. Thirteenth-century natural philosophy drew on Avicenna, in particular his *De Anima* (translated into Latin in the twelfth century), to elaborate the processes of thought enabled by the animal spirits, seen as occurring in the ventricles of the brain. Breath was intimately connected with the emotions, which were believed to occur through the movements of the vital spirits and natural heat, produced in the heart and travelling through the arteries. In joy or anger, the vital spirits and accompanying heat rushed out from the heart, marked on the body by physical responses such as blushing. In grief, distress, or fear, the vital spirits and heat withdrew into the heart, reflected in pallor or swooning. Withdrawal of the spirits equated to withdrawal of breath: at its most extreme, this could cause unconsciousness or even death. Sighs and tears were believed to carry the overburdening spirits out of the heart, purging and cooling it, though sighs might prove dangerous in excess, causing the heart to dry out and wither. The concept of the porous, breathing heart pervaded literary and theological works. As Webb emphasises, the senses were 'imagined to bring something from the outside world into the heart', stimulating the vital spirits:

> The *spiriti* generated in the medieval heart created different possibilities for the experience of the world. They extended far beyond the boundaries of a single body, reaching into other bodies and thus forming the potential for a certain porosity of selfhood that corresponded to the porosity of the heart.... The medieval heart ... was meant to be promiscuous; the world rushed into it with every breath.[15]

Love, grief, melancholy, then, were profoundly bodily, 'spiritual' in a material, bodily sense, as well as felt in the soul, treated by physicians as well as philosophers and imaginative writers.

The classical notion of *pneuma* related in complex ways to Christian conceptions of the Holy Spirit. As Thomas Hunt demonstrates in this volume, while immateriality was emphasised, physical notions of air, voice, and breath continued to resonate with conceptions of the Spirit. In the later medieval thought-world, the theory of the bodily spirits encouraged such blurring. The influential writings of St Paul depict the Spirit of divinity and life as external to the individual, moving within the cosmos, yet the Spirit is also inspiring and inspired, moving souls, breathing new life into them. St Augustine, whose works laid the foundation for later medieval theology, employs the concept of *spiritus* (*pneuma*) to explore both the immaterial quality of the divine and the life force.[16] The Holy Spirit might be understood as the divine spark, igniting the vital fire within the heart. The dual sense of *spiritus/pneuma*, as breath of God and vital spirits, gives the breath a special status in conceptions of the animating force, its ramifications both physical and spiritual. The heart of Christ and the heart of the lover were in many ways analogous, infused by the Spirit and profoundly wounded by the spirits, breathing out tears and sighs.

Fainting Hearts

Medieval romances are imbued with—and take for granted—the extreme affects of love. Such affects are fundamental to romance sensibilities, but they are also deeply embedded in medieval physiological and psychological models. Romance reworkings of the classical topos of love-sickness are invested with realism in ways not likely to be apparent to modern readers. They assume the movement of the spirits and their affective play on the body, and can also include sophisticated medical detail. Swoons, shrieks, and sighs signal emotional extremes; they also play essential narrative roles, their affects heightened by their physiological grounding. Romance treatments of love can be highly sophisticated, drawing on detailed medical knowledge, as in Chaucer's writings, but the broad outlines of physiological and psychological models to which the affective play of breath is integral are also evident in much less conspicuously learned works, demonstrating the pervasive power of such ideas from the thirteenth to the fifteenth centuries.

The thirteenth-century French dream-vision *Le Roman de la Rose* (begun by Guillaume de Lorris and completed by Jean de Meun), parts of which were translated into English by Chaucer in the late fourteenth century, offers the archetypal model of medieval love-sickness.[17] The Lover is wounded by the arrow of the God of Love: 'Than I anoon gan chaungen hewe / For grevaunce of my wounde newe, / That I agayn fell in swonyng / And sighede sore in compleynyng' (1865–68).[18] The pallor of his countenance and his ensuing faint bespeak the withdrawal of the vital spirits into the heart; his sighs send forth the overburdening spirits, while the poem too is a voicing of his 'compleyning'. In Chaucer's *Knight's Tale*, the walls of the temple of Venus are painted with images of love-sickness: 'The broken slepes, and the sikes colde, / The sacred teeris, and the waymentynge, / The firy strokes of the desirynge' (1920–22); in *The Parliament of Fowls* the lover's sighs are 'hoote as fyr', 'sikes engendered with desyr' (246, 248). The rushing out of the vital spirits in passionate desire creates hot sighs, while cold sighs reflect the absence of warmth effected by grief, when the spirits have withdrawn from the rest of the body. The Ovidian paradoxes of fire and ice, the burning and longing sighs of love, are famously taken up by Petrarch, recurring across medieval and early modern love poetry, and repeatedly used in Renaissance madrigals, where the word 'Ohime' (Alas!) with its two falling tones musically represents the sighs of the lover.

Middle English popular romance demonstrates just how widespread the physiological model of the vital spirits and the connection of breath with emotion are. The early romance *Floris and Blanchefleur* (Middle English *ca.* 1250, based on an earlier French romance), popular across Europe, exemplifies this. The stylised narrative relies on patterning, convention, and repetition, yet these are integrated with the physiological realisation of emotions at work. When the protagonist Floris, a Spanish prince whose parents are unwilling to let him marry a Christian, is told that his beloved Blanchefleur is dead, he swoons in response, meeting death with seeming death.[19] In turn, his swoon evokes the affective responses of the queen and king, whose heart is 'al in care' (253); on seeing Blanchefleur's grave, his response is heightened: 'Thre sithes Florys sownydde nouth; / Ne speke he myght not with mouth' (267–68); when he 'awoke' (269) from his swoon, a verb pointing up the death-like effect of the spirits withdrawing, his grief leads him to attempt suicide. It is the extremity of Floris' grief, lived out in his swoon, that causes his mother to fear for his life and his parents to reveal that Blanchefleur is living. On hearing that she has been taken to Babylon, Floris' spirits rise: he 'in his hert bygan to lyght' (417), and he plans to follow her there. The movements of breath, then, are carefully manipulated in relation to the plot, with swooning placed both to signal extreme distress and to effect a turning point.

A second early Middle English romance, *King Horn* (*ca.* 1225), deploys the play of the vital spirits to characterise its strikingly active female protagonist, Rymenhild. Her desire is manifest in near-madness that disorders the animal spirits generated by the vital spirits: she 'lovede so Horn child / That negh heo gan wexe wild [mad]'; 'in heorte heo hadde wo'.[20] Her rejection by Horn is dramatically reflected in the withdrawal of the vital spirits: 'Tho gan Rymenhild mislyke, / And sore gan to sike. / Armes heo gan bughe [bend]; / Adun he feol in swoghe' (425–28). Again the swoon has a pivotal narrative function, effecting in turn a change in Horn's spirits: 'Horn in herte was ful wo' (429). Rymenhild's shock at her dream warning of his departure again draws the animal spirits from the brain, affecting her inner 'wits' or faculties: she sits numbed, 'also he were of witte [out of her wits]' (652), an image repeated at the thought of her enforced marriage (1086). The deathly movement of the spirits into the heart is made explicit when she believes Horn to be dead: her heart 'bigan to chelde [grow cold]' (1150), echoing her swoon. The movement of the vital spirits, the play of breath governed by them, and their power over the wits animate the depiction of Rymenhild's remarkable agency.

Chaucer, a century later, portrays such affective responses, in which breath moves in and out of the heart, with remarkable medical detail. His engagement with physiology and conversance with works such as Trevisa's is evident across his oeuvre, investing his depictions of emotion with a realism rooted in knowledge of the complex workings of mind, body, and affect. Chaucer's early dream vision *The Book of the Duchess* (1369–72) probes the physiology of loss and grief. The narrator's unexplained melancholy, which we assume to be caused by love, has 'sleyn [his] spirit of quyknesse' (26)—afflicted the vital spirits—and deprived him of sleep. The classical story he reads, of Ceyx and Alcyone, echoes his own fleeting spirits in its account of Alcyone's grief-stricken swoon at hearing that her husband is drowned. She is 'cold as ston' (123), waking only to die within three days: the most extreme manifestation of the flight of the vital spirits into the heart. In the narrator's dream of the grieving Man in Black such a withdrawal of the spirits is elaborated in precise medical detail:

> His sorwful hert gan faste faynte
> And his spirites wexen dede;
> The blood was fled for pure drede
> Doun to hys herte, to make hym warm—
> For wel hyt feled the herte had harm.... (488–92)

No physician, 'noght Ypocras ne Galyen [Hippocrates or Galen]' (572), can treat his heart. The terms, however, are clearly reminiscent of Galenic physiology. Withdrawal of vital spirits and breath effects faintness; the flight of blood into the heart, the 'membre principal' (495) of the body, causes 'al / Hys hewe chaunge and wexe grene / And pale' (496–98). Loss of the beloved engenders loss of vital spirits and renders body and mind 'ded as stoon' (1300).

Chaucer's epic romance *Troilus and Criseyde* (*ca.* 1382–85) offers still more extended physiological detail. The narrative of Troilus' double sorrows is the narrative of his heart wounded in love and loss, and of the extreme movements of the vital spirits written on his body. Chaucer enhances and medicalises his source, Boccaccio's *Il Filostrato* (*ca.* 1335/40). He draws on both neo-Platonic convention and the kind of Aristotelian notions taken up by Avicenna to describe how, on seeing Criseyde, Troilus' heart is affected through the eyes: it is caused to 'sprede and rise' as if on fire, wounding and quickening his 'affeccioun [desire]' (I, 278, 296) as the vital spirits rush outwards. But the enduring effect is that of withdrawal of the over-burdened spirits: Troilus feels that 'with hire look [will die], the spirit in his herte' (I, 307). Again and again, the illness of the heart consumes the body. When, on seeing Criseyde weep, Troilus feels 'the crampe of deth to streyne [constrain] hym by the herte' (III, 1071), and faints, Chaucer offers careful physiological detail:

> Therwith the sorwe so his herte shette* *shut
> That from his eyen fil there nought a tere,
> And every spirit his vigour in knette,* *contracted its force
> So they astoned* or oppressed were. *surprised
> The felyng of his sorwe, or of his fere,
> Or of aught elles, fled was out of towne;
> And down he fel al sodeynly a-swowne. (III, 1086–92)

The sudden, complete withdrawal of the vital spirits and of the breath that they govern causes unconsciousness. Here, the connection between pulsation and the heart is comically reiterated as Pandarus and Criseyde chafe Troilus' pulse and palms until breath returns. But ultimately, Troilus' entire physical being is unmade: so extreme are the pallor, emaciation, and weakness caused by the affects of love that he is unrecognisable, walking with a crutch (V, 1219–25). Chaucer envisages the physical pressure of the withdrawing spirits: Troilus complains of grievous pain around his heart. Yet, ironically, Troilus' death is caused not by the final loss of breath in love but when he is slain in battle by Achilles. Now freed from the torments of the bodily spirits, his 'lighte goost' (V, 1808) looks down and laughs at the sufferings of those on earth.

Chaucer's engagement with that moment at which the vital spirits withdraw is also evident in his graphic account of the lover Arcite's death at the end of the *Knight's Tale*, written at approximately the same time, and drawing on another work of Boccaccio, *Il Teseida*. Here, Chaucer explores the possibility of physical rather than emotional pressure on the vital spirits. An injury to Arcite's chest when he is thrown from his horse prevents the clotted blood from being expelled: with 'the pipes [tubes] of his longes' swollen and the muscles in his chest 'shent [destroyed] with venym and corrupcioun [decayed matter]' (2752–54), the spirits fail in their 'vertue' of expelling the poison. The 'vital strengthe' (2802) is cut off, unable to circulate, and the 'coold of deeth' comes upon his body

(2800). Only his 'intellect' (2803), his consciousness, remains, manifest in his eyes and plea of 'Mercy' to Emilye (2808). But this too fails 'whan the herte felte deeth' (2805). The vital spirits are unable to animate the body: 'Dusked his eyen two, and failed breeth' (2806). Separated from the body, Arcite's 'spirit', now synonymous with his soul or, in this tale with its classical framework, *psyche*, changes 'hous' to another, celestial dimension (2809). The episode disentangles the complex connections between heart, spirits, intellect, breath, and soul of the lover, and probes the workings of the breathing body and mind.

Awakening Life

The movement between death and life, the visibility of the vital spirits and the possibility that breath may be latent yet present, all these coalesce in the figure of the seemingly dead lady. The motif echoes down the centuries, spanning both fairy tale—the Sleeping Beauty stories—and tragedy—*Romeo and Juliet*. It finds its most influential pre-modern treatment in the tale of Apollonius of Tyre, extant in a Latin version but probably based on a Greek prose fiction, retold in several medieval versions, most extensively by Chaucer's contemporary John Gower in Book VIII of his *Confessio Amantis*, and by Shakespeare in *Pericles*, where 'ancient Gower' is his Chorus. Marie de France's *lai* of *Eliduc* offers a partial twelfth-century Anglo-Norman analogue. Marie is specific about the absence of breath when the lady Guilliadun swoons, seemingly dead, on discovering that her lover Eliduc is already married:

> Desur sun vis cheï paumee,
> Tute pale, desculuree.
> En la paumeisun demurra,
> Que el ne revient ne suspira.
> Cil ki ensemble od lui l'en porte
> Quidot pur veir ke ele fust morte.
>
> (She fell on her face in a faint,
> all pale, colorless.
> She remained in her faint,
> she did not recover or sigh.
> He who is taking her with him
> truly believed that she was dead.)[21]

Later, Marie emphasises Eliduc's wonder, on visiting her corpse, at the fact that she has lost little of her colour. As with Juliet, death's pale flag is not yet advanced in her countenance, the clue that she only seems dead. When Eliduc's wife Guildelüec discovers her, Guilliadun's gem-like beauty is described in terms of marvel. Yet though her recovery too is wondrous, it is also medicalised, effected through the application of a herb that heals a weasel killed by the wife's servant. Marie offers realistic detail about how the herb is placed within the maiden's mouth, so that 'Un petitet i demurra, / Cele revint e suspira' ('Just a little while she waited: / the girl recovered, and sighed', 1063–64). Guildelüec emphasises Eliduc's confusion between swoon and death: the original meaning of Old French 'pasme', 'cramp', a sudden taking or spasm, suggests the suddenness of swooning, its death-like quality. This is a near-death state which medical knowledge can resolve, not a sleep to be awakened by love. But the swoon also enables the probing of Eliduc's love through his grief, and the resolution of the dilemma through Guildelüec's wonder and pity, leading her to give up her husband.

Similar emphases are found in the Apollonius story, which is reworked by Gower to probe the boundary between marvel and medicine, fantasy and reality. The classical text offers specific detail about the swoon, which occurs in childbirth: 'While they were delayed at sea … the girl gave birth in the ninth month. But the placenta failed to be discharged, her blood clotted, her breathing became constricted, and she suddenly died.'[22] The focus of the doctor's pupil who revives her is on breathing:

> The young man took a jar of unguent, went to the girl's bier, pulled aside the clothing from the upper part of her body, poured out the unguent, ran his suspicious hands over all her limbs, and detected quiescent warmth in her chest cavity. The young man was astounded to realize that the girl was only apparently dead. He touched her veins to check for signs of movement and closely examined her nostrils for signs of breathing; he put his lips to her lips, and, detecting signs of life in the form of slight breathing that, as it were, was struggling against false death, he said, 'Apply heat at four points'. When he had had this done, he began to massage her lightly and the blood that had coagulated began to flow because of the anointing. … Her blood, which had congealed because of severe cold, began to flow once heat was applied, and her previously obstructed breathing began to infiltrate to her innermost organs. (753–54)

The liberation of the vital force effected by warming the blood unlocks the breath.

Gower's narrative is rooted in intimate familiarity with medieval medical theory, taking up the idea of the vital spirits to emphasise the physiological underpinnings of what might seem magical. His account of the swoon, the absence of breath that seems to take away life itself, is briefer but acute: Apollonius' queen is 'ded in every mannes syhte', her loss inspiring him to swoon in turn.[23] Recovery is realised with precise detail, as the physician Cerymon searches and finds 'a signe of lif' in the body (VIII, 1189). His actions follow established medical procedure for reviving the vital spirits, drawing them out of the heart: he lays the queen on a soft couch, lights fires, warms her breast, anoints her joints with 'certein oile and balsme', and 'putte a liquour in hire mouth, / Which is to fewe clerkes couth'—a learned parallel to the life-giving herb in *Eliduc* (VIII, 1198–1200). While the revival of the seemingly dead lady speaks to the wish-fulfilment quality of romance, it is carefully depicted so as to appear not as magic but skilled medicine, which uses specialised, learned arts of healing, heat, aromatic scents, and 'liquour' (all cures recommended to physicians), to reawaken the vital spirits in the heart so that it begins to 'flacke and bete', and revive the 'colde brest', returning breath to the body (VIII, 1195–96). Gower enhances affect for reader and spectator by rooting marvel within the realistic possibilities of medicine, using a contemporary physiological framework to make real the possibility of waking the dead.

Engagement with the play of breath and the precise physiology of the vital spirits that govern the emotions, then, allows romance writers to dramatise the physicality of intense experience and the embodied quality of emotion; to illuminate the intimate connections between mind, body, and affect; to probe the nature of spirits, intellect, and soul; and to explore the mysterious boundary between life and death. Moments of extreme emotion, particularly enacted through the swoon, the flight of breath into the heart, are crucial turning points and foci for the revelation of the psyche. Romance structures and their conventions have their own spiritual force: they work to evoke and animate the texture of human experience, of breathing, being, and dying in the world.

Swooning into Vision

The physiology of the bodily spirits and breath plays a central part in romance; it also powerfully infuses religious writing, in particular, visionary writing with its emphasis on individual spiritual experience.[24] The long history of the concept of *pneuma*, its complex intersections with life force, spirit, and soul, and the essential connection between vital spirits and breath all played a crucial part in the bodiliness of medieval devotion. Inspiration by the spirit was not cerebral but profoundly physical: a literal inspiring. The start of Richard Rolle's *Incendium Amoris* (*The Fire of Love*, *ca.* 1343) memorably depicts the blurring between physical and spiritual: Rolle feels the heat of the flame of divine love, touching his breast to see whether his heart is literally on fire: 'it set my soul aglow as if a real fire were burning there'.[25] Such representations are intimately connected with the notion of the breathing heart, its heat infused and increased by the vital spirits. Vital spirits, soul, and Holy Spirit seem to intersect. As in romantic love, the individual responds to and with overpowering love that can manifest itself in the flight of the spirits into the heart, causing not just sighs that relieve its fire but also swoons, while the rushing out of the spirits is marked by tears and sobs, the disruption of breath. There was no clear boundary between theology and natural philosophy, and learned clerics are likely to have been broadly familiar with physiological concepts such as that of the vital spirits, and to have recognised the coincidence of this model with ideas of the Spirit as inspiring, moving force. Monastic libraries owned copies of works such as Avicenna's *Canon* and Bartholomaeus' *De proprietatibus rerum*, suggesting the interest of religious as well as lay readers in both physiology and psychology. Religious writers take up conventional images and metaphors, reworking them with new force in the light of new understandings of breath and the spirits.

The late fourteenth-century Middle English alliterative poem *Pearl* merges the conventions of romantic and divine love in its dream-vision. Here again the swoon, withdrawing breath, is the catalyst, effecting a transformation in the Dreamer-narrator. Grief at the loss of his pearl, probably his infant daughter, leads him to fall down onto the grass 'vpon a slepyng-slaȝhte': the term 'slaȝte' has the sense both of a sudden blow and of death, implying a sudden, death-like sleep.[26] From this place his 'spyryt þer sprang in space', liberated from his body, into 'sweuen' or vision (61–62). This swoon of grief becomes a form of dying to live, looking towards the redemption offered by actual death. It is not waking but the swooning sleep itself that is medicinal, opening onto a visionary, celestial world of blue and silver trees where precious pearls crunch under the feet and the Dreamer sees his daughter transformed into a beautiful bride of Christ. The change in breath-state marks the crossing of an ontological boundary, but the Dreamer cannot fully enter into the celestial world: his attempt leads him to wake at the end of the poem 'sykyng' (1170) and swooning in grief (1180) at the loss of his visionary state. Waking breath remains synonymous with loss, even while it is infused with the possibility of embodied vision.

In Julian of Norwich's *Revelations of Divine Love* too, the swoon becomes a kind of dying to live. But this instance is partially voluntary, leading beyond loss. The *Revelations* recount the visions experienced by Julian in 1373 as she lay apparently dying at the age of thirty. The illness occurs in answer to her prayer to experience three miracles, one of which is to 'have all manier peynes bodily and ghostly that I should have if I should dye, with all the dreds and tempests of the fends, and all maner of other paynes, except the outpassing of the soule'.[27] All three of her wishes are fulfilled at once: the loss of all but spiritual comfort, visual experience of Christ's Passion, and the three wounds of contrition, compassion, and 'willfull longyng to God' (ch. 2, 30). The description of Julian's illness accords with medieval physiological models. It is characterised by withdrawal of the breath, signifying the withdrawal of the vital spirits, and hence, by the failure of sight and feeling occasioned by the effect of that withdrawal on the animal spirits:

> After this the over party of my body began to dyen so ferforth that onethys I had ony feleing. My most payne was shortnes of onde [breath] and failyng of life. And than I went sothly to have passid. And, in this, sodenly all my peyne was taken fro me and I was as hole, and namely in the over party of my body, as ever I was aforn. (ch. 3, 32)

In this near-death, trance-like state, Julian experiences fifteen of her sixteen visions. Then her sickness returns, in her head 'with a sound and a dynne', and in her body, 'sodenly all my body was fulfillid with sekenes like as it was aforn, and I was as baren and as drye as I never had comfort but litil' (ch. 66, 136). Dismissing the visions as 'raving' and then repenting of her disbelief, she falls asleep, to feel the devil taking her by the throat. Sleep and waking overlap as 'anon a lytel smoke came in the dore with a grete hete and a foule stinke' (ch. 66, 137)—the breath of the devil opposing the power of the Spirit and her own vital spirits; later, the fiend returns 'with his hete and with his stinke' (ch. 69, 142). Images of air and wind recur. In the eighth revelation, inspired by Christ's words, 'I thirst', Julian sees his flesh shrivelling in death, like cloth hung out to dry, pained 'with blowing of the wynde from withowten that dryed him more, and peynd with cold, than myn herte can thynkyn' (ch. 17, 57–58). Spiritual dryness, the cold breath of despair, and the smoke of hell are countered by the breath of the Holy Spirit, which brings life and rest, 'the Holy Gost graciously inspirith [breathes] into us gifts ledand to endless life'. The soul will ultimately perceive God with all the senses: 'hym verily seand and fulsomly feland, hym gostly heryng, and hym delectably smellyng, and hym swetely swelowyng' (ch. 43, 98). He is breathed in as well as breathing into, inspired and inspiring. This is the all-consuming, true vision inspired by the Spirit, seeing face to face rather than 'through a glass darkly'. Julian's experience is fully embodied as her own breath and that of the spirit world around her intermingle.

Nothing is known about Julian's life beyond the fact of her anchoritic existence, but the theological sophistication of her interpretations suggests deep reading and knowledge, perhaps the result of spiritual instruction by her confessors. While *Pearl* ends with the sighs and swoons of grief and longing for lost vision, the Long Version of the *Revelations* combines with visionary moments the meanings contemplated for twenty years, creating a breath-world the textures of which are both inspired and deeply thought-through.

It is in *The Book of Margery Kempe* that we find the most extended treatment of the play of breath in affective piety. Though her name is frequently linked with that of Julian of Norwich, the life of Kempe (born *ca.* 1373) was far from anchoritic: married, she bore fourteen children, ran a brewing business, and travelled on pilgrimage as far as Rome and Jerusalem. She too, however, was a visionary who adopted a severely ascetic life—and her references to many devotional texts make clear the shaping of her imagination by the kind of meditative tradition that Julian would have practised, of projecting the self into, and empathising with, scenes of Christ's life. Unlike Julian's *Revelations*, Kempe's *Book* (1436–38) does not offer extended theological interpretations. Its status is complex: it is mediated by two amanuenses, its illegible first version interpreted, rewritten, and extended by a second scribe some twenty years after the earliest experiences it recounts. The process of writing and the accounts in the *Book* of Kempe's programme of reading with her priest-confessor, however, suggest an intellectual context in which physiological and theological ideas concerning breath, the spirits, and the inspiration of the Spirit would have been familiar. At the same time, the breath-world of the *Book* is uniquely and immediately Kempe's own.

Its primary focus is 'inspiration', being literally breathed into by the Holy Spirit. The images of tears, hearts, and flames signifying such inspiration that dominate the *Book* are familiar, extending back at least to St Bernard of Clairvaux and forward to the hearts and flames drawn in the margin by sixteenth-century readers. But they also have a powerful material meaning that keys into medieval understandings of the connections between mind and body, the movements of the vital spirits and the feelings they stimulate, and especially, the workings of the animating Spirit. While Kempe's down-to-earth, sometimes simplistic account of spiritual revelation and grace has been dismissed for its conventionality, Kempe's embodied experience may also be seen as shaped by contemporary physiological concepts. The emphases of the *Book* on fire in the heart and compulsive weeping are not simply performances of piety, but correspond with models of feeling that are bound up with the breath. Profound affect exacerbates the heart's heat, so that the vital spirits are emitted in sighs, tears, and cries, while extreme grief causes them to rush violently back in and leads to swoons.

From the start, Kempe's inward life is intimately associated with the body. The *Book* recounts Kempe's 'mevynggys and hyr steringgys' as she is 'enspyred of the Holy Gost', and the stirrings of others to believe in her: the breath of the Spirit works on her own breathing body.[28] The feeling and sound of divine breath is a repeated motif. Kempe's conversion is brought about by a melody 'so swet and delectable, hir thowt, as sche had ben in paradyse' (325–26), which inspires 'greet sobbyngys and syhyngys aftyr the blysse of heven' (333–34); she is 'drawt [drawn]' (335) to God. As well as further 'gret sowndys and gret melodiis' (6224) that signal heavenly merriment, she hears the noise of the bellows blowing, explicitly identified as the 'the sownd of the Holy Gost' (2968). The sound becomes that of a dove, the traditional symbol of the Holy Spirit, and then a robin, all inspiring 'gret grace' (2972). Kempe both experiences divine *pneuma* through sight, sound, and feeling and is consumed by that breath, which enters into her heart. The traditional association of the Holy Spirit with fire as well as breath is pointed up by the narrative's repeated descriptions of the fire or flame of love: 'sche had many holy teerys and wepingys, and oftyntymys ther cam a flawme of fyer abowte hir brest ful hoot and delectabyl' (7369–71). In Kempe's visionary experience, the flame is not simply a metaphor for ardent love, but a literal representation of the inspiration of the heart by the Spirit, profoundly and often painfully physical. Her heart is so far 'consumyd wyth ardowr of lofe' (929–30) that, like Troilus, she swoons: 'unmythy to kepyn hirselfe in stabilnes for the unqwenchabyl fyer of lofe whech brent ful sor in hir sowle' (3242–44). The *Book* also, however, suggests conflicting models: while the Carmelite friar Alan of Lynn and other clerics attempt to persuade Kempe's enemy, the learned Franciscan William Melton, that her experience is genuinely spiritual, Melton suspects she suffers from a disease of the heart ('a cardiakyl,' 5063).

Kempe's cryings, which commence with her first vision of heaven and continue over her entire life, are closely connected to her experience of the fire of love. While the cryings are often singled out in critical discussions of Kempe, they may better be understood as one aspect of her embodied devotion, making manifest the workings of the Spirit that connect heart and mind through the breath-play of the vital spirits. Tears are a long-standing religious topos and partial analogies may be found in the weeping of Christ, the Virgin, and Mary Magdalene, while reading of the tears of Mary of Oignies inspires belief in the priest who condemns Kempe, but the *Book* also emphasises the discomfort and wonder of her cryings. Her tears respond to the trauma of the Passion, marking her sharing of Christ's pain, but her 'synguler and ... specyal yyft' (3268) is also extreme and compulsive, transformed when she travels to Jerusalem into a 'krying and roryng' (2216), 'plentyvows terys and boystows sobbyngys, wyth lowde cryingys and schille shrykyngys' (3534–36). The repeated connection with physical 'fallyng' (2190) suggests the most extreme effect of the vital spirits drawn into the heart through profound emotion ('the fervowr of the spiryt was so meche that the body fayld', 6684–85), and hence, the violent disruption of breath. The connection with breath is made differently explicit when Kempe turns blue or the colour of lead from the intensity of her cryings (2256), a symptom often connected with epilepsy, but one that readily fits the model of the vital spirits rushing into and out of the heart in a combination of ecstasy and grief. Overcome by the passions, the heart draws in more breath through gasps and sobs, while the withdrawing vital spirits deprive the rest of the body of its animating force. Crying and falling, 'these bodily mevyngys' (2278), the effects of the rush of spirits and breath, are thus closely linked to the flame of *pneuma/spiritus* in the heart: theological and physiological models intersect in the meeting of air and fire.

While Kempe herself may not have been conversant with the detail of medical or philosophical theory, at least some members of the East Anglian clerical circle with which she engaged will have been able to offer generalised models for the workings of emotional and spiritual experience. As medieval romances show, these ideas permeated the medieval thought-world down to its popular roots. They were available in the texts and lives that influenced Kempe and shaped her spiritual experience, animating her *Book*. To place Kempe's behaviours simply as conventional or performative—or to define them in pathological terms according to modern medical disorders—ignores the physicality of contemporary understandings of divine inspiration and of emotion.[29] Breathing connects mind and body, sends the vital spirit from the heart to the brain, kindles the fire of love, and in extreme emotion also stimulates the disturbance of the vital spirits that causes severe crying and falling, a rushing out of the fire and a sudden withdrawal of the spirits into the heart. Kempe's *Book* presents an embodied spirituality that is strange, uncomfortable, extreme yet also explicable—a breathing, inspiring spirituality. Like the Dreamer in *Pearl*, she is consumed by the depth of her feeling. As with Julian, her identification with the Passion leads to revelatory, all-encompassing experience in which her entire body is caught up by the Spirit. For the visionary, human breath mingles with divine, vital spirits with the Spirit, to effect a breathing being not so unlike that of the fainting lover of romance, consumed by passion. These swoons, however, also evoked by deep desire, are not to be woken from but ardently sought after, for in them, the fire and air of the heart, spirits and Spirit meet.

Medieval breath, then, was profoundly 'spiritual'—integral to emotional experience, both romantic and divine, and to the essential force of life. The thought-world of medieval literature is rich and strange. Yet in the world of the twenty-first century, where breath has become more and more prominent, opening onto the most fundamental issues of health, belief and freedom, that strangeness of breath's spirituality resonates with uncanny relevance.

Notes

1. Heather Webb, *The Medieval Heart* (New Haven, CT: Yale University Press, 2010), see especially ch. 2, 'The Porous Heart', 50–95.

2. Armelle Debru, 'Physiology', *The Cambridge Companion to Galen*, ed. by R. J. Hankinson (Cambridge: Cambridge University Press, 2008), 263–82: 275. On the progression of Greek thought on *pneuma* and Galen's response to it, see further Geoffrey Lloyd, '*Pneuma* between Body and Soul', in *Wind, Life, Health: Anthropological and Historical Perspectives*, ed. by Elisabeth Hsu and Chris Low, *The Journal of the Royal Anthropological Institute* 13, issue s1 (2007): S135–S146.
3. Mario Vegetti, 'Between Knowledge and Practice: Hellenistic Medicine', in *Western Medical Thought from Antiquity to the Middle Ages*, ed. by Mirko D. Grmek, co-ord. by Bernardino Fantini, trans. by Antony Shugaar (Cambridge, MA: Harvard University Press, 1998), 72–103: 77–78; see further Gad Freudenthal, *Aristotle's Theory of Material Substance: Heat and Pneuma, Form and Soul* (Oxford: Clarendon Press, 1995).
4. Gerald J. Grudzen, *Medical Theory About the Body and the Soul in the Middle Ages: The First Western Medical Curriculum at Monte Cassino* (Lewiston: Edwin Mellen Press, 2007), 57; Debru, 'Physiology', 268.
5. See Galen, *On the Use of Breathing* (*De Usu Respirationis*) in *Galen: On Respiration and the Arteries*, ed. by David J. Furley, and James S. Wilkie (Princeton: Princeton University Press, 2014), ch. 5, K502, 120–2; Debru, 'Physiology', 272. See also A. A. Long's essay in this volume.
6. Debru, 'Physiology', 272.
7. Debru, 'Physiology', 275.
8. Vivian Nutton, 'The Fortunes of Galen', *The Cambridge Companion to Galen*, 355–90: 364; Nancy G. Siraisi, *Avicenna in Renaissance Italy: The Canon and Medical Teaching in Italian Universities after 1500* (Princeton: Princeton University Press, 2014), 29. For an edition of the *Canon*, see Avicenna, *The Canon of Medicine* (*al-Qānūn fī'l-ṭibb*), adapted by Laleh Bakhtiar from translations by O. Cameron Gruner, and Mazar H. Shah, 5 vols (Chicago: Great Books of the Islamic World, 1999–2014).
9. Laleh Bakhtiar, Introduction to Avicenna, *The Canon of Medicine*, l; see Fen [Part/Lecture] 1, Doctrina 6.
10. See Danielle Jacquart, 'Medieval Scholasticism', in *Western Medical Thought from Antiquity to the Middle Ages*, ed. by Grmek, 197–240, 214–16. For a comprehensive account of Avicenna's *Canon* with reference to Gerard of Cremona's translation, see Siraisi, *Avicenna in Renaissance Italy*, ch. 2, 'The Canon of Avicenna', 19–40; on the dissemination of the *Canon* in the medieval period, see Siraisi, ch. 3, 'The Canon in the Medieval Universities', 43–76, in particular, 43–47.
11. See Jacquart, 'Medieval Scholasticism', 236, and Webb, *The Medieval Heart*, 21.
12. Constantine, *Pantegni*, *Theorica* IV.19, Lyons edition, 1515, cited in Charles Burnett, 'The Chapter on the Spirits in the *Pantegni* of Constantine the African', in *Constantine the African and ʿAlī Ibn al-ʿAbbas*

al-Mağūsī: The Pantegni and Related Texts, ed. by Charles Burnett and Danielle Jacquart, Studies in Ancient Medicine 10 (Leiden: E. J. Brill, 1994), 99–120: 105–06; Grudzen, *Medical Theory About the Body and the Soul*, 63–64; 200–01.

13. John Trevisa, *On the Properties of Things*, ed. by M. C. Seymour, 3 vols (Oxford: Oxford University Press, 1975), vol. 1, III.15, 104–05; for the Latin, see Bartholomaeus Anglicus, *De rerum proprietatibus* (1601; Frankfurt: Minerva, 1964). Subsequent references are from this edition cited by book, chapter, and page number.
14. See also Pedro Gil Sotres, 'The Regimens of Health', in *Western Medical Thought from Antiquity to the Middle Ages*, ed. by Grmek, 291–318.
15. Webb, *The Medieval Heart*, 52–53, and see 50ff for a discussion of the porous heart in poetic, theological and philosophical works.
16. See G. Verbeke, *L'Evolution de la doctrine du 'pneuma' du stoïcisme à S. Augustin, étude philosophique* (Paris: Desclée, de Brouwer, 1945), 489–508.
17. For a more extended discussion of Chaucer's romance writings, see my 'The Play of Breath: Chaucer's Narratives of Feeling', in Arthur Rose, Stefanie Heine, Naya Tsentourou, Corinne Saunders, and Peter Garratt, *Reading Breath in Literature*, Palgrave Studies in Literature, Science, and Medicine (New York: Palgrave Macmillan, 2019), 17–37.
18. Chaucer, Geoffrey, *The Romaunt of the Rose*, in *The Riverside Chaucer*, ed. by Larry D. Benson, 3rd edn (Oxford: Oxford University Press, 1988), Fragment B, 1865–68 (Fragment B is less securely attributed to Chaucer). Subsequent references to the works of Chaucer are from this edition, cited by line number.
19. *Floris and Blanchefleur*, in *Of Love and Chivalry: An Anthology of Middle English Romance*, ed. by Jennifer Fellows, Everyman's Library (London: Dent, 1993), 43–72, l. 246. Subsequent references are from this edition, cited by line number.
20. *King Horn*, in *Of Love and Chivalry*, 1–41, ll. 251–52, 263. Subsequent references are from this edition, cited by line number.
21. Marie de France, *Guildelüec and Guilliadun*, or *Eliduc*, in *The Lais of Marie de France: Text and Translation*, ed. and trans. by Claire M. Waters, Broadview Editions (Peterborough, ON: Broadview Press, 2018), 300–59, ll. 853–58. Subsequent references are from this edition, cited by line number.
22. *The Story of Apollonius, King of Tyre*, trans. by Gerald N. Sandy, in *Collected Ancient Greek Novels*, ed. by B. P. Reardon, 2nd edn (Berkeley: University of California Press, 2008), 736–72: 752. Subsequent references are from this edition, cited by page number.
23. John Gower, *Confessio Amantis*, *The English Works of John Gower*, ed. by G. C. Macaulay, 2 vols, Early English Text Society, ES 81 and 82

(London: Oxford University Press, 1900, 1901), vol. 2, Book VIII, ll. 1056, 1060. Subsequent references are from this edition, cited by book and line number.

24. For a more extended discussion of breath in *The Book of Margery Kempe* with reference to Julian of Norwich's *Revelations of Divine Love* and other works, and to the connections between breath and prayer, see my article with Amy Appleford, 'Reading Women in the Medieval Information Age: The *Life of Elizabeth of Spalbeek* and *The Book of Margery Kempe*', *Studies in the Age of Chaucer* 42 (2020), 253–81.

25. Richard Rolle, *The Fire of Love*, trans. by Clifton Wolters (Harmondsworth: Penguin, 1972), Prologue, 45; for the original, see *The 'Incendium Amoris' of Richard Rolle of Hampole*, ed. by Margaret Deanesly (Manchester: Manchester University Press, 1915), Prologue, 145.

26. *Pearl*, in *The Poems of the Pearl Manuscript: 'Pearl', 'Cleanness', 'Patience', 'Sir Gawain and the Green Knight'*, ed. by Malcolm Andrew and Ronald Waldron, Exeter Medieval Texts and Studies, 5th edn (Exeter: Exeter University Press, 2007), 52–110, l. 59, and see note. Subsequent references are from this edition, cited by line number.

27. Julian of Norwich, *Revelations of Divine* Love, ed. by Barry Windeatt (Oxford: Oxford University Press, 2016), The Long Text, ch. 2, 30. Subsequent references are from this edition, cited by chapter and page number. For a translation see Julian of Norwich, *Revelations of Divine Love*, trans. by Barry Windeatt, Oxford World's Classics (Oxford: Oxford University Press, 2015).

28. *The Book of Margery Kempe*, ed. by Barry Windeatt (Cambridge: D. S. Brewer, 2004), ll. 74, 1071–72. Subsequent references are from this edition, cited by line number. The unique manuscript, a copy written by a Norfolk scribe named 'Salthows', dates to *ca.* 1450. For a translation see *The Book of Margery Kempe*, trans. by B. A. Windeatt, Penguin Classics (London: Penguin Books, 1985, repr. 2004).

29. Kempe's behaviours have been retrospectively diagnosed as manifestations of hysteria, psychosis, and temporal lobe epilepsy. See further Richard Lawes, 'The Madness of Margery Kempe', in *The Medieval Mystical Tradition: England, Ireland, and Wales. Exeter Symposium VI: Papers Read at Charney Manor, July 1999*, ed. by Marion Glasscoe (Cambridge: D. S. Brewer, 1999), 147–67: Lawes argues for a diagnosis of temporal lobe epilepsy. On the limits of psychopathological diagnoses see Alison Torn, 'Madness and Mysticism: Can a Mediaeval Narrative Inform our Understanding of Psychosis?', *History and Philosophy of Psychology* 13 (2011): 1–14; and 'Looking Back: Medieval Mysticism or Psychosis', *The Psychologist* 24/10 (2011): 788–90.

Selected Bibliography

Fanous, Samuel, and Vincent Gillespie, eds 2011. *The Cambridge Companion to Medieval English Mysticism*. Cambridge: Cambridge University Press.

Getz, Faye. 1998. *Medicine in the English Middle Ages*. Princeton: Princeton University Press.

Grmek, Mirko D., ed. 1998. *Western Medical Thought from Antiquity to the Middle Ages*. Co-ord. Bernardino Fantini. Trans. Antony Shugaar. Cambridge, MA: Harvard University Press.

Kemp, Simon. 1990. *Medieval Psychology*. New York: Greenwood Press.

Knuuttila, Simo. 2004. *Emotions in Ancient and Medieval Philosophy*. Oxford: Oxford University Press.

Kruger, Steven F. 1992. *Dreaming in the Middle Ages*. Cambridge: Cambridge University Press.

Orlemanski, Julie. 2019. *Symptomatic Subjects: Bodies, Medicine, and Causation in the Literature of Late Medieval England*. Philadelphia: University of Pennsylvania Press.

Rawcliffe, Carole. 1995. *Medicine and Society in Later Medieval England*. Stroud: Sutton.

Siraisi, Nancy G. 1990. *Medieval and Early Renaissance Medicine: An Introduction to Knowledge and Practice*. Chicago: Chicago University Press, 1990.

Webb, Heather. 2010. *The Medieval Heart*. New Haven, CT: Yale University Press.

CHAPTER 6

The Transformative Power of Breath: Music, Alternative Therapy, and Medieval Practices of Contemplation

Denis Renevey

Unlike Eastern meditative practices in yoga, Buddhist, or Hindu traditions, the Western medieval mystical tradition does not provide general technical advice about the practice of contemplation. Furthermore, it is silent about the role played by breathing in the practice of contemplation and does not address the question of the transformative power of volitional breathing. This essay aims to shed light on volitional breathing activities as performed within the context of the contemplative practices of medieval English mystics. It assesses the way in which volitional breathing, combined with socio-religious practices, contributes significantly to the alteration of consciousness, which triggers mystical experiences in the form of ecstasies or visions, or any other movement of the consciousness that transcends everyday reality.

D. Renevey (✉)
English Department, University of Lausanne, Quartier UNIL-Chamberonne Bâtiment Anthropole, Lausanne, Switzerland

D. Fuller et al. (eds.), *The Life of Breath in Literature, Culture and Medicine*, Palgrave Studies in Literature, Science and Medicine,
https://doi.org/10.1007/978-3-030-74443-4_6

In view of the absence of evidence in medieval texts, I rely on practice and theory derived from manuals for brass players, but also applicable to wind instruments in general.[1] Pranayama, the art of breath control practised in Eastern meditative traditions, can also help us speculate about the importance of breathing for medieval mystics. Luce Irigaray has shown how Western philosophical tradition has abandoned breath in its epistemological considerations.[2] Lenart Škof and Petri Berndtson also deplore the loss of thinking and breathing as 'a chiasmic relation in which the thinker and the experience of breathing somehow constantly intertwine in an essential manner, perpetually inspiring each other'.[3] It is therefore not surprising that traditional Indian spiritual practices, with their insistence on the power of breath, should attract interest from Western philosophers; nor that they should also be used by 'new age' movements, packaged for audiences interested in quick access to 'the spiritual'. One such 'new age' practice is Holotropic Breathwork, of special interest for its claim that hyperventilation offers one of the most effective means of modifying consciousness beyond the confines of the rational, a claim of particular relevance in investigating the possible function of breathing patterns in the practice of medieval mystics.

The medieval material used for this investigation is limited to the writings of two medieval English mystics, Richard Rolle of Hampole, a fourteenth-century Yorkshire hermit, and the author of the mystical treatise *The Cloud of Unknowing*, possibly a Carthusian monk active in the late fourteenth century, who remains anonymous to this day. Both writers share a common interest in the practice of contemplation based on the repetition of short prayers, and it is mainly as related to this specific spiritual practice that I offer hypotheses on the role of the transformative power of breath.

***

Producing a sound from a wind instrument depends on the player's production of wind. Arnold Jacobs (1915–1998), long-standing tuba player of the Chicago Symphony Orchestra, and perhaps one of the best brass teachers ever, offers a simple but significant statement about breath in his teaching: 'Breathing to play an instrument is clearly different from breathing to live'.[4] Jacobs' teaching focused particularly on breath. He developed simple but powerful breath exercises that led to the improvement of sound production by his students. Breathing to play requires a move from natural to volitional breathing that implies knowledge about one's own respiratory functions and abilities, as well as practice in the control of airflow, away from the instrument or, in the case of singing, without sound production.[5]

Breathing can be divided into three phases: inhalation, exhalation, followed by a brief pause. When we inhale, our respiratory system searches for oxygen; the exhalation phase deals with the expulsion of carbon dioxide. Volitional breathing for the practice of a wind instrument, a category to which brass instruments like the tuba, the French horn, the trombone, and the trumpet belong, relies on one of the three forms of breathing characterized according to physiological parameters. Diaphragmatic breathing, or deep breathing, makes use of the lower part of the lungs and full use of the diaphragm muscle. Jacobs describes the diaphragm in the following way: 'The diaphragm is a muscular partition between the thoracic and the abdominal cavities. Its location, in the front, is at the base of the sternum (breastbone) and in the back on the spine and at the base of the rib cage'.[6] Diaphragmatic breathing is central to the practice of brass instruments. It allows for the production of pressurised air which is required for an airflow that will produce notes: the higher the notes the more pressurised the air has to be. Proper use of the diaphragm keeps other parts of the body (neck, throat, face muscles, lips) free from unproductive pressure. The second form of breathing is intercostal breathing, which takes place just above the diaphragm, at the level of the rib cage. Intercostal breathing produces a movement pushing the thoracic cage outward. The third, clavicular breathing, is located at the apex of the lungs, with the air placed at the high extremity of the lungs, and therefore pushing the thoracic cage upwards. The combination of these three forms of breathing in yoga practice is called 'full breathing'. It can be practised volitionally and has therapeutic benefits, in increasing energetic potential, stimulating creativity and sensitivity. It has also the ability to help individuals reach a sustained level of relaxation and balance.[7]

Diaphragmatic breathing, also called deep breathing, is at the heart of brass playing. Michel Ricquier's *Traité méthodique de pédagogie instrumentale* contains a series of thirteen breathing exercises, all emphasising deep breathing.[8] Central to the teaching of both Ricquier and Jacobs is their emphasis on the need to use breath efficiently to create wind to play a wind instrument.[9] Deep breathing can easily be experienced if one experiments with an exercise suggested by Ricquier, which consists in first adopting an upright sitting posture, with hands joined as if praying, with elbows touching one another. Once the elbows are pressed against the abdomen, the practitioner should spread his/her legs and lean downwards and breathe. This position will automatically impose a deep breathing mode. This deep breathing, according to yogic philosophy, has positive physiological and emotional effects. It also makes possible the compression of the air required for playing a wind instrument. Deep breathing can easily be practised in that position, but is much more difficult to maintain in an upright position. The exercises that follow in Ricquier's treatise aim to help the practitioner maintain deep breathing when moving from this position to the standing position of the solo performer. The combined information offered from a medico-physiological perspective (Jacobs) and a perspective influenced by the yoga practice of pranayama (Ricquier) shows the extent to which breath participates actively in the life of both body and mind. As a way of maintaining deep breathing in all circumstances, Ricquier invites the practitioner to imagine the breathing phenomenon as taking place at the level of the substernal hole, near the diaphragm muscle, and to forget the passageways of the lips, mouth, trachea, and upper part of the lungs. In order for the compression of air to take place at this level, without the uninvited participation of these other physical organs, one should imagine the air going down when both inhaling and exhaling. This vertical downward movement, working as a piston, depends on the diaphragm as its foundation. It is called the air column, and is used for efficient brass instrument playing.[10] Although Ricquier, like Jacobs, is initially interested in providing food for thought and practice to brass players, he also realizes that the potential benefits of proper breathing go beyond the practice of wind instruments. Indeed, Ricquier links this practice to pranayama, the dynamics of breath in the yoga tradition and the desire to improve well-being.[11] There is indubitably a close connection between the practice of volitional deep breathing as part of brass instrument practice and meditation centred on breathing. For instance, in addition to the breathing exercises mentioned above that are common practice among professional brass players as part of their daily warm-ups, playing long notes for tone, pitch, stamina, and sound improvement depends completely on airflow control. Normal exhalation is usually less than four seconds, while sustained sound production may last as long as twenty to thirty seconds. This therefore depends on the controlled release of air, closely resembling the yoga practice mentioned above, and the practice of repetition of short-word prayers indicated by some medieval English mystics.[12]

While most Western philosophical and theological writing does not engage directly with the practicalities or transformative potential of breathing in contemplative practice, various schools of yoga rely on pranayama as part of their teaching.[13] One of the foundational texts on pranayama is the *Yoga Sutra* of Patanjali (*ca.* 400 BCE—400 CE), which offers a culturally constructed view of breath, from organic and animal life at one end of the spectrum, to the spiritual dimension at the other. The point of pranayama and other Eastern schools of philosophy based on breathing is that a move along that continuum, from the organic to the spiritual, can be reached by the acquisition of breathing techniques. The move from involuntary to volitional breathing allows for a change of quality in the breathing process, a change achieved through practice similar to the regular breathing exercises undertaken by brass players in their daily warm-up routines, which they keep practising as part of their musical performances. But as Cleo McNelly Kearns notes, practical techniques in matters of spirituality in the Western tradition are perceived as antithetical to infused grace, and therefore not regarded as necessary for contemplative practice. So, the absence of technical instruction concerning contemplative activity in the writing of Richard Rolle or the *Cloud*-author is culturally motivated. And yet the following aspects are paramount to efficient prayer. First, conscious breathing should be intimately linked to speaking, and therefore to praying, with much significance given to silence.[14] Second, controlled breathing (therefore controlled speaking and praying—and playing, as we have seen above) should be practised, with the expectation that it triggers spiritual transformation. Also, however ineffable and disembodied some aspects of contemplation may be, the reliance on breath in contemplative practice situates it strongly in the body.[15]

Luce Irigaray's investigation of Eastern spiritual traditions has led her to contend that transformation of our elemental breath into something more subtle and controlled helps heart, thought, and speech. Breathing is therefore at the heart of her respiratory philosophy, developed from Indian spiritual practices. *The Forgetting of Air in Martin Heidegger* (1983) represents a turning point in Irigaray's thought, leading her to propose a new age, the 'Age of the Breath'. Her deep knowledge of Western philosophy and her experiential approach to yogic breath work are shown in the conscious and thoughtful breathing which is at the heart of her reflections in *Between East and West*. Irigaray here applies breathing techniques from yoga and pranayama to Western forms of spiritual practices, such as prayer. She offers convincing evidence that 'breath itself and the kinds of religious discipline related to it are situated in the body'.[16]

The examples from both brass instrument playing and pranayama show, therefore, how a move from organic to volitional breathing can have profound transformative power in relation to the physical and spiritual make-up of practitioners. Jacobs and Ricquier both show the extent to which regular breathing exercises transform the ability to produce efficient wind for the production of sound, while the latter explores further the physical and spiritual benefits induced by conscious and wider-ranging breathing patterns that include deep breathing and full breathing. The transformation of consciousness induced by such practice is attested by the changes in the quality of sound production as well as contact with the inner self. Although the latter may not be the primary objective of musical practitioners, it is an aim comparable to those of yoga schools: Irigaray's reference to 'spirited' breath, which involves complete inhalation and complete exhalation, brings about results similar to the exercise of full breath discussed by Ricquier.[17]

If producing efficient wind leads to a state of well-being that helps an individual get in touch with the inner self, the production of sound on brass may also have unwanted effects in relation to maintaining a certain clarity of mind for proper performance. In normal situations, the concentration of oxygen and carbon dioxide in the blood that is the result of breathing varies constantly. The oxygen that we breathe is brought to the tissues by the blood, and carbon dioxide is the waste product that we exhale.[18] In some circumstances, the ratio between oxygen and carbon dioxide can be unfavourable, thus contributing to an alkaline shift in the blood's pH. This may result in hyperventilation, the effects of which Jacobs describes in the following manner:

> When you hyperventilate a bit, you actually alter the pH (potential hydrogen) of the body. The alkaline/acid relationships are affected, and a person is apt to feel a little bit peculiar with very moderate hyperventilation. Usually it will start out as dizziness. How will I say it—a little leaving of the ground—you begin to float a little bit. The symptoms of hyperventilation are due to a lack of carbon dioxide and its effect on the brain. The carbon dioxide is washed out of the blood by the heavy breathing

> (increased ventilation) of the respiratory system. Keeping the air flow into the tuba at the maximum, makes the person feel dizzy.[19]

Jacobs's scientific experiments with researchers at the Pulmonary Functions laboratory at the University of Chicago led to several major discoveries with regard to brass players' abilities to breathe without the symptoms of hyperventilation. The scientists were also aware of its effects on the brain. In the context of musical performance, hyperventilation is to be avoided at all costs as performers find themselves in situations that require control of all their faculties for a good quality performance. In the case of Jacobs, hyperventilation was avoided by inhaling back carbon dioxide through his tuba. However unwanted the experience during musical performance, what Jacobs calls 'a little leaving of the ground—you begin to float a little bit', can clearly be said to be an alteration of consciousness generated by increased ventilation. Volitional breathing and breathing performed during strenuous physical exercises can lead to hyperventilation as well, and thus to significant changes of consciousness.[20]

Though the contemporary therapeutic practice of Holotropic Breathwork has a different status from the more tried and tested practices of a traditional Eastern religion and the breathing techniques of brass teachers, its focus on hyperventilation as a trigger for transformation of consciousness offers a link with the description of one of its effects by Jacobs, that is, 'a little leaving of the ground'. While Jacobs offers suggestions as to how to avoid hyperventilation and its unwanted effects, Holotropic Breathwork offers hyperventilation as its main tool for altered consciousness. This focus on accelerated breath and the positive outlook on hyperventilation as contributing to the alteration of consciousness may be of interest in assessing the role played by breath in medieval mystical experiences. Holotropic Breathwork considers that the brain contributes to mediating consciousness, but does not actually generate it. It stipulates that we have no evidence that consciousness is produced in the brain and by the brain. Rather, like a television set, the brain absorbs and conveys material that is generated outside the brain, depending on the connections that it has made with external sources.[21] Holotropic Breathwork allows for the brain to be connected with immanent and transcendental sources that feed spiritual experiences. As the main propagators of the technique, Stanislav and Christina Grof, write:

> In holotropic states, our consciousness can reach far beyond the boundaries of the body/ego and obtain accurate information about various aspects of the material world that we have not obtained in this lifetime through the mediation of our sensory organs. We have already mentioned reliving of birth, of prenatal memories, and of conception. In transpersonal experiences our consciousness can identify with other people, with members of various species of the animal kingdom from primates to unicellular organisms, with plant life, and even with inorganic materials and processes. We can also transcend linear time and experience vivid ancestral, racial, karmic and phylogenetic sequences, and episodes from the collective unconscious.[22]

Holotropic Breathwork situates itself in a long tradition of psycho-spiritual systems that have sought to understand human nature and have 'viewed breath as a crucial link between the material world, the human body, the psyche and the spirit'.[23]

The practice of Holotropic Breathwork consists of faster breathing, evocative music, and releasing body work.[24] It emphasizes the ability of volitional breathing to influence consciousness. Hyperventilation and prolonged withholding of breath are two of the best means to generate such alteration. Holotropic Breathwork, its practitioners claim, acknowledges the many different states of consciousness that each individual may experience, allowing the brain to connect with personal, perinatal, and psycho-spiritual dimensions that cannot be accessed in a normal state of consciousness. The holotropic state is reached via accelerated breath, as is made clear by the guidance given to a breather in the context of a workshop:

> Now that your body is more relaxed, open, bringing attention to your breath ... feeling the normal, everyday rhythm of your breath ... and as you continue to be aware of your breath, beginning to increase the rhythm of your breathing ... so your breath becomes faster than usual ... allowing the breath to travel all the way to your fingertips, all the way to your toes, as the breath fills the entire body ... the breath is quite a lot deeper and faster than usual ... Now increasing the rhythm of the breath even more ... so that it becomes faster than usual, and deeper ... and deeper and faster ...
>
> At this point, if you have a source of inspiration in your life, you may try to connect with it asking for guidance ...
>
> So by now, the breath is quite a lot deeper and faster than usual ... As the music begins, allowing the music to support your breathing ... finding

> your own rhythm ... Opening yourself to the breath, to the music, and whatever experience that presents itself.[25]

Holotropic Breathwork, which consists mainly of accelerated and deep breathing, leads, its practitioners suggest, to a wide spectrum of experiences, some of them mystical. This 'new age' practice demonstrates the continuing desire for altered states of consciousness. The results it claims to achieve by inducing hyperventilation suggest the possibility that the effects of unsolicited hyperventilation may have been causal factors in the transformation of consciousness in medieval contemplative practice.

***

Although Western medieval mystics are silent about techniques of contemplation in general and breathing in particular, it is justifiable to assume that breathing was part of their spiritual routines and to assess whether volitional breathing led them to move to the spiritual end of the breathing spectrum. Also, even if their volitional breathing did not tend towards accelerated breathing leading to hyperventilation, it is worth considering whether some of their experiences may have been generated by unsolicited hyperventilation. The aim of the second part of this essay is to consider these possibilities in the context of the practice of short one-word prayers as advocated by Richard Rolle and the anonymous *Cloud*-author. The early fourteenth-century Yorkshire mystic Richard Rolle is one of the staunchest proponents of ceaseless prayer in the form of devotion and prayer to the Name of Jesus. The prayer was well-known in both Western and Eastern Christian traditions.[26] When reduced to its simplest form, it consisted in both traditions in the repetition of the name, in Middle English 'Ihesu'. Examples of the use of the prayer to the Name of Jesus abound in Rolle's corpus.[27] But more importantly, statements about the presence of the name 'Ihesu' in Rolle's mind show the degree to which it became an essential component in the shaping of his state of consciousness. For instance, Rolle writes the following about the Name in his *Melos amoris*:

> Dum vigilo, non vaco a voce vitali: si legerem, si scriberem, si ambulem, si agerem, continuo canorem concentu cupitum; si comedam, si bibam, non cadit a corde nomen notandum solum in quo salvari me sencio securum. Non obliviscar Illius qui animum assumpsit ut ardens anhelem Ipsi adesse.

Sine Iesu non iubilo qui genitum me gessit a gemitu et gelu ut germiner ingenuus ierarchie. Et cum sedeo inter suspirantes ad seculare solacium, dum non garriunt ut clangentes michi in secreto quod non sciunt, acsi essem in requiem restauratus, resultat organum angelicum et sonat, sed subtilissime, celica symphonia.

[When I watch, this powerful voice does not leave me. Whether I read or write, whether I walk or work, the harmonious song I love continues. Whether I eat or drink, from my heart the only Name which is worthwhile mentioning stays with me; I am confident about salvation. I cannot forget the one who has taken possession of my soul that I long to be in his presence.

Without Jesus I do not jubilate. From birth he protected me from the weeping and from the cold, so that I could grow, introduced as I had been among the angelic line. Even when I am in the middle of those who long for the pleasures of this century, as long as they do not break my ears shouting secrets they ignore, the angelic music resounds within me as if rest was given back to me, and the celestial symphony resounds very delicately.][28]

In Rolle's *Oleum effusum*, the Name serves to fight off the demon, disguised in the shape of a woman.[29] In *Incendium amoris,* Rolle states on several occasions how much his unceasing devotion to the Name contributes to his spiritual state.[30] References to devotion to the Name of Jesus in these treatises indicate that it was used by Rolle in different ways and for different purposes. Using his preferred sitting posture, which he mentions on several occasions, Rolle would have repeatedly pronounced the word 'Ihesu' during controlled and sustained exhalation phases, thus forcing a pattern of breathing that he would have closely scrutinized. The breathing rhythm imposed by such practice would necessarily be diaphragmatic breathing, leading to an initial state of well-being that would bring about a change in state of consciousness. The intoning of this short one-word prayer in one's own familiar language invites a focus on the sound of the word itself, on the quality of silence between each intonation, and on the quality of breath.[31]

Rolle is also unique among medieval English mystics in recommending the sitting posture as the preferred one for his contemplative activity. Two significant moments that are part of his conversion begin with reference to his being seated in a particular chapel:

> Sedebam quippe in quadam capella, et dum suauitate oracionis uel meditacionis multum delectarer, subito sentiui in me ardorem insolitum et iocundum…
>
> Dum enim in eadem capella sederem, et in nocte ante cenam psalmos prout potui decantarem, quasi tinnitum psallencium uel pocius canencium supra me ascultaui… Puto tamen neminem illud accepturum, nisi specialiter nomen Ihesum diligat, et eciam in tantum honoret ut ab eius memoria numquam, excepto sompno, recedere permittat.
>
> [I was sitting in a certain chapel, delighting in the sweetness of prayer or meditation, when suddenly I felt within myself an unusually pleasant heat…
>
> While I was sitting in that same chapel, and repeating as best I could the night-psalms before I went into supper, I heard, above my head it seemed, the joyful ring of psalmody, or perhaps I should say, the singing … All the same I fancy that no one will receive them (celestial joys) unless he has a special love for the Name of Jesus, and so honours it that he never lets it out of his mind, except in sleep.][32]

In the following chapter Rolle describes his solitary state as a sitting posture ('*sedens in solitudine*') and, more significantly, he projects his heavenly stay as being seated in Paradise ('*in paradise positus*').[33] The sitting posture becomes a hallmark of Rolle's spirituality, one that he even includes in the lyric that concludes *Ego dormio*, one of his three epistles, written for the attention of a nun of Yedingham: 'I sytt and syng of luf langyng þat in my breste es bredde/ Jhesu, Jhesu, Jhesu when war I to þe ledde?'[34]

Ricquier recommends the sitting posture in order to experience deep breathing. In view of the significance of breathing to the practice of short prayers, it is not impossible that Rolle's preference for that posture was linked to the ease it allowed in finding a good breathing pattern for the practice of the unceasing prayer devoted to the name 'Jhesu'. Interestingly, the practice of the Jesus prayer in orthodox spirituality, which can be reduced to the simple utterance of the word 'Jesus', also offers the sitting posture as the best for its practice. But unlike Rolle, who seems to have kept to a traditional sitting posture, with the torso upright, the hesychast method practised by mystics of the Eastern Orthodox Church on Mount Athos proposes a position very similar to that of the diaphragmatic breathing postulated by Ricquier:

> St Gregory of Sinai advises sitting on a low stool, about nine inches high; the head and shoulders should be bowed, and the eyes fixed on the place of the heart. He recognizes that this will prove exceedingly uncomfortable after a time. Some writers recommend yet more exacting posture, with the head held between the knees, following the example of Elijah on Mount Carmel.[35]

Modern-day practitioners of the Jesus prayer provide very useful information about coordination with heartbeat and breath in order to integrate it with oneself:

> Prayer is not meant to be a sporadic occurrence. It must be assured of continuity. It must be given room to live, as surely as there must be room in a person's life for his heart and his breathing. As important as my breathing and heartbeats is my steady prayer to Jesus. To practise this constant praying by mechanically co-ordinating it with the beat of the heart or the work of the lungs is just one way of really providing prayer with this room of its own.[36]

The unceasing quality of the Jesus prayer discussed here matches some of the comments made by Rolle about his own practice, as well as reflecting the number of lyrics or lyrical moments in his prose treatises that integrate the name 'Ihesu' with his devotional outbursts.

An example of the accommodation of the name 'Jhesu' occurs within the lyric that concludes *Ego dormio*, at a moment when Rolle provides tools to reach the highest level in the contemplative life, according to three degrees of love. This lyric is offered as part of the third and highest degree of love, called 'contemplatife lyfe'. The lines beginning with 'Jhesu' appear to function according to the same breathing paradigm described above for intoning the name 'Ihesu'. Each verse, preceded by a quick inhalation, is uttered during the course of one exhalation, thus creating a breathing pattern that is regular and under the control of the will. Although no technical instruction is provided for this iteration, or for the reading of the whole epistle, the rhythmical pattern invites breathing that very much follows the deep and volitional breathing pattern discussed earlier:[37]

> Jhesu, my dere and my drewry [sweetheart], delyte ert þou to syng.
> Jhesu, my myrth and melody, when will þow com, my keyng?
> Jhesu, my hele [cure], and my hony, my whart [joy] and my comfortyng,
> Jhesu, I covayte for to dy when it es þi payng [pleasure].[38]

Unlike Irigaray, I would not want to disentangle meaning from intoning. They work superbly together here to facilitate a change of consciousness that will bring the reader to the highest and final degree of love. However, insistence on the role of breathing as part of these highly performative exercises gives them a strong physical grounding. This spiritual medicine has been tested by Rolle himself who mentions elsewhere that he practised devotion to the Name ceaselessly, day and night, with the effect that he was brought to a state of consciousness which made it possible for him to hear heavenly song continuously. Reading is therapeutic in part because it forces volitional breathing, which often leads one to develop deep breathing, which is known to have therapeutic properties.[39]

Rolle is not alone in proposing short one-word prayers. The *Cloud*-author also shows interest in the practice of short prayers. His most well-known text, *The Cloud of Unknowing*, stands out in medieval English writing as the most representative of the apophatic tradition. This tradition provides instructions on how to experience contact with the divine by suggesting a self-emptying, a noughting of the mind's images and concepts in order to let the divine pour itself in. This approach asserts that no image, object, or concept derived from the created world can successfully convey any aspect of the divine. It resists any form of comparison of the created world with divinity. This text and its companion piece, *The Book of Privy Counselling*, are unique in Middle English in using the apophatic tradition for contemplative practice.

Considering that annihilation and avoidance of analogies made to the divine and the created world are at the heart of the apophatic approach, it is not surprising that the *Cloud*-author proposes a set of short prayers for the practice of contemplation. Chapters thirty-seven to forty-one provide very useful information about the practice of short prayers. The *Cloud*-author asserts with great conviction his belief in their efficacy. He writes:

> A man or a womman, affraied wiþ any sodeyn chaunce of fiir, or of mans deeþ, or what elles þat it be, sodenly in þe heiȝt of his speryt he is dreuyn upon hast & upon nede for to crie or for to prey after help. Ȝe, how? Sekirly not in many woordes, ne ȝit in o woorde of two silabes. & whi is þat? For hym þinkeþ to longe tariing, for to declare þe nede & þe werk of his spirit. & þerfore he brestiþ up [bursts out] hidously wiþ a grete spirit, & cryeþ bot a litil worde of o silable, as is þis worde FIIR or þis worde OUTE.[40]

This passage gives prominence to the need, in situations of emergency, of voicing as quickly as possible the will of the speaker. In this particular case, the passage emphasizes the power of breath in voicing the words 'fiir' or 'oute'. As the *Cloud*-author writes, 'he brestiþ up hidously wiþ a grete spirit'. The idea of bursting out takes into account the physiology of urgent speech that forces one to take a quick deep breath in order to utter powerfully the words of warning that need to be heard by those who are in danger. I also contend that the *Cloud*-author employs 'sperit' with multiple meanings, such as 'vital breath', 'emotion', or simply 'breath or air'.[41] His words insist on the physical and physiological dimension of uttering short words with conviction and power, and it is this very physiological aspect that he transfers to the practice of praying with one-syllable words. Indeed, like 'fiir' or 'oute' in the above example, they express powerful urgency and desire for transformation, unmediated by unnecessary affect and metaphorical language. Before suggesting specific words, he offers detailed information about the role of prayer, as a 'deuoute entent directe vnto God, for getyng of goodes & remowyng of yuelles'.[42] 'Synne' and 'God' are two of the words proposed for prayer by the author, although one may also be stirred by God to use certain words, such as 'love'. Although the semantic range of each word plays an important function in the efficacy of each prayer, I would like to emphasize the physiological process, more particularly the breathing patterns, required for its performance. The *Cloud*-author is quite specific about the way in which such prayers should be performed. Indeed, they are part of what he, like Rolle, calls ceaseless praying. The *Cloud*-author is very precise when he writes that 'it schuld neuer sees tyl þe tyme were þat it had fully getyn þat þat it longid after'.[43] Mention is made of the possibility of creating formulaic prayers based on the repetition of a combination of one-syllable words. For instance, the following sequence is mentioned: 'Synne, synne, synne; oute, oute, oute'. The *Cloud*-author specifies that the prayer can be voiced or practised inwardly.[44] In both cases, silent or vocal performance would need to work in synchrony with a controlled breathing pattern that, if practised with the sense of urgency described in the first passage, would rely on deep breathing.

While the *Cloud*-author does not insist as much as Rolle on the posture that should be adopted while contemplating, *The Book of Privy Counselling* nevertheless associates the intoning of a short word with the love of sitting alone:

> Fewe ben þi wordes, bot ful of frute & of fiir. A schorte worde of þi mouþ conteneþ a woreld ful of wisdam, ȝit semeþ it bot foly to hem þat wonen [dwell] in here wittis. Þi silence is softe, þi speche ful speedful [profitable], þi preier is preue [personal], þi pride ful pure [guiltless], þi maners ben meek, þi mirþe ful mylde, þi list [craft, desire] is likyng to pleye wiþ a childe. Þou louest to be only & sit by þi-self;...[45]

The *Cloud*-author advocates restraint in many areas of the contemplative life: few words, short words, silence, private prayer, nothing more ambitious than to play with a child, and, importantly, to love to sit in solitude. Affective devotional restraint is also one of the characteristics of the apophatic tradition to which these texts belong. In this respect, *The Cloud of Unknowing* and *The Book of Privy Counselling* contrast significantly with the affective spirituality of Richard Rolle. But in other respects, the desire for solitude, the sitting posture, and prayers based on a few short words point to a world of silence in which breath finds its organic place and is shaped at will for the contemplative to reach his or her inner being in communion with the divine.

***

Breathing techniques associated with brass instrument playing, yoga, and Holotropic Breathwork offer interesting clues concerning the role played by breath in these different practices. In all these cases, breath is both essential and strongly embodied, showing that alteration of consciousness is physiologically triggered. Since information about breathing in the context of medieval contemplative practice is lacking, this discussion of modern breathing practices in complementary contexts can help us understand better the transformative power that breath had in the practice of one-syllable word prayers. The fact that Rolle and the *Cloud*-author do not offer practical instructions for their one-word prayers should not detract from investigations of the power breath has as part of contemplative practice. Breathing to live spiritually implies volitional breathing that fulfils therapeutic functions and leads to the discovery of realms of experience that go beyond the rational horizon of expectations. Combined with divine grace, volitional breathing contributed to some of the most inspiring lived experiences of medieval English mystics.

Notes

1. As a brass player (trumpet and French horn), I practise these exercises on a regular basis. Although I have never taken part in Holotropic Breathwork sessions, I have practised meditation based on a one-syllable word.
2. See Luce Irigaray, *Between East and West: From Singularity to Community*, trans. by Stephen Pluhácek (New York: Columbia University Press, 2002); also Lenart Škof and Emily A. Holmes, *Breathing with Luce Irigaray*, Bloomsbury Studies in Continental Philosophy (London: Bloomsbury, 2013).
3. See Lenart Škof and Petri Berndtson, 'Introduction', in *Atmospheres of Breathing*, ed. by Lenart Škof and Petri Berndtson (Albany: SUNY, 2018), ix–xxvii (xi).
4. Briand Frederiksen, *Arnold Jacobs: Song and Wind*, ed. by John Taylor (Gurnee, IL: WindSong Press, 2012), Kindle, loc. 2943.
5. For further information on the exercises advocated by Jacobs, see 'Arnold Jacobs *Almost Live:* Breathing Exercises', YouTube, <https://www.youtube.com/watch?v=KrNmbaRql0Y> [accessed March 3, 2019].
6. Frederiksen, *Arnold Jacobs*, loc. 3028.
7. See Michel Ricquier, *Traité méthodique de pédagogie instrumentale* (St-Jean d'Arvey: T.M.P.I., 1976), 99–106.
8. See Ricquier, *Traité méthodique*, 15–46.
9. I am paraphrasing Ricquier, *Traité méthodique*, 15.
10. Ricquier, *Traité méthodique*, 37.
11. Ricquier, *Traité méthodique*, 25. Interestingly, Ricquier's second book moves somewhat away from his initial concern with brass instrument playing. His *L'Utilisation de vos ressources intérieures: Dans votre activité instrumentale, artistique, sportive,* etc. (Paris: Billaudot, 1984) deals with the pursuit of fulfilment of one's own full being.
12. Frederiksen, *Arnold Jacobs*, loc. 3055–61.
13. Cleo McNelly Kearns, 'Irigaray's *Between East and West*: Breath, Pranayama, and the Phenomenology of Prayer', in *The Phenomenology of Prayer*, ed. by Bruce Ellis Benson and Norman Wirzba (New York: Fordham University Press, 2005), 103–18. For various explorations of a 'respiratory philosophy', see Škof and Berndtson (eds), *Atmospheres of Breathing*.
14. On the process of silence as part of contemplative practice, see Maggie Ross, *Silence: A User's Guide*, Vol. 1: Process (London: Darton, Longman and Todd, 2014).
15. McNelly Kearns, 'Irigaray's *Between East and West*', 104–05.
16. McNelly Kearns, 'Irigaray's *Between East and West*', 105.
17. McNelly Kearns, 'Irigaray's *Between East and West*', 116.
18. Frederiksen, *Arnold Jacobs*, loc. 3110.

19. Frederiksen, *Arnold Jacobs*, loc. 3117. I have experienced this kind of dizziness on a few occasions when playing the trumpet.
20. During summer hiking camps in the Swiss Alps friends occasionally suffered from hyperventilation when going up a mountain top. The way to stop the associated dizziness was to breathe into a plastic bag, so as to take in carbon dioxide in order to balance the ratio with oxygen. This is a very similar method to that described by Jacobs with his tuba. This technique is also described by Grof as a means of suppressing spontaneous hyperventilation; see Stanislav Grof and Christina Grof, *Holotropic Breathwork: A New Approach to Self-Exploration and Therapy* (Albany, NY: State University of New York Press, Excelsior Editions, 2010), Kindle, loc. 923–30.
21. Grof and Grof, *Holotropic Breathwork*, loc. 812–23.
22. Grof and Grof, *Holotropic Breathwork*, loc. 823–28.
23. Grof and Grof, *Holotropic Breathwork*, loc. 885.
24. Grof and Grof, *Holotropic Breathwork*, loc. 869.
25. Grof and Grof, *Holotropic Breathwork*, loc. 1600.
26. For a summary of Western tradition, see Denis Renevey, 'The Emergence of Devotion to the Name of Jesus in the West', in *Aspects of Knowledge: Preserving and Reinventing Traditions of Learning in the Middle Ages*, ed. by Marilina Cesario and Hugh Magennis (Manchester: Manchester University Press, 2018), 142–62.
27. See, for instance, the *Incendium amoris*, *Emendatio vitae*, *The Form of Living*, and *Ego dormio*, as examples of the inclusion of the devotion to the Name within Rolle's mystical system. For a brief description of Rolle's spiritual and textual legacy, see Tamás Karáth, *Richard Rolle: The Fifteenth-Century Translations*, Medieval Church Studies 40 (Turnhout: Brepols, 2018), 13–33.
28. Richard Rolle, *Le Chant d'amour*, 2 vols, ed. by E. J. F. Arnould, intro. by François Vandenbroucke, Sources Chrétiennes 168–9 (Paris: Editions du Cerf, 1971), vol. 1, 134.
29. For a Middle English version of this text, see *Richard Rolle: Uncollected Prose and Verse*, ed. by Ralph Hanna, EETS 329 (Oxford: Oxford University Press, 2007), 9–11.
30. For instance, chapters 36, 38, 40, 41, and 42 show the extent to which Richard Rolle has integrated the prayer to the Name of Jesus: see Richard Rolle, *Incendium amoris*, ed. by Margeret Deanesly (Manchester: Manchester University Press, 1915), 248, 250, 267, 272–73, and 278.
31. See McNelly Kearns, 'Irigaray's *Between East and West*', 109; Irigaray discusses the intoning of prayers in non-familiar languages, thus displacing discursive meaning and drawing attention to sound, silence, and breath.

32. Richard Rolle, *Incendium amoris*, 189–90; Richard Rolle, *The Fire of Love*, 2nd edn, trans. by Clifton Wolters (London: Penguin, 1988), 93–94.
33. See Richard Rolle, *Incendium amoris*, 193.
34. See *English Writings of Richard Rolle, Hermit of Hampole*, ed. by Hope Emily Allen (Gloucester: Alan Sutton, 1988), 71.
35. Bishop Kallistos of Diokleia, *The Power of the Name: The Jesus Prayer in Orthodox Spirituality*, rev. edn (Oxford: SLG Press, 1986), 25.
36. See Per-Olof Sjögren, *The Jesus Prayer*, rev. edn, trans. by Sydney Linton (London: Triangle, 1996), 44.
37. *English Writings*, ed. by Allen, 69.
38. *English Writings*, ed. by Allen, 71.
39. These exercises participate in what Daniel McCann considers to be therapeutic reading: see Daniel McCann, *Soul-Health: Therapeutic Reading in Later Medieval England*, Religion and Culture in the Middle Ages (Cardiff: University of Wales Press, 2018); see also Daniel McCann, 'Blood and Chocolate: Affective Layering in *Swete Ihesu, Now wil I Synge*', in *Middle English Lyrics: New Readings of Short Poems*, ed. by Julia Boffey and Christiania Whitehead (Cambridge: Brewer, 2018), 45–56.
40. *The Cloud of Unknowing and The Book of Privy Counselling*, ed. by Phyllis Hodgson, EETS 218 (Oxford: Oxford University Press, 1944), 74.
41. See 'sperit', *Middle English Dictionary*, <https://quod.lib.umich.edu/m/middle-english-dictionary/dictionary/MED42245> [accessed July 9, 2019].
42. *The Cloud*, 77.
43. *The Cloud*, 78.
44. *The Cloud*, 78.
45. *The Cloud*, 166–67.

Selected Bibliography

Arnold Jacobs. *Almost Live:* Breathing Exercises. https://www.youtube.com/watch?v=KrNmbaRql0Y. Accessed March 3, 2019.

Frederiksen, Brian. 2012. *Arnold Jacobs: Song and Wind*. Ed. John Taylor. Gurnee, IL: WindSong Press.

Grof, Stanislas, and Christina Grof. 2010. *Holotropic Breathwork: A New Approach to Self-Exploration and Therapy*. Albany, NY: State University of New York Press, Excelsior Editions.

Kallistos, Bishop of Diokleia. 1986. *The Power of the Name: The Jesus Prayer in Orthodox Spirituality*. Rev. edn. Oxford: SLG Press.

McCann, Daniel. 2018. *Soul-Health: Therapeutic Reading in Later Medieval England*. Religion and Culture in the Middle Ages. Cardiff: University of Wales Press.

McNelly Kearns, Cleo. 2005. Irigaray's *Between East and West*: Breath, Pranayama, and the Phenomenology of Prayer. In *The Phenomenology of Prayer*, ed. Bruce Benson and Norman Wirzba, 103–18. New York: Fordham University Press.

Renevey, Denis. 2018. The Emergence of Devotion to the Name of Jesus in the West. In *Aspects of Knowledge: Preserving and Reinventing Traditions of Learning in the Middle Ages*, ed. Marilina Cesario and Hugh Magennis, 142–62. Manchester: Manchester University Press.

Ricquier, Michel. 1976. *Traité méthodique de pédagogie instrumentale*. St-Jean d'Arvey: T.M.P.I.

Ross, Maggie. 2014. *Silence: A User's Guide*. Vol. 1: Process. London: Darton, Longman and Todd.

Sjögren, Per-Olof. 1996. *The Jesus Prayer*. Rev. edn., trans. Sydney Linton. London: Triangle.

CHAPTER 7

A Breath of Fresh Air: Approaches to Environmental Health in Late Medieval Urban Communities

Carole Rawcliffe

In 1450, Thomas Cornwaleys brought a prosecution in the Court of King's Bench against six London butchers for dumping 'dung, intestines and other foetid and disgusting matter' in the garden of his house at Aldgate. Since the King's Bench was England's premier common law court, which dealt with cases of treason, murder, and other major felonies, it seems strange on the face of things that the justices should have entertained litigation about the medieval equivalent of fly-tipping. But this was no ordinary nuisance. The butchers were accused of endangering Cornwaleys' life and the lives of his servants with the lethal stench (*fetor*) of their waste, and of making it impossible for him to remain in his home without the greatest physical risk (*absque maximo corporum suorum periclo*). After several delays, the accused appeared in court and were committed to prison on 2 December, by which point, not coincidentally, plague had broken out in London.[1] The outcome of the suit, in which

C. Rawcliffe (✉)
School of History, University of East Anglia, Norwich, UK

D. Fuller et al. (eds.), *The Life of Breath in Literature, Culture and Medicine*, Palgrave Studies in Literature, Science and Medicine,
https://doi.org/10.1007/978-3-030-74443-4_7

Cornwaleys sought damages of £40,[2] is not known, but complaints of this kind (especially from high status individuals) were treated very seriously. They also had a long history, predating the Black Death of 1348–50 by many years.

At the start of the fourteenth century, for example, the staff and students of Oxford University had submitted a battery of petitions to the crown about the filthy state of the town's streets and the environmental pollution that ensued. In 1310 they singled out Oxford's butchers as particular offenders on this score, since they caused 'so great a corruption' when slaughtering animals that 'many people' fell ill and some even died through the inhalation of toxic air.[3] On other occasions the king himself might intervene directly, as happened in 1332 when Edward III reprimanded the rulers of York over the 'abominable smell abounding in the said city [...] from dung and manure and other filth and dirt', ordering them to clean the streets before Parliament assembled there and lives were put at risk.[4] It is thus hardly surprising that he should insist upon the permanent removal of slaughterhouses from London during the second plague epidemic of 1361–62, renewing the order when pestilence returned in 1368–69. His efforts to confine these noisome activities to Knightsbridge (in the west) and Stratford (in the northeast) none the less had unwelcome consequences and encountered widespread resistance. Whereas some butchers raised their prices to allow for the cost of transport, thereby occasioning disturbances in the City, others took the easy option of butchering cattle just outside the walls and leaving their offal to decompose in the watercourses, fields, ditches, and gardens of people like Cornwaleys.[5]

The outcry made in 1379, another plague year, by a group of influential courtiers and other prominent residents living in Holborn reflects a timeless desire on the part of affluent householders to preserve the value of their property. Yet a real fear of infection is also apparent from the petition for redress which they submitted to Parliament, the highest court in the land, protesting that:

> [...] because of the great and horrible stenches and deadly abominations which arise there from day to day from the corrupt blood (*sank corrupt*) and entrails of cattle, sheep and pigs killed in the butchery next to the church of St Nicholas in Newgate and thrown in various ditches in two gardens near to Holborn Bridge, the said courtiers, frequenting and dwelling there, contract various ailments, and are grievously exposed to

disease (*trope grevousement mys a disease*) as a result of the infection of the air, the abominations and stenches above said, and also by many evils that notoriously ensue.[6]

Protests of this kind eventually gave rise to a parliamentary statute of 1388 that comprehensively forbade the deposit of butchers' waste and similar refuse in or near *any* English towns or cities because of the threat to public health posed by miasmatic (noxious) air.[7] The butchers sued by Cornwaleys in 1450 were thus not only potential homicides, but were also in breach of statute law, which constituted another serious criminal offence.

The English were certainly not alone in demonstrating acute sensitivity to the stench of rotting carrion. In 1416 the advisors of King Charles VI of France had, for instance, ordered the demolition of Paris's Great Butchery (les Halles), which stood alarmingly near the royal palace of the Châtelet, and its replacement by four new flesh markets outside the walls. The adjacent slaughter- and scalding-houses were likewise to be moved to a site beyond the Tuileries, 'less dangerous to the public health of our said city and less likely to corrupt the air of the same'. The entire operation was justified on medical grounds, it being deemed necessary 'to provide and take precautions against the infections and corruptions noxious to the human body' engendered by so much potentially toxic waste.[8] When the relocation of slaughterhouses proved impractical, other precautions, such as rebuilding on hygienic principles, would be adopted, preferably with improved water supplies for more effective cleansing.[9] As a last resort, regulations issued by the Dublin Assembly in 1484 for the safer management of the 'flesshambles' insisted that a keeper should 'locke and steke the dorres and wyndouus of the saide shambles and so kepe them fast at all tymes but when the said bouchers selleth their flesshe'.[10] Hefty fines penalised any infringements that might contaminate the surrounding area.

Such measures were not confined to butchers, nor were they, as we have already seen in the case of Oxford and York, simply a knee-jerk response to the panic which understandably gripped urban communities in the aftermath of the Black Death. Recent work on medieval public health has stressed the comparatively early date of environmental legislation that explicitly reflects the connection made by urban magistrates between bad smells and the spread of disease.[11] In 1283–84 the rulers of Treviso ordered that dung and waste should be removed quickly from the streets because they 'infect the air and create a pestilence (*aeram infliciunt et faciunt pestilentam*) through which human bodies succumb to illness and suffer death'.[12] Ordinances for the expulsion of noxious trades such as tanning from Pistoia, dated 1296, were likewise justified specifically because 'it is civil and expedient for the preservation of people's health that the city [...] be cleared of stenches from which the air is corrupted and pestilential diseases ensue (*ex quibus aer corumpitur et pestilentiales egretudines oriuntur*)'.[13] Chemical as well as organic processes were also beginning to cause anxiety. It was in the seventeenth century that John Evelyn compared London to 'the face rather of *Mount Ætna*, the *Court of Vulcan*, *Stromboli*, or the Suburbs of Hell', but already by the 1280s lime-burning with sea coal, which produced an 'intolerable stench', appeared to spread 'infection and corruption of the air' as it wafted into the homes of prominent citizens and even the queen herself.[14]

Where did these ideas originate? A Latin version of the pseudo-Hippocratic text *Airs, Waters, Places* had been in circulation from the twelfth century, underscoring the importance of fresh, temperate air for communal health. Its influence is apparent in the urban panegyrics that proliferated during the later Middle Ages, praising not only the cleanliness but also the invigorating environments of various European cities.[15] Thus, for example, Bonvesin de la Riva's paean to Milan, *De magnalibus Mediolani* of 1288, asks with a fine rhetorical flourish: 'Are there no putrid pools or lakes corrupting the air with their damp or stench? Certainly not! Here one finds clear fonts and fertile rivers'.[16] Similarly, the Franciscan friar Francesc Eiximenis begins the regimen that he composed in 1383 for the magistrates of Valencia by extolling first among the city's many attractions its 'fine and clear' air, uncontaminated by any of the fumes or turbidity that blighted the climate of England, France, and Germany.[17]

More significant in the present context, however, was the work of the celebrated Greek physician Galen of Pergamum (d. 216), which underpinned most medieval ideas about human physiology. Since the effectiveness of all physical and mental processes initially depended upon what one ate, health was, to a notable extent, determined by diet, 'the first instrument of medicine'.[18] Having been cooked in the oven of the stomach, partially digested food was conveyed to the liver, where it was converted into humoral matter: blood (hot and wet), phlegm (cold and wet), black bile (cold and dry), and yellow bile (hot and dry). From the liver, the blood and other humoral matter, known as the natural spirit, travelled along the veins to the organs and extremities, being absorbed as essential nourishment (Plate 7.1). During this third, final stage of the

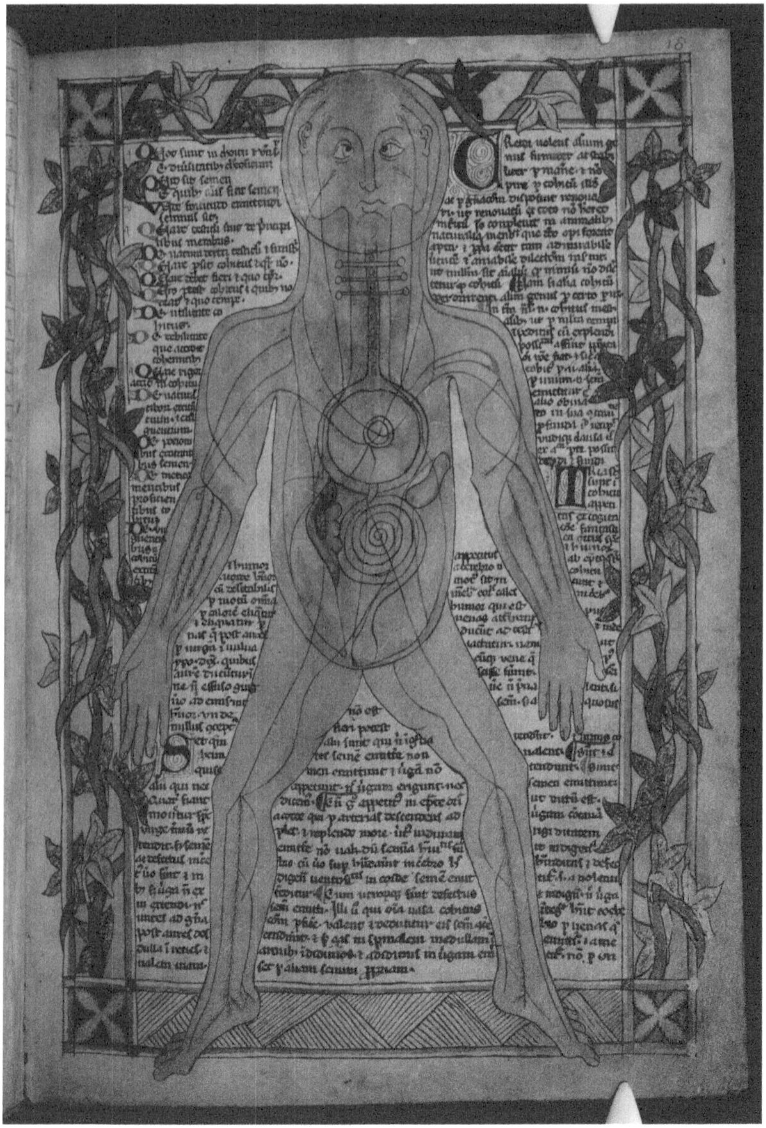

Plate 7.1 From an English-owned medical treatise of *ca*. 1292, this diagram of the venous system depicts the stomach, intestines, and liver (the leaf-shaped organ on the figure's right side), where the natural spirit, or humoral matter, is generated from food and transported along the veins to the extremities. (Bodleian Library, Oxford, MS Ashmole 399, f. 18r)

digestive process, the natural spirit might easily be affected by the quality of the air drawn into the body through the open pores. For this reason, anything likely to raise one's temperature and encourage perspiration, including sexual activity, rich food, and excessive amounts of alcohol, was to be avoided during epidemics. Exemplifying the symbiotic relationship between morality and medicine characteristic of so much late medieval advice literature, one vernacular plague tract warned that 'men that abusen them self with wymenn, or vsen ofte times bathes, or men that be hote with labour or grete angre' would inevitably be 'more disposed to this sekenes'.[19]

Although the concept of circulation as we know it today was not clearly understood until the seventeenth century, it was assumed that some blood passed directly to the heart, whose function was to generate heat, the source of life itself. Flowing through the septum, from right to left, it mingled with cooling air from the lungs and entered the arterial system as a frothy substance, known as vital spirit or *pneuma*, because it transported enlivening warmth throughout the body.[20] The nature of the external environment played a crucial role at this stage, since corrupt air, such as that thought to carry plague, could rapidly poison the entire system, while fresh breezes and fragrant aromas could strengthen it (Plate 7.2). Many authorities regarded odours as corporeal entities or 'smoky vapours', somewhere between water and air, which transported 'the prynte and likenes' of the thing from whence they came directly into the bodies of those who inhaled or absorbed them.[21] As both Hippocrates and Galen had observed, they could therefore play a supplementary role in nutrition, bypassing the first two stages of the digestive process altogether, and providing a valuable source of nourishment for delicate patients with poor appetites. Writing in 1489, the Italian physician Marsilio Ficino recommended the application of poultices of warm bread mixed with mallow wine and mint powder to the abdomens 'and sometimes just the noses' of elderly or emaciated individuals, who would be comforted by the heat and sustained by the fortifying smell.[22]

The vital spirit that reached the brain was, according to Galen, filtered through a network at the top of the spinal cord called the *rete mirabile*. Once mixed with air inhaled through the nostrils this purified blood—known as the animal spirit—quite literally animated both body and mind, influencing behaviour in part according to information derived from the senses. It responded with great acuity to sights and smells, both of which were believed to have a powerful impact upon mental as well as physical

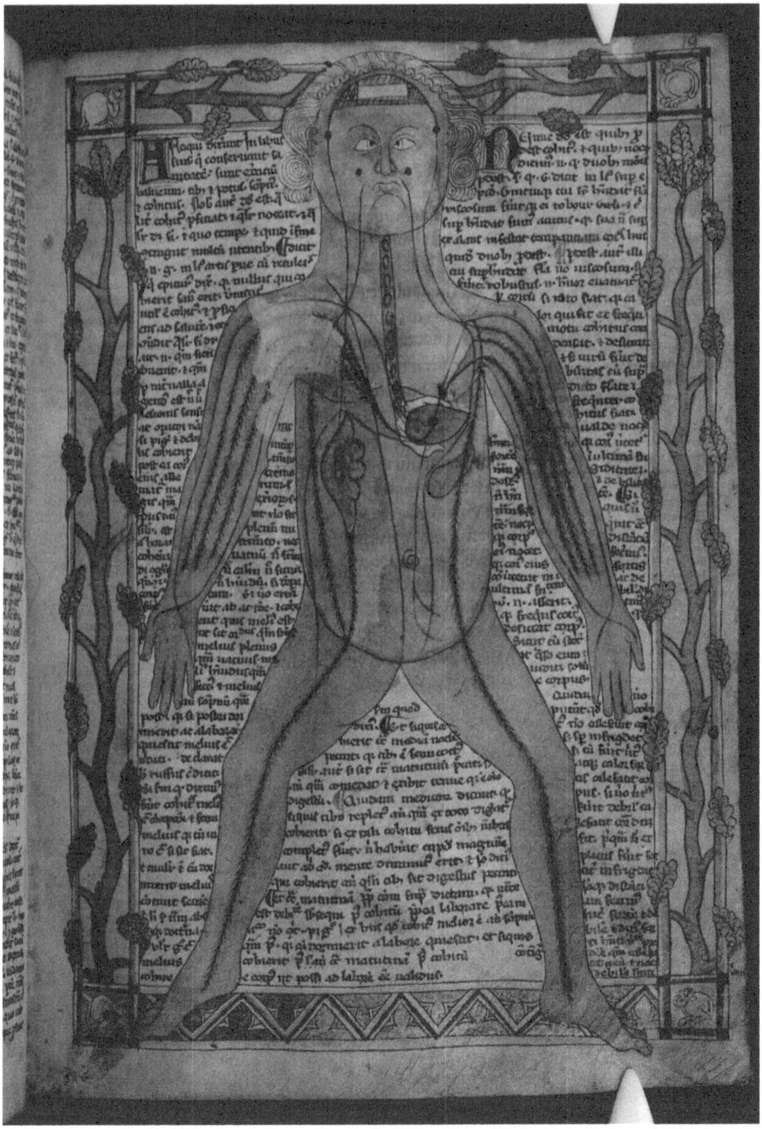

Plate 7.2 From the same medical treatise, this diagram of the arterial system depicts the process whereby some venous blood passes through the heart (the lozenge on the left of the figure's chest), where it is warmed and purified. When mixed with air from the lungs it becomes *pneuma*, or vital spirit, which is carried along the arteries. (Bodleian Library, Oxford, MS Ashmole, f. 19r)

health. Graded according to the relative quantities of heat, cold, moisture, and aridity that they possessed, aromatic plants could be prescribed to rectify the humoral balance and soothe or stimulate the spirits. Since the scent of marjoram was, for instance, warm and dry in the third degree, it served to counterbalance excessive phlegm in the stomach and intestines, to purify the blood and to counteract occlusions of the brain.[23] Conversely, however, repellent odours had both a contaminating *and* destabilising effect. Once they penetrated the inner recesses of the brain, the 'derknesse and stenche' of acrid smoke, such as that generated by lime-burners and plumbers, threatened the processes of thought and movement, and thus seemed particularly dangerous.[24]

A clear exposition of these theories, which had initially been developed in several different Galenic texts, might be found in the *Canon* of Avicenna (Ibn Sīnā, d. 1037). This great work, which systematised and embellished Galen's medical teachings, formed the bedrock of the European medical syllabus from the thirteenth century onward. An influential passage in book four describes the toxic effect of miasmatic exhalations upon the human body:

> Vapours and fumes rise [into the air] and spread in it and putrefy it with their debilitating warmth. And when air of this kind reaches the heart, it corrupts the complexion of the spirit that dwells within it; and, surrounding the heart, it then putrefies it with humidity. And there arises an unnatural heat; and it spreads throughout the body, because of which pestilential fever will occur, and will spread to a multitude of men who likewise have vulnerable dispositions.[25]

It was only to be expected that the university authorities in Cambridge, who knew their Avicenna, should insist in the aftermath of the fourth national pestilence of 1374–75 that all 'putrid flesh' and other noxious waste should be removed from the market every morning and evening.[26] Members of the intellectual elite did not, however, exercise a monopoly in this regard, and might well find themselves at the receiving end of complaints from aggrieved neighbours. In an ongoing dispute with Christ Church priory, residents of Canterbury accused the monks in 1425 of destroying the city ditch with 'ordure and felthe' from their dormitory, while recklessly endangering the health of local people. Whether or not many of them had, indeed, been 'enfectyt gretly' by the contaminated air and become 'grievously […] dyssesyd', the fact that petitions of this kind increasingly assumed an awareness of the underlying medical theory is itself highly significant.[27]

Guidance about the preservation of health had by then begun to circulate widely in the more accessible format of the *Regimen sanitatis*, or Regimen of health, which explained how to avoid sickness through the careful management of six external factors known as 'non-naturals'. The earliest of these texts were commissioned by royalty and other influential figures, being customised to meet their personal needs, although the quality of the air and the salubriousness of the environment invariably commanded particular attention.[28] The physician Aldobrandino of Siena (d. 1287) produced one such regimen for the use of Eleanor of Provence, queen to Henry III, which was written in her native French rather than Latin. It stresses the importance of avoiding 'the corruption of the air', while also drawing attention to the dangers of fumes and smoke,[29] and may well have influenced her decision in 1257 to leave Nottingham for the purer air of Tutbury because the burning of sea coal made life there profoundly disagreeable.[30] Both Latin and vernacular versions of Aldobrandino's regimen attracted a wide readership in the later Middle Ages, confirming that much of the advice on offer was applicable to less privileged individuals and, indeed, entire communities. A regimen devised by the eminent Catalan physician, Arnald of Villanova (d. 1311), for King James II of Aragon dealt at great length with the complex mechanisms of respiration. Yet its underlying message was commendably clear and concise:

> The first item or consideration with regard to the preservation of health concerns the choice of air. For among the things which, by necessity, affect the human body nothing changes it more than that which, inhaled by the mouth and nostrils [...] and mixed with the spirit of the heart, travels along all the arteries and by which means all the activities of daily life are accomplished.[31]

In a society which positively encouraged plagiarism, works by eminent authorities such as Aldobrandino and Arnald were enthusiastically copied, translated, abridged, and simplified to suit new urban audiences.

John Mirfield, a priest who lived at St Bartholomew's priory in London, near the unsavoury purlieus of the Smithfield butcheries, produced two *regimina* for the benefit of the resident canons at the turn of the fourteenth century. Drawing heavily, but without acknowledgement, upon the *De conservatione vite humane* of the French physician, Bernard Gordon (fl. 1305), he describes the salutary effects of a brisk walk in the countryside, far away from polluted streets and overflowing cesspits. As he explains, exercise offers 'pure recreation of the soul and body when it is performed in the open; for then a man is exposed to wholesome air (*bono aeri*), and he rejoices in gazing far and near, and upon the sky, the sea and the green landscape'.[32] Access to fresh air was a major preoccupation of religious living in cramped urban monasteries, as we can see from complaints repeatedly voiced by the Durham Benedictines. They objected at various times to the polluted environment (*aer corruptus*) of their 'intolerably overcrowded' precinct and asked for a garden in which they could enjoy the benefits of healthy recreation in wholesome air.[33]

On the pragmatic basis that 'protecting [people] is a much more powerful and secure response than treating them once they are actually sick',[34] vernacular self-help manuals began to proliferate after the Black Death, when the demand for advice about preventative medicine naturally grew. Indeed, a more specialist type of guide or *consilium* achieved widespread popularity from the 1340s onward by concentrating upon the avoidance of epidemics and highlighting the risks posed by such hazards of urban life as dung heaps, butchers' waste, and stagnant drains. Many of these works were produced, initially at least, for the assistance of magistrates. The physician Jacme d'Agramont (d. 1348) wrote a concise *Regiment de preservacio a epidemia* in his native Catalan for the rulers of Lerida, explaining how 'particular' sources of airborne corruption might be contained by sanitary measures.[35] Tommaso del Garbo (d. 1370) applied himself to 'the well-being and health of the men who live in the city of Florence', while the *Regimen contra pestilentiam* drawn up by Jean Jacme (d. 1384) for the benefit of the 'commonwealth' of Montpellier took on a new life in the fifteenth century as the much-copied and revised 'Canutus' plague treatise.[36]

During the devastating epidemic of 1407, physicians teaching at the University of Oxford supplied the mayor and aldermen of London with a *consilium* (copied verbatim from an earlier tract by John of Burgundy),[37] although their advice may by then have seemed unnecessary. The number of orders for cleansing the streets and water courses of noxious waste recorded in the city's Letter Books had already risen fourfold from just sixteen between 1300 and 1349 to at least sixty-five during the second half of the century. Since the population of London fell by at least half during the same period and thus generated far less garbage, these figures seem particularly striking. Practical steps had also been taken to improve refuse collection: in the aftermath of the 1368–69 epidemic, for example, the authorities invested in twelve new tumbrels and twenty-four more draught horses to facilitate the regular removal of household waste from wards that needed additional resources.[38]

The spread of literacy among urban populations, and with it the ability to read simple vernacular texts, further expanded the market for accessible literature. Like many other late medieval religious, John Lydgate (d. 1449/50), a Benedictine monk of Bury St Edmunds in Suffolk, considered it his Christian duty to provide essential information in a format that ordinary laymen and women could understand. Written in verse, so that his recommendations could more easily be remembered by those with little or no learning, his 'Dietary and Doctrine against the Pestilence' has been described as 'a medieval best-seller'. It survives in at least fifty-seven manuscript copies and was published in prose versions by England's first three leading printers.[39] Lydgate focussed principally on diet and the importance of moderation, but he, too, emphasised the benefits of 'clere ayre', warning his readers to avoid 'mystys blastys', 'infect placys' and 'the aire of pestylens' at all costs. Longstanding connections with the aldermanic elite of London gave him a sound grasp of the practical problems faced by magistrates in the struggle against disease, while inspiring him to imagine what might be achieved through the systematic application of the regimen to communal life. His *Troy Book* (1412–20) was composed at a time when the Londoners were beginning to replace and extend the system of pipes and conduits that had brought fresh, unpolluted water within the walls since the thirteenth century.[40] Did he intend to offer advice as a medically informed 'proto-city planner' or simply to reflect more generally upon the importance of effective sanitation?[41] It is certainly worth noting that his evocation of Troy as a utopian community celebrates King Priam's ingenious use of hydraulics to flush away 'ordure & fylthes' from the streets through subterranean channels:

> Wher-by the toun was outterly assured
> From engenderyng of al corrupcioun,
> From wikked eyr & from infeccioun,
> That causyn ofte by her violence
> Mortalite and gret pestilence.[42]

Shorn of their theoretical underpinnings, these ideas began to percolate downwards through society, their transmission being hastened by repeated outbreaks of plague from the 1360s onwards.[43] Public proclamations, whereby the texts of official measures would be read aloud in the streets, helped further to spread information among the illiterate.[44] Royal directives, such as Henry VII's stern missive of 1489 to the rulers of Dublin, were clearly intended for this purpose. Having warned that 'dung-heaps, swine, hog-sties, and other nuisances in the streets, lanes and suburbs [...] infect the air and produce mortality, fevers and pestilence throughout the city', the writ went on to explain that a mandatory programme of street cleaning would 'prevent loss of life from pestilential exhalations', while also encouraging trade.[45] Receptive or not, few residents could claim complete ignorance of the rationale behind these pronouncements, or deny the fact that an unhealthy environment could have serious economic consequences.

In English towns and cities, a network of local courts (often known as 'leets'), where infringements of bylaws and other perceived nuisances could be reported and penalised, offered the public an opportunity to protest about antisocial behaviour. The lists of notifiable offences compiled by magistrates for the guidance of juries invariably included such threats to survival as noxious privies and stinking dung heaps. These documents and the indictments that followed provide a fascinating insight into the way that medical ideas about the airborne spread of disease could inform proceedings.[46] Presumed lepers had, for example, long been obliged to live 'outside the camp', although their exclusion had little to do with overt fears of infection and was far from absolute. The belief, clearly enunciated in many plague tracts, that 'pestilence sores be contagious because of enfect humours' and that 'the reke or smoke of suche sores is venomous and corrupteth the ayer', led, however, to an inevitable hardening of attitudes toward segregation. So too did warnings about the pollution exhaled by anyone with tainted breath.[47] Cases involving the identification and removal of suspect *leprosi* refer increasingly from the 1360s onward to the danger that they posed as sources of unpleasant odours (*fetor*), and thus as potential vectors of plague as well as leprosy.[48] An analysis of 113 such presentments made between 1369 and 1501 to leet courts in the port of Great Yarmouth, which suffered badly from epidemics, reveals a distinct correlation between the identification of suspects and outbreaks of pestilence.[49] Significantly, when the rulers of London recorded a directive of 1472 (a plague year) from Edward

IV ordering them to confine anyone 'enfecte with the contagious and perilous siknes of lepour' in an appropriate institution, lest they contaminate persons 'of clene compleccon', they entered it in the city journal next to a copy of the above-mentioned statute of 1388 for the safe disposal of butchers' waste.[50]

Complaints about insalubrious living conditions came from the bottom up as well as the top down. The wording of presentments made at neighbourhood level reflects an obvious awareness on the part of ordinary people that bad air was not only potentially lethal but also a source of shame that reflected badly upon the entire community. Collective repute was a powerful weapon in the battle against environmental pollution. In 1438–39 one of the Yarmouth leets that were on the look-out for suspect lepers presented four men for depositing 'filth and the blood of beasts and blubber near the town gates in a public thoroughfare, causing a great stench to the grave nuisance of locals and outsiders and the *disparagement* of the vill'.[51] Not surprisingly, Londoners were particularly sensitive on this score, in part because they were subject to constant surveillance by members of the royal household and government officials travelling to and from Westminster. In 1421 residents of St Martin's parish, Ludgate, presented one William ate Wode:

> [...] for making a great nuisance and discomfort to his neighbours by throwing out horrible filth on to the highway, the stench of which is so odious and infectious that none of his neighbours can remain in their shops, which is *a great reproof* [my italics] to all this honourable city, because of the lords and other gentlemen and men of the court who go and pass there.[52]

Two years later, householders in the ward of Farringdon Without, which lay beyond the walls and was thus rather more prone to insanitary nuisances than those at the centre, gave vent to their exasperation at years of official inertia and neglect. It is interesting to note that they too referred to the malign effects of contaminated air upon the city's reputation as well as its health:

> The comune prive [privy] of Ludgate is ful defectif and perlus and the ordur ther-of rotith the stone wallys, and makith other while a orrible stench and a foul sight, to grete desese and nusans of all folk therabout dwellyng and therby passing, and a *desclaunder to all this Citee* [my italics],

> that so foul a nusans shuld be so nygh so comune an hy wey, and oft hit hath be presentyd and no remedye yit is ordeined.[53]

Protests of this kind did not necessarily fall upon deaf ears. Municipal schemes for the preservation of public health, often launched in the aftermath of plague outbreaks, attracted growing support from private individuals, who regarded them as an extension of the 'comfortable works' incumbent upon all devout Christians [Matthew 25:32–36]. As a further inducement to almsgiving, many plague tracts maintained that charity was itself a prophylactic against epidemic disease.[54] From this standpoint, investment in street cleaning or the construction of purpose-built suburban slaughterhouses brought immediate physical benefits to the community, as well as promising long-term spiritual rewards to the donor.[55] Quite probably in response to appeals such as those from the householders of Dowgate ward, where lanes leading down to the Thames were rendered impassable by heaps of 'stinking privy filth',[56] the wealthy London mercer Richard Whittington (d. 1423) left money for the construction of a public lavatory with the capacity to seat 128 men and women. Malodorous waste could now be rapidly flushed away from the aptly named 'long house' directly into the tidal waters of the Thames, rather than contaminating the entire neighbourhood with its stench.[57] As his many bequests reveal, Whittington sought to improve the daily lives of his fellow citizens, while also assiduously providing for the health of his immortal soul. In response to the shockingly high death rate among prisoners exposed to 'the infected and corrupt air and other perils and horrible diseases' in 'the hateful gaol of Newgate', he paid for the entire complex to be demolished and rebuilt on sanitary principles.[58] The new prison, which opened in 1432, was lighter and better ventilated. Some chambers boasted their own privies and chimneys; there were facilities for indoor and outdoor recreation; and, thanks to the generosity of another eminent Londoner, Thomas Knolles, freshly piped water was freely available.[59] Whittington merits attention because of the scale and variety of his benefactions, but in other respects he was but one of many successful merchants driven by a combination of piety, civic pride, and compassion to tackle the environmental problems so clearly identified in contemporary *regimina* and plague tracts.

Despite the wealth of evidence to the contrary, the assumption that medieval people were incapable of conceptualising disease in any other than theological terms, and thus of taking practical steps to improve communal health, remains entrenched.[60] It is easy to see why the sanitary reformer and author, Charles Kingsley (d. 1875), should bewail the tragic failure of medieval magistrates to recognise that 'foul air', 'foul water', and 'unclean backyards', rather than the 'seemingly supernatural arrows of [the] angel of death', gave rise to epidemics.[61] He was, after all, the product of a society that believed in the inexorable march of scientific progress. But Victorian attitudes linger on, even today, when the word 'medieval' is still all too often employed as synonym for ignorance and superstition. The profoundly held belief that pestilence (and, indeed, all human suffering) was ultimately an act of God did not preclude a conviction that more immediate environmental hazards made people sick and might be eliminated, or at least contained, through the rational application of medical theory.[62] We should also bear in mind that the provision of public works, such as regular refuse collections, paved streets, covered drains, and piped water, all of which helped to clear the air, ranked as an act of Christian compassion that would speed the donor on his or her way to paradise. Cleanliness was, in this instance, quite literally a manifestation of godliness; and although the implementation of sanitary measures may often have been an uphill battle, the reasoning behind them ought surely to defy our condescension.

Notes

1. The National Archives, Kew, KB27/758, rot. 51. All translations into English from the original Latin or French in manuscripts and printed editions are my own.
2. Then equivalent to the annual salary of a senior member of the royal household.
3. Herbert Edward Salter, ed., *Munimenta civitatis Oxonie*, Oxford Historical Society 71 (Devises: G. Simpson, 1920), 13–14.
4. *Calendar of Close Rolls, 1330–1333* (London: HMSO, 1898), 610. N.J. Ciecieznski, 'The Stench of Disease: Public Health and the Environment in Late-Medieval English Towns and Cities', *Health, Culture and Society* 4 (2013): 92–104, lists some further examples, but fails to engage with the medical rationale behind them.
5. For the background to much of what follows, see Ernest L. Sabine, 'Butchering in Mediaeval London', *Speculum* 8 (1933): 335–53; and Carole Rawcliffe, '"Great Stenches, Horrible Sights and Deadly Abominations": Butchery and the Battle against Plague in Medieval English Towns', in *Plague and the City*, ed. by Lukas Engelmann, John Henderson, and Christos Lynteris (London: Routledge, 2018), 18–38.

6. Geoffrey Martin and Christopher Given-Wilson, eds, *The Parliament Rolls of Medieval England, VI: Richard II 1377–1384* (Woodbridge: Boydell Press, 2005), 181. Residents of Holborn had been obliged to contend with airborne pollution since at least 1290: Marjorie Honeybourne, 'The Fleet and Its Neighbours in Early and Medieval Times', *London Topographical Record* 19 (1947): 13–87, on 51–54.
7. Andres Luders, et al., eds, *Statutes of the Realm*, 11 vols (London: Record Commission, 1810–28), vol. 2, 59–60.
8. René de Lespinasse, ed., *Les métiers et corporations de la ville de Paris I: XIVe–XVIIIe siècle, ordonnances générales metiers de l'alimentation* (Paris: Imprimerie Nationale, 1886), 274.
9. Rawcliffe, 'Butchery and the Battle against Plague', 24–26.
10. John Thomas Gilbert, ed., *Calendar of Ancient Records of Dublin*, 1 (Dublin: J. Dollard, 1889), 366.
11. See, for example, Guy Geltner, *Roads to Health: Infrastructure and Urban Wellbeing in Late Medieval Italy* (Philadelphia: University of Pennsylvania Press, 2019), 27–31, 47–67; and the various contributions in *Policing the Urban Environment in Premodern Europe*, ed. by Carole Rawcliffe and Claire Weeda (Amsterdam: Amsterdam University Press, 2019).
12. Bianca Betto, ed., *Gli Statuti del comune di Treviso (sec. XIII–XIV)*, 2 vols (Rome: Instituto Storico Italiano per il Medio Evo, 1984, 1986), vol. 1, 192–93. The authorities additionally banned the disposal of butchers' waste in the city's watercourses: 195. See also Guy Geltner, 'Urban *Viarii* and the Prosecution of Public Health Offenders in Late Medieval Italy', in *Policing the Urban Environment*, ed. by Rawcliffe and Weeda, 104.
13. Lodovico Zdekauer, ed., *Statutum Potestatis Comunis Pistorii (1296)* (Milan: Hoepli, 1888), 150–51.
14. Carole Rawcliffe, *Urban Bodies: Communal Health in Late Medieval English Towns and Cities* (Woodbridge: Boydell Press, 2013), 165–66; Peter Brimblecombe, *The Big Smoke: A History of Air Pollution in London Since Medieval Times* (London: Methuen, 1987), 6–16.
15. Claire Weeda, 'Cleanliness, Civility and the City in Medieval Ideas and Scripts', in *Policing the Urban Environment*, ed. by Rawcliffe and Weeda, 49–55.
16. Bonvesin de la Riva, *Le meraviglie di Milano*, ed. and trans. by Giuseppe Pontiggia (Milan: Bompiani, 2015), 29.
17. Francesc Eiximenis, *Regiment de la cosa pública* (Valencia: Cristofol Cofman, 1499), f. 5r.
18. For what follows, see Carole Rawcliffe, 'The Concept of Health in Late Medieval Society', in *Le interazioni fra economia e ambiente biologico nell'Europa preindustriale. Secc. XIII–XVIII*, ed. by Simonetta Cavaciocchi (Florence: Florence University Press, 2010), 317–34.

19. Johannes Jacobi, *A Litill Boke Necessarye and Behouefull agenst the Pestilence* (London: W. de Machlinia, *ca.* 1485), f. 3r.
20. Heather Webb, *The Medieval Heart* (New Haven, CT: Yale University Press, 2010), 26–31, 96–107.
21. Richard Palmer, 'In Bad Odour: Smell and Its Significance in Medicine from Antiquity to the Seventeenth Century', in *Medicine and the Five Senses*, ed. by William F. Bynum and Roy Porter (Cambridge: Cambridge University Press, 1993), 61–68; Rawcliffe, *Urban Bodies*, 120–21.
22. Marsilio Ficino, *The Book of Life*, ed. and trans. by Charles Boer (Irving, Texas: Spring Publications, 1980), 52, 73–78.
23. Luisa Cogliati Arano, ed., *The Medieval Health Handbook: Tacuinum Sanitatis* (New York: George Braziller, 1974), nos XIX and 183.
24. Bartholomaeus Anglicus, *On the Properties of Things: John Trevisa's Translation of Bartholomaeus Anglicus'* De proprietatibus Rerum, ed. by M. C. Seymour, et al., 3 vols (Oxford: Clarendon Press, 1975–88), vol. 1, 561–62.
25. Avicenna, *Liber Canonis Medicine* (Lyon: Jacques Myt, 1522), liber IV, f. 329r. This passage is cited in one of the earliest printed vernacular English plague tracts, Jacobi, *A Litill Boke*, f. 2r.
26. Charles Henry Cooper, ed., *Annals of Cambridge*, 5 vols (Cambridge: Warwick & Co., 1892–1908), vol. 1, 114.
27. Ernest F. Jacob, 'Chichele and Canterbury', in *Studies in Medieval History Presented to Frederick Maurice Powicke*, ed. by R. W. Hunt, W. A. Patin, and R. W. Southern (Oxford: Clarendon Press, 1969), 400–01.
28. Marilyn Nicoud, *Les régimes de santé au Moyen Âge: Naissance et diffusion d'une écriture médicale*, 2 vols (Rome: École française de Rome, 2007); Pedro Gil-Sotres, 'The Regimens of Health', in *Western Medical Thought from Antiquity to the Middle Ages*, ed. by Mirko D. Grmek (Cambridge, MA: Harvard University Press, 1998), 291–318, and for air esp. 303–4.
29. Aldobrandino of Siena, *Le Régime du corps de Maître Aldebrandin de Sienne*, ed. by Louis Landouzy and Roger Pépin (Paris: H. Champion, 1911), 59–61; Peter Murray Jones, *Medieval Medical Miniatures in Illuminated Manuscripts* (revised edn, London: The British Library, 1998), 103–7.
30. Henry Richards Luard, ed., *Annales monastici*, 5 vols (London: Rolls Series, 1866), vol. 3, 203–04.
31. Arnald de Villanova, *Regimen sanitatis ad regem Aragonum*, ed. by Luis Garcia-Ballester and Michael R. McVaugh (Barcelona: Fundació Noguera, 1996), 423.
32. Sir Percy Horton-Smith Hartley and Harold Richard Aldridge, *Johannes de Mirfeld of St Bartholomew's Smithfield: His Life and Works* (Cambridge: Cambridge University Press, 1936), 140–41.

33. Richard Barrie Dobson, *Durham Priory 1400–1450* (Cambridge: Cambridge University Press, 1973), 35; J. A. Twemlow, ed., *Calendar of Papal Registers: Papal Letters, XI, 1455–1464* (London: HMSO, 1921), 5–6.
34. Luke Demaitre, *Medieval Medicine: The Art of Healing, from Head to Toe* (Santa Barbara, CA: Praeger, 2013), 64, citing Valesco de Tharanta (d. 1417).
35. Jon Arrizabalaga, 'Facing the Black Death: Perceptions and Reactions of University Medical Practitioners', in *Practical Medicine from Salerno to the Black Death*, ed. by Luis García-Ballester, et al. (Cambridge: Cambridge University Press, 1994), 237–88; John Aberth, ed., *The Black Death: The Great Mortality of 1348–1350* (Boston: Bedford/St Martin's, 2005), 50–55.
36. Demaitre, *Medieval Medicine*, 65–66; George R. Keiser, 'Two Medieval Plague Treatises and Their Afterlife in Early Modern England', *Journal of the History of Medicine* 58 (2003), 292–324; John Henderson, 'The Black Death in Florence: Medical and Communal Responses', in *Death in Towns: Urban Responses to the Dying and the Dead, 100–1600*, ed. by Steven Bassett (Leicester: Leicester University Press, 1992), 136–50.
37. Dorothea Waley Singer and Annie Anderson, *Catalogue of Latin and Vernacular Plague Texts in Great Britain and Eire in Manuscripts Written before the Sixteenth Century* (London: William Heinemann, 1950), 27–28. For a modern translation of John of Burgundy's influential tract, see Rosemary Horrox, ed., *The Black Death* (Manchester: Manchester University Press, 1995), 184–93.
38. Ernest L. Sabine, 'City Cleaning in Mediaeval London' *Speculum*, 12 (1937): 19–43, on 23–24, 27–28.
39. George Shuffelton, ed., *Codex Ashmole 61: A Compilation of Popular Middle English Verse* (Kalamazoo, MI: Medieval Institute Publications, 2008), no. 31. Shuffelton's observation that plague tracts derived from a 'folk' rather than an academic tradition is, however, erroneous.
40. Caroline Barron, *London in the Later Middle Ages: Government and People 1200–1500* (Oxford: Oxford University Press, 2004), 256–57, and ch. 10 generally for the urban environment.
41. Paul Strohm, 'Sovereignty and Sewage', in *Lydgate Matters: Poetry and Material Culture in the Fifteenth Century*, ed. by Lisa H. Cooper and Andrea Denny-Brown (Basingstoke: Palgrave Macmillan, 2008), 58–70.
42. Henry Bergen, ed., *Lydgate's Troy Book, Part I* (Early English Text Society ES 97, 1906), Bk II, 166. Significantly, Thomas More, who was involved in pioneering measures for the introduction of quarantine in London and Oxford during the plague of 1518, stressed the cleanliness of his own model community. On the island of Utopia noxious waste was also carried away by running water, 'for fear that the air, tainted by putrefaction,

should engender disease': *Utopia*, in *The Complete Works of St. Thomas More, Volume 4*, ed. by Edward Surtz and Jack H. Hexter (New Haven, NJ: Yale University Press, 1965), 138–39, 418.
43. For a chronological survey of national and regional epidemics up to 1530, see Rawcliffe, *Urban Bodies*, 360–74.
44. In London, for example, whatever ordinances were currently in force would be proclaimed in English throughout the city every year: Henry Thomas Riley, ed., *Munimenta gildhallae Londoniensis: Liber albus* (London: Longman, 1859), 260, note 1. See also Rawcliffe, *Urban Bodies*, 45–50.
45. Gilbert, *Calendar of Ancient Records of Dublin*, 139–40.
46. Carole Rawcliffe, 'The View from the Streets: The Records of Hundred and Leet Courts as a Source for Sanitary Policing in Late Medieval English Towns', in *Policing the Urban Environment*, ed. by Rawcliffe and Weeda, 69–95.
47. Jacobi, *A Litill Boke*, ff. 3v, 4v; Margaret Healy, 'Discourses of the Plague in Early Modern London', in *Epidemic Diseases in London*, ed. by Justin A. I. Champion (London: Centre for Metropolitan History, 1993), 21.
48. Avicenna maintained that the inhalation of 'air corrupted by itself or because of the proximity of lepers' might be sufficient to infect a vulnerable person with the disease: *Liber canonis*, Bk IV, f. 442v. Even so, although writers such as Bartholomaeus Anglicus reiterated this warning and advised against prolonged association with the leprous (*On the Properties of Things*, I, 424, 426), their recommendations had little discernible impact upon English urban communities until after the Black Death: Carole Rawcliffe, *Leprosy in Medieval England* (Woodbridge: Boydell Press, 2006), 92–93.
49. Rawcliffe, *Leprosy in Medieval England*, 274–84.
50. London Metropolitan Archives, COL/CC/01/01/008, ff. 21v–22r.
51. Norfolk Record Office, Y/C4/147, rot. 16r. In this instance the word 'vill' refers to the entire town.
52. Arthur Hermann Thomas, ed., *Calendar of Plea and Memoranda Rolls of the City of London, 1413–1437* (Cambridge: Cambridge University Press, 1943), 129.
53. R. W. Chambers and Marjorie Daunt, eds, *A Book of London English 1384–1425* (Oxford: Clarendon Press, 1931), 134–35; Thomas, ed., *Calendar of Plea and Memoranda Rolls of the City of London, 1413–1437*, 157. For the importance of cleanliness as a civic ideal and the shameful connotations of dirt, see Douglas Biow, *The Culture of Cleanliness in Renaissance Italy* (Ithaca, NY: Cornell University Press, 2006), 75–94.
54. As, for example, Thomas Forestier's tract of 1485: British Library, Additional MS 27582, p. 71.
55. Rawcliffe, *Urban Bodies*, 6, 61–62, 99–100, 283–89, 357–58.

56. Thomas, ed., *Calendar of Plea and Memoranda Rolls of the City of London, 1413–1437*, 133–34.
57. P. E. Jones, 'Whittington's Long House', *London Topographical Record* 23 (1974 for 1972), 27–34.
58. Henry Thomas Riley, ed., *Memorials of London and London Life in the XIIIth, XIVth, and XVth Centuries* (London: Longmans, Green and Co., 1868), 677; Sir Nicholas Harris Nicolas, ed., *Proceedings and Ordinances of the Privy Council of England*, 7 vols (London: Eyre and Spottiswoode, 1834–7), vol. 3, 79–80. This was not the first attempt to safeguard the health of London's prisoners. In 1356, the moat around the Fleet prison in Holborn was cleared of noxious filth from sewers, tanneries and illegally sited latrines following a complaint by Edward III that 'by reason of the infection of the air, and the abominable stench which there prevails, many of those there imprisoned are often affected with various diseases and grievous maladies': Riley, ed., *Memorials*, 279–80; Honeybourne, 'The Fleet and its Neighbours', 39–40.
59. Margery Bassett, 'Newgate Prison in the Middle Ages' *Speculum*, 18 (1943), 233–46: 239.
60. Rawcliffe, 'Butchery and the Battle against Plague', 19; Geltner, *Roads to Health*, 7–9.
61. He expressed these views in a lecture on 'Great Cities and their Influence for Good and Evil' first given in 1857: Charles Kingsley, *Sanitary and Social Lectures and Essays* (London: Macmillan, 1889), 194–96. Ironically in 1850 Charles Dickens had personified the miasmatic air arising from England's slums as a 'perversion of the Holy Ghost', or 'profane spirit, breeding and brooding' over the entire country: Alexandra Harris, *Weatherland: Writers and Artists under English Skies* (London: Thames and Hudson, 2015), 298.
62. The authors of plague tracts generally recognised three levels of causation: the ultimate, namely God; the intermediate or universal, which involved the movement of winds and heavenly bodies; and the proximate, among which stagnant waste, rotting food and similar sources of miasmatic air ranked prominently. See, for example, Horrox, ed., *Black Death*, 159–61.

Selected Bibliography

Arrizabalaga, Jon. 1994. Facing the Black Death: Perceptions and Reactions of University Medical Practitioners. In *Practical Medicine from Salerno to the Black Death*, Ed. Luis García-Ballester, Roger French, Jon Arrizabalaga, and Andrew Cunningham, 237–88. Cambridge: Cambridge University Press.

Brimblecombe, Peter. 1987. *The Big Smoke: A History of Air Pollution in London since Medieval Times*. London: Methuen.

Englemann, Luke, John Henderson, and Christos Lynteris, eds 2018. *Plague and the City*. London: Routledge.
Geltner, Guy. 2019. *Roads to Health: Infrastructure and Urban Wellbeing in Late Medieval Italy*. Philadelphia: University of Pennsylvania Press.
Healy, Margaret. 1993. Discourses of the Plague in Early Modern London. In *Epidemic Diseases in London*, Ed. Justin A. I. Champion, 19–34. London: Centre for Metropolitan History.
Henderson, John. 1992. The Black Death in Florence: Medical and Communal Responses. In *Death in Towns: Urban Responses to the Dying and the Dead, 100–1600*, Ed. Steven Bassett, 136–50. Leicester: Leicester University Press.
Palmer, Richard. 1993. In Bad Odour: Smell and Its Significance in Medicine from Antiquity to the Seventeenth Century. In *Medicine and the Five Senses*, Ed. William F. Bynum and Roy Porter, 61–68. Cambridge: Cambridge University Press.
Rawcliffe, Carole. 2013. *Urban Bodies: Communal Health in Late Medieval English Towns and Cities*. Woodbridge: Boydell Press.
Rawcliffe, Carole, and Claire Weeda, eds 2019. *Policing the Urban Environment in Premodern Europe*. Amsterdam: Amsterdam University Press.

PART III

The Early Modern Period

CHAPTER 8

'Being Breathed': From *King Lear* to Clinical Medicine

Katharine A. Craik and Stephen J. Chapman

I

In Act 2 Scene 2 of Shakespeare's *King Lear* (1605–06), a servant, Oswald, appears onstage gasping and scarcely able to utter. To the Earl of Kent, the spectacle of Oswald's breathlessness seems inseparable from his cowardly loyalty to his mistress, the king's daughter, Goneril:

> Oswald: I am scarce in breath, my lord.
> Kent: No marvel, you have so bestirred your valour, you cowardly rascal; nature disclaims in thee—a tailor made thee.
> Cornwall: Thou art a strange fellow—a tailor make a man?
> Kent: Ay, a tailor, sir; a stone-cutter or a painter could not have made him so ill, though they had been but two years o' the trade.[1]

K. A. Craik (✉)
Faculty of Humanities and Social Sciences, School of English and Modern Languages, Oxford Brookes University, Oxford, UK

S. J. Chapman
Oxford University Hospitals, Oxford, UK

D. Fuller et al. (eds.), *The Life of Breath in Literature, Culture and Medicine*, Palgrave Studies in Literature, Science and Medicine,
https://doi.org/10.1007/978-3-030-74443-4_8

Oswald's breathlessness reflects not only his moral compasslessness but also the loss of his natural personhood. As Kent puts it, 'nature disclaims in thee—a tailor made thee'. If Oswald is man-made, he is not even properly man-made. He is so ill put together, in fact, that even an apprentice painter or stonemason could not have produced such an assemblage. Only a tailor, proverbially rough and dishonest, would be capable of delivering such a botched job. Later in the same scene, Kent's recollection of Oswald's 'reeking post, / Stewed in his haste, half breathless, panting forth' (2.2.220–21) is in keeping with his drubbing of him as

> A knave, a rascal, an eater of broken meats; a base, proud, shallow, beggarly, three-suited-hundred-pound, filthy, worsted-stocking knave; a lily-livered, action-taking knave; a whoreson, glass-gazing, superserviceable, finical rogue. (2.2.14–18)

Kent imagines Oswald assembled from a series of scraps no better than the 'broken meats' he eats. He is a patchwork person—not whole, not integral—and his breathlessness seems part of the general unboltedness that eventually prompts Kent to threaten, terribly, to trample Oswald into a filthy human paste: 'I will tread this unbolted villain into mortar and daub the wall of a jakes with him' (2.2.63–65). Losing altogether a sense of bounded personhood, Oswald comes apart at the seams and is trodden into something compound. Now he is a nothing more than a 'zed', an 'unnecessary letter' (2.2.62). To Kent, Oswald's wheezing breathlessness, rank materiality and gross thingness mark him out as the play's 'lowest and most dejected thing'.[2]

This brief and easily overlooked episode suggests the importance of breath, and breathlessness, as a means of articulating early modern ideas about natural personhood and ethical responsibility. Oswald's disintegration also signals Shakespeare's more specific interest in *King Lear* in breath's fundamental role in making us persuasively real—not least to ourselves. This chapter begins by considering *King Lear* as Shakespeare's most extended and subtle study of the relationship between breath and identity. Our second and perhaps more ambitious aim is to offer a novel perspective on present-day breathlessness by considering this within the unfamiliar context of early modern literature and culture. Literary and cultural historians interested in the history of the body and *soma* have generally confined their attention to one historical period at a time. Those working with Renaissance sources have recently explored breath's place in the formation of late sixteenth- and seventeenth-century subjectivity, uncovering how breath felt, and what it meant, prior to the theorization of respiration through modern science. Focusing especially on the early modern passions or humours, these critics have shown how breath discloses the porous relationship between self and surroundings, revealing an ecology of Renaissance personhood beyond the boundaries of the skin.[3] Our work is indebted to this important research but goes one step further by proposing that accounts of breath, as a lived experience from the past, can shed light on the conceptual categories which shape medical approaches to breathless patients now. Working together as a literary historian (Craik) and a respiratory physician (Chapman), we propose that recovering the existential and social meanings of breath in the Renaissance can reveal unexpected insights for today's medical science. This process also works in the opposite direction since medical science can illuminate some of the important aspects of human and ethical experience which Shakespeare uncovers when he attends closely to breath, breathing, and breathlessness.

Breathlessness (or dyspnoea) is clinically defined as 'a subjective experience of breathing discomfort that consists of qualitatively distinct sensations that vary in intensity'.[4] The unpleasant sensation and distress associated with breathlessness may be all-encompassing: 'when we are healthy, we take our breathing for granted... But when our lung health is impaired, nothing else but our breathing really matters'.[5] Breathlessness is a highly complex, multifactorial symptom and its underlying pathophysiological mechanisms remain poorly understood. Many recent studies of physiology and neuroimaging have subdivided breathlessness into sensory and affective components. The sensory component of breathlessness describes the intensity and quality of the neurological sensation itself, whereas the affective domain refers to the unpleasantness and distress associated with the symptom.[6] While disease-focused approaches have furthered our understanding of pathophysiology, they have not yet been translated into effective treatments for breathlessness. It remains unclear why breathlessness differs so significantly between individuals with the same disease and indeed why the extent of physiological lung function impairment is such a poor predictor of 'real life' breathlessness severity.[7] More recent medical approaches have therefore favoured what has been described as a 'total breathlessness' model with increased focus on dimensions of breathlessness such as suffering, indignity, shame, guilt, stigma, memory, disability, and social isolation.[8] Such approaches recognise that, while a lack of holism might be considered characteristic of present-day western healthcare in general, this may have particular consequences for those suffering from breathlessness. The very recent proposal of a 'chronic breathlessness syndrome' as a medical entity emphasises the disability and functional consequences of breathlessness over and above its purely sensory and affective dimensions.[9]

Building on this recent clinical research, our central interest lies in how breath is experienced and felt. Phenomenological work in the medical humanities has already identified important differences between disease (as a set of symptoms, or a diagnosis) and illness (as experienced by patients). While acknowledging the value of scientific empiricism and objectivism, this work offers a balance from the perspective of lived experience. As a theoretical approach, phenomenology tends to disturb neat distinctions between people and things, bracketing '*noumena*—things in themselves—in order to attend to *phenomena*: things as appearances, things as apprehended by the subject'.[10] In the universalist, presentist approaches of Merleau-Ponty, et al., such insights are claimed for everyone. But our own phenomenological project is both historical *and* presentist. *King Lear* provides a particularly rich starting point as one of western culture's most powerful explorations of human suffering and ethical relationality from the distant past which still carries powerful cultural weight today. We focus on two of the play's most important scenes: Gloucester's encounter on the cliffs at Dover with his son Edgar, disguised as the 'spirit' Poor Tom; and Lear's anguished response to his daughter Cordelia's death at the play's conclusion. In both examples, breath emerges as a distinctly early modern experience which also suggests new avenues for clinical practice now—and, more broadly but not unconnectedly, for our continuing ethical relationship to one another.

II

The pagan landscape of *King Lear* often resonates with Christian theology, and Shakespeare would have had firmly in mind Genesis 2:5 where God's breath animates man to enter a life of faith:

> And the Lord God formed man [of] the dust of the ground, and breathed into his nostrils the breath of life; and man became a living soul.

Breath was the divine inspiration which infused the body and soul with the Holy Spirit.[11] 'Spirit' indeed simply denotes God himself in John 4:24, and early modern subjects were familiar with the idea of being breathed by God so that to be in breath implied not only individual aliveness but also membership of a fellowship where everyone shared one 'common breath among Christians'.[12] In *Paradise Lost* (1667), Milton describes how Adam was shaped through the 'breath of life' into an image of God's likeness.[13] Renaissance theology and philosophy, looking back to the New Testament and early Christian theologians, regarded breath as indistinguishable from animate life so that a person's ability to breathe was more or less inseparable from their possession of a spirit kindled into a life of faith.[14] Breath was considered by some as an important participatory aspect of the Eucharist: 'the breath of their owne mouthes togither with the signe of the Crosse, may touch the bread and the Cuppe'.[15] Early modern breath therefore encoded the faithful subject's dependence upon and service to God, as well as the shared commitment, responsibility and protection involved in belonging to a Christian community. To breathe as an early modern subject was always to be breathed by, with and among others: 'our breath is not our owne'.[16]

Philosophical and mechanistic Renaissance theories of aliveness were also expressed through breath, looking back to classical and medieval conceptions. The fundamental connections between breath, life, and being were well-established, as Gina Bloom has written: '*Anima*, like the terms *pneuma* and *spiritus*, signifies a range of ideas we associate with living creatures—including mind, soul, feeling, and living being more generally—but is most fundamentally connected to breath'.[17] Air continued to be seen as one of the six 'non-naturals' necessary for life, the others being motion and rest; sleep and waking; food and drink; excretions; and passions or emotions.[18] As Aristotle had written, inhalation and exhalation 'control life and death; for when respiring creatures can no longer respire, then destruction comes to them'.[19] As Long shows in this volume, although different ancient writers used the term *pneuma* in different ways, this 'wind' or 'breath' was generally regarded as essential to all living organisms and often closely connected to *psyche* ('breath' or 'soul').[20] In Hippocrates' *On Regimen*, for example, *psyche* refers to the body's 'vital stuff' but also to a more abstract principle of 'animation'.[21] Later Galen described the body's vital *pneuma* which, distributed through the arteries and processed by the brain into psychic *pneuma*, nourished the functioning of the soul.[22] All of these theories, together and separately, informed early modern medical understanding of breath. As James I's physician Helkiah Crooke would remember, 'in inspiration and expiration life doth consist' since 'the pulse and respiration… serue one faculty that is the Vitall; for they were both ordained onely for the heart which is the seate of the vitall faculty'.[23] Breath emerges in such accounts as synonymous with being rather than simply supporting or sustaining it, binding the body together with the mind and soul into an integral whole. It is this sense of integral personhood—together with its implied correlative, moral integrity—which the fragmented, unbolted, and breathless Oswald so conspicuously lacks.

Early modern breath *was* life, then, encompassing vitality and sensitivity in the mind, body, and soul. In *The Winter's Tale* (1611), a play deeply concerned with the hazy boundaries between corporeal and spiritual aliveness, Paulina tells Leontes that he may re-marry only 'when your first queen's again in breath'. Later, when Leontes encounters what he believes is a startlingly life-like statue of his late wife, Hermione, his first question probes, through breath, the difference between aesthetic and bodily liveliness: 'What fine chisel / Could ever yet cut breath?'.[24] Earlier, however, Shakespeare had given breath's ability to articulate these same boundaries more sustained attention in *King Lear*. At the start of Act 4 Scene 6, Gloucester has lost his eyes at the hands of Lear's daughter Regan and the Duke of Cornwall and has retreated to Dover, planning to commit suicide. Here the blinded Gloucester, all 'dark and comfortless' (3.7.84), describes his mutilated body as the stub of a candle whose dim, continued smouldering might be painfully borne:

> If I could bear it longer…
> My snuff and loathed part of nature should
> Burn itself out. (4.6.37–40)

His loathsome body, now merely a fragmented 'part' of what it used to be, will consume air slowly towards its natural end unless Gloucester finds a way to snuff it out first. Towards the end of the scene, however, he finds himself persuaded at last that his allotted span is worth living:

> You ever gentle gods, take my breath from me;
> Let not my worser spirit tempt me again
> To die before you please. (4.6.212–14)

Now Gloucester recognises that his desire to stop his own breath evinces his earlier possession by the 'worser spirit' of human frailty. As we will see, the intervening lines contain an unsparing study of breath at the faultlines between life and death, or salvation and damnation.

Here the blinded Gloucester speaks to his own son Edgar, disguised as the beggar Poor Tom, who persuades Gloucester that he is walking along the top of a precipitous cliff, although in reality he is in no physical danger. Poor Tom has earlier been explicitly identified as a spirit, albeit by Lear's Fool:

> Edgar [*within*]: Fathom and half, fathom and half: Poor Tom!
> Fool: Come not in here, nuncle, here's a spirit. Help me, help me!
> Kent: Give me thy hand. Who's there?
> Fool: A spirit, a spirit. He says his name's Poor Tom. (3.4.39–42)

The exposed, wretched Poor Tom may be, as Lear says, an 'unaccommodated man' (3.4.105), but he is also a breath, or a spirit—a semi-carnate agent, either malign or beneficent.[25] And like breath, Poor Tom has the power to prolong or extinguish life, since the elaborate tableau he sketches in the blinded Gloucester's agonised mind of an imaginary and 'horrible steep' (4.6.3) cliff proves as powerfully dangerous (because as powerfully real in Gloucester's imagination) as any 'headlong' (4.6.3) plunge to the ground.[26] Later in the same scene, Edgar encounters his father at what Gloucester believes is the foot of the cliff. Abandoning his Poor Tom persona, Edgar fears for a moment that his father has actually died:

> Edgar: [*to Gloucester*] Alive or dead?
> Ho, you sir! Friend! hear you, sir? Speak!
> [*aside*] Thus might he pass indeed. Yet he revives. -
> What are you, sir?
> Gloucester: Away and let me die.
> Edgar: Hadst thou been aught but gossamer, feathers, air,
> So many fathom down precipitating,
> Thou'dst shivered like an egg; but thou dost breathe,
> Hast heavy substance, bleed'st not, speak'st, art sound.
> Ten masts at each make not the altitude
> Which thou hast perpendicularly fell.
> Thy life's a miracle. Speak yet again.
> Gloucester: But have I fallen, or no? (4.6.45–56)

If Gloucester had been made of anything heavier than gossamer, feathers, or air, his body would surely have 'shivered' (fragmented) into pieces like an egg. But Gloucester's substance is neither fragile nor immaterial, for he is not made of breath—unlike the fiction spun earlier by Poor Tom. The miracle is that Gloucester has survived, despite being made of substantial stuff: 'the clearest gods, who make them honours / Of men's impossibilities, have preserved thee' (4.6.73–74). Nevertheless Edgar's attention to his father's breath plays an important part in this newly realised vitality: 'thou dost breathe, / Hast heavy substance, bleed'st not, speak'st, art sound.'[27] Breath is one of the holy things (like speaking) which has prised Gloucester away from precarity and back to authentic, felt reality. In fact Edgar's intervention goes further, redeeming his father from the humiliating curtailments imposed by mortal life or what he calls 'men's impossibilities'. Gloucester is, miraculously, still in breath—although his spirit, Poor Tom, has departed from him:

> Edgar: Upon the crown o'the cliff what thing was that
> Which parted from you?
> Gloucester: A poor unfortunate beggar. (4.6.67–68)

To Gloucester, the unexpected continuance of his breath at first exposes only the traumatic cracking of his being: 'Is wretchedness deprived that benefit / To end itself by death?' (4.6.61–62). But in time he recognises his life must continue: 'Henceforth I'll bear / Affliction' (4.6.75–76). Gloucester's life, such as it is, is for the time being salvaged through Edgar's determined witnessing and affirming of his father's bodily and spiritual *pneuma*.

Early modern breath has often been understood to signal vulnerability or pliability, standing in metaphorical relation to the fragility of life itself. As the New Testament Epistle of James puts it, life is but 'a vapour, that appeareth for a little time, and then vanisheth away'.[28] Shakespeare does indeed sometimes describe breath in the same or related ways in *King Lear*, as when the Fool dismisses his own song as a worthless nothing like 'the breath of an unfee'd lawyer' (1.4.127).[29] Even when breath's insubstantiality points towards something assertive, early modern cultural historians have tended to emphasise its 'unpredictable movements and subtle material nature', assuming that breath resists our regulation or control.[30] In *Hamlet*, Carla Mazzio has argued, 'th'incorporeal air' becomes part of the play's 'eerie atmospheric haunt', suggesting the limits of knowledge, perception, and observation.[31] And insofar as

Shakespearean breath has suggested inter-relationality beyond the hapless individual subject, it has been recognised mainly as the spreading kind of infectious, 'all-taynting breath' which threatens to disrupt the political commonwealth.[32] In *Coriolanus*, for example, Martius dreads having to beg the 'stinking breath' of the plebeians by seeking their votes for his consulship of Rome: 'You common cry of curs whose breath I hate / As reek o' th' rotten fens'.[33] In the above example from *King Lear*, however, the stakes seem far higher than individual human agency or indeed the vanishing coordinates of a collapsed political structure. Here, in Shakespeare's most uncompromising tragedy, breath involves nothing less and nothing more than one person recognising another's aliveness. As such, witnessing the breath of someone else involves the powerful injunction to stand before and respond to them.[34] Edgar's affirmation of Gloucester's breath in this way recalls the frankness of Cordelia's statement of love for *her* father at the start of the play: 'I love your majesty / According to my bond, no more nor less' (1.1.92–93). Lear cannot recognise Cordelia's words for what they are, or at least not yet; but Gloucester is remade, by what Edgar says, into something more than a dim flame waiting to 'burn itself out'. As the preacher William Perkins wrote, also in 1606, when one man encounters another it is 'breath… [which] sheweth him to be alive'. Breath's mediation of individual agency and identity is most powerfully realised when it is seen, understood, and acknowledged by others.[35] When Edgar observes his father's breath, then, he recreates, reinvents, and redeems him. Even if, as the rest of the play will reveal, this redemption must remain tragically incomplete, Shakespeare's interest in breath allows us to see *King Lear* as something other than a play about 'the melancholy perception of a life no longer recoverable'.[36]

Breath emerges in *King Lear* as synonymous with existential being, fundamental to early modern personhood—and a hopeful marker of our shared aliveness beyond mortal precarity. How might these connections between breath and selfhood be relevant to modern medical practice? Insights have come from a recent series of qualitative interviews with patients suffering severe forms of the incurable lung condition chronic obstructive pulmonary disease (COPD), as well as their informal and professional carers. Patients with COPD describe severe symptoms that cause significant disruption to their day-to-day life in terms implying a passive acceptance or 'weary resignation'; their lung disease is depicted as a 'way of life' rather than an 'illness'.[37] When describing their condition, patients tell a 'chaos narrative' of their illness characterised by unpredictable and uncontrollable events with little sense of restitution or quest. Unlike in conditions such as cancer, patients with chronic lung disease often tell a disease story that is indistinguishable from their life story.[38] A 'culture of normalcy' of breathlessness has been described, in which this persistent symptom is distressing and disabling, but not perceived by patients as disruptive to their sense of being.[39] This may reflect patients adjusting their sense of self and identity in order to adapt and accommodate the all-encompassing sensation of chronic breathlessness. For some patients this acceptance may be an important coping strategy,[40] but on the other hand it may represent the stigma experienced by many and also lead to harm by disempowering patients and restricting their access to healthcare, as they are less likely to report their symptoms. An example of the latter is the significant underuse of palliative care services in the United Kingdom by COPD patients when compared to cancer patients,[41] resulting in a lack of support and potentially unnecessary suffering at end-of-life for patients with chronic lung disease.

Insights from early modern culture, where breath was central to integrated personhood and mind–body consciousness, suggest that there may be therapeutic value in actively challenging this modern passive acceptance of breathlessness. Particular benefits may arise from a revised understanding of the intimate link between the breath and a sense of the 'whole' self. Raising awareness among patients and public that breathlessness is not a natural consequence of ageing, for example, may result in earlier presentation and diagnosis. Attempts should be made to empower patients with chronic lung disease to have higher expectations and improve their access to health care, including palliative care. An increased understanding among clinicians, too, of patients' frequent passive acceptance of breathlessness will facilitate their role as patient advocate. There may also be value in directly addressing the concept of 'breath as being' through psychological approaches that focus on how patients disabled with breathlessness might retain a stronger sense of integral personhood. Patients' own views and ideals should be at the heart of such research strategies.

At the start of Act 4 Scene 6, as we have seen, Gloucester's life is reduced to something barely tolerable: 'if I could bear it longer'. If the airy spirit Poor Tom is in league with Gloucester's despairing 'worser spirit', in seeming to assist his self-destruction, it is also Gloucester's own breath, recognised by Edgar, which shunts him back into life. To be sure, what Gloucester experiences—lying crumpled, as he thinks, far below the cliff's 'dread summit' (4.6.57)—is neither a restitution nor the accomplishment of a quest. His restored breath cannot be said to restore any unproblematic sense of identity, Christian or otherwise. But it does stage a modest intervention into a life that had looked inevitably lost; and relatedly, a move away from breath as the register of unavoidable vulnerability or fragility. To be out of breath, in the early modern period, is to be out of life; to be out of breath, in *King Lear*, is to be exiled into the barren psychic and environmental spaces that this play anatomises so unsparingly. But when Edgar raises his father back up, re-affirming the connection between breath and aliveness, he is also raising the possibility of an altogether different and more promising future. In this way, Shakespeare reminds us of breath's fundamental place in existential awareness, and in our lives among others. He returns to this same inter-relationality in the final scene of the play, the second episode thoroughly concerned with the boundaries between living and dying.

III

Utterly humiliated by Act 5 Scene 3, Lear holds his daughter Cordelia in his arms. She is already dead, executed on the orders of Gloucester's bastard son Edmund, but Lear cannot bear to embrace this reality:

> Lear: I know when one is dead and when one lives;
> She's dead as earth. [*He lays her down.*]
> Lend me a looking-glass;
> If that her breath will mist or stain the stone,
> Why then she lives.
> Kent: Is this the promised end?
> Edgar: Or image of that horror?
> Albany: Fall, and cease.
> Lear: This feather stirs, she lives: if it be so,
> It is a chance that does redeem all sorrows
> That ever I have felt. (5.3.258–65)

Cordelia's breath might become visible through mist on a mirror, or through the movement of a feather in the air. These are ways of dramatising—visibly and theatrically, through material props—the transition from life to death. But like breath, and like Poor Tom, the mist and the movement are almost but not quite seen. Once again breath straddles visible and invisible, material and immaterial, organic and inorganic, body and spirit. Around forty lines later, Lear returns to Cordelia's breath:

> No, no, no life!
> Why should a dog, a horse, a rat have life
> And thou no breath at all?...
> Do you see this? Look on her: look, her lips,
> Look there, look there! [*He dies*]. (5.3.304–310)

The last two lines of this important passage appear in the folio (1623), but not in the version printed in the first quarto (1608). As the play's Arden editor R. A. Foakes points out, the folio addition 'allows us to suppose Lear may die in the joyful delusion of thinking Cordelia is still alive'.[42] The play's heightened emotional climax, in the later version of the play, therefore hinges on whether or not Cordelia is breathing; or, rather, whether or not her breath can be witnessed. Here the play comes full circle in its exposure of the lie that something immeasurable can be measured; this began in the first scene of the play when Lear asked his three daughters 'Which of you shall we say doth love us most?' (1.1.51). The two extra folio lines in Act 5 Scene 3 increase the pathos, and emphasise Lear's own infirmity, but also engage a broader set of questions in keeping with breath's meanings across the play as a whole. The agony of the play's conclusion lies in its collapsing of the difference between empirical reality on the one hand (if the feather *is* moving, this can't be because of Cordelia's breath); and phenomenological or theatrical possibility on the other (Cordelia might still be alive if Lear, and we, were only prepared to believe it). In the folio, Lear opts for what he sees rather than what is. Lear dies in the joyful belief that Cordelia is still alive, raising the prospect of a different version of the future where it is still possible, despite everything, for breath—signalling the possibility of life itself, among those we love—to 'redeem all sorrows'.

King Lear focuses intensely on what it means to look at the breath of someone else; and the consequences of this for both the breathing subject and the person who recognises the breath of the other. Although written more than four hundred years ago, these descriptions of what it means to see, hear, and witness the breath (and breathlessness) of another are highly relevant to contemporary clinical care. Qualitative studies exploring chronic breathlessness from the perspective of patients' informal carers (typically life partners or family members) illustrate the shared impact of lung disease, with one carer remarking that it is 'harder on the spouse than the patient in some ways'.[43] Strikingly, informal carers describe their partner's disease using the first person plural: '*We* were diagnosed… *We* both have a lot of fear'.[44] Very recent research has gone further, describing a phenomenon of 'vicarious dyspnoea' in which healthy volunteers themselves experienced breathlessness on viewing images depicting breathless people.[45] Interestingly viewers with higher degrees of empathy experienced greater breathlessness. These findings suggest there is a need to study breathlessness in caregivers and to examine the shared experience of breathlessness between patients and their informal and professional carers. In *King Lear*, western culture's most canonised drama about suffering, the stakes involved in witnessing the breath of others, particularly beloved others, could scarcely be higher. Tracing the legacy of breath's cultural meanings into the present might lead us to suggest that modern medicine has failed to recognise and address the wider social context of breathlessness. Modern medicine is moving towards an acceptance of a 'whole body' approach to patient care, including the care of breathless individuals; but perhaps what is really needed is a greater commitment to a 'whole family' or even a 'whole society' approach to breathlessness. Such a commitment might do better justice to breath's relational dimensions, and the wider phenomenological possibilities involved in feeling, perceiving, and articulating the experience of breath in the world. After all, as we remember, 'our breath is not our owne.'[46]

IV

Tracing through *King Lear* the historical roots of the close connection between breath and selfhood, we have aimed to recover the forgotten legacy of this connection in the present. As Jamie McKinstry and Corinne Saunders have recently argued, 'breathing and breathlessness can only be understood fully by drawing not only on physiological and pathological evidence, but also on cultural, historical, and phenomenological sources'.[47] Our chapter has sought to spark a mutually enriching dialogue between early modern literary studies and very recent research into chronic breathlessness with the twofold aim of illuminating in new ways Shakespeare's anatomisation of self, *soma,* and ethical relationality; and of informing and improving clinical approaches to breathless patients today.

With these goals in mind, we have focused on two particular synergies. The first relates to the inseparability of breath (and breathlessness) from affective life. We have seen how early modern medical, philosophical, and spiritual ideas about subjectivity were often articulated through descriptions of what it meant, and how it felt, to be in or out of breath. Breathing goes far beyond the mechanical operation of inhalation and exhalation—although the pathos of Lear's feather, at the play's conclusion, momentarily reduces the limitlessness of life and love to exactly that. Nor does breath merely signify life's precarity. Instead early modern breath, particularly as it is revealed in *King Lear*, stands for a broader, more vivid and more promising version of aliveness. At this time, and in this place, aliveness meant the fulfilment of Christian hope and the possibility of redemption. Today, in clinical practice, patients still experience and starkly express the inseparability of breath and aliveness. This tends, however, to emerge through 'chaos' narratives where breathlessness seems inseparable from the story of life itself, as it is lived. Being breathless is 'just the way it is', and becomes something of 'a way of life', as patients adapt their sense of what is possible in order to accommodate their condition. We propose that, for breathless patients now, the continuing inseparability of breath from affective life has led towards a too-hasty acceptance of loss, limitation, and the curtailment of freedom. The next question for clinicians, and surely for us all, is whether this must be so. The connection between breath and aliveness, if more fully historicised and carefully understood, might allow for the recovery of breathless patients' sense of personhood and potential—which might, in turn, open avenues towards the improvement of their quality of life.

Secondly, and building on recent phenomenological work in the critical medical humanities, we have considered how breath is perceived, witnessed, and experienced by others. Breath is bestowed upon Adam in the Book of Genesis, inaugurating a reciprocal relationship between God and humankind. *King Lear* retains a powerful sense of breath's spiritual ramifications, not least through the abject Oswald whose gasping breathlessness suggests, at least to Kent, his utter abandonment of justice and integrity as he descends into mere materiality. But through Edgar's testimonial witnessing of Gloucester's breath, and Lear's wishful revival of Cordelia's, Shakespeare draws attention to the profound implications of breath's inter-relationality. Acts of witnessing and describing the breath of beloved others are integral to these two scenes in which Gloucester, Edgar, Lear, and Cordelia are disclosed to one another in new ways. The transformation works in both directions, since those who imagine or describe the breath of others are seldom themselves unchanged. It is this mutually transformative exchange, rather than the fact of breathing itself, which is, as Lear says, 'a chance that does redeem all sorrows'. Redemption is withheld in this bleakest of tragedies where a different ethical realm, and the possibility of human relationality, remain out of reach. Shakespeare nevertheless shows their unfulfilled potential as breath carries the weight of responsiveness and responsibility involved in all human relations.[48] In clinical practice today, breathing again emerges as a shared rather than an individual experience—'*we* were diagnosed'—and breathlessness precipitates further breathlessness among healthy volunteers. Just as four hundred years ago, then, breathing involves not a singular, isolated self but is instead more accurately understood as a reciprocal or shared exchange. Attending carefully to breath in *King Lear* allows us to suggest something of the continuing existential possibility—as well as the risks—of the togetherness implied by breath's intimate and forgotten relationality.

Acknowledgments The authors would like to thank David Fuller, Jane Macnaughton, and Corinne Saunders for the opportunity to pursue this work. Thanks are also due to Sarah Waters for research assistance.

Notes

1. *King Lear*, 2.2.51–58. All references are taken from R. A. Foakes' Arden edition (London: Arden Shakespeare, 1997; repr. 2003).
2. This is Edgar's line, spoken of himself in the person of Poor Tom, as he prepares to embrace the 'unsubstantial air' upon the heath (4.1.7).
3. See, for example, Gail Kern Paster, *Humoring the Body: Emotions and the Shakespearean Stage* (Chicago: University of Chicago Press, 2004), especially at p. 41 on the 'ambiguity of breath'; and Mary Floyd Wilson and Katherine Rowe, *Reading the Early Modern Passions* (Philadelphia: University of Pennsylvania Press, 2004). Mary Floyd Wilson's chapter in this volume, 'English Mettle', describes how the air people breathed was understood as a marker of early modern ethnic distinction, 130–46, at 133.
4. Louis Laviolette and Pierantonio Laveneziana (on behalf of the ERS Research Seminar Faculty), 'Dyspnoea: A Multidimensional and Multidisciplinary Approach', *European Respiratory Journal* 43 (2014), 1750–62.
5. Capucine Morélot-Panzini, et al., 'Breathlessness Despite Optimal Pathophysiological Treatment: On the Relevance of Being Chronic', *European Respiratory Journal* 50 (2017), 1701159.

6. Robert W. Lansing, Richard H. Gracely, and Robert B. Banzett, 'The Multiple Dimensions of Dyspnea: Review and Hypothesis', *Respiratory Physiology and Neurobiology* 167 (2009), 53–60.
7. Rebecca Oxley and Jane Macnaughton, 'Inspiring Change: Humanities and Social Science Insights into the Experience and Management of Breathlessness', *Current Opinion in Supportive and Palliative Care* 10 (2016), 256–61.
8. Anja Hayen, Mari Herigstad, and Kyle T. S. Pattinson, 'Understanding Dyspnoea as a Complex Individual Experience', *Maturitas* 76 (2013), 45–50.
9. Miriam J. Johnson, et al., 'Chronic Breathlessness: Re-thinking the Symptom', *European Respiratory Journal* 51 (2018), 1702326.
10. Kevin Curran and James Kearney, 'Introduction', *Criticism: Shakespeare and Phenomenology* 54/3 (2012), 353–64, at 358. For an account of phenomenology's usefulness in the study of the experience of disease, including breathlessness, see Havi Carel, *Phenomenology of Illness* (Oxford: Oxford University Press, 2016), esp. 35–39.
11. Genesis, 2:7. All biblical quotations follow the King James version (1611). As Gina Bloom notes, 'The Latin term *spiritus* derives from the Proto-Indo-European base *(s)peis*, 'to blow' and comes into English as spirit primarily through the Vulgate, where it serves as a translation for the Greek *pneuma* and the Hebrew *ruach*—used in the Hebrew Bible to refer to the breath of life that animated Adam in the book of Genesis.' See Bloom, *Voice in Motion: Staging Gender, Shaping Sound in Early Modern England* (Philadelphia: University of Pennsylvania Press, 2007), 80.
12. Philip Stubbes, *A Motive to Good Works* (London, 1593), 181.
13. *Paradise Lost*, ed. by Alastair Fowler (Harlow: Longman, 1991), Bk 7, l. 526.
14. For discussion of earlier developments of these ideas, see Thomas E. Hunt's essay in the present volume.
15. John Willoughbie, *Mnemosynon Kyrio-eucharistion* (Oxford, 1603), 121.
16. John Dod, *The Bright Star Which Leadeth Wise Men to our Lord Jesus Christ* (London, 1603), 30.
17. Bloom, *Voice in Motion*, 81.
18. As earlier essays in the present volume have shown, continuities can be found in the development of ideas about breath and breathing through classical, medieval, and early modern sources.
19. Aristotle, *On the Soul. Parva Naturalia. On Breath*, trans. by W. S. Hett (Cambridge, MA: Harvard University Press, 1935), 443.
20. See also Sylvia Berryman's article on 'Pneuma' in *The Encyclopedia of Philosophy*, ed. by Donald M. Borchet, 2nd edn, 10 vols (London: Routledge, 2006), vol. 7, 649.

21. Elizabeth M. Craik, 'Holism of Body and Mind in Hippocratic Medicine and Greek Tragedy' in *Holism in Ancient Medicine and its Reception*, ed. by Chiara Thumiger (Leiden: Brill, 2020) 184–200, at 184.
22. Berryman, 'Pneuma', 649.
23. Crooke, *Mikrokosmographia: A Description of the Body of Man* (London, 1615), sigs. 2O1^{v}–2O2^{r}; 422–23.
24. *The Winter's Tale*, ed. by Susan Snyder and Deborah T. Curren-Aquino (Cambridge: Cambridge University Press, 2007), 5.1.83 and 5.3.78–79.
25. On *pneuma* as discarnate personal agency, see Berryman, 'Pneuma', 649.
26. Edgar again makes clear the power of the imagination in an aside: 'I know not how conceit may rob / The treasury of life when life itself / Yields to the theft' (4.6.42–44).
27. James Kearney argues that 'a world of ethical promise' lies in the play's deferral of Gloucester's recognition of Poor Tom as his son; it is this deferral which stages 'the possibility—or impossibility—of an ethical relation to the stranger'. See '"This is above all strangeness": *King Lear*, Ethics, and the Phenomenology of Recognition', *Criticism*, 54/3 (2012), 455–67, at 456, 465.
28. The Epistle of James, 4:14. According to Calvin's *An Abridgement of the Institution of Christian Religion* (1585), 'the soule or spirit of man is only a breath of power inspired or poured into the bodie', 74.
29. Compare *The Rape of Lucrece* where the 'lust-breathed' Tarquin describes the imagined ravishment as 'A dream, a breath, a froth of fleeting joy'. See *Shakespeare's Poems*, ed. by Katherine Duncan-Jones and H. R. Woudhuysen (London: Arden Shakespeare, 2007), ll. 3 and 212. The Duke addresses life itself as 'a breath... / Servile to all the skyey-influences' in *Measure for Measure*, ed. by Brian Gibbons (Cambridge: Cambridge University Press, 1991), 3.1.8–9.
30. Bloom, *Voice in Motion*, 85.
31. *Hamlet*, ed. by Ann Thomson and Neil Taylor (London: Arden Shakespeare, 2006), 3.4.14. See Carla Mazzio, 'The History of Air: *Hamlet* and the Trouble with Instruments', *South Central Review* 26/1–2 (2009), 153–96 (159). Compare Carolyn Sale's account of breath's importance in *Hamlet*'s generally 'relentless materializing of everything it discusses'. Sale argues that Shakespeare envisages the early modern theatre itself 'as a place for the exchange of breath'. See 'Eating Air, Feeling Smells: *Hamlet*'s Theory of Performance', *Renaissance Drama* 35 (2006), 145–68, at 148, 161.
32. John Marston, *The Scourge of Villanie* (1598), 4. For an account of 'perceptions of smoky air'—as both a cultural construct in art and literature, and a material reality in Shakespeare's London—see William M. Cavert, 'Airs: Smoke and Pollution, 1600–1775' in *The Smoke of London: Energy*

and Environment in the Early Modern City (Cambridge: Cambridge University Press, 2016), 32–39, at 33.

33. *Coriolanus*, ed. by Peter Holland (London: Arden Shakespeare, 2013), 2.1.229 and 3.3.119–20. See Hristomir A. Stanev, '"A Plague's the Purge to Cleanse a City": Harmful Touch, Rotten Breath, and Infectious Urban Strife in *Coriolanus* and *Timon of Athens*' in *Sensory Experience and the Metropolis on the Jacobean Stage, 1603–1625* (Farnham: Ashgate, 2014), 161–82, esp. 166–78.
34. Here our reading is indebted to Emmanuel Levinas' influential account of the ethical relation between self and other in *Entre Nous: Thinking of the Other*, trans. by Michael B. Smith and Barbara Harshav (London: Continuum, 2006), esp. 124–26 and 131.
35. William Perkins, *A Godlie and Learned Exposition upon the Whole Epistle of Jude* (1606), 27.
36. David Bevington, '"Is This the Promised End?": Death and Dying in *King Lear*', *Proceedings of the American Philosophical Society*, 133.3 (1989), 404–15, at 404.
37. Hilary Pinnock, et al., 'Living and Dying with Severe Chronic Obstructive Pulmonary Disease: Multi-perspective Longitudinal Qualitative Study', *British Medical Journal* 342 (2011), d142; Ann Hutchinson, Natalie Barclay-Klingle, Kathleen Galvin, and Miriam J. Johnson, 'Living with Breathlessness: A Systematic Literature Review and Qualitative Synthesis', *European Respiratory Journal* 51 (2018), 1701477.
38. Pinnock, et al., 'Living and Dying with Severe Chronic Obstructive Pulmonary Disease', d142.
39. Oxley and Macnaughton, 'Inspiring Change', 256–61.
40. Gunvor Aasbø, Kari Nyheim Solbrække, Ellen Kristvik, and Anne Werner, 'Between Disruption and Continuity: Challenges in Maintaining the 'Biographical We' When Caring for a Partner with a Severe, Chronic Illness', *Sociology of Health and Illness* 38/5 (2016), 782–96.
41. Chloe I. Bloom, et al., 'Low Uptake of Palliative Care for COPD Patients within Primary Care in the UK', *European Respiratory Journal* 51 (2018), 1701879.
42. 'Introduction' in *King Lear*, ed. by Foakes, 139.
43. Amanda Belkin, Karen Albright, and Jeffrey J. Swigris, 'A Qualitative Study of Informal Caregivers' Perspectives on the Effects of Idiopathic Pulmonary Fibrosis', *BMJ Open Respiratory Research* 1 (2013), e000007.
44. Belkin, et al., 'A Qualitative Study of Informal Caregivers' Perspectives on the Effects of Idiopathic Pulmonary Fibrosis', e000007.
45. Michaela Herzog, et al., 'Observing Dyspnoea in Others Elicits Dyspnoea, Negative Affect and Brain Responses', *European Respiratory Journal* 51 (2018), 1702682.
46. Dod, *The Bright Star*, 30.

47. 'Medievalism and the Medical Humanities' in *postmedieval: A Journal of Medieval Cultural Studies*, 8/2 (2017), 139–46 (143).
48. Compare Derrida's account of 'a form of involvement with or relation to the other that is a venture into absolute risk, beyond knowledge or certainty'. 'The Gift of Death' in *The Gift of Death & Literature in Secret*, trans. by David Wills (Chicago: University of Chicago Press, 2008), 7–8.

Seleted Bibliography

Bloom, Gina. 2007. *Voice in Motion: Staging Gender, Shaping Sound in Early Modern England*. Philadelphia: University of Pennsylvania Press.

Hayen, Anja, Mari Herigstad, and Kyle T. S. Pattinson. 2013. Understanding Dyspnoea as a Complex Individual Experience. *Maturitas* 76: 45–50.

Herzog, Michaela, et al. 2018. Observing Dyspnoea in Others Elicits Dyspnoea, Negative Affect and Brain Responses. *European Respiratory Journal* 51: 1702682.

Johnson, Miriam J., et al. 2018. Chronic Breathlessness: Re-thinking the Symptom. *European Respiratory Journal* 51: 1702326.

Kearney, James. 2012. 'This Is Above All Strangeness': *King Lear*, Ethics, and the Phenomenology of Recognition. *Criticism* 54 (3): 455–67.

Lansing, Robert W., Richard H. Gracely, and Robert B. Banzett. 2009. The Multiple Dimensions of Dyspnea: Review and Hypothesis. *Respiratory Physiology and Neurobiology* 167: 53–60.

Laviolette, Louis, and Pierantonio Laveneziana. 2014. Dyspnoea: A Multidimensional and Multidisciplinary Approach. *European Respiratory Journal* 43: 1750–62.

Mazzio, Carla. 2009. The History of Air: *Hamlet* and the Trouble with Instruments. *South Central Review* 26/1–2: 153–96.

McKinstry, Jamie, and Corinne Saunders. 2017. Medievalism and the Medical Humanities. *postmedieval: A Journal of Medieval Cultural Studies* 8/2: 139–46.

Oxley, Rebecca, and Jane Macnaughton. 2016. Inspiring Change: Humanities and Social Science Insights into the Experience and Management of Breathlessness. *Current Opinion in Supportive and Palliative Care* 10: 256–61.

CHAPTER 9

'Let Lovers Sigh Out the Rest': Witnessing the Breath in the Early Modern Emotional Body

Naya Tsentourou

To argue that the breathing body is an emotional body runs the tautological risk of attempting to define the shared, universal experience of breathing with the shared, universal experience of feeling. After all, we breathe, ergo we feel. Or is it that we feel, thus we breathe? Except for cases of mechanical ventilation supporting bodies beyond the capacity of emotional or other communication, breath and emotions are vital in how we engage with the world around us and with each other. Everyday discussions of breath and emotions in restorative contexts, such as medicine and yoga, centre around the rhetoric of inside/out as in calls to take a deep breath and 'let it flow out gently' in order to negotiate negative feelings like stress, anxiety, and panic.[1] Breath is essential to well-being as it can effect relaxation: 'with a bit of practice changing how you breathe can make a real difference to how you feel'.[2] The onus in such

N. Tsentourou (✉)
University of Exeter, Exeter, UK

D. Fuller et al. (eds.), *The Life of Breath in Literature, Culture and Medicine*, Palgrave Studies in Literature, Science and Medicine,
https://doi.org/10.1007/978-3-030-74443-4_9

contexts falls on the patient's breathing body to help alleviate the symptoms of the emotional body and reinstate the patient as a physically and emotionally regulated being. The rhetoric of these therapeutic guidelines therefore serves to produce two bodies: the one in need and the one in agency. This essay is concerned with the history of this division and its implications for current perspectives on the 'usefulness' of breath in articulating and managing emotions. Focusing on instances of breathing and suffering in cases of love melancholy in early modern literary and medical texts, I seek to complicate the notion that breath helps us manage feelings by arguing instead that breath itself *is* a feeling that registers the emotional displacement of the self, rather than serving as the self's founding principle. The essay concludes by showing how displaced breath points to the body's emotional entanglement with its surroundings and its observers, an entanglement that confuses the patient and physician binary.

Versions of a displaced self are common in seventeenth-century religious poetry where the speaker often tries to reconcile his or her current fallen and ignorant state with the promise of divine grace and fulfilment. George Herbert's 'Love Unknown' is, as the title suggests, a poem about the speaker's inability to understand God's love, presented to him in three quasi-allegorical episodes where the speaker's heart is acted upon via water, fire, and thorns. Published in the 1633 collection, *The Temple*, the poem charts a tale 'long and sad' (l. 1) that allows the speaker to dwell in the pain of his encounter with the servant, a proxy for God, who patiently educates him into understanding that it is the heart's renewal that God seeks, not the works and rituals professed to accomplish that.[3] The narrative is divided into four sections: the first three list the steps the speaker takes to approach God and how these are corrected by the servant. The ceremonial, almost detached, offering of the heart on a plate of fruit is interrupted by the servant seizing the heart and throwing it in a font to be washed and violently wrung (l. 17). Next, the speaker's sacrifice *from* the heart turns into a sacrifice *of* the heart when the servant intervenes again to throw the heart 'into the scalding pan' (l. 35). Finally, returning to his bed and the comfort of insipid formal worship, the speaker finds that the servant/God disrupts the temporary solace with painful, yet productive, moments of self-reflection in the shape of thorns (l. 52). The fourth section is spoken in the voice of the interlocutor and spells out the lessons learnt in the three episodes. The overall message of the poem seems to be that 'God does not work upon man to get him to do certain things but to get him to be a certain way [...] in a state, that is, of complete and total responsiveness'.[4] While, as Strier argues, the 'aim and mark of the whole discourse is joy', the dominant emotion that shapes the speaker's responses to the events is pain.[5] Pain is captured and reimagined in embodied terms such as 'the very wringing yet / Enforceth tears' (ll. 17–18), or in paratextual cues, such as the capitalization of *AFFLICTION* (l. 28), pointing to the 'predisposed melancholy' that permeates Herbert's verse.[6] Melancholic pain interjects in the poem, and in the overall experience of coming to know God's 'love unknown', with dysregulated, aching breaths at key moments in the devotional situation. The three episodes of correction follow a parallel format: the speaker is under the impression he is performing his religious duties right or is enjoying momentary relief after the servant's intervention, only to walk into his next disorienting trial.

> To him I brought a dish of fruit one day,
> And in the middle plac'd my heart, But he
> (I sigh to say)
> Lookt on a servant, who did know his eye
> Better then you know me, or (which is one)
> Then I my self. The servant instantly
> Quitting the fruit, seiz'd on my heart alone. (ll. 6–12)
>
> But you shall hear. After my heart was well,
> And clean and fair, as I one even-tide
> (I sigh to tell)
> Walkt by my self abroad, I saw a large
> And spacious fornace flaming, and thereon
> A boyling caldron, round about those verge
> Was in great letters set *AFFLICTION*. (ll. 22–28)
>
> … I hasted to my bed.
> But when I thought to sleep out all these faults
> (I sigh to speak)
> I found that some had stuff'd the bed with thoughts,
> I would say thorns. Deare, could my heart not break,
> When with my pleasures ev'n my rest was gone? (ll. 48–53)

While the transitions demarcate the speaker's gradual awakening to the centrality and intimacy of the heart, the reader cannot but notice the visual effect of the lines. If, in Herbert's poetry, 'words are to be seen as well as heard, and the shape of a poem on the page can contribute significantly to its meaning', what the reader sees each time connecting the two emotional extremes—temporal complacency and painful encounter—are the bracketed exhalations: '(I sigh to say)', '(I sigh to tell)', and '(I sigh to speak)'.[7] To 'say', to 'tell', to 'speak' of pain is not enough without the—literally and metaphorically—side-lined breath that makes it all possible. Breath becomes materially grafted on the text as interruption, a bracketed interjection that visualizes a new model of readerly engagement: by pausing the narrative to take a bracketed breath, the text turns the reader into the friend present in the speaker's account. In these aside moments the poet momentarily abandons his role of 'an empiricist conducting research', witnessing the suffering body from a privileged, external point of view, and we come to see and hear the exhalations in real time.[8] The displaced heart at the emotional centre of Herbert's poem co-exists with and manifests itself in the displaced breath. Moreover, the displaced sighs become a threshold where the speaker meets the divine and meets the reader as well, encountering the other in a space simultaneously inside and outside discourse. Christian love is about the suffering body but, in its tormented breathing, that body is not passively reflecting an emotional state. Via its exhalations, it connects to the world.

Even though the 'emotional turn' in early modern studies has so far completely bypassed Herbert, his poetry not only speaks to the field's pursuit of particularizing emotions in their cultural and historical milieu (especially in relation to seventeenth-century theological and literary contexts), but, as is evident in the example of 'Love Unknown', it has the potential to expand this pursuit to include considerations of the breath, especially in relation to the study of melancholy.[9] The study of the history of emotions has found ample ground in the early modern period's blend of humoral medicine with emerging discourses of embodied subjectivity, whether these relate to religion, theatre, material culture, or environment.[10] While the body's humours have been exhaustively revisited and documented either in favour of or against Galenic models of selfhood, and while blood has been the subject of multiple publications and scholarly interest, the body's pneumatic consistence has been less well-recognized, to the point where we can extend to historians of emotion Irigaray's accusation of Heidegger, of forgetting the air.[11] Discussions dedicated explicitly to early modern breathing are sparse, and tend to examine breath alongside humoralist macro–micro analogies and conventional (i.e., Galenic) medical frameworks that pathologize the body.[12] Breath and air often become indistinguishable in such analyses, and understandably so. Herbert, however, in trying to formulate divine–human love, does not refer to the air; he refers to the breath. This is a breath that is not just released in pain but does something to the text and to the world outside that text, in communicating love as displacement.

The performative displacement of the self in love that we encounter in Herbert was a common feature of early modern medical literature on the passions.[13] Like the friend in 'Love Unknown' who diagnoses the speaker's condition and explains the moral lesson, medical writers recorded their observations of lovesickness by turning the patient's breathing, among other symptoms, into an object for study. The remainder of this essay engages directly with references to the respiration of the melancholic lover in these publications to establish how the breath is seen to choreograph the responses of the body in a way that distances it socially and (inter)textually.[14]

Robert Burton's magnum opus, *The Anatomy of Melancholy* (1620), lists the symptoms of love melancholy with the air of detached and disinterested observation that characterizes the majority of the citations employed so profusely in this dizzyingly encyclopaedic work:

> Symptomes are either of Body or Mind; of body, palenesse, leannesse, drinesse, &c. [a] *Pallidus omnis amans, color hic est aptus amanti*, as the Poet describes lovers: *fecit amor maciem*, Love causeth leannesse. [b] *Avicenna de Ilishi cap. 23. makes hollow eyes, drinesse, Symptomes of this disease, to goe smiling to themselves, or acting, as if they saw or heard some delectable object. Valleriola lib. 2. observat. cap. 7. Laurentius cap. 10. Ælianus Montaltus de Her. amore. Langius epist. 24. lib. 1. epist. med.* deliver as much, *corpus exangue pallet, corpus gracile, oculi cavi*, leane, pale, *ut nudis qui pressit calcibus anguem*, hollow-eyed, their eyes are hidden in their heads, [c] *Tenerque nitidi corporis cecidit decor*, they pine away, and looke ill with waking, cares, sighes, *Et qui tenebant signa Phœbeæ facis / Oculi, nihil gentile nec patrium micant*. With groanes, griefe, sadnesse, dulnesse, [d] *Nulla jam Cereris subit / Cura aut salutis* want of appetite, &c. A reason of all this, [e] *Jason Pratensis* gives, *because of the distraction of the spirits the Liver doth not performe his part, nor turnes the aliment into bloud as it ought, and for that cause the members are weake for want of sustenance, they are leane and pine, as the hearbs in my garden doe this month of May, for want of raine.*[15]

The body in love is a body left to waste, as its complexion and basic functions, like breathing, eating, and digesting food, are suspended as a result of distraction by the love object. The symptoms of love melancholy entail a paradoxical performance of embodied yet at the same time 'anti-bodied' gestures. Leanness, paleness, dryness, hollow eyes, and sighs are defined by what they are not: not enough fat, blood or moisture, eyes that look but do not see, sighs instead of regular breaths. In the tradition of melancholy writings sighing is evidence of an overheated, overemotional heart, as well as the body's natural way of restoring balance by cooling the heart through larger than normal inhalations. Sighs and sobs, as Timothy Bright writes, respond to the heart's 'greater necessity of breathing, which being not answered through imbecility of the breathing parts, discharges the office of respiration by sobbes, which should be performed by one draught of breath'. 'Differing onely in that sobbes are sighinges interrupted, and sighes sobbes at large', excessive breathing is a response to the feebleness caused by melancholy.[16]

In the extract from the *Anatomy* Burton casually places sighing on the list of symptoms, but refrains from the clinical language of Bright; after all the subject of the third partition is love melancholy and, in his usual repetitive and relentless style, Burton has already catalogued medical opinion on the symptoms more extensively elsewhere (see 1. 381–384, 'Symptomes, or Signs of Melancholy in the Body'). The third partition is instead taken up with numerous verses that interrupt the flow (if there is a flow) of the treatise to reference examples of lovers and their predicament. Section 2, Member 2, Subsection 4 is a case in point. The subject matter of this section centres around enticement and amorous exchange, covering voice, singing, kissing, dancing, promising, and weeping, as strategies lovers employ to fulfil their desires. Burton references a wide spectrum of mythical and literary tales to support his discussion, acknowledging their fictional status: 'These are tales you will say, but they have most significant Moralls, and doe well expresse those ordinary proceedings of doting Lovers' (3.118). Analysing the 'progresse of this burning lust', he invites the reader to 'observe but with me those amorous proceedings of *Leander* and *Hero*', before quoting the following passage in Latin and English:

> With becks and nods he first beganne,
> To try the wenches minde,
> With becks and nods and smiles againe

> An answere he did finde.
> And in the darke he tooke her by the hand,
> And wrung it hard, and sighed grievously,
> And kissed her too, and wo'd her as he might,
> With pitty me sweet heart or else I dye,
> And with such words and gestures as there past,
> He wonne his Mistris favour at the last. (3.117)

Leander's performance of courtship shares all the features Burton has been listing, such as becking, nodding, smiling, sighing, and kissing. The 'grievous sighing' is part of the vocabulary of the melancholic lover's diagnosis: 'And surely this Aphorisme is most true, *If a lover be not jealous, angry, waspish, apt to fall out, sigh & sweare, he is no true lover*' (Burton 3.119), a convention that reduces sighs to a trope famously embodied by Romeo in Shakespeare's *Romeo and Juliet*, as in this description by his father: 'Many a morning hath he there been seen / With tears augmenting the fresh morning's dew, / Adding to clouds more clouds with his deep sighs' (1.1.129–131). As the audience is invited to imagine a Romeo in isolation, releasing heavy breaths in response to his love, the *Anatomy* reader is asked to observe from a safe distance the unfolding of Leander's passion. The spatial exclusion of the melancholic lover is to a large extent self-imposed, the result of the body withdrawing and excluding itself from its everyday functions. The respiring lover appears to be constantly under observation as he chooses to place himself outside social circles. André Du Laurens's influential *A Discourse of the Preservation of the Sight; of Melancholike Diseases; of Rheumes and of Old Age* (published in French in 1597, and translated into English in 1599) presents to the reader the lover's extreme isolation in similar lines:

> You shall finde him weeping, sobbing, sighing, and redoubling his sighes, and in continuall restlesnes, auoyding company, louing solitarines, the better to feed & follow his foolish imaginations; feare buffeteth him on the one side, & oftentimes dispayre on the other; he is (as *Plautus* sayth) there where indeede he is not.[17]

The distance in the lover's case renders sighing theatrical, and if not insincere, then a superficial indication of a superficial love. The act of witnessing from the physician's perspective the fantastical plight of the patient turns comical in the prefatory material of the 1640 translation of Jacques Ferrand's treatise, *Of Lovesickness* (1610). Edmund Chilmead's translation, titled *Erotomania*, is prefaced with a compilation of dedicatory verses addressed 'To the Author on his Love-Melancholy', playfully mocking the lover's appearance:[18]

> Pray y' pitty him, Lady! How you make him looke!
> His cloathes he weares, as if he had mistooke
> One peice for t'other; and you may safely sweare,
> Though he seeme drest, yet they still scatter'd are.
> His buttons, (like Tarquins Poppy heads) fall down,
> Some halfe a dozen at a sigh; and's Crowne
> Is grown bald with scratching Tunes out. (Ferrand 1640, np)

The dishevelled image is accentuated by the constant sighing, which is imagined as requiring a double inhalation, stretching the chest, and thus the clothes of the lover, to the limit, undoing buttons with every breath. Sighing performatively discovers the body not only in the act of releasing air but in undressing it. It is one of the tricks the lover as actor puts on and which another poet dismisses as inauthentic:

> You that still sigh, not breath; and fondly dote
> On every Black-bagge, and new Petticoate.
> Playing your sad and Melancholy tricks,
> Like devout Iesuits 'fore a Crucifixe.
> Being All things, but your selfe. (Ferrand 1640, np)

In Ferrand's own account, sighs are the result of forgetfulness. Forgetful, the melancholic body attempts to carve a space for itself that goes against its physiology as the lover is seen to interfere with the natural process.

> Sighs come to melancholy lovers because they forget to breathe due to the absorbing fantasies they feed upon, whether in looking upon the beloved or in contemplating her absence, in meditating upon her winning qualities or in searching for the means to enjoy the desired object. Once the lack is realised, nature is constrained to draw in the quantity of air in a single gasp that is taken in normally in two or three breaths: that form of respiration is called a sigh, which is in fact a doubling of the breath. (Ferrand 1623, 280)

To be the melancholic lover means to be fully 'absorbed', to be so consumed by the passion that there is no more space for air to circulate in the heart. The beloved party has succeeded in literally pushing out from the lover's heart any other matter. Seeking to cool the heart's suffocating heat by inhaling more than usual, sighs are the body's attempt to dislocate the love object from the heart and to reinstate some air. The deep breath restores some equilibrium, but this balancing act inevitably privileges nature over imagination, body over mind. Nature is perceived to rescue the body that is threatened by the mind or soul's obsessive meandering. The duality is perplexing: sighing is a product both of nature and 'art' (in the sense of its being a staple of the lover's repertoire). What creates this division is perspective, and the observer's view. Burton, Du Laurens, and Ferrand as observers on the one hand invite the reader to single out the patient and diagnose his breathing as a theatrical tool of wilful displacement, and on the other undermine such conscious involvement on the part of the patient by describing sighing as a reflex. Unlike Herbert's poetic blend of pneumatic subjectivities of reader, speaker, and author, passionate breathing in writings on lovesickness imposes boundaries between author and subject of study, boundaries that appear to devalue or in the best case pathologize the experience of the emotion.

In trying to cordon off emotional breathing and relegate it from nature to art, however, the authors cannot escape the artifice of their supposedly objective, or scientific, discourse and the multiplicities inherent in it. For instance, Burton's Partition 3, Section 2, Member 2, Subsection 4 in its inclusion of 'tales' with 'most significant Moralls' is the most digressive section of the *Anatomy*, not only thematically, as it moves from one method of enticement and one love gesture to the next, but also structurally, as the flow is continuously interspersed with verses in Latin and English. The visual outcome is pages that look more like a selection of quotations of poetry, akin to a commonplace book, where the clinical and detached style of the voluminous treatise is abandoned for the heartfelt verses of Leander, Cleopatra, Venus, from poems by Theocritus and Apuleius among much else. By the time the sixth edition, 'corrected and augmented by the author', appears in 1651, the section is bursting with numerous other verses, including lines spoken by Orpheus, Dido, Euryalus, Phaedra, Myrrha, and Medea. Famous love stories from antiquity and up to the late medieval period, including Petrarch's pleas to Laura, are represented with passionate verses that bring emotion to the *Anatomy*'s otherwise passionless project. If we accept that the *Anatomy*'s textuality embodies the 'melancholy performativity' it seeks to record, and that Burton 'divorces citation from endorsement' 'to keep authority itself permanently in question', Partition 3, Section 2, Member 2, Subsection 4 is a prime example of how the medical veneer that keeps the melancholic lover at arm's length is interrupted and undercut by the voice of that lover.[19] Seeing the verses as short vigorous outbursts in a dull scientific context, the section embodies in its fragmented textuality the palpitation of the heart and shortness of breath the writers associate with love melancholy. The *Anatomy* here executes what Burton instructed earlier, referring to love as madness in his prefatory epistle 'Democritus Junior to the Reader': 'I shall dilate this subject apart, in the meane time let Lovers sigh out the rest' (1.103). The lovers indeed take centre stage and make the point for and instead of the author. The emotion speaks from the margins, from intertextual references, that complicate the reader–author dynamic as they seek to immerse both in their indented reality. Despite their claim to scientific impartiality, early modern writings on melancholy like Burton's can be seen to operate on an emotional axis similar to Herbert's bracketed verses: letting in and expressing passionate love while cordoning it off as their subject matter.

A similar blend of intertextual emotionality infuses Ferrand's Chapter XIV on 'Diagnostic Signs of Love Melancholy' (Ferrand 1623, 269–73). Explaining how lovers 'sigh at frequent intervals and complain without reason', Ferrand follows his statement with a reference to Sappho who 'could no longer weave her cloth' (Ferrand 1623, 270). The rest of this section is dominated by the common link between the tale of Antiochus and Stratonice and Sappho's poetry, the embodied symptoms of love melancholy. 'The verses of the learned and amorous poet Sappho' are cited in Greek, Latin, and French translations, with the latter being the most extensive (Ferrand 1623, 271). As in the *Anatomy*, the description of lovesickness requires the medical narrative to be interrupted by verses of love.[20] The lover's fragmented speech and inability to breathe regularly are reflected in the text's disjointed appearance. The rendition of the passionate inarticulacy that accompanies love in multiple different languages and words mirrors the 'babbling that comes from the surfeit of the heart' (Ferrand 1623, 270). It is as if Ferrand cannot help indulging in Sappho's poetry of the ineffable. 'Speechless', 'my tongue falters', 'my tongue becomes heavy', this is Sappho, but also Ferrand in love with Sappho. Ending the poetic infusion with a rhetorical question, Ferrand revels in the poet's subjective telling of the classical physician's objective knowledge: 'Does it not appear that Sappho was as wise and as experienced in this art as our Greek, Latin, and Arab physicians in light of the fact that they mentioned no indisputable signs that this lady did not already know?' (Ferrand 1623, 272). Although the revised treatise is considered to be 'more sober and vigilant, more disciplined by the procedures of the scholastic medical treatise' compared to the 1610 edition, in moments like this, Sappho, Ferrand, and the lover blend into one, retaining the Ovidian elements of the first treatise (Ferrand 1623, 23). Approximating Burton's philosophical and literary style, Chapter XIV challenges the perceived differences between the 1610 readership of 'the lovers themselves, captivated by the beauty of their ladies' and the 1623 audience of 'the medical profession concerning these lovers' (Ferrand 1623, 23).

Unruly emotion does not only produce the self-unravelling breaths that are sighs but also affects pulse fluctuation. The most famous example is the story of Antiochus and Stratonice, which remained popular in the Western literary and medical imagination from its early versions in Valerius Maximus and Plutarch to its re-imagining by Petrarch, right through the time these treatises of passions were produced and beyond.[21] It is a tale that depicts the extreme physical ailment of unrequited love and the dangers of lovesickness, while it celebrates the unconditional love of a father for his son. Burton, Du Laurens, and Ferrand all engage with the story. Du Laurens summarizes the tale:

> By all these tokens the great Phisitian *Erasistratus* perceiued the disease of *Antiochus* the sonne of *Seleucus* the king, who was readie to dye for the loue of *Stratonica* his mother in law. For seeing him to blush, to waxe pale, to double his sighes, and change his pulse so oft at the very sight of *Stratonica,* he deemed him to bee troubled with this eroticke passion; and so advertised his father.[22]

Erasistratus relates his diagnosis to King Seleucus who in an act of self-sacrificial parental love hands his wife Stratonice to his son. From a medical point of view the story is evidence of how intensity of feeling accelerates the pulse and how this symptom is a safe indicator of an internal passionate condition. Antiochus, who in most versions of the story is said to have been so desperate that he decided to perish by refusing to eat, has his body and his secret betrayed by his pulse as it is read by the physician without his knowledge and consent. Erasistratus in this case might not have been acting only as the king's physician, but might have had a personal interest in experimenting with a lover's pulse; he has been credited with discovering the role of the diaphragm in breathing, and belongs to the group of Hellenistic philosophers interested in determining how respiration worked and the role of blood and *pneuma* in the human body.[23] The presence of the lover was necessary to throw the pulse off its regular rhythm. Burton cites another physician, Josephus Struthlus, who speaks of 'this particular love pulse', and determines the object of his patient's love just by naming him: 'hee named many persons, but at the last when he named him whom he suspected, her pulse began to varie and to beat swifter, and so by often feeling her pulse, he perceiued what the matter was' (Burton 3.142–143). Ferrand goes even further and instead of reciting the tale of Antiochus and Stratonice, he records a version of it with himself as the protagonist. He includes an anecdote from 'May 1604, when I was just beginning my practice in Agen' (Ferrand 1623, 273). In the story, Ferrand 'diagnosed…the love madness of a young scholar' who, having fallen into despair and resorted to every physician and a mountebank in the city, cannot be cured (Ferrand 1623, 273). Ferrand narrates how he observed his symptoms, and taking into account his scholarly profession, 'I concluded, for my own part, that he was lovesick' (Ferrand 1623, 273). As another Erasistratus, he succeeds in identifying the object of the scholar's love when 'an attractive girl of the house came in with a lamp as I was taking his pulse, which from that moment went through a series of changes' (Ferrand 1623, 273). The continuities with the classical tale do not end here: Ferrand is entreated by the scholar to negotiate the match with the maid despite his father's objections to his marrying someone from a lower class, since 'his life depended on' it (Ferrand 1623, 273). The scholar's father is no Seleucus, however; the match falls through and the scholar only survives following Ferrand's 'medical remedies', which appear much later in the book and are unknown to the reader at this stage (Ferrand 1623, 273). The tale of Antiochus's self-sacrificial love turns in Ferrand's narrative into a parody, especially as the lover misguidedly quotes elegiac poetry as evidence his father will agree to his marriage plans (Ferrand 1623, 273).[24]

The beating pulse, like breathing, establishes a body's natural rhythm, a rhythm that is disrupted in cases of love melancholy. The disruption of that rhythm is for the physician to observe, record, and diagnose by experimenting with and producing knowledge at the expense of the lover but for the lover at the same time. The lover both performs and is performed by the symptoms, the sigh and pulse both ridiculed and taken seriously. Trying to account for the body's disrupted rhythms, the physician creates a narrative of displacement that works on multiple levels. First, the love object allows no space in the heart for the air to circulate; love has displaced breath. Second, as the lover's sighs are evidence of an extra, a double breath to account for the missed opportunity to breathe regularly, the displaced air returns forcefully to the body, throwing it out of tune. Third, sighing becomes synonymous with verse recital: the body vents its love through poetry, regaining the balance and rhythm (poetic in this case) that it has lost in material terms. Finally, the physician identifies the melancholic body's isolation yet isolates it further by keeping a perceived objective distance from it. Observing inevitably includes a process of distancing and narrativizing that keeps the writers separate from the condition they describe.

Yet the external point of view at times collapses, bringing ill and healthy bodies into conversation and conspiration with one another. This is most obvious in poetry, as in Herbert's case, where speaker, poet, and reader meet in the brackets, in an example of conspiration that works not only vertically, but, as Peter Adey claims, horizontally.[25] The coordinates of breath, however, do not only have two dimensions but expand to map a depth of transhistorical and transliterary points of reference. Conspiration in this respect is not only between bodies or between bodies and the cosmos, but also between bodies and texts. This level of interdependent breathing finds its way through poetry in Burton's text too: the quick, short, and successive verses suggest the tempo of the writing changes to capture the intensity and multivocality of love melancholy. The lover's rhythm supersedes and encompasses the author's and the two meet not in brackets as such but in printed demarcations that introduce voices other than the speaker's. Similarly, Ferrand's tribute to Sappho and his creative reimagining of the Antiochus and Stratonice story expose a playful and inventive tendency to be part of the passions he records. The treatise's intertextual references make for a fluid and interpersonal experience that transcends the physician–patient binary. In the above cases, the emotional body that suffers is displaced as its air is displaced, but the form in which this body is written complicates pathological readings. The authorial subjectivity of the observer in the above cases is shaped in tandem with the patient's emotional load.

Breath, therefore, is not just a useful reflex mechanism allowing the body to restore its balance. It facilitates the move from the aesthetic realm into which love transposes the patient back to the natural rhythms of existence. Yet in doing so it complicates this binary and the broader binary to which the performative sighing versus healthy body points: that of objective measurement and knowledge and subjective experience. Relating to the lover, but also to the author, relating to medicine, but also to poetry, the representation of breath in the texts studied in this essay destabilizes the role of the presumably external, disengaged, and disinterested witness. It shows points of connection and alternative interpretations that challenge the 'rift' that Havi Carel describes in the clinic, and specifically in respiratory medicine: 'there is a substantial difference between the objective domain of physical bodies, disease, and physiological dysfunction, and the subjective experience of it. There is a gap between symptom and experience, and between physiological disease and how it is lived by people'.[26] This gap is evident in the observer-patient cases of Herbert, Burton, and Ferrand, where bodies in love and their breathing are reduced to their diagnostic symptoms. This essay has tried to show that if we expand our understanding of disrupted rhythms and patterns from the body outwards towards observers and the language they use to record disruption we might approach the phenomenological perspective Carel envisions for clinical practice in a new light. Carel's call for phenomenology in the clinic invites health professionals to consider first-person subjective accounts of breathlessness so they can build on the patient's 'insider' perspective.[27] The poetic and non-discipline-specific vocabulary of early modern treatises on the passions may suggest ways in which the language physicians use might be (consciously or not) informed by the experience of the patients, allowing for a deeper level of empathy and insight.[28] Seen as producing and re-enacting an entangled mess of bodies and stories, emotional breathing can cast either patient or observer as the subject of Du Laurens' claim that in illness, 'he is there where indeede he is not'.

Acknowledgements The author would like to thank the editors of the collection for their generosity, insight, and patience, and the TRU3910 Feeling Bodies seminar groups (University of Exeter, Penryn, Spring 2019) for their inspiring and provocative readings of emotions in early modern literature and culture.

Notes

1. 'Breathing Exercise for Stress', NHS Moodzone (2018) https://www.nhs.uk/conditions/stress-anxiety-depression/ways-relieve-stress/# [accessed 12th April 2019].
2. 'ReConnect2Life—Creating Skills for the Future: Deep Breathing Exercise', NHS Torbay and South Devon (2019) https://www.torbayandsouthdevon.nhs.uk/services/pain-service/reconnect2life/creating-skills-for-the-future/learning-relaxation-skills/deep-breathing-exercise/ [accessed 12th April 2019].
3. George Herbert, 'Love Unknown', in *The Works of George Herbert*, ed. by F. E. Hutchinson (Oxford: Clarendon, 1941), 129–31. All references to 'Love Unknown' are to this edition and are listed parenthetically and by line number in the text.
4. Richard Strier, *Love Known: Theology and Experience in George Herbert's Poetry* (Chicago: Chicago University Press, 1983), 164.
5. Strier, *Love Known*, 163. On the emotional rollercoaster of Protestant experience, see also Alec Ryrie, *Being Protestant in Reformation Britain* (Oxford: Oxford University Press, 2013), 17–98.
6. On Herbert's melancholy see Anne-Marie Miller Blaise, 'George Herbert's Distemper: An Honest Shepherd's Remedy for Melancholy', in *John Donne and the Metaphysical Poets*, ed. by Harold Bloom (New York: Bloom's Literary Criticism, 2010), 142.
7. Helen Wilcox, *George Herbert: 100 Poems* (Cambridge: Cambridge University Press, 2016), x.
8. Angela Balla, 'Baconian Investigation and Spiritual Standing in Herbert's *The Temple*', *George Herbert Journal* 34/1 (2010): 65 https://doi.org/10.1353/ghj.2010.0001.
9. Richard Meek and Erin Sullivan, eds, *The Renaissance of Emotion* (Manchester: Manchester University Press, 2015), 3. Meek and Sullivan usefully summarize key studies in fn. 3 of the cited page.
10. For quick reference to the wealth of perspectives, see the contents page in Susan Broomhall, ed., *Early Modern Emotions: An Introduction* (London: Routledge, 2017).
11. Predominantly Gail Kern Paster, *Humoring the Body: Emotions and the Shakespearean Stage* (Chicago: The University of Chicago Press, 2004); Luce Irigaray, *The Forgetting of Air in Martin Heidegger* (London: Athlone Press, 1999).
12. See Carla Mazzio, 'The History of Air: *Hamlet* and the Trouble with Instruments', *Shakespeare & Science* 26/1 (2009): 153–96 https://doi.org/10.1353/scr.0.0039. Notable exceptions that take into account the poetics of breath and its rhetorical force include Gina Bloom, *Voice in Motion: Staging Gender, Shaping Sound in Early Modern England*

(Philadelphia: University of Pennsylvania Press, 2007), Leah Knight, *Reading Green in Early Modern England* (London: Routledge, 2014), and Rebecca Totaro, '"Revolving This Will Teach Thee How to Curse": Lessons in Sublunary Exhalation', in *Rhetorics of Bodily Disease and Health in Medieval and Early Modern England*, ed. Jennifer C. Vaught (New York: Routledge, 2010): 135–51.

13. I am referring to these texts as treatises of the passions rather than emotions here, following their titles and early modern terminology. Nevertheless, the essay does not adopt rigid distinctions between these terms. I agree with Thomas Dixon that 'there are important differences in nuance to all these terms [he cites emotions, passions, affections, or sentiments] that should not be effaced', but I take issue with curtailing the use of these terms. In reference to emotion specifically, Dixon argues that 'the word "emotions" is currently used too liberally by historians of psychology and its reference needs to be narrowed'; he suggests any application of the term to pre-nineteenth-century contexts is anachronistic [see Thomas Dixon, *From Passions to Emotions: The Creation of a Secular Psychological Category* (Cambridge: Cambridge University Press, 2003), 11–12]. Scholars of emotions in the early modern period, however, counter-argue that 'during the seventeenth century, it [emotions] is a word whose meaning is in flux' [see David Thorley, 'Towards a History of Emotion, 1562–1660', *The Seventeenth Century* 28/1 (2013): 15, https://doi.org/10.1080/0268117X.2012.757943] and acknowledge there is always 'a certain amount of arbitrariness built into the project of choosing which current terms to use to describe the past' and 'the term "emotions" begins to look quite attractive as an option that covers the variety of equally ambiguous terms in these texts' [see Kirk Essary, 'Passions, Affections, or Emotions? On the Ambiguity of 16th-Century Terminology', *Emotion Review* 9/4 (2017), 368 https://doi.org/10.1177/1754073916679007].
14. A full discussion of the passion of lovesickness and its appeal to the early modern medical and literary imagination is beyond the scope of my analysis in this essay. For a firm grounding in its cultural and philosophical contexts throughout the centuries, see the Introduction by Beecher and Ciavolella in their edition of Ferrand's treatise, *Of Lovesickness*. For lovesickness in an early modern English context see Lesel Dawson, *Lovesickness and Gender in Early Modern English Literature* (Oxford: Oxford University Press, 2008).
15. Robert Burton, *The Anatomy of Melancholy*, ed. by Thomas C. Faulkner, Nicolas K. Kiessling, and Rhonda L. Blair, with commentary by J. B. Bamborough with Martin Dodsworth, 6 vols (Oxford: Clarendon Press, 1989–2000), vol. 1, 139. All references to Burton are to this edition and

are cited parenthetically in the text, with volume number followed by page numbers.

16. Timothy Bright, *A Treatise of Melancholie* (London: Printed by Thomas Vautrollier, 1586), 160. For a wider discussion of sighing and the regulation of breath in the early modern period see Naya Tsentourou, 'Wasting Breath in *Hamlet*', in *Reading Breath in Literature*, by Arthur Rose, et al., (Palgrave, 2018): 39–62.
17. André Du Laurens, *A Discourse of the Preservation of the Sight; of Melancholike Diseases; of Rheumes and of Old Age*, trans. by Richard Surphlet (London: Imprinted by Felix Kingston, 1599), 118.
18. Jacques Ferrand, *Erotomania, or a Treatise Discoursing of the Essence, Causes, Symptomes, Prognosticks, and Cure of Love*, trans. by Edmund Chilmead (Oxford: Printed by L. Lichfield, 1640). See also Ferrand, *A Treatise on Lovesickness*, ed. and trans. by Donald A. Beecher and Massimo Ciavolella (New York: Syracuse University Press, 1990). Beecher and Ciavolella's translation is based on the second edition of the treatise, published in French in 1623. The first edition, significantly shorter, appeared in 1610. For a selection of potential sources, the editors have identified for Ferrand's examination of sighing in lovers, including Du Laurens, see pages 452–53, fn. 4. Subsequent references to Ferrand's treatise are cited parenthetically in the text, by year of edition and page numbers.
19. Drew Daniel, *The Melancholy Assemblage: Affect and Epistemology in the English Renaissance* (New York: Fordham University Press, 2013), 161. On Burton's style see also Mary Ann Lund, *Melancholy, Medicine and Religion in Early Modern England: Reading the Anatomy of Melancholy* (Cambridge: Cambridge University Press, 2010).
20. Editors Beecher and Ciavollela suggest that, although Burton references Ferrand in the fourth edition of the *Anatomy* (1632), no direct influence can be established. 'After Ferrand, only Burton had the energy to peruse the vast documentation traditionally attached to the *topos* [of *amor hereos*], but for reasons different to Ferrand's' (Ferrand 1623, 16).
21. For a concise summary of the story's influence see Ferrand 1623, 48–51. For the story's wider historical context and its place in medical tradition see Daniel Ogden, *The Legend of Seleucus: Kingship, Narrative and Mythmaking in the Ancient World* (Cambridge: Cambridge University Press, 2017), ch. 5.
22. Du Laurens, *A Discourse*, 118–19.
23. Jean-William Fitting, 'From Breathing to Respiration', *Respiration* 89 (2015), 82–87 https://doi.org/10.1159/000369474.
24. The line is from Propertius's *Elegies*, Bk 1, Elegy 5: 'Love knows no pedigrees, will cede to no old portraits'. It is spoken by Propertius to Gallus warning him against courting Cynthia, the poet's own love object. Propertius, *Elegies*, ed. and trans. by G. P. Goold (Loeb Classical Library

18. Cambridge, MA: Harvard University Press, 1990), 51–52 https://doi.org/10.4159/dlcl.propertius-elegies.
25. For my understanding of conspiration I am very much indebted to Phillip Horky's presentation, "Our Common Breath: συμπνοία from Early Christianity to Neopythagoreanism". The Life of Breath: History, Texts, Contexts Conference, 10–12 July 2018, Durham University. Peter Adey made the point about horizontal and vertical breathing as part of his closing presentation at the same conference.
26. Havi Carel, 'Breathlessness: The Rift between Objective Measurement and Subjective Experience', *Lancet Respiratory Medicine* 6/5 (2018), 332 https://doi.org/10.1016/S2213-2600(18)30106-1.
27. Havi Carel, 'Invisible Suffering: The Experience of Breathlessness', in *Atmospheres of Breathing: The Respiratory Questions of Philosophy*, ed. by Lenart Škof and Petri Berndtson (New York: SUNY Press, 2018), 241.
28. In this light, a literary work like Paul Kalanithi's *When Breath Becomes Air* (London: Bodley Head, 2016) can be seen as a modern-day equivalent to the aspects of Burton and Ferrand's work on which this essay has focused.

Selected Bibliography

Carel, Havi. 2018. Invisible Suffering: The Experience of Breathlessness. In *Atmospheres of Breathing: The Respiratory Questions of Philosophy*, ed. Lenart Škof and Petri Berndtson, 233–46. New York: SUNY Press.

Dawson, Lesel. 2008. *Lovesickness and Gender in Early Modern English Literature*. Oxford: Oxford University Press.

Kalanithi, Paul. 2016. *When Breath Becomes Air*. London: Bodley Head.

Lund, Mary Ann. 2010. *Melancholy, Medicine and Religion in Early Modern England: Reading the Anatomy of Melancholy*. Cambridge: Cambridge University Press.

Meek, Richard, and Erin Sullivan, eds 2015. *The Renaissance of Emotion*. Manchester: Manchester University Press.

Paster, Gail Kern. 2004. *Humoring the Body: Emotions and the Shakespearean Stage*. Chicago: The University of Chicago Press.

CHAPTER 10

What Is 'the Breath of Our Nostrils'? *Ruach* and *Neshamah* in John Donne's 1622 Gunpowder Day Sermon

Patrick Gray

In the early seventeenth century, as fragile religious pluralism on the Continent degenerated into all-encompassing, international sectarian conflict, and England itself lurched rapidly towards what would become a civil war, John Donne found himself engaged in high-stakes domestic peace-making between the Crown and restive Puritans. With fellow Calvinists under threat in Bohemia, Protestant nationalists proved increasingly unwilling to accept King James's protracted efforts to secure Infanta Maria Anna of Spain, a Habsburg and a Catholic, as a wife for his son Charles, the English heir apparent.

Correspondence between Englishmen at the time, as well as reports from foreign diplomats, give a picture of a nation on the verge of open rebellion. As Thomas Cogswell observes, 'Never before in James's reign had a single political issue so deeply divided the kingdom'.[1] Sermons, in particular, proved occasions for public protest. In April of 1622, John

P. Gray (✉)
Department of English Studies, Durham University, Durham, UK

D. Fuller et al. (eds.), *The Life of Breath in Literature, Culture and Medicine*, Palgrave Studies in Literature, Science and Medicine,
https://doi.org/10.1007/978-3-030-74443-4_10

Knight preached a sermon at Oxford that could be construed as arguing in favour of the assassination of tyrannical monarchs, in keeping with emerging Huguenot theories of a right to resist political suppression of the Reformation. Knight was arrested and his books burnt, but James remained alarmed. A Protestant version of the 1605 Gunpowder Plot was no longer an unthinkable prospect.[2]

On 4 August 1622, in a letter to the Archbishop of Canterbury, James issued *Directions concerning Preachers*, setting out strict limits to what the Crown would thenceforth allow as the content of sermons. No preacher 'under the degree of a Bishopp or deane (at the least)' was henceforth to 'presume to preache in any popular auditory the deepe poynts of *Predestinac[i]on Election Reprobac[i]on* or of the *universality Efficacity Resistability or Irresistabillity of gods grace*', but instead was directed to 'leave theise themes to be handled by learned men', 'as being fitter for the Schooles & universities than for simple auditories'. 'Noe preacher', moreover, 'of what title or denominac[i]on soever' was henceforth to 'fall into any bitter Invectives & undecent rayleinge speeches against the persons of either Papists or Puritanes'. Above all, 'no preacher of what title or denominac[i]on soever' was henceforth to 'presume ... to declare, limit, or bound out ... the power prerogative jurisdic[ci]on Authority or Duty of Soveraigne princes, or otherwise meddle with theise matters of State'.[3]

Despite the unpopularity of James's foreign policy, Donne had reason to feel indebted to the king. As a younger man, an intellectual prodigy with a taste for libertine poetry, Donne seems to have dreamed of a career in civil service, perhaps as a diplomat on the Continent. As the son of Catholic recusants, such a position would have allowed him to avoid awkward questions about the precise nature of his own theological commitments. These hopes for a secular future overseas were dashed, however, when he eloped with his patron's niece. Following many years of unemployment, Donne finally took Holy Orders only after repeated encouragement from James, as well as what seems to have been a crisis of conscience.[4] The king was also instrumental in his appointment to the Deanship of St Paul's.

By royal command, Donne gave an initial sermon in defence of James's *Directions* at St Paul's on 15 September 1622. Contemporary accounts reveal great anticipation of the event among his contemporaries, as well as some measure of approval afterwards, although one, at least, John Chamberlain, reports in a private letter that Donne 'gave no great satisfaction, or as some say spake as yf himself were not so well satisfied'.[5] James for his part was delighted; he praised the sermon as 'a piece of such perfection as could admit neither addition nor diminuition' and recommended publication.[6] Hoping, perhaps, for an even more spectacular encore, James then asked Donne to preach a second such sermon at St Paul's on 5 November 1622: Gunpowder Day.

At liberty to choose his own text for the occasion, Donne singled out a cryptic verse from Lamentations: 'The breath of our nostrils, the anointed of the Lord, was taken in their pits' (4:20). Although the passage today may seem obscure, the choice in its original context is well-informed and artful, allowing Donne room for Janus-like equivocation. As his audience was keenly aware, whether the phrase in Lamentations, 'the anointed of the Lord', refers to a good king, Josiah, or a bad king, Zedekiah, was at the time a notorious textual crux. The Geneva Bible glosses 'the anointed' as Josiah; Calvin in his commentary as Zedekiah.

The phrase 'the breath of our nostrils' in Donne's chosen text, an allusion to Genesis 2:7, introduces yet another deep-rooted enigma. What does 'breath' refer to here, metaphorically speaking? Like St Augustine's exposition of Genesis 2:7 in his *Commentary on Genesis against the Manichees*, Donne's learned exposition of Lamentations 4:20 rests on fine distinctions between 'breath,' 'soul', and 'spirit' in the languages of Scripture: *neshamah* vs *nepesh* vs *ruach* (Hebrew); *pnoē* vs *psychē* vs *pneuma* (Greek).

In sum, as Donne seems to have intended, making sense of his argument in this high-stakes, highly visible sermon, much less his own theological and political commitments, is no easy task. Following his reasoning requires understanding, not only several scholarly disputes, but also what Donne's audience at St Paul's would have been likely to perceive as at stake in the possible resolution of these controversies. Before turning to the nub of the matter, to wit, whether hot Protestants should obey a king who was seeming cooler by the day, Donne works through and dismisses several less obviously topical debates which then serve as precedents.

Is Lamentations a history or a prophecy? Does Lamentations 4:20 refer to Josiah or to Zedekiah? Does the word 'pits' in this verse refer to a grave, a prison, or a pitfall? Does the circumlocution 'the breath of our nostrils' in Genesis 2:7, as well as Lamentations 4:20, refer to human breath, to the soul, or to God himself, as the Holy Spirit? Throughout Donne's sermon, these and other recondite, seemingly harmless brainteasers serve as proxies for more pressing, obviously political, and dangerously provocative questions that he is reluctant to approach more directly.

Is James more like Josiah or Zedekiah, that is to say, is he a good king or a bad king? Is the Protestant Reformation in England analogous to the destruction of Jerusalem described in Lamentations, or should the contemporary resonance of Lamentations be re-imagined with the Catholic Church in the role of Babylon? Finally, and most importantly, just how closely is King James's authority to be identified with that of God himself? In terms of the two rival meanings of 'the breath of our nostrils', is he more like *ruach* or like *neshamah*? In each case, Donne works assiduously not so much to resolve the controversy as to complicate it beyond repair, so that in the end, he can wave it aside as unimportant.

James himself did not attend Donne's sermon at St Paul's Cross, but he did commission a scribal copy, which Donne corrected in his own hand before sending on.[7] No intellectual slouch in matters theological, and a man of much stronger opinions, James seems to have been, insofar as we can tell, altogether less pleased with Donne's second, more ambiguous defence of his *Directions*. No record of any subsequent royal praise has been found, and no recommendation of publication either, which, Jeanne Shami speculates, 'may indicate his less than enthusiastic reception'. 'After 5 November', Shami notes, 'we find no more special commissions for Donne in the pulpit. Nor is Donne promoted either by James or by Charles to the bishopric that the "absolutist" reading of his politics predicts'.[8]

Izaak Walton's seventeenth-century life of Donne emphasizes the role of divine providence in guiding a man of manifest genius towards Protestantism.[9] Reacting against Walton's quasi-hagiography, twentieth-century biographers R. C. Bald and John Carey proposed a rival vision of Donne as a calculating careerist whose conversion to the Church of England was motivated less by any genuine change of heart than by the mercenary impulse of worldly ambition.[10] This sense of Donne as concerned above all with patronage and advancement was then taken up by critics in the 1980s such as Jonathan Goldberg as an explanation for his sermons in defence of the absolutism of James I.[11]

In the 1990s, critics such as David Norbrook and Annabel Patterson, as well as Shami, began to question this revisionist account of Donne's life and motives.[12] Dennis Flynn, for example, shows that Donne 'behaved without much ambition in the early years of the reign of James I,' even turning down several attractive offers of employment.[13] Annabel

Patterson draws attention to his close friendships with parliamentarians strongly opposed to James's efforts to consolidate power.[14] Richard Strier mocks the idea of an 'oppositional' Donne; even Strier, however, sees Donne's service to the established church and state as a matter of 'principled loyalty,' rather than status-driven 'toadying'.[15]

As Achsah Guibbory explains, 'critical disagreement has centered on the question of what Donne's position was in the 1620s, when the English church was the site of escalating religious conflict that would eventually lead to civil war'.[16] Still more specifically, the focus of the controversy remains still today Donne's two sermons in 1622 justifying James's *Directions*, which Norbook concedes have become 'notorious', and Strier describes more vividly as 'quite a camel for the oppositionists to swallow'.[17] One appealing compromise between these two camps is Arthur Marotti's more recent emphasis on what he identifies as Donne's 'obvious ambivalence'. Donne was 'sometimes ill at ease with the role of official apologist', he observes, 'both before and after his appointment as dean of St Paul's'. His defence of James's edict is fascinating because it is 'at odds with itself'.[18]

In the case of Donne's Gunpowder Day sermon, such doublemindedness is especially pronounced. Starting straightaway with his choice of text, Donne seems to move at times in two opposite directions at once. 'The breath of our nostrils, the anointed of the Lord, was taken in their pits / Of whom we said, Under his shadow we shall live among the heathen': for Donne, this passage would have been more than usually familiar. He translated the entirety of Lamentations into poetry on more than one occasion, working from the Latin version of Tremellius, a convert from Judaism to Protestantism. In his own translation, Donne writes:

> Th' anointed lord, breath of our nostrils, he
> Of whom we said, under his shadow we
> Shall with more ease under the heathen dwell,
> Into the pit which these men digged, fell.[19]

'Pits' becomes more clearly here a pitfall; otherwise, however, Donne sticks remarkably close to his source. The poem, moreover, seems to be his only translation directly from Scripture. Given Donne's praise for the Sidney translation of the Psalms, Graham Roebuck finds himself compelled to wonder 'why he completed only one such exercise'.[20]

Then again, Donne's singular interest in Lamentations may not be entirely surprising. As John Klause lays out in some detail, for Elizabethan Catholics such as Donne's family, the prophet Jeremiah's description of the destruction of Jerusalem by the Babylonians held a special resonance. 'Bare, ruined Jerusalem was, simply, Catholic England'.[21] In the 1580s and 1590s, the recusant composer William Byrd, for example, wrote numerous motets grieving for '*Sion ... deserta*' and '*Ierusalem ... deserta*'. Donne signals his departure from this tradition by foregrounding in a subtitle that his source is Tremellius, a Protestant, rather than the Catholic Vulgate. In his Gunpowder Day sermon, Donne casts Catholics in the role of Babylonians and his fellow Protestants as imperilled Jews.[22]

For contemporary Protestants, interest in Lamentations arose more immediately out of a different tradition of politicized interpretation. As Graeme Murdock explains, 'Many Calvinists found the life of King Josiah to be an excellent model for contemporary magistrates'.[23] Josiah was responsible for what is now sometimes known as the Deuteronomic Reform, a set of thorough-going changes to the practice of worship in Judah inspired by the high priest Hilkiah's discovery of 'a book of the law of the Lord given by Moses', presumably, the Book of Deuteronomy, during a renovation of the Temple of Solomon (2 Chron. 34:14).

In the second Book of Kings, when Josiah hears this 'book of the law of the Lord' read to him, he tears his clothes and proclaims, 'Great is the wrath of the Lord that is kindled against us, because our fathers have not hearkened unto the words of this book'. Calling the people of Judah together, 'he read in their ears all the words of the book of the covenant which was found in the house of the Lord' (2 Kings 22:13, 23:2). We might compare here Donne's description of King James in his Gunpowder Day sermon as 'a most perfect Text-man, in the Book of God' (4:256). Seized with renewed religious zeal, Josiah purged the Temple of Solomon of 'all the vessels' dedicated to the worship of anything other than Yahweh, destroyed all sites of pagan worship throughout Judah, and banished or killed all 'idolatrous priests' (2 Kings 23:3–24; *cf*. 2 Chron. 24:2).

In the sixteenth century, following the death of Henry VIII, Protestants such as Thomas Cranmer, Hugh Latimer, and Martin Bucer cited Josiah in particular, anointed king of Judah at the age of eight, as an illustration of their hopes for his successor, Edward VI, crowned at the age of nine, and England's first monarch to have been raised as a Protestant. Calvin himself wrote to Edward from Geneva to second their opinion.[24]

For John Knox and the other authors of the 1560 Scots Confession of Faith, the example of Josiah shows that kings are 'not only appointed for civil government but also to maintain true religion and to suppress all idolatry and superstition'.[25]

The story of Josiah encapsulates what contemporary English Protestants hoped for from James. In contrast to Edward VI, however, James I seemed to them more like Josiah's successor, the bad king Zedekiah, who 'did that which was evil in the sight of the Lord his God, and humbled not himself before Jeremiah the prophet speaking from the mouth of the Lord'. Following his example, 'all the chief of the priests, and the people, transgressed very much after all the abominations of the heathen; and polluted the house of the Lord which he had hallowed in Jerusalem' (2 Chron. 36:12, 14). Against Jeremiah's advice, Zedekiah led a rebellion against Babylonian rule which resulted in disaster. King Nebuchadnezzar razed Jerusalem to the ground, including the Temple, and took many of its people back to Babylon as slaves. Zedekiah himself was caught attempting to flee and carried off to life imprisonment.

Lamentations mourns the loss of one of these kings, but which one? Zedekiah or Josiah? And is this king the same king whom Jeremiah describes as 'the anointed of the Lord' in the specific verse Donne singles out? As Donne explains at the outset of his sermon, in the second Book of Chronicles we learn that after the death of King Josiah, '*Jeremy lamented for Josiah, and all the people speake of him, in their Lamentations*' (4:240; *cf*. 2 Chron. 35:25). By this light, Donne proposes, the 'anointed of the Lord' may be taken to refer to Josiah, 'for so Saint *Hierome*, and many of the *Ancients*, and many of the Jewes themselves take it' (4:240).[26] But not everyone would agree. Calvin, in particular singles out Jerome as 'grossly mistaken' for thinking that the Book of Lamentations is 'the Elegy which Jeremiah composed on the death of Josiah'.[27]

The Geneva Bible of 1560, which had been displaced by the Bishops' Bible of 1568, followed by the authorized translation of 1611, now known as the King James Version, glosses Lamentations 4:20 as referring to Josiah. As Donne concedes, however, in his 1622 sermon, it is 'more ordinarily, and more probably held by the Expositours' to refer to 'the transportation and the misery of an ill King, of *Zedekiah*' (4:240). 'This verse has been ignorantly applied to Josiah, who fell in battle long before the fall of the city', Calvin insists, and 'cannot be understood except of King Zedekiah'. 'He had fled by a hidden way into the desert, and he thought that he had escaped from the hands of his enemies; but he was soon seized, and brought to king Nebuchadnezzar. As, then, he had unexpectedly fallen into the hands of his enemies, rightly does the Prophet say metaphorically, that he was taken in their snares'.[28]

Throughout the sermon, Donne minimizes the importance of the identity of the king in question, Josiah or Zedekiah, and instead emphasizes his status as 'the anointed of the Lord', as well as, especially, Jeremiah's description of him as 'the breath of our nostrils'. 'We argue not, we dispute not now,' Donne insists; 'we imbrace that which arises from both, That both good Kings, and bad Kings, *Josiah* and *Zedekiah*, are the *anointed* of the Lord, and the *breath of the nostrils*, that is, The life of the people' (4:240). The phrase itself, 'the life of people', reveals the likely influence of the gloss on this passage in the base text for the King James Bible, the 1602 edition of the Bishops' Bible. Unlike the Geneva Bible, which specifies that 'breath' here refers to Josiah, the Bishops' Bible is more ambiguous: 'the breath' is 'the lyfe, meaning the king, who is called the lyfe of the people, for that the people are like an headlesse body that hath no lyfe in it, when they be left without a king or governour'.

What does it mean, however, to say that the king is '*Spiritus narium*', that is, '*the breath of our nostrills*'? In keeping with Genesis 2:7, which it cites as precedent, the gloss in the Bishops' Bible suggests a relatively mundane interpretation. Breath as 'lyfe', comparable to a human head, corresponds to the Hebrew *neshamah* (*cp.* Greek *pneuma*, 'breath'). Donne, however, chooses a more exalted reading: breath as the Hebrew *ruach* (*cp.* Greek *psychē*, 'spirit'), which he identifies with God. Later in the sermon, Donne walks back this line of argument. But it is striking here, not least as an extravagant compliment to James.

'First', Donne argues, '*Spiritus*' refers to the Third Person of the Trinity, the Holy Spirit. 'He is called by this Name, by the word of this Text, *Ruach*, even in the beginning of the Creation, God had created Heaven and Earth, and then *The Spirit of God*, *sufflabat*, saith *Pagnins* translation, (and so saith the *Chaldee Paraphrast* too) it *breathed upon the waters*' (4:252).[29] 'The Attribute of the Holy Ghost and his Office, which is, to apply to man the goodnesse of God, belongs to Kings also; for, God gives, but they [that is, kings] apply all blessings to us'. King James is like the Holy Spirit insofar, in particular, as he 'moves upon *the waters*, by his royall and warlike *Navy at Sea* (in which he hath expressed a speciall and particular care)' (4:252).

As the Bishops' Bible notes in its gloss, Lamentations 4:20 alludes to Genesis 2:7: 'the Lord God formed man of the dust of the ground, and breathed into his nostrils the breath of life; and man became a living soul.' Omitting all mention of Genesis 2:7, however, at least for the moment, Donne turns instead to Genesis 1:2: 'the Spirit of God moved upon the face of the waters'. He also cites a 1528 Latin translation, Sante Paganini's, esteemed for its literal translation of the Hebrew, which allows him to translate the Hebrew verb לַעתְפֶּחֲרְמ־ (*al-marahaefeth*; *lit.*, 'hovering over, brooding over') that the King James version translates as 'moved' as 'breathed' (Latin, *sufflabat*) instead, emphasizing the connection to literal breath. Part of the rationale behind this decision is that the Hebrew term typically translated in English as 'spirit,' חוּר_(*lit.*, 'wind'), that Donne transliterates as *ruach,* and that appears in Lamentations 4:20, as well as Genesis 1:2, is not the same as the Hebrew term הָמָשְׁנ (*lit.*, 'breath') that Donne transliterates as *neshamah* and that appears in Genesis 2:7.

How much Hebrew did John Donne know? Judith Scherer Herz notes that Donne frequently translates and explains single Hebrew words throughout his sermons. Anthony Raspa, by contrast, points out that 'literally all of Donne's references to past and present Hebrew writers are to the Latin translations of German, Italian, and Spanish translators'.[30] Borrowing Matt Goldish's concept of a 'third-order Hebraist', Chanita Goodblatt makes a convincing case for some truth to both perspectives: Donne was someone 'who could read *some* Hebrew, but who knew and used significant amounts of Jewish literature in Latin and vernacular translation'.[31] In this case, Donne seems likely to have known enough about Hebrew to know that the difference between *ruach* and *neshamah* is theologically significant.

Briefly put, the crux of the contrast between the two terms has to do with agency. *Ruach,* like wind, comes from without. *Neshamah,* like breath, comes from within. This distinction is not as pronounced in post-exilic Hebrew, but nonetheless still apparent. Part of what can make it difficult to grasp for English speakers is not only the evolution of biblical Hebrew, however, wherein figurative language gradually weakens the force of the original distinction, but also the assumption throughout the Old Testament that strong emotions can be understood, like madness, as more closely analogous to wholly external forces such as the weather or divine providence than to actions such as speech or internal deliberation that we today tend to see as more susceptible to our own individual control. As H. Wheeler Robinson explains, the term *ruach* 'was originally applied both to the "blowing" of the wind and to the "blowing" or panting of men and animals in distress or excitement'. Thus, it became by extension 'strong passion (anger, zeal, impatience)'.[32]

In his sermon, Donne initially translates *neshamah* in contrast to *ruach* as 'the *soule*, the immortal soule' (4:256). This translation, however, is questionable. Notably, Donne departs here from standard contemporary practices of English translation. In the Geneva Bible, the Bishops' Bible, and the King James Bible, the word used to translate *ruach* is indeed 'spirit', in keeping with the Septuagint, which uses πνεῦμα (*pneuma*; *cp.* Latin, *spiritus*). The English word 'soul', however, is used to translate a different Hebrew term, נֶפֶשׁ (*nepesh*), again in keeping with the Septuagint, which uses a different Greek term, ψυχή (*psychē*; *cp.* Latin, *anima*).

In the Geneva Bible, the Bishops' Bible, and the King James Bible, as well as Patristic commentary, the Hebrew term *neshamah* is typically translated not as 'soul' but as 'breath'. In his commentary on Genesis 2:7 in his *City of God*, St Augustine, in particular, takes pains to emphasize that in the Septuagint, *neshamah* is translated as πνοή (*pnoē; cp.* Latin, *spiraculum*, in the Vulgate; *flatum*, in Tremellius) rather than *pneuma*.[33] Donne himself eventually suggests that *neshamah* might be better translated as 'temporal life,' rather than 'soul' (4:256). The closer equivalent in Hebrew, however, of this concept, 'temporal life', would be *nepesh*.

Whereas the primitive, concrete, site-specific definition of *ruach* is 'wind', and *neshamah*, 'breath', the original meaning of *nepesh* is 'throat' or perhaps 'gullet'. As Paul MacDonald explains, 'A further meaning is 'desire or longing', that is, to satisfy the need for food, drink, or air taken in through the throat'.[34] By extension, over time, *nepesh* comes to mean something like life itself. As Horst Seebass puts it, *nepesh* is not 'simply 'life', however, 'but rather the individuation of life'.[35] MacDonald clarifies the distinction: 'An individual does not *have* a *nepesh* in the sense of a separate or separable possession; rather, an individual *is* a *nepesh*'.[36] This soul or self, moreover, seems to be seen in the Old Testament as inseparable from embodiment; especially, breathing.[37] *Nepesh* is like what Aristotle means by *psychē*: a form that can be separated from matter only in concept.

In his commentary on Genesis against the Manichaeans, St Augustine compares human beings to mud. 'Just as water,' he explains, 'collects earth and sticks and holds it together when mud is made by mixing it in, so too the soul by animating the material of the body shapes it into a harmonious unity, and does not permit it to fall apart into its constituent elements'.[38] In his second sermon on Genesis 1:26, Donne cites and agrees with this commentary of St Augustine's and offers a further elaboration, drawn from the *Thesaurus* of St Cyril of Alexandria. As the soul is the form of the body, so also the image of God is the form of the human soul. 'This Image is in our soule, as our soule is the wax, and this Image is the seale' (9:79).[39]

St Cyril's choice of metaphor recalls, not coincidentally, Aristotle's in his *De anima* ('On the soul'), where he dismisses the possibility that the soul could be separated from the body: 'one need no more ask whether body and soul are one than whether the wax and the impression it receives are one'.[40] In addition to Aristotle's more familiar categories of form and matter, and in keeping with the elaborate hierarchies of form characteristic of Neo-Platonism, St Cyril's adaptation of this conceit introduces, in effect, a further layer of existence, the image of God, which stands in relation to form as form does to matter.[41]

Turning back to Donne's 1622 Gunpowder Day sermon, an analogous sense of simultaneous differentiation and entanglement seems to inform Donne's description of the relation between spirit (*ruach*) and breath (*neshamah*). As a lower category in the 'chain of being', so to speak, *neshamah* is presented as inseparable from *ruach,* albeit distinct. Donne not only mistranslates *neshamah* as 'soul', however, rather than as 'breath', but also equivocates on its relation to *ruach*. Are these two concepts different or the same?

Donne begins his exposition of the phrase 'the breath of our nostrils' by comparing King James to *ruach* in the sense of the Holy Spirit, that is, God himself, the Third Person of the Holy Trinity. As the sermon goes on, however, Donne backs away from this flattering gloss, probably because he knows it is tenuous at best and at worst a flagrant departure from longstanding interpretive consensus. 'We carry not this word *Ruach*, Spirit, so high'. Instead, Donne proposes, he will 'take the word literally, as it is in the text [*sc.* in Lamentations 4:20]; *Ruach, spirit,* is *the Breath* that *we breathe,* the *Life* that *we live*'. 'The King', Donne maintains, 'is that *Breath*, that *Life*, and therefore that belongs to him' (4:252).

Here at last we arrive at the core of Donne's argument, which he takes up thereafter as a preacherly refrain. King James, he insists, 'is our Breath; *our Breath is his*' (4:253). As in the case of contemporary doctrines of predestination, the logic here remains, however, somewhat elusive. Donne's case rests, in effect, on a conflation of what would normally be considered two very different types of possession. Is my breath the king's in the sense that his own breath is, or his hand, or his foot? If so, then what need to exhort me to obey him? Or, contrariwise, if I am, by contrast, capable of independent agency, in what sense am I his?

Donne's argument is in keeping, nonetheless, with a surprisingly sophisticated sense of the relation in the Old Testament between *ruach* and *neshamah,* as well as *nepesh.* As Paul MacDonald explains, whereas *neshamah* means 'the breath of an animal being', *ruach* refers to 'the power or force behind the wind or breath'. 'No human can control the wind; only the divine power of Yahweh can do so; thus, by analogy, it is the breath of God which instills or *inspires* the life-force in living things, though *this* breath (not the animal's breathing) is only observable through its effects'.[42]

Walter Eichrodt makes a similar observation as regards the relation between *ruach* and *nepesh.* 'If *nepesh* is the individual life in association with a body, *ruach* is the life-force present everywhere and existing independently against the single individual. We might say that the same force is considered from different points of view'. Whereas *nepesh* is 'individual', however, 'and comes to an end with the death of the individual,' '*ruach* is universal, and independent of the death of the creature; it does not die'.[43]

Was Donne aware of the resonance between his argument and Hebrew assumptions about the relation between 'breath' (*neshamah*) and 'spirit' (*ruach*)? The connection is more likely to have been indirect: an effect of the influence of this connection on the Christian concept of the Holy Spirit. Donne's former naval commander, Sir Walter Raleigh, for example, tackles some of the same concerns in the opening to his *History of the World.* As a young man, Donne travelled with Raleigh to the Azores on an expedition against the Spanish, and it is likely that he would have been familiar with this work, Raleigh's *magnum opus.* Here, after citing a wide range of Christian authorities, Raleigh defines 'the Spirit of God' as 'the infinite active power of God, which then formed and distinguished, and which now sustaineth and giveth continuance to the Universal'.[44]

In any case, howsoever derived, whether directly or indirectly, in his 1622 Gunpowder Day sermon, Donne's core conceit turns out to correspond to a well-informed Jewish sense of the relation between the divine as 'spirit' and the human as 'breath'. As he repeats several times, the king is our breath, and our breath is his. 'First our *Breath*, that is, *sermo*, our *speech* belongs to him' (4:252–53). In fact, the Latin word *sermo* neither itself means 'breath' nor derives from any older word for breath, but instead from the proto-Indo-European root *ser-, meaning approximately to line up, bind, or join together.[45] Donne's angle in this first example, however, is not so much etymological as topical. By re-imagining sermons as *sermo*, 'speech', and speech as a form of breath, he is able to connect his chosen text more readily to his commission: a defence of James's *Directions concerning Preachers*.

Following the same through-line of argument, Donne then turns to various other possible definitions of 'breath'. 'Princes' in general 'are our breath'; '*Our breath is theirs*', and 'it is theirs especially, *in our prayers*'. Breath is not only 'speech', but '*Breath is life too,* and so *our life is his*'. Our lives are the King's, not only to die for him, if need be, but also to 'live for him', in the sense of living '*peaceably*', '*honestly*', and '*industriously*' (4:253–55). Donne then stops short, however. Our bodies may indeed be subject to the Crown. But what about our souls? Here Donne stands on more precarious footing. As Shakespeare's Henry V says, giving voice to a pervasive belief, 'Every subject's duty is the king's; but every subject's soul is his own'.[46]

As if for support at this most delicate juncture, Donne now turns at last to Genesis 2:7. 'When the breath of life was first breathed into man, there it is called by another word, *Neshamah,* and that is the *soule*, the immortall soule'. 'Is the King the breath of that life?' Donne asks. That is to say, 'Is he the soule of his subjects so, as that their *soules* are *his*; so, that as they must sinne *towards men,* in doing unjust actions, or sinne *towards God*, in forsaking, and dishonouring him, if the King will have them?'. James himself, Donne assures his audience, 'would be the first man, that would say, *No, No*; your souls are not mine, so' (4:256). In more straightforward, abstract language, what Donne asserts here is that King James does not claim that his right to demand obedience from his subjects supersedes their more fundamental obligation to revere and worship God, as well as their responsibility to preserve justice in their actions towards each other.

Once again, however, Donne introduces a distinction only to collapse it forthwith. 'He claimes not your souls so', he insists, 'It is *Ruach* here, it is not *Neshamah*; your life is his, your soule is not his, in that sense'. 'But yet, beloved', he goes on,

> these two words are promiscuously used in the Scriptures; *Ruach* is often the soule; *Neshamah* is often the temporall life; And thus farre, the one, as well as the other, is the Kings, That hee must answer for your soules; so they are his; for hee is not a King of bodies, but a King of men, bodies and

> soules; nor a King of men onely, but of Christian men; so your Religion, so your soules are his; his, that is, appertaining to his care, and his account. (4.256)

Donne's conclusion here builds upon the peculiar nature of his argumentation up until this point. Again and again, Donne invokes a notorious scholarly controversy only to dismiss it as unimportant: whether Lamentations is prophecy or history; whether Lamentations 4:20 refers to Josiah or Zedekiah; whether *ruach* means anything substantively different from *neshamah*. These abstruse, inconclusive inquiries lay the groundwork for his analogous response to the burning political question of the moment: whether James's subjects should obey him with regards to religion.

> Therefore, though you owe no obedience to any power under heaven, so as to decline you from the true God, or the true worship of that God, and the fundamentall things thereof, yet in those things, which are, in their nature but circumstantiall, and may therefore, according to times, and places, and persons, admit alterations, in those things, though they bee things appertaining to Religion, submit your selves to his directions; for here, the two words meet, *Ruach*, and *Neshamah*, your lives are his, and your souls are his too; His end being to advance Gods truth, he is to be trusted much, in matters of indifferent nature, by the way. (4:256–57)

In closing, Donne's solution, once again, is to try to minimize the significance of the division in question, drawing this time more explicitly, if only in translation, on the Stoic notion of ἀδιάφορα (*adiaphora*; Greek, 'things indifferent', i.e., ethically irrelevant).[47]

Donne is not simply apologizing for King James, however, by whatever means he can. Instead, his argumentation in this, his most high-profile performance, can be understood as of a piece with his perspective throughout his sermons. As Jeanne Shami observes, Donne consistently rejects 'the rhetorical style of either-or, of separation and division'.[48] In a 1620 sermon on Matthew 18:17, for example, 'Woe unto the world because of offences', Donne urges his audience 'to detest' the polarizing assumption 'that there is nothing done, if all bee not done; that no abuses are corrected, if all be not removed; that there's an end of all Protestants, if any Papists bee left in the world'. Instead, Donne proposes, 'reduce all to the precious mediocrity' (3:182–83).

Contesting Donne's supposed lack of principle, Joshua Scodel proposes, like Jeanne Shami, that Donne can be better understood as deliberately and consistently pursuing the *via media* or 'middle course' that since Elizabeth I has come to be considered a defining feature of Anglicanism, even against the current of his historical moment.[49] In his first sermon preached at St Paul's, for example, Donne proclaims straightaway, 'we decline wranglings, that tend not to edification' (3:365). In his sermon on Matthew 18:17, Donne unabashedly begs his countrymen not to be 'so ombragious, so startling, so apprehensive, so suspicious'. 'Leaving fundamentall things, and necessary truths', he laments, 'we wrangle uncharitably about Collaterall impertinencies' (3:175). Here in his 1622 Gunpowder Day sermon, Donne's dazzling, sometimes dubious argumentation as regards Josiah and Zedekiah, as well as *ruach* and *neshamah*, can be understood by this light as neither unprincipled flattery nor unchecked absolutism, but instead as a paradigmatic example of his lifelong aversion, like Erasmus before him, to rarefied theological polemic as a 'thing indifferent'. For Donne, power politics is not a temptation, but instead an unwelcome distraction from the very different demands of Christian ethics.

Notes

1. Thomas Cogswell, *The Blessed Revolution: English Politics and the Coming of War, 1621–1624* (Cambridge: Cambridge University Press, 1991), 52.
2. Richard Serjeantson, 'Preaching Regicide in Jacobean England: John Knight and David Pareus', *The English Historical Review* 134/568 (2019), 553–88.
3. King James VI and I, *Directions Concerning Preachers*, in *King James VI and I: Selected Writings*, ed. by Neil Rhodes, Jennifer Richards, and Joseph Marshall (New York: Routledge, 2016), 379–84, at 383.
4. Dennis Flynn, 'Donne's Politics, "Desperate Ambition", and Meeting Paolo Sarpi in Venice', *The Journal of English and Germanic Philology* 99/3 (2000), 334–55.
5. John Chamberlain, *The Letters of John Chamberlain*, ed. by Norman Egbert McClure (Philadelphia: The American Philosophical Society, 1939), 2:451. On the reception of this first sermon on James's *Directions*, see Annabel Patterson, *Censorship and Interpretation: The Conditions of Reading and Writing in Early Modern England* (Madison, WI: University of Wisconsin Press, 1984), 98–100; Jeanne Shami, *John Donne and Conformity in Crisis in the Late Jacobean Pulpit* (Cambridge: D. S. Brewer, 2003), 110–15; and Brian Cummings, *The Literary Culture of the Reformation: Grammar and Grace* (Oxford: Oxford University Press, 2002), 390–92.

6. Recounted by the Earl of Carlisle in a congratulatory letter to Donne; quoted in R. C. Bald, *John Donne: A Life*, ed. by W. Milgate (Oxford: Clarendon Press, 1970), 435.
7. British Library MS Royal 17.B.XX; reprinted in *John Donne's 1622 Gunpowder Plot Sermon: A Parallel-Text Edition*, ed. by Jeanne Shami (Pittsburgh, PA: Duquesne University Press, 1996).
8. Jeanne Shami, *Conformity in Crisis*, 204, 138.
9. Izaak Walton, 'The Life of Dr. John Donne', in Walton, *The Lives of Dr. John Donne* [et al.] (London: Thomas Newcomb, 1670), 12–81.
10. Bald, *Life of Donne*, and John Carey, *John Donne: Life, Mind, and Art* (London: Faber, 1981).
11. Jonathan Goldberg, *James I and the Politics of Literature: Jonson, Shakespeare, Donne, and Their Contemporaries* (Baltimore, MD: Johns Hopkins University Press, 1983). *Cf*. Debora K. Shuger, *Habits of Thought in the English Renaissance: Religion, Politics, and the Dominant Culture* (Berkeley, CA: University of California Press, 1990).
12. David Norbrook, 'The Monarchy of Wit and the Republic of Letters: Donne's Politics', in *Soliciting Interpretation: Literary Theory and Seventeenth-Century English Poetry*, ed. by Elizabeth D. Harvey and Katharine Eisaman Maus (Chicago: University of Chicago Press, 1990); and Annabel Patterson, '*Quod oportet versus quod convenit*: John Donne, Kingsman?' in Patterson, *Reading Between the Lines* (Madison, WI: University of Wisconsin Press, 1993), 157–206.
13. Flynn, 'Donne's Politics', 339.
14. Annabel Patterson, 'All Donne', in *Soliciting Interpretation*, 37–67.
15. Richard Strier, 'Donne and the Politics of Devotion', 94, in *Religion, Literature, Politics in Post-Reformation England, 1540–1688*, ed. by Donna B. Hamilton and Richard Strier (Cambridge: Cambridge University Press, 1996), 93–114.
16. Achsah Guibbory, 'Donne's Religion: Montagu, Arminianism, and Donne's Sermons, 1624–1630', *English Literary Renaissance* 31/3 (2001), 412–39, at 413.
17. Norbrook, 'Monarchy of Wit', 22; Strier, 'Politics of Devotion', 94.
18. Arthur Marotti, 'John Donne's Conflicted Anti-Catholicism', *Journal of English and German Philology*, 101/3 (2002), 358–79, at 369; *cf*. Patterson, *Censorship and Interpretation*, 98–100.
19. John Donne, 'The Lamentations of Jeremy', in *John Donne: The Divine Poems*, ed. by Helen Gardner (Oxford: Oxford University Press, 1952), 35–48, at 46.
20. Graham Roebuck, 'Donne's *Lamentations of Jeremy* Reconsidered', *John Donne Journal* 10 (1991), 37–44, at 38.
21. John Klause, 'The Two Occasions of Donne's *Lamentations of Jeremy*', *Modern Philology* 90/3 (1993), 337–59, at 345.

22. John Donne, 'A Sermon upon the Fifth of November 1622: Being the Anniversary Celebration of our Deliverance from the Powder Treason', in *The Sermons of John Donne*, ed. by G. R. Potter and E. M. Simpson, 10 vols (Berkeley: University of California Press, 1953–1962), 4:235–63, at 239–40.
23. Graeme Murdock, 'The Importance of Being Josiah: An Image of Calvinist Identity', *The Sixteenth Century Journal* 29/4 (1998), 1043–59, at 1048.
24. Murdock, 'Josiah', 1047–51.
25. Article 24, citation 5, on 1 Peter 2:14; quoted in Murdock, 'Josiah', 1048.
26. *Cf*. St Jerome, *Commentary on Zechariah*, 12:11–14.
27. Calvin, 'Preface' to *Commentary on Lamentations*, in *Calvin's Commentaries*, 22 vols (Grand Rapids: Baker Books, 2005), vol. 11, *Lamentations*, trans. by John Owen.
28. Calvin, *Commentary on Lamentations*, 4:20.
29. The source that Donne, like Calvin, calls the 'Chaldee paraphrast' is now known as the Targum Onkelos. *Cf*. Sir Walter Raleigh on Genesis 1:2: 'Some of the Hebrews convert it to this effect, *Spiritus Dei volitabat*: the Spirit of God did flutter; the Chaldaean Paraphrast in this sense, *Ventus à conspectus Dei sufflabat*, or, as others understand the Chaldaean, *Flabat*'. Quoted from Raleigh, *The History of the World in Five Books*, 1.1.6 (London: Thomas Basset, et al., 1687).
30. Anthony Raspa and Judith Scherer Herz, 'Response', *Renaissance and Reformation* 20 (1996), 97–98.
31. Chanita Goodblatt, *The Christian Hebraism of John Donne: Written with the Fingers of Man's Hand* (Pittsburgh: Duquesne University Press, 2010), 22.
32. H. Wheeler Robinson, *The Christian Doctrine of Man* (Edinburgh: T. & T. Clark, 1913), 17.
33. Augustine, *The City of God*, 8.24.
34. Paul S. MacDonald, *History of the Concept of Mind: Speculations About Soul, Mind and Spirit from Homer to Hume* (Aldershot: Ashgate, 2003), 2.
35. Horst Seebass, '*nepesh*', in *Theological Dictionary of the Old Testament*, ed. by G. J. Botterweck and H. Ringgren, 15 vols (Grand Rapids, MI: Eerdmans, 1974–90), 9:498–519, at 9:510.
36. MacDonald, *Concept of Mind*, 6.
37. For a thorough review of the scholarly consensus that the Hebrew concept of the self is monist, as well as an argument that it might sometimes nonetheless be considered dualist, see Richard Pleijel, 'To Be or to Have a *nephesh*?' *Zeitschrift für die alttestamentliche Wissenschaft* 131/2 (2019), 194–206.

38. Augustine, *On Genesis Against the Manichees*, 2.7.9; quoted in Gerald P. Boersma, *Augustine's Early Theology of the Image: A Study in the Development of Pro-Nicene Theology* (Oxford: Oxford University Press, 2016), 213.
39. For further discussion of variations on this conceit in Donne's poetry and prose, see James Jaehoon Lee, 'John Donne and the Textuality of the Two Souls', *Studies in Philology* 113/4 (2016), 879–918. For a contemporary review of commentary on the *imago Dei*, see Raleigh, *History*, 1.2.1.
40. Aristotle, *On the Soul*, trans. by W. S. Hett (Cambridge, MA: Harvard University Press for the Loeb Classical Library, 1957), 412b.
41. Cyril of Alexandria, *Thesaurus* 34 and *In Joannem* 11; quoted in Walter J. Burghardt, *The Image of God in Man According to Cyril of Alexandria* (Washington, DC: Catholic University of America Press, 1957), 72.
42. MacDonald, *Concept of Mind*, 7.
43. Walther Eichrodt, 'The "soul" (*nepeš*)', in Eichrodt, *Theology of the Old Testament*, 2 vols (London: SCM Press, 1967), 2:131–45, at 136.
44. Raleigh, *History*, 1.1.6.
45. August Schleicher, *A Compendium of the Comparative Grammar of the Indo-European, Sanskrit, Greek, and Latin Languages*, trans. by Herbert Bendall, 2 vols (Cambridge: Cambridge University Press, 1877), 2:206.
46. William Shakespeare, *Henry V*, ed. by T. W. Craik (London: Routledge, 1995), 4.1.175–77.
47. On ἀδιάφορα in the Epistles of St Paul, see James L. Jaquette, 'Life and Death, "*Adiaphora*", and Paul's Rhetorical Strategies', *Novum Testamentum* 38/1 (1996), 30–54. On *adiaphora* in Calvin's *Institutes*, see Edward F. Meylan, 'The Stoic Doctrine of Indifferent Things and the Conception of Christian Liberty in Calvin's *Institutio religionis christianae*', *Romanic Review* 28/2 (1937), 135–45. On the concept of 'things indifferent' as a recurring influence in medieval and early modern Christian intellectual history, see G. R. Evans, '*Sancta Indifferentia* and *Adiaphora*: "Holy Indifference" and "Things Indifferent"', *Common Knowledge* 15/1 (2009), 23–38.
48. Shami, *Conformity in Crisis*, 80.
49. See, e.g., Gale H. Carrithers, Jr., *Donne at Sermons: A Christian Existential World* (Albany, NY: State University of New York Press, 1972), as well as Joshua Scodel, 'John Donne and the Religious Politics of the Mean', in *John Donne's Religious Imagination: Essays in Honor of John T. Shawcross*, ed. by Raymond-Jean Frontain and Frances M. Malpezzi (Conway: University of Central Arkansas Press, 1995), 45–80.

Bibliography

Boersma, Gerald P. 2016. *Augustine's Early Theology of the Image: A Study in the Development of Pro-Nicene Theology*. Oxford: Oxford University Press.

Burghardt, Walter J. 1957. *The Image of God in Man according to Cyril of Alexandria*. Washington, DC: Catholic University of America Press.

Donne, John. 1996. *John Donne's 1622 Gunpowder Plot Sermon: A Parallel-Text Edition*. Ed. Jeanne Shami. Pittsburgh, PA: Duquesne University Press.

MacDonald, Paul S. 2003. *History of the Concept of Mind: Speculations About Soul, Mind and Spirit from Homer to Hume*. Aldershot: Ashgate.

Mix, Lucas John. 2018. *Life Concepts from Aristotle to Darwin: On Vegetable Souls*. Cham: Springer for Palgrave Macmillan.

Pleijelm, Richard. 2019. To Be or to Have a *Nephesh*? *Zeitschrift für die alttestamentliche Wissenschaft* 131/2: 194–206.

Richardson, W. F. 1976. Pneuma in the Septuagint. *Prudentia* 8/2: 99–114.

Shami, Jeanne. 2003. *John Donne and Conformity in Crisis in the Late Jacobean Pulpit*. Cambridge: D. S. Brewer.

van Kooten, George H. 2008. *Paul's Anthropology in Context: The Image of God, Assimilation to God, and Tripartite Man in Ancient Judaism, Ancient Philosophy and Early Christianity*. Tübingen: Mohr Siebeck.

van Ruiten, Jacques T. A. G. M., and George H. van Kooten, eds 2016. *Dust of the Ground and Breath of Life (Gen. 2:7): The Problem of a Dualistic Anthropology in Early Judaism and Christianity*. Leiden: Brill.

PART IV

The Eighteenth Century

CHAPTER 11

Breathscapes: Natural Environments in Eighteenth-Century Physiology and Psychosomatics of Breathing

Rina Knoeff

> The common air, that element in which we live, that invisible fluid which surrounds the whole earth [...] never engaged so much the attention of the learned as it has of late years. This fluid, diffused everywhere, the *breath of life*, deserves so much the more attention and investigation of philosophers, as it is the only substance without which we can scarce subsist alive a single moment, and whose good or bad qualities have the greatest influence upon our constitution.[1]

Thus begins the preface of *Experiments upon vegetables, discovering their great power of purifying the common air in the sun-shine, and of injuring it in the shade and at night* (1779), written by Dutch physician Jan Ingenhousz (1730–1799). Building on Stephen Hales' experiments with pneumatic troughs and Joseph Priestley's speculations that plants have the ability to improve rotten air, Ingenhousz (who has been generally forgotten by historians of science) discovered that sunlight is an

R. Knoeff (✉)
History Department, University of Groningen, Groningen, The Netherlands

D. Fuller et al. (eds.), *The Life of Breath in Literature, Culture and Medicine*, Palgrave Studies in Literature, Science and Medicine,
https://doi.org/10.1007/978-3-030-74443-4_11

essential ingredient of photosynthesis.[2] As a result of numerous experiments he argued that during the night plants breathe just like animals and so putrefy the air. In daylight, however, they restore the quality of air again. For this reason, Ingenhousz argued, the excretory ducts of plants and trees are placed at the underside of leaves, so dephlogisticated air (oxygen), which is heavier than common air, falls downwards 'as a beneficial shower for the use of the animals'.[3]

Following Ingenhousz' experiments (and those of Hales and Priestley), other eighteenth-century physicians also busied themselves with experiments and speculations on photosynthesis. Towards the end of the eighteenth century it was generally agreed that beneficial air rained down during the day, but that the breath of vegetation was extremely damaging at night. This led to a growing (medical) awareness of the (dis)advantages of plants and trees in houses and urban areas. For instance, Ingenhousz carefully considered the benefits of indoor plants and he argued that the practice of keeping orange trees in a room, 'by way of ornament and to keep the air of the room wholesome', could 'contribute somewhat to purify the air', as long as they were placed close to a window where they could profit maximally from daylight. Yet he strongly advised taking out the plants by night as plants 'absolutely tend to foul the air, principally when they are in flower'.[4] In a similar vein, the Dutch physicians Rudolph Deiman (1743–1808) and Adriaan Paets van Troostwyk (1752–1837) argued that 'plants certainly are very useful in stuffy rooms', which made them wonder how nature, during the winter months, when the trees and lands were bare, improved the air.[5] Willem van Barneveld (1747–1826) and Joachim Muller (dates unknown), a Utrecht apothecary and an Amsterdam gentleman, concluded after many experiments on different plants that the plant *Sedum Telephium* (sometimes appropriately known as 'live forever') could restore great quantities of air and was therefore suitable to place at the windows of almshouses, hospitals, and sickrooms.[6] Ingenhousz' discoveries were soon known throughout Europe and it became generally known that 'the leaves of plants growing in light purify or renovate the air, by supplying it with fresh oxygen'.[7]

With respect to the improvement of air through plants and trees, and the subsequent step to include 'nature' in urban spaces, another important question was how many trees should be planted and what the distance should be between them. It was generally acknowledged that trees were necessary for purifying the air, but on account of their foul breath at night it was said that they should not be placed too close to bedroom windows. Against the planting of too many trees, it was also argued that autumn leaves on the streets and in the canals would breathe extra unhealthy vapours. Moreover, the presence of too many trees in narrow streets would hinder the winds necessary for blowing away harmful particles in the air. Not surprisingly, an article in the Dutch journal *Vaderlandsche Letteroefeningen* in 1810 argued that the Dutch and English had been wrong in planting too many trees in close vicinity to their mansions: it is better, the article stated, 'to plant small forests some distance from the house, they break the strength of the winds, give shadow, lessen humidity and increase the purity of the air'.[8]

Although the photosynthesis experiments of Ingenhousz and his circle have been largely forgotten in the wake of Lavoisier's (chemical) discoveries concerning the role of oxygen and hydrogen in combustion, they were an important part of late eighteenth-century medical thinking about the importance of natural environments and clean air for respiratory physiology. From Antiquity onwards, air, the first of the six Hippocratic non-naturals, referred not only to the air we breathe, but also to the natural environments that produce (breathe) and determine the quality of air. During the 1760s, it was discovered that air was an amalgam of different kinds of air. This led to the revival of an old (Hippocratic) discussion of whether bad air (i.e., bad natural environments) can cause contagious diseases. Moreover, this discussion added to Ingenhousz' earlier notion that plants imbibe and release 'the breath of life'. As a result (natural) environments became an essential part of respiratory theories.

This essay is concerned with the question of how the medicine of respiration is bound up with natural environments, not only in a literal sense, concerned with the quality of air, but also in the layout of parks in urban spaces and the material culture of households, in particular in the design of chimneys and ventilators, as well as in the widespread presence of landscape paintings in living rooms. With respect to the latter, it was thought that the viewing of spacious landscapes would stimulate the imagination and literally give breath. The inclusion of natural environments in medical ideas and advice about breathing was facilitated by a changing focus in eighteenth-century physiologies of respiration. In the wake of new (chemical) theories about the nature of air, experiments in photosynthesis, and speculations concerning oxygen and combustion, the focus of respiratory medicine changed from *anatomy* and an emphasis on the *organs of respiration* to the *physiology and pathology of breathing* as a process that was as much linked to individual organs as it was to the mind and (natural) environments.[9] The effects of this change were considerable. The Scottish physician Robert Menzies argued that 'physicians were, till very lately, totally ignorant of the effects of respiration on the system: Modern chemistry, by discovering the component principles of atmospheric air [...] induced philosophers to examine the subject with the degree of attention it is merited'.[10] The new focus on respiratory physiology, in short, turned breathing into one of the big topics of eighteenth-century physiology.

Respiration in Eighteenth-Century Medicine

The changing physiology of respiration can be very well observed in the academic lectures of the famous medical teacher Herman Boerhaave (1668–1738) and the pathology of Jerome Gaub (1705–1780), Boerhaave's pupil and follower in the chairs of medicine and chemistry at Leiden University.[11] Boerhaave linked breathing to the mechanical motion of the circulation of the blood and emphasized that the quality of breathing is indicative of the health of the underlying organs. Gaub, on the contrary, considered the physiology and pathology of respiration as an autonomous process that was to a certain extent independent of the organs of breathing.

Boerhaave discussed respiration in Book IV of his *Academical Lectures* 'containing the economy of the external and internal senses, sleep and respiration' and again in Book VI on the 'signs of health'. He argued that 'what respiration is, or how it is performed, and why it continues without the influence of the mind' is difficult to understand. He concluded that it can best be studied by considering the 'infinite number of organs' involved in the process.[12] In Boerhaave's mechanical teaching, respiration, that is, the inflation and deflation of the lungs and thorax (which also affects the movement of the abdomen), facilitated by the elasticity of air, assists the heart and blood vessels in propelling the blood forward. Boerhaave proved his point with reference to a surgical procedure: when a vein has been opened, if a patient is ordered to cough, the blood 'instantly flies forth in a full stream [...] therefore coughing must evidently accelerate the motion of the blood'.[13] As a result, Boerhaave's medicine envisioned a relation between the contraction of the heart and the number of breaths. The anatomy of organs, muscles, and bones involved was a key concern. The nature and quality of air were only important in relation to its weight and pressure on the lungs. So Boerhaave argued that light air did not give enough pressure, so the resulting constriction of the lungs would suffocate asthmatic patients; moist air was not elastic enough; and cold air was too heavy and squeezed the blood vessels.

In other words, in Boerhaave's system respiration was an important sign (together with the pulse and urine), indicating the (un)healthy functioning of internal organs. He formulated an extensive semiotics related to respiration, arguing that 'there are hardly ten people in a hundred having the same respiration'. A painful respiration denoted an internal inflammation; a small or short respiration signalled that the lungs or windpipe were full of blood or mucus; a quick respiration referred to injured organs of breathing; a suffocating respiration was a sign of inflamed lungs or asthma. By contrast, Boerhaave considered 'an easy, large, slow, equal, and refreshing respiration, performed only by a gentle motion of the intercostal and abdominal muscles, with the diaphragm, the most healthy'. According to Boerhaave even Galen had admitted that no one ever yielded to any disease while manifesting these signs.[14] In short, for Boerhaave difficult breathing was mainly related to the heart and blood circulation and almost certainly signalled organic trouble. He argued that it 'is therefore one of the worst presages'. Boerhaave nowhere considered respiration as an independent physiological process that depends as much on the condition of organs as it does on passions and emotions and the quality of the surrounding air.

During the second half of the eighteenth century, Boerhaave's ideas on respiration came under attack. Robert Harrington (1751–1837) blandly stated that 'Boerhaave is demonstrably wrong'.[15] London physician Hugh Smith (d. 1790) rejected Boerhaave's ideas, calling them absurd, and rather than pointing to the organs of respiration, he argued that the noxious effects of impure air infused in the blood were devastating (in particular in the case of pox and other epidemic diseases).[16] Most criticism was the result of a general shift away from the organs of respiration towards the nature and quality of air in animal and 'vegetable' breathing. As a result, breathing was no longer solely seen as a *sign* pointing to the 'true' condition of underlying organs. Instead, breathing was increasingly considered a *symptom* of respiration which could become a distemper itself.[17]

An excellent example of this shift can be observed in the work of Jerome Gaub. His *Institutiones pathologiae medicinalis* (1758) was widely read throughout Europe. The Edinburgh professor of medicine William Cullen (1710–1790) recommended Gaub's work as 'the best system of pathology'. Rudolph Virchow still used Gaub's pathology as a medical student in Berlin in the 1830s and later called it 'the world's first textbook of general pathology and the standard until far into the nineteenth century'.[18]

Gaub placed the symptoms of respiration among the symptoms of the vital motions, together with the symptoms of the heart's motion, the pulsation of the arteries, and the natural functions. He defined a symptom as a 'visible alienation from the natural state, which arises in the body from disease'. It is visible to the senses of both the patient and doctor and should not be confused with the disease itself. The latter is important: Gaub insisted that symptoms 'arise in the body from disease, but *in such a manner as to be considered distinct from it*.'[19] This means that although symptoms are often caused by the disease, they just as often originate in the cause of the disease or from other symptoms. As a result, Gaub's list of the numerous and various injuries to which respiration is liable is long and diverse, and includes causes as different as the quality of air, troubles with blood, and mental disturbances leading to anxiety. The kinds and degrees of breathing difficulties are even more numerous: 'shortness of breath, asthma, short, creaking, loud, and deep respiration, strangulation &c.'[20] In addition, Gaub argued that symptoms of breathing are often morbid as well as salutary. Thus coughing and sneezing can be symptoms of a disease, but these violent movements also remove noxious particles from the nose, mouth, and lungs. Morbid laughing is similarly caused by a tickling of the diaphragm, but can be uplifting at the same time; sighing is caused by grief, but eases the anxiety that goes with it; yawning has the salutary purpose of waking up drowsy people.

Additionally, Gaub argued that symptoms might arise from the mind (in particular the power of excitement and the urge to avoid pain) or the 'circumambient air [...] things ingested or applied externally'.[21] In fact, Gaub considered respiration the most important process mediating between the body's interior and external surroundings. All this means that in Gaub's rational pathology symptoms do not have a one-on-one connection to disease, but instead must be considered individually and to a certain extent as independent from the body and its organs. In viewing respiratory trouble as a symptom, rather than a sign as Boerhaave had done, Gaub signalled that troubled breathing could be considered a medical problem in itself, rather than merely as a sign indicative of an organic problem. Moreover, the idea that breathing has its own dynamics opened up the possibility of including natural environments, the nature of air, and the 'government of the mind' in the physiology of respiration. So, in addition to obstructions in the respiratory organs (inflammations, pustules, ulcers, tubercles, congestions of blood, pain, etc.), Gaub included 'the air offending by heat, humidity, gravity, levity, stagnation and corruptions' as well as

> celestial bodies, meteors, climates, seasons of the year, diversity of soils, mountains, seas, lakes, marshes, rivers, vegetables, animals, subterraneous caverns, the number of inhabitants of a place, manner of life, aliments, drink, fuel of the fire, employments, arts, commerce, and many other things, [which] all contribute their share.[22]

In other words, surrounding atmospheres were made an integral part of the physiology of respiration and as such were widely discussed in terms of health regimens, medical treatment, and climate control.[23]

As a result, atmospheric health not only stimulated chemico-medical research, but also asked for technological innovations and interventions in natural environments. The building of chimneys and fireplaces was reconsidered as they were thought to be very unhealthy: 'some of them very fatal in their consequences to health' which 'cost the lives of thousands every year in this island' [of Ireland].[24] Particularly widespread were Stephen Hales' ventilators which significantly improved the air in enclosed spaces such as prisons, workhouses, ships, hospitals, and mines.[25] One of the best-known inventions was the eudiometer, an instrument to measure the breathability of air. The instrument dovetailed with worries about climatic susceptibility and the idea that aerial pathologies were indicative of the contemporary moral decline into laxity and effeminacy. As a result, eudiometry became a key element in a general programme of economic and social reform.[26] Apparently, the British believed they were particularly susceptible to the prevailing dullness and dampness of the air, which had a depressing effect on the spirits of the population.[27] In the Netherlands, medical writers advocated drying the marshes in order to improve the air and by extension the (moral) character of the Dutch.[28]

Outdoor exercise was similarly promoted as a way to improve breathing.[29] John Sinclair wrote in his *Code of Health and Longevity* (1807) that 'the great object of training, is to obtain the benefit of a free respiration'.[30] As a result he advised regular exercise, which he appropriately labelled the 'breathing hour' of each day.

> A free and powerful respiration is most essential to a fresh colour of the face, to lively spirits, and cheerful feelings and to the healthy and vigorous actions of the body. "It is my breathing hour of the day," says Hamlet to Osric [...] there is little doubt, that if all those who linger away their hours in luxurious and indolent relaxations, were to assign a regular portion of their time, to the hardy and manly exercises of walking, riding, fencing, &c. and would take *their breathing hour*, they would breathe long and well.[31]

Walking and horseriding were favoured as exercises *par excellence*. They literally afforded 'various changes of the air', as well as changes of scenery which 'amuse the mind'.[32] Asthmatics and consumptives were particularly advised to undertake outdoor exercise in addition to 'talking out loud' in order to improve their breathing.[33] Exercise was best carried out in the morning, for the morning air 'braces and strengthens the nerves'. In addition, the bad air of towns must be avoided 'like the plague'.[34]

As a result of the increasing emphasis on 'taking the air', health committees and garden theorists advocated public parks in urban spaces 'for the health of rich and poor'. These parks were designed according to the latest theories about the air-improving qualities of vegetation. For instance, with reference to Ingenhousz, the Amsterdam Committee of Medical Supervision decided that trees should be carefully spaced 'in order that the earth's vapors beneath them would be substantially purified and dispersed in the atmosphere'.[35] Indeed representations of late eighteenth-century parks in Britain and on the Continent all depict people leisurely strolling below carefully spaced trees (Plate 11.1).

The parks provided the beneficial air needed for healthy respiration and by extension the vitality of the body. Not surprisingly, by the beginning of the nineteenth century there were fears that the English Royal Parks—what William Pitt (1708–1778) had allegedly called the 'lungs of London'—were in danger of being gobbled up by a rapid expansion of the surrounding buildings, seriously threatening the health of the population.[36]

By the end of the eighteenth century, 'taking the air' was considered so important that physicians generally warned that bad weather should not detain one from venturing outside. William Cadogan, author of the controversial thesis that most chronic diseases (most notably consumption and gout) are caused by lack of outdoor exercise, argued that 'in bad weather I can see no great evil in throwing a cloak round his [the

Plate 11.1 View of the Canal in St. James's Park, London. Basset, Paris, 1700–1799 (Courtesy of the Rijksmuseum Amsterdam)

patient's] shoulders and walking even in the rain; the only difficulty is to summon resolution enough to venture out'. Yet he also admitted, like many physicians in his day, that sometimes people were too weak or it was simply too cold and wet to venture outside. In these cases he advised the use of an indoor exercise machine.[37] Staying indoors significantly reduced the benefits of 'getting fresh air', but, as I argue below, decorations of idyllic and breathy landscapes in paintings and on wallpaper were meant to stimulate the impression of healthy air and give the viewer the illusion of being outside. The emerging eighteenth-century field of 'psychosomatics' further legitimized the representation of outside air in indoor settings in aid of a better respiration.[38]

The Psychosomatics of Indoor Breathing

Ever since ancient Hippocratic writings and the medical works of Galen, physicians and natural philosophers have acknowledged that troubled breathing and other bodily discomforts often originate in disturbances of the soul. As a result of the eighteenth-century revival of Hippocratic medicine, these concerns were high on the medical agenda again. According to the historian of medicine L. J. Rather, eighteenth-century doctors 'ascribed as much or more in the way of bodily change to the emotions or "power of imagination" than would all but the most convinced proponents of the psychological causation of diseases today'.[39] Consequently, treating the mind as well as the body became a serious medical concern. The French physician Pierre Jean Georges Cabanis (1757–1808) wrote in 1795, 'woe to the medical man, who has not learned to read the human heart as well as to recognize the febrile state', and Gaub argued that 'the careful management of the mind [...] is the concern of everyone and above all that of the physician'.[40] For Gaub, this was most definitely the case with respiratory trouble: 'respiration is indeed subject to the government of the mind, as it performs various functions, which are regulated by the will'.[41]

Respiration in relation to emotional balance was a prominent theme in Gaub's two influential orations on the regimen of the mind. Gaub delivered these orations as Vice Chancellor of Leiden University in 1749 and 1763. At the time they were well known for their clarity and elegance in expressing the 'truly philosophical' notions and principles of mind–body interactions.[42] Yet although the orations were popular in Gaub's own time, they were quickly forgotten afterwards. In the second oration (1763) Gaub argued that the mind actively partakes in the maintenance of bodily health: 'the mind can variously alter the state of the body, so that absolutely none of the motive powers, functions, interior organs and humors can resist its force'.[43] He explicated his argument with a reference to respiration: 'To begin at birth, what is more important to the life of the newborn infant when it first comes to the light and thereafter to man at every stage of growth, than breathing? Do you hold that the body of its own accord initiates breathing? Even though an answer to this question is still lacking, it is more likely the mind'.[44] When respiration is impaired, leading to anxiety and distress, the mind actively intervenes. For instance, when a man is troubled by respiratory difficulty, the mind strengthens the voluntary muscles and adjusts the position of the body, so that it breathes more easily.[45] The assistance of the mind in restoring health is often effected through the unconscious mediation of (violent) emotions. Again, Gaub referred to respiration:

> Sighing, by the assistance of the voluntary muscles, draws in the breath very deep, even to the bottom of the breast, and, more forcibly than usual, expels it upwards. Thus nature, by a great and long draught of respiration, increases the beneficial effects of the air, procures refrigeration to the suffocating heat, excitement to the pulmonary circulation, a remedy to difficulty of breathing, and ease to anxiety.[46]

Not surprisingly, Gaub closely linked respiration to anxiety: 'impediments of respiration furnish the most frequent cause of anxiety', and vice versa, through imaginary fear, anger, terror, excessive joy, or lovesickness 'the mind is often thrown into extreme anguish', affecting respiration. Thus, anxiety is 'a monitor of the vital system', and 'even more than pain deserves to be attended to'.[47] An important way to ease anxiety was through taking exercise, getting fresh air, and contemplating nature outside in landscapes and parks as well as inside through viewing idyllic and pastoral paintings. In fact, eighteenth-century writers on art invoked medical ideas in contemplating the effects of viewing beautiful landscapes on the body's fibres and (blood) vessels.

For instance, the well-known London writer Uvedale Price (1747–1829) in his *Essay on the Picturesque* (1796) argued that the viewing of beautiful landscapes alleviated anxiety, calmed the mind, and steadied breathing. In his explanation he referred to contemporary physiological ideas on fibres, the elementary and vital building-blocks of the body, in stating that beauty excites passion and relaxes the fibres.[48] Although some authors would claim that pleasure stimulates rather than relaxes the fibres, Uvedale conceded that it did not matter: whether stretching or relaxing, the 'active agency [of the picturesque which leads to curiosity] keeps them [the fibres] to their full tone'.[49] Vice versa, 'wrong' colours or representations could lead to illness. The Dutch physician Lambertus Bicker (1732–1801) reports on a gentleman who always suffered from breathlessness and swooning as soon as he entered a room with blue wallpaper.[50] Likewise, Diderot's idea that paintings should represent the colours of nature in order not to offend the organs of the body ('the tender and weak eye will never find pleasure in lively and strong colours') was a familiar theme.[51] Precisely for this reason, it was advised never to decorate bedrooms in white or red, but to follow nature and use green paints and wallpaper.[52] Speculations about the effect of vision and colours on passions, organs, and fibres dovetailed with Gaub's psychosomatics, as well as with the work of William Cullen and other eighteenth-century medical writers who argued that there was a direct connection between ‘the energy of the brain’ and the fibrous strength of vital organs.[53]

The art historian Frances Gage has similarly argued that landscape paintings in early modern Italian domestic spaces had therapeutic effects:

> Landscape painting [...] replicated many of the perceived therapeutic and preservative effects of actual landscape and could function as a substitute for the direct experience of nature [...] through its compositional structures and varied depictions of nature’s most desirable properties, landscape painting also promised a more completely healthful experience than did nature herself.[54]

In particular, Italian galleries, placed away from the polluted streets and infused with good air, were meant as places of health and exercise. They were decorated with landscape paintings and gave early modern gentry the sense of outdoor exercise while taking a stroll inside.[55] Italian doctors were convinced that ‘the viewing of beautiful perspectives’ improved the experience of breathing good air. Through gazing at such paintings, the beholder would ‘follow a visually cathartic process of purification’ in moving from bad air in his direct environment to the wholesome air of the painting.[56] In other words, imagining good air would improve breathing experiences.

During the eighteenth century, landscape painting also boomed in Northern Europe. Artists and physicians began emphasizing the health benefits of their paintings in a similar way. In particular the art of French painter Claude Lorrain (1600–1682) became famous across Europe and was appreciated for the ‘calm and dignified repose and harmony of tint’, ‘the calm beauties of antiques’, the ‘warmth’ they transmitted to the beholder, and the illusion of becoming inhabitants of the mythical Elysium.[57] From the 1750s the comforts of landscape painting were widely recommended as they made the viewer ‘imagine himself either hunting, or taking the air, or walking, or sitting, and giving himself up to agreeable meetings’.[58] The well-known English painter Joshua Reynolds (1723–1792) argued that Lorrain’s paintings ‘lead us to the quiet rest of Arcadian scenery or the country of the fairies’ and that ‘the purpose of these paintings is to impress the senses and imagination; the imitation of nature often achieves this goal’.[59] In the Netherlands, it was increasingly important to identify with Dutch landscapes. In order to get the most out of getting fresh air indoors, people had to relate to familiar outdoor views. So, while seventeenth-century poet Joost van den Vondel (1587–1679) observed that the arcadic landscape paintings of Herman Zachtleven (1609–1685) allowed one to take the air while conveniently staying at home, from the 1750s this gradually changed to an appreciation for the simple and familiar views of Dutch nature.[60] This meant the enjoyment of particular Dutch landscapes with low horizons, vast skies, rivers, grassy meadows, and cows: 'though Italy may delight its inhabitants' senses with the splendor of its most beautiful scenery, all the splendid views of a Vesuvius cannot compare to the sight of a rich meadow full of thriving cattle'.[61]

The close connection between medical and artistic ideas shows that paintings—of idyllic landscapes in particular—were considered an important part of the material culture of eighteenth-century households, conducive to the mental *and* physical health of their residents. This may also explain the revival and popularity of scenic wall decoration during the second half of the eighteenth century.[62] In Britain and the Netherlands wallpaper manufacturers thrived. From the 1780s newspaper articles advertising cheap or second-hand wallpaper (the latter originating from household effects) doubled. Wallpaper factories not only produced luxurious, expensive wallpaper, primarily depicting naturalistic and idyllic landscapes, made according to the individual tastes of wealthy buyers and printed on linen, but also cheaper wallpaper printed on cotton or paper for a growing market of middle-class homes.[63] The former was meant for manor houses, townhalls, and theatres (Plate 11.2).

The history of eighteenth-century wallpaper remains to be written, but art historians have recently linked the popularity of floor-to-ceiling 'painted rooms' to the eighteenth-century popularity of natural history and idyllic gardening. They have suggested that 'painted rooms' secured the enjoyment of nature inside the house. In particular during the long and harsh winter months, the 'green tree crowns and a summer atmosphere' would invigorate the weary soul.[64] The (landscape) painter Arnold Houbraken (1660–1719), in a poem on a painted room in the house of statesman Cornelis Backer in Amsterdam, contrasted the chill of the barren and snowy trees outside with the impression of summer provided by leafy trees inside: 'Here he [the weary statesman] can relax and string his bow again'.[65] Indeed, although taking the air indoors was a familiar motif in early modern Europe, it flourished during the eighteenth century and kept pace with a mounting appreciation for healthy breathing in relation to natural environments (Plate 11.3).

Plate 11.2 Wallpaper advertisement of Jan Smit, wallpaper entrepreneur in Amsterdam. Adolf van der Laan, 1734 (Courtesy of the Rijksmuseum Amsterdam)

Plate 11.3 Design for wallpaper with a view of a park. Anonymous, 1700–1800. The viewer would be at the same height as the boy playing with the dog, giving the sensation of being in the park (Courtesy of the Rijksmuseum Amsterdam)

Conclusion

One of the most beautiful painted rooms can be found in the luxurious interior of the doll's house of Petronella Oortman (1656–1716).[66] The doll's house, which cost as much as a real Amsterdam canal house, was a popular tourist attraction during the eighteenth century. For the purpose of this essay it is particularly interesting to have a closer look at the drawing room or *Zaal*, which is decorated with floor to ceiling landscape paintings. Significantly, the paintings continue in a painted ceiling with birds flying in a serene and slightly clouded light blue sky. To eighteenth-century viewers this must have looked familiar. Not only were painted ceilings with clouds and birds a popular motif in early modern art, but also a translucent sky visually represented the most healthful air.[67] In this case, the impression inside such a room must have been of being outside, the ceiling giving the sensation of being in open air so breathy, pure, and light that the birds simply float on it (Plates 11.4 and 11.5).

In a way the mini-drawing room summarizes the argument of this essay. The healthy air provided by idyllic landscapes—outdoors as well as imagined through paintings and murals—literally gave air and assisted

Plate 11.4 Petronella Oortman's doll's house. The drawing room is located on the left side on the second floor. Anonymous, *ca.* 1686–1710 (Courtesy of the Rijksmuseum Amsterdam)

healthy breathing. Supported by new physiologies of respiration as well as a renewed emphasis on the imagination as a means to relieve anxiety-related breathing trouble, eighteenth-century people quite literally believed that nature breathes as we do. This means that taking a walk in the park, along the trees that exhale and inspire the atmosphere with beneficial air, would surely improve respiration and health. Likewise, as Frances Gage has argued, the viewing of landscapes with airy forms, indistinct contours, and softened colors produced 'a visual analogue to the softening and purifying effects that trees and foliage were understood to exert upon the air and upon the substance of the body in return'.[68] In all cases, breath was an important, yet often invisible, go-between linking body, mind, and environment. So, in Oortman's doll's house, the drawing room's landscapes signify more than pretty paintings meant to divert the eye. They were meant to animate the house, transcending the threshold

Plate 11.5 Detail of the painted ceiling in Petronella Oortman's doll's house. Attributed to Nicolaes Piemont, *ca.* 1690–1709 (Courtesy of the Rijksmuseum Amsterdam)

between life and death, giving the viewer the impression of gazing at a household invigorated with the breath of life, perhaps even feeling better for it.

Notes

1. Jan Ingenhousz, *Experiments Upon Vegetables, Discovering Their Great Power of Purifying the Common Air in the Sun-Shine, and of Injuring It in the Shade and at Night* (London, 1779), xiii.

2. Geerdt Magiels, 'Dr Jan IngenHousz, or Why We Don't Know Who Discovered Photosynthesis', paper for the 1st Conference of the European Philosophy of Science Association (Madrid, 15–17 November 2007) http://philsci-archive.pitt.edu/3790/1/IngenHousz_EPSA07.pdf [last consulted 14 March 2019]. See also: Norman and Elaine Beale, 'The Life of Dr Jan IngenHousz (1730–1799), Private Counsellor and Personal Physician to Emperor Joseph II of Austria', *Journal of Medical Biography* 13 (2005), 15–21; 'Jan Ingenhousz: Plant Physiologist,' *Chronica botanica* 11 (1949), 285–396.
3. Ingenhousz, *Experiments*, 9–10.
4. Ingenhousz, *Experiments*, 67–69.
5. Jan Rudolph Deiman and A. Paets van Troostwijk, 'Proefneming omtrent de verbetering der lugt, door middel van de groeij der planten', *Vaderlandsche Letteroefeningen* (1778): 338–49, 437–46, 482–90, 485. See also Joost Kloek and Wijnand Mijnhardt, *Dutch Culture in a European Perspective: 1800, Blueprints for a National Community* (London: Palgrave, 2004), 321–22.
6. *Verhandelingen van het Provinciaal Utrechtsch Genootschap van Kunsten en Wetenschappen*, 1 (1781), 467–68.
7. Clement Archer, *Miscellaneous Observations on the Effects of Oxygen on the Animal and Vegetable Systems* (Bath, 1798), 79.
8. 'Iets over den invloed van woonplaats of verblijf op de gezondheid', *Vaderlandsche Letteroefeningen* (1810), 373. See also William Buchan, *Domestic Medicine, or a Treatise on the Prevention and Cure of Diseases* (Philadelphia: Richard Folwell, 1797), 70.
9. Two simultaneous traditions of respiratory medicine, respiratory anatomy and respiratory chemistry, have been identified by historian Leon Goddlieb in his *A History of Respiration* (Springfield, IL: Charles Thomas, 1964).
10. Robert Menzies, *A Dissertation on Respiration* (Edinburgh, 1796), iii.
11. Jerome Gaub also employs the Latin version of his name, Hiëronymus David Gaubius. I here use his name in the vernacular.
12. Herman Boerhaave, *Dr. Boerhaave's Academical Lectures on the Theory of Physic*, 6 vols (London, 1757), vol. 4, 341, 343.
13. Boerhaave, *Lectures*, vol. 4, 402.
14. Boerhaave, *Lectures*, vol. 6, 205, 210–11.
15. Robert Harrington, *A Philosophical and Experimental Enquiry into the First and General Principles of Animal and Vegetable Life* (London, 1781), 413.
16. Hugh Smith, *Philosophical Inquiries into the Laws of Animal Life* (London, 1780). The British Library catalogue ascribes this book to Hugh Smith the Younger (1735/1736–1789). However, as the *Philosophical Inquiries* are 'founded on the principles delivered in a course of

philosophical lectures in the beginning of the year 1778', I assume Hugh Smith (d. 1790), who gave very popular lecture courses on the theory and practice of physic (ODNB), authored this publication.

17. Today the distinction between a sign and a symptom is understood in terms of the doctor's or the patient's experience: a sign is considered objective evidence of disease that can be observed by others, most notably the doctor. A symptom refers to the subjective patient experience of disease. By contrast, Boerhaave defined the sign of a disease as 'an appearance perceptible by the senses, from which the physician discovers the true cause of the disease'. A symptom is 'the preternatural appearance in a diseased body, which flows from the distemper as its cause, but is distinguishable from the disease itself' and can become a distemper itself (Boerhaave, *Lectures*, vol. 4, 1, 113).
18. Virchow, quoted in L. J. Rather, *Mind and Body in Eighteenth-Century Medicine. A Study Based on Jerome Gaub's 'De regimine mentis'* (London: Wellcome Library, 1965), 23.
19. Hieronymus David Gaubius, *The Institutions of Medicinal Pathology by H. D. Gaubius, Translated from the Latin by Charles Erskine* (London, 1778), 23–24. My italics.
20. Gaubius, *Institutions*, 284–85.
21. Gaubius, *Institutions*, 27 and 30.
22. Gaubius, *Institutions*, 284 and 146.
23. Rina Knoeff, 'The Body is a Barometer. Dutch Doctors on Healthy Weather and Strong Constitutions', in *Lifestyle and Medicine in the Enlightenment: The Six Non-Naturals in the Long Eighteenth Century*, ed. by James Kennaway and Rina Knoeff (London: Routledge, 2020), 19–42.
24. Benjamin Rumford, *An Essay on Chimney Fire-Places; with Proposals for Improving Them to Save Fuel; to Render Dwelling-Houses More Comfortable and Salubrious, and Effectually to Prevent Chimnies from Smoking* (Dublin, 1796), 1.
25. Stephen Hales, *A Description of Ventilators: Whereby Great Quantities of Fresh Air May with Ease Be Conveyed into Mines, Goals* [sic], *Hospitals, Work-Houses and Ships* (London, 1743).
26. Simon Schaffer, 'Measuring Virtue. Eudiometry, Enlightenment and Pneumatic Medicine', in *The Medical Enlightenment of the Eighteenth Century*, ed. by Andrew Cunningham and Andrew Wear (Cambridge: Cambridge University Press, 1990), 281–318. On eudiometry in the Netherlands see also Huib J. Zuidervaart, 'An Eighteenth-Century Medical-Meteorological Society in the Netherlands: An Investigation of Early Organization, Instrumentation and Quantification. Part 2', *BJHS* 39 (2006), 49–66.
27. Jan Golinski, *British Weather and the Climate of Enlightenment* (Chicago: University of Chicago Press, 2007), 137–40.

28. See for instance Iman J. van den Bosch, *Over de Ziekten van het Vaderland* (Amsterdam, 1794).
29. James Kennaway and Rina Knoeff, '"For it is the debilitating fibres that exercise restores": Movement, morality and moderation in Eighteenth-Century Medical Advice Literature', in Kennaway and Knoeff, *Lifestyle and Medicine in the Enlightenment*, 111–38.
30. John Sinclair, *The Code of Health and Longevity. Or, a Concise View, of the Principles Calculated for the Preservation of Health, and the Attainment of Life*, 2 vols (Edinburgh: A. Constable and Co., 1807), vol. 1, 708.
31. Sinclair, *Code of Health*, vol. 1, 708; vol. 2, Appendix, 'On Athletic Exercise', 87.
32. George Cheyne, *An Essay of Health and Long Life* (Bath: G. Strahan, 1724), 49.
33. See for instance Cheyne, *Essay of Health*, 50.
34. Buchan, *Domestic Medicine*, 69, 73.
35. Kloek and Mijnhardt, *Blueprints*, 322–24.
36. Paul Rabbits, *Hyde Park. The People's Park* (Stroud: Amberley, 2015), introduction.
37. William Cadogan, *A Dissertation on the Gout: And All Chronic Diseases, Jointly Considered, as Proceeding from the Same Causes* (London, 1772), 94.
38. For convenience I use 'psychosomatics' for the emerging field of 'science of the soul' or *zielkunde*, the 'new' field where physicians were encouraged to treat the mind as well as the body.
39. Rather, *Mind and Body*, 17.
40. Gaub in Rather, *Mind and Body*, 10, 194–195.
41. Gaubius, *Institutions*, 283.
42. Ernst von Feuchtersleben, quoted in Rather, *Mind and Body*, 24.
43. Gaub in Rather, *Mind and Body*, 155.
44. Gaub in Rather, *Mind and Body*, 162.
45. Gaub in Rather, *Mind and Body*, 164.
46. Gaubius, *Institutions*, 289.
47. Gaubius, *Institutions*, 252. See also Rather, *Mind and Body*, 58.
48. Uvedale Price, *An Essay on the Picturesque* (London, 1796), 104. On the importance of fibres in eighteenth-century physiology see Hisao Ishizuka, *Fiber, Medicine, and Culture in the British Enlightenment* (New York: Palgrave Macmillan, 2016); Tobias Cheung, 'Omnis Fibra Ex Fibra: Fibre Oeconomics in Bonnet's and Diderot's Models of Organic Order', in *Early Science and Medicine* 15 (2010) 66–104; Alexander Berg, 'Die Lehre von der Faser als Form- und Funktionselement des Organismus. Die Geschichte des biologisch-medizinischen Grundproblems vom kleinsten Bauelement des Körpers bis zur Begründung der Zellenlehre', in *Virchows Archiv* 309 (1942), 333–460.

49. Price, *Essay on the Picturesque*, 105.
50. Lambertus Bicker, *Natuur- en geneeskundige verhandeling over de oorzaken, den aart, en de genezing der zenuwziekten* (Utrecht, 1793). Presumably Bicker referred to the colour 'Berlin or Prussian blue', invented in 1704 and cheaply available from 1724. See C. Willemijn Fock, ed., *Het Nederlandse Interieur in beeld 1600–1900* (Zwolle: Waanders, 2001), 195.
51. Diderot's *Essai sur la Peinture,* discussed in *Nieuwe Algemene konst- en letter-bode* 14 (1794), 51–52.
52. George Adams, *Verhandeling over het zien* (Amsterdam, 1792), 110–11.
53. John Thomson, et al., *An Account of the Life, Lectures and Writings of William Cullen*, 2 vols (Edinburgh: Blackwood, [1832] 1859), vol. 1, 296–97. Thomson quotes William Cullen, *First Lines of the Practice Physic* (1784), vol. 2, 363.
54. Frances Gage, *Painting as Medicine in Early Modern Rome. Giulio Mancini and the Efficacy of Art* (Pennsylvania: Pennsylvania University Press, 2008), 85.
55. Gage, *Painting as Medicine*, 82–85.
56. Frances Gage, 'Chasing "good air" and Viewing Beautiful Perspectives: Painting and Health Preservation in Seventeenth-Century Rome', in *Conserving Health in Early Modern Culture. Bodies and Environments in Italy and England*, ed. by Sandra Cavallo and Tessa Storey (Manchester: Manchester University Press, 2017), 237–61, 254.
57. Antony Pasquin, *Memoirs of the Royal Academicians; Being an Attempt to Improve the National Taste* (London, 1796), 20, 121, and 139.
58. Roger Piles, *The Principles of Painting* (London, 1743), 123.
59. Joshua Reynolds, 'Redevoering', in *Algemeen Magazyn van Wetenschap, Konst en Smaak* (Amsterdam, 1785), 132 and 138.
60. *De Werken van Vondel*, 10 vols (Amsterdam: De Maatschappij voor goede en goedkope lectuur, 1927), vol. 9, 300.
61. Rheinvis Feith, quoted in Kloek and Mijnhardt, *Blueprints*, 510.
62. Ineke Strouken, *Tussen plafond en print. Cultuurgeschiedenis van het behang* (Utrecht: Nederlands Centrum voor Volkscultuur, 1992).
63. Fock, *Het Nederlandse Interieur*, 81–82; J. de Loos-Haaxman, 'De behangselfabriek der Vaderlandsche Maatschappij te Hoorn', *Nederlands Kunsthistorisch Jaarboek* 12 (1961), 149–92, 172.
64. Inge Verslype, Margriet van Eikema Hommes and Annemieke Heuft, 'De oorspronkelijke gedaante van de geschilderde kamer in het Martenahuis te Franeker', in *Bulletin KNOB* (2017), 101–22, 113. See also C. Willemijn Fock, 'Het interieur in de Republiek 1670–1750: (g)een plaats voor schilderkunst?', in ed. by Ekkehard Mai, Sander Paarlberg, and Gregor J. M. Weber, *De kroon op het werk. Hollandse schilderkunst 1670–1750* (Köln: Locher, 2006), 63–86, 70.

65. Jet Pijzel-Domisse, *Het Hollandse poppenhuis. Interieur en huishouden in de 17de en de 18de eeuw* (Zwolle: Waanders, 2000), 96.
66. For details on the dollhouse and the painted room see Pijzel-Domisse, *Het Hollandse poppenhuis*, 247–346.
67. Gage, *Painting as Medicine*, 49; Fock, *Het Nederlandse Interieur*, 91, 187.
68. Gage, *Painting as Medicine*, 81.

Selected Bibliography

Fock, Willemijn, ed. 2001. *Het Nederlandse Interieur in beeld 1600–1900*. Zwolle: Waanders.

Gage, Frances. 2008. *Painting as Medicine in Early Modern Rome. Giulio Mancini and the Efficacy of Art*. Pennsylvania: Pennsylvania University Press.

Golinski, Jan. 2007. *British Weather and the Climate of Enlightenment*. Chicago: University of Chicago Press.

Ishizuka, Hisao. 2016. *Fiber, Medicine, and Culture in the British Enlightenment*. New York: Palgrave Macmillan.

Kloek, Joost, and Wijnand Mijnhardt. 2004. *Dutch Culture in a European Perspective: 1800, Blueprints for a National Community*. London: Palgrave.

Kennaway, James and Rina Knoeff, eds 2020. *Lifestyle and Medicine in the Enlightenment: The Six Non-Naturals in the Long Eighteenth Century*. London: Routledge.

Rather, L. J. 1965. *Mind and Body in Eighteenth-Century Medicine: A Study Based on Jerome Gaub's 'De regimine mentis'*. London: Wellcome Library.

Schaffer, Simon. 1990. Measuring Virtue. Eudiometry, Enlightenment and Pneumatic Medicine. In *The Medical Enlightenment of the Eighteenth Century*, ed. Andrew Cunningham and Andrew Wear. Cambridge: Cambridge University Press.

Vila, Anne C. 1998. *Enlightenment and Pathology. Sensibility in the Literature and Medicine of Eighteenth-Century France*. Baltimore: Johns Hopkins University Press.

CHAPTER 12

'Spoken from the Impulse of the Moment': Epistolarity, Sensibility, and Breath in Frances Burney's *Evelina*

Gillian Skinner

'Every word seemed spoke from the impulse of the moment'[1]: thus the heroine, newly arrived in London for the very first time, describes her elation after the experience of watching David Garrick act, in a letter early on in Frances Burney's only epistolary novel, *Evelina* (1778). What impresses her so much is his capacity to perform spontaneity—to speak a written part as if 'from the impulse of the moment'. The comment is itself a performance by the novelist of a performance of spontaneity on the part of the letter-writer—of her enthusiasm, scribbled down in the heat of the moment, on her return from the theatre—and such an impression of immediacy is a key quality of epistolary fiction of the period. Samuel Richardson's famous phrase, 'writing, to the moment', encapsulated the idea of the epistolary form as peculiarly suited to the rendition of human emotion as it was felt, minute by minute, apparently unmediated by any process of recollection, revision, or censorship.[2] Evelina's description of

G. Skinner (✉)
Department of English Studies, University of Durham, Durham, UK

D. Fuller et al. (eds.), *The Life of Breath in Literature, Culture and Medicine*, Palgrave Studies in Literature, Science and Medicine,
https://doi.org/10.1007/978-3-030-74443-4_12

Garrick's performance focuses attention specifically on the act of speech and what prompts it (or appears to do so); in the use of the word 'impulse' she hints at an important nexus of relationships between breath, the heart, and the motivations behind human action that this essay will explore. The meanings of 'impulse' encompass the beating of the heart, the idea of sudden and unreflective action, and, more archaically, the influence of the spirit.[3] Questions of physiology, emotion, and moral decision-making are brought together in the word, as they are frequently in the sentimental epistolary novel of the period and in *Evelina* itself.

Drawing out these aspects of Burney's novel illuminates the neglected role of breath in the mapping and understanding of eighteenth-century sensibility. Sensibility has been studied in terms of the senses, the significance of the heart has been examined, and, above all, the centrality of developing ideas about the nervous system has been scrutinised, but breath in its multiple aspects and significances has not received the specific attention it deserves.[4] Yet in even a cursory consideration of the distinguishing features of the sentimental protagonist, her (and, though less frequently, his) sighs, tears, and swoons would suggest that breath and its disruptions are central to the physiological manifestations and emotional impact of sensibility. The sentimental text is typically written with acute regard for the ebbs and flows of its protagonists' feelings, fluctuations that are often mirrored in the speed or loss of breath, factors which are rendered textually in a range of typographical signs. The letter, widely popular both as narrative form and narrative device in the fiction of the period, is itself a sign of breath in its role as a substitute for speech. This role is powerfully expressed early in the century by Eloisa in Pope's verse epistle *Eloisa to Abelard*:

> Heav'n first taught letters for some wretch's aid,
> Some banish'd lover, or some captive maid;
> They live, they speak, they breathe what love inspires,
> Warm from the soul, and faithful to its fires,
> The virgin's wish without her fears impart,
> Excuse the blush, and pour out all the heart,
> Speed the soft intercourse from soul to soul,
> And waft a sigh from *Indus* to the *Pole*.[5]

Standing in for the 'banish'd lover' or 'captive maid', letters are said to 'live' and 'speak': most importantly for Eloisa, 'breath[ing] what love inspires', they communicate desire directly and fully, more fully indeed than might be the case if the lovers met in person. The letter, in its love-inspired breath, pouring out the heart and wafting sighs around the globe, moves from 'soul to soul' transparently. Eloisa's description conjures up a kind of ideal letter, a letter that can ease the pain of a 'wretch' parted from his or her loved one by expressing desire directly, faithfully and without fear: this sentimental ideal of transparent epistolary sincerity is shown to be more complicated in practice by both Eloisa's own agonised verses and by Evelina's correspondence. Composition, it turns out, is not quite so natural as breathing, just as breathing, however natural it may be, is itself a culturally weighted process.

Tension between a conception of letter-writing as offering unparalleled direct access to the heart and a concern that such sincerity and transparency were potentially dangerous and needed regulation can be found frequently in eighteenth-century letter manuals, often resulting in contradictory advice. '[W]rite what your heart shall dictate', urges a correspondent to her niece in Lady Dorothea Du Bois' *The Lady's Polite Secretary* (1771), before giving 'another caution, which one would hardly think necessary did not daily experience convince us of the contrary, and that is, *never take up the pen till you have considered what you are going to write.*'[6] As Mary Favret has commented, 'When common consent granted the letter the power to convey "passion" in language, it also accused the letter of a lack of intrinsic control', and she goes on, 'In the words of a 1780 rhetoric book, the letter allowed its writer "to be at ease and give vent occasionally to the overflowings of his heart." At the same time, the rhetorician felt obliged to rein in the antisocial impulses of any correspondent [...]'.[7] The book in this case is *Lectures on Rhetoric and Belles Lettres* by the literary scholar Hugh Blair, whose 'overflowings' of the heart recall the 'pouring out' of Eloisa's. Yet as Dorothea Du Bois' cautionary comment implies, there may be many reasons why writing the heart is problematic. In Evelina's case, the transparency of the letter is always compromised by the impossibility of explicitly acknowledging her own desire, which must be inferred by the reader but unclaimed by the letter-writer herself. When Lord Orville finally declares his love, Evelina writes:

> I attempt not to describe my sensations at that moment; I scarce breathed; I doubt if I existed,—the blood forsook my cheeks, and my feet refused to sustain me: Lord Orville, hastily rising, supported me to a chair, upon which I sunk, almost lifeless.

It is fitting that her collapse comes at the moment when Lord Orville protests the inadequacy of speech—she is 'dearer to [him] than speech has the power of telling!'—and significant that the remainder of their conversation, once Evelina has sufficiently recovered, is not given to the reader. 'I cannot', she tells Mr Villars, 'write the scene that followed, though every word is engraven on my heart' (351–52). Intimacy is unrepresentable, while the lifelessness that overtakes Evelina at the moment of Orville's declaration—the consummation of her desire—relates breathlessness and sex in the only way such a novel could (although with plenty of precedent in the experiences of earlier sentimental heroines, such as Eloisa and Pamela). As Bradford K. Mudge writes, romance novels in the late eighteenth century are 'dependent on the displacement and/or deferral of sexual passion' so that 'passion is at once central to the drama and almost entirely absent from the field of vision'.[8] Breathlessness in such novels is the repeated signifier of such an absent presence.

Just as Eloisa's heart is closely associated with breath and Evelina's breathlessness exposes the state of her heart, so Blair's use of the word 'vent' brings heart and lungs together and suggests the letter's important function of release, both emotional and physiological. In Johnson's *Dictionary*, the first meaning of 'vent' is 'A small aperture; a hole; a spiracle; passage at which any thing is let out'. 'Spiracle' is worth some attention: the *OED* defines it as

> 1. Breath, spirit. *Obs.* (the last example of its use is in 1654)
> 2. a. A small opening by which a confined space has communication with the outer air; *esp.* an air-hole or air-shaft.
> *b. spec.* An opening in the ground affording egress to subterranean vapours or fiery matter; a volcanic vent-hole.

Johnson's examples of the use of 'vent' with the meaning of 'spiracle' include the moment in *The Rape of the Lock* when Umbriel opens the 'wondrous bag' over the dejected heroine Belinda, slumped in the arms of her friend Thalestris:

> Full o'er their Heads the swelling Bag he rent,
> And all the Furies issued at the Vent. (IV, 91–92)

This bag, significantly, contains 'the Force of Female Lungs, / Sighs, Sobs, and Passions, and the War of Tongues' (IV, 83–84): breath, voice, and tears.[9] The particular resonance of 'vent' in its relation to pent-up emotion, figured by the holding and releasing of the breath (and potentially by the writing and sending of letters), becomes clear. As the eponymous heroine of Samuel Richardson's *Pamela* (1740) laments to her parents: '...to whom but you can I vent my Griefs, and keep my poor Heart from bursting!', while in his *Clarissa* (1748) the heroine's mother writes to Mrs Norton, 'My heart is full. Writing may give some vent to my griefs, and perhaps I may write what lies most upon my heart [...]'[10] Evelina also needs to give 'free vent' to her feelings at one of the novel's key crisis points.

As with Blair's comment above, Pamela's lament and Mrs Harlowe's complaint point vividly to the relation between lungs and heart towards which Evelina's appreciation of Garrick's acting gestured. In classical physiology, there was an 'all-important linkage ... of the two systems of blood flow and respiration' and the lungs acted in tandem with the heart, as Robert Erickson explains, quoting Plato's *Timaeus*:

> The lungs gave 'coolness and the power of respiration', alleviating the heat of the heart, and 'when passion was rife within, the heart, beating against a yielding body', was cooled and more ready to join with passion in the service of reason.[11]

The symbolic and physiological importance of the lungs had been recognised since the classical theories of Galen,[12] and classical ideas and associations still pertained in the eighteenth century, notwithstanding developments in medical understanding: for Blair, Pamela, and Mrs Harlowe, giving vent to emotions is presented as the lungs easing pressure on the heart, providing much-needed relief to the inflammations caused by various kinds of passion. Comprehending the operation of the breath physiologically was still a challenge at the time *Evelina* was written in 1778. Joseph Priestley's essay 'Observations on Respiration' of 1776 still laboured under the misconceptions caused by the phlogiston theory, not to be finally disentangled until Antoine Lavoisier published his *Traité élémentaire de chimie* in 1789. Still, whatever the confusions in the detail of their descriptions, what Edmund Goodwyn's 1788 study referred to as *The Connexion of Life with Respiration* encapsulated one of the key reasons for the fascination with the topic: breath and breathlessness were the essential symptoms of life and death. In a sample letter included in

The Universal Letter-Writer (1770), a clergyman reflects that 'The partition which separates time from eternity is nothing more than the breath of our nostrils, and the transition may be made in the least particle of time'.[13] Epistolary and sentimental literary techniques in the period were peculiarly attuned to the transition from breath to breathlessness and the challenges of representation it presented to writers.

Writing of Coleridge and Keats, Francis O'Gorman has commented that 'Poetry naturally has a close relationship with breath, and the late eighteenth-century writers who celebrated how poetry, even as it was printed, mimicked the spoken, the breath-born, were part of a long tradition'.[14] This tradition of interest in the way print mimics speech includes not only poetry but also the prose styles of the sentimental novel and epistolary fiction—very often, of course, overlapping genres. In their prioritisation of immediacy and their intent to engage the feelings of their readers, such texts made liberal use of punctuation to imitate the way lived experience and emotional pressure fractured and broke up speech. This applied particularly to the use of the letter in narrative, given its function as a substitute for the human voice. Richardson's *Pamela* illustrates this facet of sentimental fiction vividly. Early in the novel, after Mr B has sprung unexpectedly from the closet in the bedroom, Pamela writes to her parents:

> Wicked, wicked Man!—I have no Patience left me!—But yet, don't be frighted—for,—I hope—I hope, I am honest!—But if my Head and my Heart will let me, you shall hear all.—[15]

The dash, much favoured by sentimental novelists, here becomes an intensifier that adds extra weight to the already-intense exclamation mark. An indicator of breathless gaps in narration, the dash mirrors the disturbing gap in Pamela's consciousness, a gap that means she can only repeatedly 'hope', rather than 'know', that she is 'honest'.

The opening of Evelina's first letter from London is another, if rather different, case in point: 'This moment arrived. Just going to Drury-Lane theatre. The celebrated Mr. Garrick performs Ranger. I am quite in extacy. So is Miss Mirvan. How fortunate, that he should happen to play!' (27). Clipped and abbreviated, the sentences capture Evelina's excitement and hurry; the theatre beckoning, the letter draws to an end rapidly: 'I can write no more now. I have hardly time to breathe—only just this, the houses and streets are not quite so superb as I expected' (27). (The slight bathos in that observation stands as a hint that London may not always come up to the young Evelina's high expectations.) It is not only Evelina's own letters that convey the breathlessness of strong emotion. Writing early in the novel, Mr Villars, Evelina's guardian, exclaims to his old friend Lady Howard, 'This letter will be delivered to you by my child, – the child of my adoption, – my affection!' (22). The dash here is again instrumental in raising the emotional temperature of the utterance, drawing attention to the complexities of Villars and Evelina's relationship in the qualification of 'child' by 'child of my adoption', qualified in its turn by 'my affection!'

The dash is only one of a group of ellipses increasingly employed in print as the eighteenth century progressed. In *Ellipsis in English Literature*, Ann Toner lists dots, dashes, series of hyphens and asterisks as the main ways texts in the period denoted varieties of pause, interruption or aposiopesis. Her book draws attention both to the way 'the novel has given rise to especially varied and innovative uses of the ellipsis' and to the particular relationship between sensibility and ellipsis:

> Sensibility became a pan-European language in that it rejected words where possible for the vocabulary of sighs, sobs and tears. That which 'punctures or interrupts speech', as John Mullan has described the vocabulary of sensibility, is needy of a visible counterpart to document such interruptions.[16]

Toner points out that '[A]ll punctuation provides guidance as to breathing', but in the literature of sensibility especially the ellipsis became a symbol of emotionally freighted breath—the sob, the sigh, the gasp, the swoon.

In *Evelina*, as in *Pamela*, moments of high emotional stress are consistently associated with breathlessness, figuring the extreme physiological pressure such emotion puts the body under, often in the form of multiple dashes. The stresses these signify are not simply emotional, however, but also and inextricably moral. One of the novel's most intense crisis points, Evelina's first meeting with her estranged father, Sir John Belmont, illustrates the role of ellipsis in the rendering of sentimental breathing with particular clarity. Her likeness to her wronged mother triggers an acute reaction in Sir John, whose emotion is produced by his intense feelings of guilt at his mistreatment of the dead Caroline Evelyn:

> 'Yes, yes,' cried he, looking earnestly in my face, 'I see, I see thou art her child! she lives—she breathes—she is present to my view!—Oh God, that she indeed lived!—Go, child, go,' added he, wildly starting, and pushing me from him, 'take her away, Madam,—I cannot bear to look at her!' And then, breaking hastily from me, he rushed out of the room. (372)

Such moments in the novel not only mimicked the breathlessness of the characters represented, they also induced breathlessness in their readers, as contemporary reports made clear to Burney: 'Thou hast made thy old Father Laugh & Cry at thy pleasure', Dr Burney told her;[17] 'Lady Hales told Susan [Burney's sister Susanna] how she had "been reading Evelina to Madame de Ferre, the Governess, & the children, & that the meeting with the Father made them all *sob* so much, she was obliged to leave it off"'.[18] Such effects can be seen as typifying what Annamarie Jagose describes as 'that sympathetic connection between character and reader that underwrites the eighteenth-century sentimental novel's program of improvement'.[19] In its approving review of *Evelina*, the *Critical Review* implied just such a direct link between identification with a character and beneficial moral effects. It imagined a father recommending the novel to his daughters for its 'knowledge of the world and lessons of experience':

> they will weep and (what it is not so commonly the effect of novels) will laugh, and grow wiser, as they read; the experienced mother will derive pleasure and happiness from being present at its reading; even the sons of the family will forego the diversions of the town or the field to pursue the entertainment of Evelina's acquaintance, who will imperceptibly lead them, as well as their sisters, to improvement and to virtue.[20]

Thus the process of moral improvement is explicitly bound up with physiological reactions to the sentimental text ('they will weep and [...] will laugh, and grow wiser'), underlining the continuity between mind and body that was becoming of such importance both in medical and philosophical thinking and in the popular imagination as the eighteenth century progressed.[21]

This intertwining of the moral and the physiological—again, specifically the operation of breath—is especially clear in *Evelina* when the heroine believes that Mr Macartney, the despised and impoverished lodger of her vulgar relations the Branghtons, is about to attempt suicide. She tells Mr Villars:

> In a moment, strength and courage seemed lent me as by inspiration: I started, and rushing precipitately into the room, just caught his arm, and then, overcome by my own fears, I fell down at his side, breathless and senseless. (183)

Her recovery from this swoon is 'almost instantaneous', however, and she seizes the pistols and takes them away:

> The moment I reached again the room I had so fearfully left, I threw away the pistols, and flinging myself on the first chair, gave free vent to the feelings I had most painfully stifled, in a violent burst of tears, which indeed, proved a happy relief to me. (184)

Here we see again the way in which pent-up emotion needs release. As before, physiology is intimately bound up with emotion, and emotion leads to morally significant action: it is 'inspiration'—breath as spirit—that leads Evelina to act as she does, and the force of this lends her courage to act to protect Macartney, resulting in the brief breathless swoon. Once she has disposed of the pistols, the operation of breath in relation to emotion is graphically expressed in the 'vent' given by tears to her feelings.

The rapid transitions in this episode and the fears Evelina has for Macartney return us to the intrinsic involvement of breath and breathlessness in matters of life and death. 'Breathless' here means in a faint, but it was as commonly used in eighteenth-century sentimental literature as a synonym for 'dead'.[22] When, for example, Pamela contemplates suicide by the pond's edge, she imagines that 'when they see the dead Corpse of the unhappy *Pamela* dragg'd out to these slopy Banks, and lying breathless at their Feet, they will find that Remorse to wring their obdurate Hearts, which now has no Place there!'.[23] In Henry Mackenzie's *The Man of Feeling* of 1771, Old Edwards gazes 'on the breathless remains' of Harley at the novel's close.[24] Eighteenth-century usage included the meaning 'out of breath', and 'breathless' is also frequently employed in this way, but the potential danger of breathlessness is brought closer by the dual uses of the word, as *The Man of Feeling* shows. Upset on hearing that his beloved Miss Walton is to be married, Harley ruminates in his garden and is then concerned to see 'a servant, with a knot of ribbands in his hat, go into the house'.[25] Thinking this is a messenger from the Waltons with news of the engagement, he hastens to find out, but

> When he approached the door of the kitchen where he supposed the man had entered, his heart throbbed so violently, that when he would have called Peter, his voice failed in the attempt. He stood a moment listening in this breathless state of palpitation: [...][26]

The inter-relation of heart and lungs is illustrated beautifully in this passage—the throbbing of Harley's heart prevents speech. Again, the way voice and breath are bound up together is key. In his susceptibility to such suspension of action, the man of feeling's early demise is anticipated.

One of the defining features of the sentimental protagonist, as with Harley in this passage, is the degree to which sensibility disables them from taking action: the death-like swoon that overtakes Pamela, Evelina, and many other heroines of sensibility (as well as some heroes) is the ultimate example of this powerlessness. The effect of breathlessness on the voice, preventing or distorting utterance, is part of this suspension of the will, an effect with potentially dangerous consequences in certain circumstances. Evelina, led into the 'dark walks' at Vauxhall Gardens by the foolish and imprudent Branghton sisters, runs panic-stricken from one party of intimidating gentlemen into another:

> I was almost distracted with terror, and so breathless with running, that I could not speak, till another [man] advancing, said, I was as handsome as an angel, and desired to be of the party. I then just articulated, 'For Heaven's sake, Gentlemen, let me pass!'
>
> Another, then, rushing suddenly forward, exclaimed, 'Heaven and earth! what voice is that?—'
>
> 'The voice of the prettiest little actress I have seen this age,' answered one of my persecutors.
>
> 'No,—no,—no,—'I *panted out*, 'I am no actress, pray let me go,—pray let me pass—.' (197–98)

Barely articulate, panting, Evelina's breathless entreaties seem unlikely to move her persecutors, but she is rescued by the fortuitous appearance of Sir Clement—a persecutor more familiar to her. Sir Clement takes her away from the threatening crowd, but attempts to steer her into 'another of the dark alleys', '"Where"', as he explains, '"we shall be least observed"' (198). As she realises what he is doing, Evelina's feelings undergo a metamorphosis from terror to anger, physiologically experienced: she tells how her 'heart beat with resentment', and she demands 'how he dared treat [her] with such insolence'. Part of Sir Clement's problem throughout the novel is that he cannot place Evelina socially: is she a young lady of birth or is she a nobody? Can he assault her with impunity or does she have an army of protectors at her back? Finding her alone in the dark walks, he implies, is tantamount to confirming her social nullity and sexual availability. Evelina tells her reader that she is '[e]xtremely offended at this speech', and doesn't 'deign' to answer, instead walking on 'towards that part of the garden whence I perceived lights and company':

> 'So you will not explain to me your situation?' said he, at length.
> 'No, Sir,' answered I, disdainfully.
> 'Nor yet—suffer me to make my own interpretation?—'
> I could not bear this strange manner of speaking; it made my very soul shudder,—and I burst into tears. (199)

The pattern of high emotion followed by the relief of tears that we saw in the scene of Macartney's supposed attempted suicide, just a little earlier in the novel, is repeated, but the tears she sheds in this later scene are tears of mortification and of anger, not of fear or grief—a different kind of 'venting'—and her indignation has the effect of successfully reducing Sir Clement to the position of breathless suppliant:

> 'O Miss Anville—loveliest of women—forgive my—my—I beseech you forgive me;—if I have offended,—if I have hurt you—I could kill myself at the thought!—' (199)

The power dynamics of this scene are always interesting: Evelina's deliverer becomes her persecutor, from whom she effectively delivers herself. Mary Favret regards *Evelina* as 'the epitome' of the sentimental epistolary novel in English, using the term 'memoir-letter' to characterise its method.[27] Of the form more generally, she writes: 'the memoir-letter encourages resignation. The form produces and foregrounds the letter-writer's vulnerability, her lack of self-determination, and her ineffectual role in events [...] The form thus seems to foster the victimization of the ingenuous letter-writer.'[28] Given Evelina's precarious social position and the way she seems frequently to be at the mercy of characters more forceful, selfish, and powerful than she, be it Captain Mirvan, Sir Clement, or Madame Duval and the Branghtons, such a reading of the form can seem persuasive. Yet Betty Rizzo urges that 'Too much has been made of the "passivity" of both Burney and Evelina', contending that Evelina 'has a distinguished sense of self, the ability [...] to act against convention'.[29] Rizzo identifies 'at least three times' when Evelina takes independent action, in relation to Macartney and her father; her successful management of Sir Clement at Vauxhall could be added to this list, as she moves from breathless inarticulacy to a combination of indignant self-defence and deliberate withholding of speech. There is admittedly a distinct limit to Evelina's power in this case, however. Sir Clement characteristically turns suppliance into coercion and, she tells Mr Villars, 'besought me to forgive him, with such earnestness of supplication, that, merely to escape his importunities, I was forced to speak', though 'with a very ill grace' (200)—but nevertheless by this time she has almost got herself back to the safety of the 'general crowd', and the energy she exhibits in the scene is very far from the stasis experienced by Harley in *The Man of Feeling*, just as the rapidity with which she recovers from her swoon enables her to remove the pistols from Macartney's room. Furthermore, her energy in both these cases is morally directed: where the sexual threat of the amorphous group of young men elicits a visceral 'terror' that renders Evelina breathless and powerless, Sir Clement's duplicity and lack of honour arouse her sense of justice and her pride, while the imminence of the apparent threat to Macartney requires immediate action to preserve life.

It is no coincidence that the three episodes identified by Rizzo in which Evelina acts for herself are all also episodes of intense emotional investment and moral significance in which breath plays a notable role. One of these is Macartney's apparent suicide attempt. The other two are Evelina's decision to meet Macartney at the garden gate in Clifton and her refusal to leave her father when he commands her to do so, in order to effect a reconciliation with him. These instances of independent action are far from straightforward, however, and the conflicting pressures involved are signalled by the fractured and breathless exchanges that take place between the principal actors in the scenes.

Having met Macartney by accident outside Mrs Beaumont's house in Clifton early one morning, Evelina impulsively agrees to meet him there again the following day, partly hurried into doing so by the unexpected presence of Lord Orville, who has come to bring her in to breakfast. It is a decision she rapidly regrets, especially on seeing Orville's surprise, and one she attempts to reverse: she writes cancelling the appointment as politely as she can, mindful of Macartney's 'peculiar situation, his misfortunes, his sadness, and, [...] the idea I knew he entertained of what he calls his obligations to me' (300). When the letter cannot be delivered, she reluctantly concludes it is 'incumbent upon me to keep my word' and sets out to meet Macartney, but the meeting never takes place because Lord Orville—deliberately, we are to understand—is also out for an early morning stroll:

> [...] uncertain whether I was wrong or right, it was with fear and trembling that I opened the garden-gate,—judge, then, of my feelings, when the first object I saw was Lord Orville!—he, too, looked extremely disconcerted, and said, in a hesitating manner, 'Pardon me, Madam,—I did not intend,—I did not imagine you would have been here so soon,—or,—or I would not have come.'—And then, with a hasty bow, he passed me, and proceeded to the garden.
>
> I was scarce able to stand, so greatly did I feel myself shocked; but, upon my saying, almost involuntarily, 'Oh my Lord!'—he turned back, and, after a short pause, said, 'Did you speak to *me*, Madam?'
>
> I could not immediately answer; I seemed *choaked*, and was even forced to support myself by the garden-gate. (302)

Evelina's anxiety about whether she is doing the right thing becomes paralysing here. The consternation, however, is to some extent mutual, and the net result of this encounter is that Evelina does *not* go through with her plan to meet Macartney. It shows her trying to think for herself in order to discern the right course of action, but this does not result in a clear-cut decision. Instead, the episode underscores both the interconnections between emotional, moral, and physiological discernment and the difficulties involved in properly interpreting their signals.

The third of Rizzo's decisive actions—Evelina's refusal to leave her father's presence despite his commands—is similarly complicated. The first meeting between father and daughter ended as we have seen above, with Sir John Belmont rushing in disorder from the room. The second meeting almost ends in the same way, with Sir John crying 'passionately', '"go, go! [...] in pity—in compassion,—if thou valuest my senses, leave me,—and for ever!"' (383). Evelina, 'greatly terrified', cries '"I will, I will!"', moving 'hastily towards the door', but

> stopping when I reached it, and, almost involuntarily, dropping on my knees, 'Vouchsafe,' cried I, 'oh, Sir, vouchsafe but once to bless your daughter, and her sight shall never more offend you!' (383)

This action, which leads to a 'soften[ing]' of Sir John's voice and an extended, highly charged conversation in which father and daughter are properly reconciled, is 'almost involuntary' and its breathlessness is signalled in the punctuation—in the dashes of Sir John's passionate cry and in the exclamation marks and commas of Evelina's response. This barely intentional action suggests an instinctive moral discernment at an emotional and physiological level, as is the case, indeed, with the Macartney episode.

Many have noted that *Evelina* can be seen as a species of *Bildungsroman*, and Rizzo sees it as presenting a process of maturation that makes Evelina worthy of Orville, a process encouraged by Villars' exhortation that she must 'learn not only to *judge*, but to *act*' for herself.[30] Evelina's actions at key moments do not seem to allow for a division of human behaviour into judging and acting, however; rather they involve a visceral impulse that links them to the notion of a moral sense as developed by Francis Hutcheson, for whom such a sense operated precisely in advance of rational reflection. In his *Inquiry into the Originals of Our Ideas of Beauty and Virtue* (1725), he instances celebrated English and German philosophers to ask rhetorically,

> must a Man have the Reflection of CUMBERLAND or PUFENDORF, to admire Generosity, Faith, Humanity, Gratitude? Or reason so nicely to apprehend the Evil in Cruelty, Treachery, Ingratitude? Do not the former excite our Admiration, and Love, and Study of Imitation, wherever we see them, almost at first View, without any such Reflection, and the latter, our Contempt, and Abhorrence?[31]

Rather than providing evidence of maturation, then, Evelina's actions to save Macartney and to reconcile with her father could be argued to encapsulate such an almost instantaneous recognition, perceiving 'almost at first View, without any [...] Reflection' the essential moral value of both Macartney's life and the bond between father and daughter. The problem of whether to meet Macartney at the garden gate is more genuinely a dilemma (Orville calls it 'a point so delicate' [299]), since there are competing priorities—letting down the distressed and vulnerable

Macartney and the imperative of keeping his secrets versus the impropriety of meeting a man illicitly and unchaperoned. In this case, Evelina's natural impulses pull her in opposing directions, resulting in paralysis, while in the other two instances fundamental values—the preservation of life and the filial bond—override all other considerations and spur her to act.

Considering the representation of breath and breathlessness in *Evelina* leads to a greater understanding of their centrality, hitherto underappreciated, in the construction of the essential features of the sentimental protagonist in the eighteenth century. It also draws attention to the particular role breath plays in the protagonist's capacity to act. Breathlessness becomes not simply the standard marker of sentimental incapacity, of paralysis, but also the signifier of intense and instinctive moral discernment of the kind described by Hutcheson and which, in certain circumstances of crisis, is the vital condition for action. Literary sensibility has long been seen by critics as bound up with philosophical developments of the late seventeenth and eighteenth centuries; the link between breathlessness and moral discernment is a further and useful refinement of this connection.[32] It also contributes to a revision of the view that the heroine of epistolary fiction more generally, and Evelina in particular, is purely passive—done to, rather than doing. Instead, she emerges as actively involved in numerous scenarios that at once challenge her capacity for moral conduct and allow her to demonstrate her power to act.

Notes

1. Frances Burney, *Evelina*, ed. by Edward A. Bloom and intro. by Vivien Jones (Oxford: Oxford University Press, 2002), 27.
2. Samuel Richardson, letter to Lady Bradshaigh, 14 February 1754, quoted in Tom Keymer, *Richardson's 'Clarissa' and the Eighteenth-Century Reader* (Cambridge: Cambridge University Press, 1992), 73. For discussions of 'writing to the moment' and epistolary immediacy, see for example Janet Altman, *Epistolarity: Approaches to a Form* (Columbus: Ohio State University Press, 1982), 123–24; Janet Todd, *Sensibility: An Introduction* (London: Methuen, 1986), 85–87; Keymer, *Richardson's Clarissa and the Eighteenth-Century Reader*, 1–15, and Joe Bray, *The Epistolary Novel: Representations of Consciousness* (London: Routledge, 2003), 54–59.
3. See *OED* meanings 1(b) 'The shock felt on the chest-wall when the heart beats'; 3(a) 'Formerly, esp., A strong suggestion supposed to come

from a good or evil spirit' and 3(c) 'Sudden or involuntary inclination or tendency to act, without premeditation or reflection'.

4. See for example Ann Jessie Van Sant, *Eighteenth-Century Sensibility and the Novel: The Senses in Social Context* (Cambridge: Cambridge University Press, 1993), Robert Erickson, *The Language of the Heart, 1600–1750* (Philadelphia: University of Pennsylvania Press, 1997), G. J. Barker-Benfield, *The Culture of Sensibility: Sex and Society in Eighteenth-Century Britain* (Chicago: University of Chicago Press, 1992) and George Rousseau, *Nervous Acts: Essays on Literature, Culture, and Sensibility* (Basingstoke: Palgrave Macmillan, 2004).
5. Alexander Pope, *Eloisa to Abelard* (51–58), in *The Poems of Alexander Pope*, ed. by John Butt (1963; London: Routledge, 1989). Further references to Pope are to this edition. Laura Alexander points out a similar passage in John Hughes's *Letters of Abelard and Heloise* (1713), a great influence on Pope ('"Breathings of the Heart": Reading Sensibility in Pope's *Eloisa to Abelard*', *New Perspectives on the Eighteenth Century* 3/1 (Spring 2006): 34–35).
6. Lady Dorothea Du Bois, *The Lady's Polite Secretary; or, the New Female Letter Writer* (London: J. Coote and T. Evans, 1771), 8.
7. Mary Favret, *Romantic Correspondence: Women, Politics and the Fiction of Letters* (Cambridge: Cambridge University Press, 1994), 24, quoting Hugh Blair, *Lectures on Rhetoric and Belles Lettres*, 14th edn (London: T. Caddell, et al., 1825), 496.
8. Bradford K. Mudge, *The Whore's Story: Women, Pornography, and the British Novel, 1684–1830* (Oxford: Oxford University Press, 2000), 218.
9. A sob is defined in the *OED* as 'a convulsive catching of the breath under the influence of grief'.
10. Samuel Richardson, *Pamela*, ed. by Thomas Keymer and Alice Wakely (Oxford: Oxford University Press, 2001), 60; *Clarissa*, ed. by Angus Ross (London: Penguin Books, 1985), 583.
11. Erickson, *The Language of the Heart*, 1–2.
12. As Clark Lawlor explains in *Consumption and Literature: The Making of the Romantic Disease* (Basingstoke: Palgrave Macmillan, 2006), 'Galen conceptualised the lungs as the part of the body where food, transformed by the liver into blood with natural spirit, became infused with air or vital spirit (a form of *pneuma* or life force). Vital spirits would then flow to the brain—which would use it to make animal spirits to enable motion—and the rest of the body. Breath was life, both spiritual and physical. Galen's ideas appealed to the Christian ages to follow because his vitalistic ideas of an animating life force were easily transformable into a non-pagan concept of the soul' (18).
13. Thomas Cooke, *The Universal Letter-Writer; or, New Art of Polite Correspondence* (London: J. Cooke, 1770[?]), 147.

14. Francis O'Gorman, 'Coleridge, Keats, and the Science of Breathing', *Essays in Criticism* 61/4 (October 2011), 366.
15. Richardson, *Pamela*, 60.
16. Ann Toner, *Ellipsis in English Literature: Signs of Omission* (Cambridge: Cambridge University Press, 2015), 21 and 84, quoting John Mullan, *Sentiment and Sociability: The Language of Feeling in the Eighteenth Century* (Oxford: Clarendon Press, 1988), 61. As Toner points out, one of the challenges in interpreting ellipses in printed works is that 'Decisions relating to these features are more often in the domain of publishers and printers than authors. Punctuation seems precariously exposed to non-authorial management in a way that word choices are not' (15). For an example of an eighteenth-century author's close involvement with the minutiae of presentation, see Joe Bray's '"Attending to the *minute*": Richardson's revision of italics in *Pamela*' in *Ma(r)king the Text: The Presentation of Meaning on the Literary Page*, ed. by Joe Bray, Miriam Handley and Anne C. Henry (Aldershot: Ashgate, 2000). A further difficulty in the literary analysis of many eighteenth-century texts is that manuscripts were commonly disposed of by the printer after publication. In the case of *Evelina*, however, a portion of an early manuscript does survive. Joyce Hemlow's comparison of passages of the manuscript with the printed text suggests a reduction of the use of the dash in keeping with the general trend of revisions that included the 'tightening of the sentence structure [and] replacing of direct, artless, and informal effect by one of greater finish, dignity, and elegance' (Joyce Hemlow, *The History of Fanny Burney* [Oxford: Clarendon Press, 1958], 80). While it is impossible to be certain, the placing of dashes in the published novel does therefore seem likely to be Burney's own.
17. Madame D'Arblay [Frances Burney], *Memoirs of Doctor Burney* (London: Edward Moxon, 1832), vol. 1, 148, quoted in Hemlow, *The History of Fanny Burney*, 97.
18. *The Early Journals and Letters of Fanny Burney*, ed. by Lars E. Troide and Stewart Cooke (Montreal: McGill-Queen's University Press, 1994), vol. 3, 31, quoted in Hemlow, *The History of Fanny Burney*, 97. This anecdote recalls Aaron Hill's account of the little boy whose 'Succession of heart-heaving Sobs' on hearing read aloud the scene from *Pamela* when the heroine contemplates suicide at the pond's edge confirmed him as 'the youngest of *Pamela's Converts*' (Richardson, *Pamela*, Appendix 1, 515).
19. Annamarie Jagose, '"Critical Extasy": Orgasm and Sensibility in *Memoirs of a Woman of Pleasure*', *Signs: Journal of Women in Culture and Society* 32/2 (2007), 473.
20. *Critical Review*, 46 (1778), 202–203.
21. See Van Sant, *Eighteenth-Century Sensibility and the Novel*, 8–15, for a discussion of the often loose relationship between developing scientific

understandings of physiology and the popular deployment of physiological terms in the literature of sensibility.

22. The *Oxford English Dictionary* records the meanings of 'breathless' as '1. a. Without breath; b. Lifeless, dead; c. *Grammar* Unaspirated; 2. a. Breathing with difficulty, panting (as a result of swift running or violent exercise); out of breath, exhausted, spent; b. Holding one's breath, as with awe, expectation, excitement; 3. Unstirred by a breath of wind'.
23. Richardson, *Pamela*, 172.
24. Henry Mackenzie, *The Man of Feeling*, ed. by Brian Vickers (Oxford: Oxford University Press, 1987), 131.
25. Mackenzie, *The Man of Feeling*, 109.
26. Mackenzie, *The Man of Feeling*, 110.
27. Favret (*Romantic Correspondence*, 229, n. 24) explains that the term derives from Vivienne Mylne, *The Eighteenth-Century French Novel: Techniques of Illusion* (Manchester: Manchester University Press, 1965), 150–51 and François Jost, '"L'Evolution d'un genre: le roman epistolaire dans les lettres occidentales"', in *Essais de littérature comparée* (Urbana: University of Illinois Press, 1968), vol. 2, 124–25.
28. Favret, *Romantic Correspondence*, 63.
29. Betty Rizzo, *Companions Without Vows: Relationships Among Eighteenth-Century British Women* (Athens, GA: University of Georgia Press, 1994), 88, 109.
30. Rizzo, *Companions Without Vows*, 88; Burney, *Evelina*, 166. For considerations of *Evelina* as *Bildungsroman*, see for example Susan Fraiman, *Unbecoming Women: British Women Writers and the Novel of Development* (New York: Columbia University Press, 1993) and Emily Allen, 'Staging Identity: Frances Burney's Allegory of Genre', *Eighteenth-Century Studies* 31/4 (Summer 1998): 433–51.
31. Francis Hutcheson, *An Inquiry into the Original of Our Ideas of Beauty and Virtue* (1725; London: R. Ware, et al., 1753), 120. Hutcheson is one of the key philosophers of the Benevolist School whose writings underpin many aspects of the culture of sensibility. His belief in a moral instinct was developed from Shaftesbury's work, but it was rejected by later thinkers such as David Hume and Adam Smith. See John Dussinger, *The Discourse of the Mind in Eighteenth-Century Fiction* (The Hague: Mouton, 1974), 47–49.
32. See for example John Mullan, *Sentiment and Sociability: The Language of Feeling in the Eighteenth Century* (Oxford: Clarendon Press, 1988) and Adela Pinch, *Strange Fits of Passion: Epistemologies of Emotion, Hume to Austen* (Stanford: Stanford University Press, 1996).

Selected Bibliography

Alexander, Laura. 2006. 'Breathings of the Heart': Reading Sensibility in Pope's *Eloisa to Abelard*. *New Perspectives on the Eighteenth Century* 3/1: 32–45.

Bray, Joe, Miriam Handley, and Anne C. Henry, eds 2000. *Ma(r)king the Text: The Presentation of Meaning on the Literary Page*. Aldershot: Ashgate.

Favret, Mary. 1994. *Romantic Correspondence: Women, Politics and the Fiction of Letters*. Cambridge: Cambridge University Press.

Hemlow, Joyce. 1958. *The History of Fanny Burney*. Oxford: Clarendon Press.

Lawlor, Clark. 2006. *Consumption and Literature: The Making of the Romantic Disease*. Basingstoke: Palgrave Macmillan.

Mullan, John. 1988. *Sentiment and Sociability: The Language of Feeling in the Eighteenth Century*. Oxford: Clarendon Press.

Rizzo, Betty. 1994. *Companions Without Vows: Relationships Among Eighteenth-Century British Women*. Athens, GA: University of Georgia Press.

Todd, Janet. 1986. *Sensibility: An Introduction*. London: Methuen.

Toner, Ann. 2015. *Ellipsis in English Literature: Signs of Omission*. Cambridge: Cambridge University Press.

Van Sant, Ann Jessie. 1993. *Eighteenth-Century Sensibility and the Novel*. Cambridge: Cambridge University Press.

CHAPTER 13

'Eloquence and Oracle': Tobacco in Eighteenth-Century Life and Literature

Andrew Russell

Human beings need breath to function, to speak, and to do many things that are essential to life. They also use breath to inhale products that are variously pleasurable or efficacious. One such product is tobacco. From its origins in the Americas, tobacco arrived on European shores in substantial quantities from the early seventeenth century onwards. While some was imbibed as snuff or chewing tobacco, the bulk was added, as smoke, to breath. What were the consequences of the high and intense levels of tobacco consumption, particularly among the male literati of the time, on human life and thought? Can we draw parallels between the influence of tobacco in the eighteenth century and its place in indigenous societies in lowland South America where its evolutionary and human histories are longest? To answer such questions I bring anthropology, science, and technology studies into communication with literature and the other creative arts, based on the assumption that the latter can reveal 'those dimensions of personal and social life we may have little access to when using other research strategies'.[1] Using an interdisciplinary approach I

A. Russell (✉)
Department of Anthropology, Durham University, Durham, UK

D. Fuller et al. (eds.), *The Life of Breath in Literature, Culture and Medicine*, Palgrave Studies in Literature, Science and Medicine,
https://doi.org/10.1007/978-3-030-74443-4_13

argue through comparative analysis for the significance of tobacco as a hitherto unacknowledged agent in the development of the epoch we now call 'the Enlightenment'.

Tobacco is an example of things which, for ethnographers, 'may appear as material objects; as practices or concepts; as events, institutions or beliefs; as gifts, mana, traps, actants, spirits or dividuals; or as structures, perspectives, networks, systems or scales'.[2] Tobacco has multiple dimensions of 'thingness', through the multiple hybrid relationships it establishes with people. Lupton et al. use the example of 'the swimmer, who while wearing the goggles, embodies the ability to swim differently, and the goggles, who when wearing the swimmer, acquire the ability to move'.[3] We can think similarly about, and reflect on, the relationship between people and tobacco. For example, the person who inhales tobacco smoke embodies an altered state of consciousness, while tobacco, incinerated and penetrating the lungs on the breath of the smoker, acquires the ability to increase the spread and reach of its species.[4] Geismar distinguishes between objects with impact (like falling meteorites) and objects with agency (which, to be effective, must be 'entangled within social relations and indeed within our own humanity').[5] Tobacco, I argue, is both—the agency it garners through hybrid relationships with people is matched by the impact/agency it develops in its own right. My essay is thus part of a broader framework, traversing both anthropology and literary studies, associated with notions such as actor-networks, 'material agency', and 'thing power'. It invites us to think differently about tobacco, both temporally and spatially—to consider its arrival on European shores as a mere blip in time, and its transmogrification into cigarettes as an unseemly aberration of twentieth- (and now twenty-first-) century capitalism.

In what follows, I first consider some significant features of the 'deep history' of tobacco in lowland South America—its role in constituting persons, changing perspectives, and generating dualisms. I then explore the possibility that the impact of tobacco in eighteenth-century European life was associated with similar manifestations of material agency, and how its entanglements with people can be discerned through the art and literature of the period. Finally I identify two eighteenth-century literary genres where tobacco's influence may particularly reverberate, namely the popular object- or 'it'-narratives of the time, and the less commonly recognized or celebrated 'poetry of attention'.[6]

A Deep History of Tobacco in Lowland South America

Tobacco Constituting Persons, Shifting Shapes, and Changing Perspectives

Tobacco has a 10,000-year history of engagement with humanity in lowland South America. For the Muinane, Witoto, and Andoke, the self-styled 'People of the Centre' living along the Brazilian/Colombian border, tobacco actively constitutes persons, exemplified in an elder's remark that 'we all have the same hearts, made from the same tobacco'.[7] Among the Xié of Amazonia the newborn is a 'little fish': a creature fluid and vulnerable, in need of forming into a fully-fledged human being.[8] Tobacco smoke serves to cool, dry out, and firm up—to humanize the less than human. So great is this sense of tobacco constituting persons that among the People of the Centre differences that westerners might consider cultural are attributed to differences between their tobaccos.[9]

In contrast to western notions of the 'natural body', regarded as a generally stable substrate for the range and diversity of multiple human cultures, for Amazonian peoples culture is the stable constant, a 'soul essence' that underpins and is shared with a multiplicity of natural forms of which humans are only one. The 'chronically unstable'[10] boundaries within and between the variety of human and non-human animals, plants, and objects require careful nurturing, and tobacco is a key component in this process. Another manifestation of Amazonian body instabilities is belief in the propensity of bodies to alter and transform, a property known as shape-shifting.[11] This shape-shifting may also occur when, through ingestion of prodigious quantities of tobacco, shamans become were-animals (particularly jaguars), combative heroes against opposing 'spirits, sorcerers, sickness, and death'.[12] Tobacco, a mainstay of shamanic practice, is strongly implicated in this shape-shifting agenda.[13]

Shape-shifting is not always an easy matter for Amerindians to deal with. 'Metamorphosis is something that haunts the native imagination', writes Vilaça about the Wari' in the Rodonia region of Brazil.[14] A Wari' story relates the luring of a child into the forest by her mother to pick fruit. One day the child realizes they have spent an inordinate amount of time away from home and notices a tail discreetly hidden between her mother's legs. She cries out in alarm and the jaguar flees, leaving a trail of paw-prints. The anthropologist who recorded this story relates that one woman, telling her about this event, said that the girl's true mother hereafter warned the girl always to distrust other people: 'Whenever she went far from home, either with her mother or father, she should take along a brother or sister as company (in order, I assume, to secure her point of view)'.[15]

This extends into what the anthropologist Eduardo Viveiros de Castro describes as 'the conception, common to many peoples of the continent, according to which the world is inhabited by different sorts of subjects or persons, human and non-human, which apprehend reality from different points of view'.[16] So marked is this recognition of perspectives other than one's own, and so profound its consequences, that Viveiros de Castro has coined the term 'cosmological perspectivism' to encompass it. Perspectivism leaves 'indelible, though elusive, marks'.[17] A compelling example is a pattern of speech common to many lowland South American groups, reflected in phrases such as '*this is beautiful to me*', '*for him, the animal turned into a jaguar*', and '*to us, there appeared prey while we were making the canoe*', as Stolz-Lima records in the Portuguese spoken by the Juruna of Brazil.[18]

This grammatical relativization of 'self' and 'non-self' offers a constant perspectival reminder that, while this is how it was for me (or us), others might see things differently. The ability to shift perspective in this way is a recurrent trope in lowland South American folklore. I suggest there are discernible parallels in eighteenth-century Enlightenment life, literature, and the creative arts.

Tobacco Generates Dualisms

Tobacco's ability to generate dualisms is exemplified by the structuralist paradigm of Claude Lévi-Strauss, which in turn derives from his claims about deep, fundamentally binary structures of the human mind. The myths analyzed by Lévi-Strauss in his four-volume *Mythologiques* come from across the Americas, and correspond with 'the limits of tobacco distribution in the New World'.[19] Lévi-Strauss's fundamental categorical premise, indeed what has been described as a core Amerindian cosmological concern,[20] is the opposition of nature and culture, whatever these two poles might contain in specific circumstances. Volume Two of *Mythologiques* (*From Honey to Ashes*) presents one such mythic system, one which, significantly enough, 'revolves round the central theme of tobacco'.[21]

Where they are acknowledged, the binaries inherent in much structural theory seem to fit the ethnographic data in North and South America surprisingly well. This is precisely the same hemisphere marked by what Wilbert calls an ubiquitous 'tobacco ideology' which, he suggests, indicates 'the extraordinary power of this agent of diffusion'.[22] Could it be that tobacco constituting persons, bending perspectives, and possibly even generating the binary oppositions on which Lévi-Strauss was able to build his theories may have had something to do with the long-term relationship between plant and people in the region? I argue for its influence on the life and thought of the people living there over the millennia. Can similar influences be discerned in some of the intellectual changes that helped precipitate what we now call the 'Enlightenment'?

Tobacco in Enlightenment Life and Thought

Various reasons have been posited to explain the emergence of the philosophical and cultural epoch known as the 'Enlightenment' in Europe. One element that I feel has been ignored or overlooked in such analyses is the arrival of tobacco on European shores. It is suggestive that the word Enlightenment is first recorded in the English language in 1621, at around the same time as tobacco was starting to make substantial inroads into European life. Given the impact and agency of tobacco, a case can be made for including this powerful substance as a significant but hitherto neglected component in the history of the period.

The Stupefying Pleasures and Material Sociality of Tobacco

At the turn of the seventeenth century, the average price of Spanish American tobacco in London was about £1.10s. per pound; labourers' wages were about 8d. per day. However, consumption increased as the supply of tobacco from the nascent Virginian colonies started to grow from 1640 onwards. By the start of the eighteenth century, Davies estimates that tobacco imports into England totalled around 26 million pounds annually. Sixty per cent of this was re-exported to other parts of Europe, but the remaining 10.4 million pounds still permitted a generous one and a half pounds of tobacco per inhabitant per year—enough for twenty-five per cent of the adult population to enjoy a pipe and a half every day.[23] The Dutch were consuming tobacco in similar proportions.[24]

Social spaces such as coffee houses constituted a growing 'public sphere' offering opportunities for people across classes to meet (and smoke) together. Tobacco in eighteenth-century Britain was deeply imbricated in whatever conversations took place in these settings. For Porter, 'to be enlightened, a gentleman had to be sociable'.[25] This was also, he argued, an era marked by 'a *philosophy* of expediency, a dedication to the art, science and duty of living well in the here and now'.[26] Such 'materialistic worldliness' was fostered in and by the 'bubbling commercial atmosphere' of eighteenth-century England. In such a world, people's concerns were with 'the here-and-now, in matters tangible, buyable, disposable'. Tobacco, sugar, and tea were 'the first objects within capitalism that conveyed with their use the complex idea that one could become different by consuming differently…closely connected to England's fundamental transformation from a hierarchical, status-based medieval society to a social-democratic, capitalist, and industrial society'.[27] The melding of human and non-human through consumption, the 'constituting of persons', was a prominent trope in the life and literature of eighteenth-century England. In the process, tobacco was recognized, as the character Tobacco itself states in the seventeenth-century play *Wine, Beere, Ale and Tobacco: Contending for Superiority*, as a substance whose 'divine breath…doth distill eloquence and oracle upon the tongue'.[28] The capacity for eloquence and oracle was a significant contribution to the development of the 'Enlightenment'.

Sahlins sees the pleasure principle as a fundamental part of the Enlightenment, when 'self-pleasing came out of the shadow of its sinful ancestry to assume a moral position nearly 180 degrees removed. The individual's singular attention to his own good turned out to be the basis of society rather than its nemesis—as well as the necessary condition of the greatest wealth of nations'.[29] Tobacco played an important role in helping 'to frame a distinctively early modern culture in which the pursuit of pleasure was thereafter more public, routine, and unfettered'.[30]

Israel is critical of those who would make too much of 'new eighteenth-century social spaces and practices…in generating Enlightenment ideas'.[31] Yet if anything was uniting people from across the social classes, it was tobacco, 'the Old Man's Solace, and the Student's Aid', as a poem published in *The London Medley* in 1731 put it.[32] The 'Convert to Tobacco', in *A Collection of Merry Poems* published in 1736, describes tobacco offering contentment to the Welsh farmer trudging barefoot through the snow ('With thee he warms his dripping Nose, / And scrubs, and puffs, and on he goes'), as well as to the 'Justice grave', who partners tobacco with ale to hold court at his table after dinner 'Whilst sober whiff fills each Hiatus'.[33] 'The Triumph of Tobacco over Sack and Ale' similarly emphasizes the social communion tobacco engendered:

> Tobacco engages
> Both Sexes, all Ages,
> The Poor as well as the Wealthy,
> From the Court to the Cottage,
> From Childhood to Dotage,
> Both those that are sick and the healthy.[34]

Snuff, later to become the *sine qua non* of distinction and refinement, was performing a similarly unifying role in Parisian society at this time. One commentator wrote:

> at court as well as in the city: princes, lofty lords, and the people all take snuff. It ranks among the favourite occupations of the noblest ladies, and the middle-class women who imitate them in everything follow them in this activity as well. It is the passion of prelates, abbés, and even monks. Despite the papal prohibition, priests in Spain take snuff during the Mass. The snuffbox lies open before them on the altar.[35]

Yet smoking remained the primary means of tobacco consumption. As Brook argues, 'whenever tobacco showed up, a culture that did not smoke became a culture that did...Not all the original meanings of Native smoking made the jump to other cultures...but many did, including the notion that tobacco opened a door to the spiritual realm'.[36] The ways in which tobacco was consumed in eighteenth-century Europe, however, were very different from those of the twenty-first century. An amusing and somewhat moralistic account of 'Sam Scot's Smoaking Club' in mid-eighteenth-century London describes its gentlemen members—a linen draper turned dancing-master, a city musician, an engraver, and a Scottish writer—as a group that 'had acquired such an expeditious Way of consuming a Pipe of Tobacco, that when they were met together, they would make no more of smoaking a Pound in an hour'.[37] This is tobacco consumption of potentially shamanic proportions. Simon Schama describes the Dutch artist Adriaen Brouwer who, in the 1620s and 1630s, 'took great care to record the expressions of deep inhalation or drowsy puffing peculiar to the serious pipe smoker. Some of their figures appear so stunned and insensate with smoke that it has been argued—speculatively—that their tobacco might have been spiked with some sort of opiate or narcotic'.[38] Such exotic speculations are unnecessary, however. The amount of tobacco being consumed was enough to generate the same kind of narcotic intoxication experienced by shamanic practitioners in lowland South America through the heavy ingestion of *N. tabacum* to this day. The stupefied silence frequently engendered was captured by a Russian historian who visited England in 1790:

> I have dropped into a number of coffeehouses only to find twenty of thirty men sitting around in deep silence, reading newspapers, and drinking port. You are lucky if, in the course of ten minutes, you hear three words…'Your health, gentlemen!'[39]

Tobacco and Embodied Cognition

Nowhere was tobacco use more prominent than among the literati, as is brilliantly evoked in Isaac Hawkins Browne's six satirical poems in praise of tobacco. Every one of them is written in the style of a different contemporary poet. For Browne, the 'light from smoke' upon which they all rely is tobacco. He sees tobacco as either a disreputable cousin to the classical Muses or the divine inspiration of the Christian tradition. 'While offering a sublime heightening of perception and thoughts, the inspiration she [tobacco] offers is debased by fleshly cravings, foul-smelling fumes, and addiction'.[40] Oxfordshire poet John Philips, while revelling in the ability of inhaled tobacco to 'suck new Life into my Soul', likewise noted the inferiority of a 'Muse from smoke' to a classical Muse, since even as it elevates the mind, tobacco weakens the body.[41] Tobacco could also, though, be described as 'Assistant Chief' to 'Country vicar': 'If text obscure perplex his Brain, / He scratches, thinks, but all in vain; / Till lighted Pipe's prevailing Ray, / Like *Phoebus*, drives the Fog away'.[42] In some cases 'to think and smoke tobacco'[43] became a profoundly spiritual experience, described by one Scottish Secessionist minister as bringing the mind closer to God, with tobacco's effects on the body likened to the Holy Spirit's even more powerful action on the soul. Inspiration was also attributed to snuff, which was becoming recognized for its stimulant properties: when pulverized into 'smart Rappee' (a coarse snuff) it was said to invigorate 'Sir Fopling's Brain, if Brain there be', giving him the ability to shine 'in Dedications, Poems, Plays'.[44]

Heretofore there has been little recognition or acknowledgement of the breadth and profundity of tobacco's influence on the 'embodied cognition' of individuals who regularly subjected themselves to such high levels of nicotian stupefaction. One fundamental effect, I surmise, was the development of binary notions of 'self' and 'non-self' or 'other', a prerequisite, if one were needed, for the generation of individuality. The narcotic effect of tobacco can only encourage the formation of such relationships. Some scholars see these changes in terms of the Renaissance humanist idea of Man boldly stepping out from behind God's shadow and becoming a unique, all-powerful entity.[45] For Riesman and co-authors, though, there are two distinct periods in the development of modern individualism: its fifteenth- and sixteenth-century emergence from older 'tradition-directed' social forms, and what they call an 'inner-directed' stage of intense individualism which they see as developing between the sixteenth and nineteenth centuries. This of course was precisely the period when tobacco was making such great inroads into Europe.[46] Martin points out significant regional variations between and within countries in the increasing production of 'ego documents' in the sixteenth and seventeenth centuries—variations he describes as 'baffling'.[47] Although pre-tobacco late Medieval Italy was similarly full of people writing letters, diaries, memoirs, journals, and such like,[48] the eighteenth-century effulgence of ego documents took place predominantly in England and the maritime provinces of the Dutch Republic (Friesland, Zeeland, Holland), the two areas of Europe where tobacco use was particularly heavy at that time.

Both the 'individual as expressive, self-reflective subject' in Martin's formulation of the performative type of selfhood, and the shades of cosmological perspectivism inherent in a self 'increasingly conscious about the need to assume different roles in different contexts', would have been encouraged by nicotian influences (see Table 13.1). However, as well as 'self' and 'non-self', other dualisms appear to have become prominent in European thought at a time corresponding with the arrival of tobacco.

Tobacco and the Scientific Separation of 'Facts' and 'Values'

Francis Bacon, for example, was one of the new breed of seventeenth-century philosopher-scientists whose arguments exemplified the trend towards the bifurcation of 'facts' and 'values', 'reality' and 'fantasy'. According to Bacon, 'traditional' modes of knowledge continually mixed

Table 13.1 Martin's three basic types of selfhood in Renaissance Europe (from Martin, 'The Myth of Renaissance Individualism', 210–11)

Type	*When?*	*Characteristics*	*Comments*
Communal or civic	Throughout period	Group or collective identity	Often based on family or lineage
Performative or prudential	Appeared quite suddenly in early sixteenth century	Individual as expressive, self-reflective subject	Increasingly conscious of need to assume different roles based on context
Porous or open	Late medieval and early modern	Body as porous, open to strong 'spiritual' forces from outside	Often through witchcraft or possession

up 'the reality of the world with its configurations in the minds of men'.[49] The 'cracks and deformities' of the mind, he argued, prevented it reflecting 'the genuine ray of things'. The only option, in Bacon's view, was to 'dissect the nature of this very world itself'. Withington highlights the 'perennial tensions' (dualisms all) that came into play as words such as 'modern', 'society', 'company', and 'commonwealth' sidled into use—all of them at much the same time as tobacco became a significant feature of the European intoxicant landscape. These include the tensions between received wisdom and personal experience; between reform and resistance; between public service and private profit; between the common good and its political organization; between idealism and power; between the social and the natural worlds.[50] We could add Bacon's 'reality' and 'imagination' to this list.

For Bruno Latour, attempting to champion one binary and ignoring the other results in the development of an even stronger hybrid relationship between the two. Through these means, one half of the polarity comes to assume some of the attributes of the other, be it nature–culture, reality–imagination, object–subject or tobacco–people. This is because as soon as one tries to become 'modern' by splitting the world into binaries such as 'nature' and 'culture', so these hard-fought distinctions are breached and the things we would call things become people, and the people we call people become things. We can see this development in the extraordinary eruption of non-human 'object'- or 'it'-narratives on the eighteenth-century English literary scene which I shall now go on to explore. The 'poetry of attention' likewise is a genre in which I argue nicotian influences are strong.

'It'-Narratives and the Poetry of Attention—Changing Perspectives?

Although there are isolated instances of objects being given the power of speech in earlier literary history, in eighteenth-century English literature things started vocalizing as never before. 'Can the thing speak?' the anthropologist Martin Holbraad asks with an avant-garde rhetorical flourish in 2011.[51] Well, in eighteenth-century English literature things were positively gabbling. The object- or 'it'-narratives featuring a non-human voice were satirical pieces of prose, the protagonists and narrators of which were 'mundane material objects such as banknotes, corkscrews, shoes, and coins that circulate through human society, commenting upon and damning it as they go'.[52] While we might associate such a genre today with literature for children, the initial audience was overwhelmingly adult. The genre's popularity is exemplified by the fact that one of the first examples, Charles Johnstone's *Chrysal: Or the Adventures of a Guinea*, went into a third edition within three years of its first publication in 1760.[53]

These non-human 'it'-narratives are a fascinating genre, particularly given the potential of tobacco to influence cosmological perspectivism and in the context of Latour's remarks about modernity. The complex modern relationship that was developing between man and nature, the human and non-human, is reflected in the philosophical dissolution of 'nature' and 'culture' in Locke's *Essay Concerning Human Understanding*, published in 1690. Using arguments reminiscent of Lucretius, Locke argues that we cannot determine whether thought occurs in our souls, or in 'some systems of matter fitly disposed', such as our brains. It is perhaps no coincidence, given this shift to more material connections, that James Arbuckle commented in 1719 on the likeness between the tobacco leaf and the human brain.[54] John Hill, cautioning against the immoderate use of snuff in 1761, argued that the nose offered a direct path to the brain, its nostrils 'covered, in a manner, with branches of nerves: and these so thinly guarded from the air, that the brain itself may be said to lie almost naked there'.[55]

The 'sentient matter' question, as it came to be known, revolved around whether sentience derived from a soul (as the immaterialists believed), or whether man was nothing more than matter whose thought occurred in and through the brain (as the so-called 'free thinkers', or materialists, argued). The immaterialists, attempting 'to maintain the ontological privilege of humanism',[56] could only accept matter as passive. An early lecture series endowed by the chemist Robert Boyle featured a sermon delivered by Richard Bentley entitled 'Matter and Motion cannot Think: or, a confutation of atheism from the faculties of the soul'. Bentley aimed to reconcile science with the rational belief in God and was strongly opposed to the 'vulgarly received notion of nature' espoused by the materialists. 'Sensation and Perception are not inherent in matter as such', Bentley argued, 'for if it were so; what monstrous absurdities would follow? Every Stock and Stone would be a percipient and rational creature'.[57]

Locke's views, rather than Bentley's, seem to have been the imaginative touchstone for the surge in popularity of 'it'-narratives. 284 published works are recorded between 1700 and 1900.[58] With them came terms and ideas which seem highly consonant with contemporary notions of non-human agency. Indeed the term 'material agency' was first used in William Jones' essay *First Principles of Natural Philosophy* (1762) in a manner remarkably similar to how it is used by Science and Technology Studies scholars today.[59]

A diverse range of things speak in object narratives, and they go to many places. Mobile objects are at an advantage. A banknote circulates 'within the space of five pages from a milliner, to a bishop's wife, to the bishop, to a bookseller, to a printer, to a pastry cook, and to a seller of dead dogs'.[60] Alternatively static places such as a Covent Garden pub or the Bank of England are invoked, but watch and comment on the human characters that pass by them. The rupee in Helenus Scott's *Adventures of a Rupee* (1782) spends most of its time in a pawnbroker's, the novel presenting a series of portraits of visitors to the shop, whose stories are told to the rupee by the spirits of gold.[61] Like the diverse explanations for the origins of Enlightenment, the explanations posited for the eruption of the 'it'-narratives and their success with the reading public largely reflect the disciplinary interests of those who propose them. Lynch, for example, suggests they were an attempt to soften (through humanizing) the new market system 'which made English men and women uneasy'.[62] According to Hudson, they were a means of placing the old aristocracy and the new mercantile classes on an equal footing, narratively speaking.[63] Less convincing is Flint's argument that speaking objects reflect authors' anxieties concerning the public exposure of their books.[64] However, as well as reflecting the impact of the growing class of *nouveau riche*, markets or books, 'it'-narratives offer important creative explorations of some of the philosophical preoccupations arising from the Enlightenment, such as people's increasing sense of separation from, and consequent objective interest in, the material world. This trend is predicated on an increasingly strong dissociation of the human from the non-human from the seventeenth century onwards. Insofar as tobacco is strongly implicated in the hybridization that ensued, tobacco is another explanation for the development of the 'it'-narrative form.

Another example of tobacco's influence in eighteenth-century literature is afforded by what one commentator calls 'the poetry of attention'. Koehler identifies this as an emergent poetic genre marked by 'a commitment in much of the period's poetry to teaching readers how to attend closely', a concern related 'to a more widely recognized impulse in eighteenth-century poetry to describe details and to proliferate objects... to focus on the minute, the miscellaneous, the detailed, the domestic'.[65] This description is also apt for the still life paintings so popular in the Dutch Republic at the time. One wonders whether tobacco might have had an influence on this artistic form, since one of the well-documented cognitive effects of tobacco is as a stimulus barrier, enhancing attentiveness through screening out extraneous and distracting stimuli.

The fascination with the ordinary and everyday evinced by the poetry of attention stems from what Koehler calls a 'methodical, experimental attentiveness'. The examples she uses are telling. 'How does the world look from a cat's vantage? What response does the ringing of a bell evoke from hungry sparrows? [...] In these poems the ordinary is defamiliarized; it is particularized according to the perception of one attentive viewer.'[66] Note the perspectivism reverberating in the notion of animals' vantage points. As in the 'it'-narrative form, there are smoky whiffs here of the cosmological perspectivism characteristic of lowland South American life and thought. Might the attention to detail and the perspectivism that accompanies it have been heightened by the prodigious consumption of tobacco? I suggest that the insinuation of tobacco into human affairs offers an explanation for the changes we can observe in literature and the arts in Europe at this time as a result of its embodied, cognitive effects. However the poetry of attention, unlike the 'it'-narrative form, does not seem to have caught the imagination of those working in literary studies, possibly because of a disinclination among scholars to grapple with some of its psychological theories and overtones. Tobacco may serve to 'fill the gaps' with a material substance that helps explain the cognitive developments and perspectival shifts Koehler observes.

The Silence of Tobacco

Given the literary fashion for things, plants, and animals to tell stories from particular vantage points, the apparent lack of any extant 'it'-narratives delivered by tobacco or its paraphernalia is paradoxical. Despite the acculturation of things ('nature') through narrative, tobacco, although an eminent candidate for such a task, appears to have remained steadfastly silent, rather than making its own voice heard, at least in prose. Perhaps the reason is a practical one. Unlike a coin of the realm (for example) tobacco is an inherently unreliable speaking object, since it is less likely to be passed from person to person than to go up in smoke (in the hands of a smoker) or, as became increasingly popular during the eighteenth century, to go up the nostril, as snuff. Nevertheless, a pipe, tobacco box, pouch, or snuff box would appear to be ideal objects to offer perspicacious observations in the manner so evidently enjoyed by readers of the 'it'-narrative genre.

Tobacco's 'it'-narrative silence seems doubly surprising considering there was ample precedent for its appearance as a speaking character in at least two early seventeenth-century plays. Rather than its 'eloquence and oracle', championed in the Dutch play above, it is the power of tobacco that is emphasized in the 1604 play *Lingua* (attributed to Thomas Tomkis). In this play, tobacco is variously described as 'God', 'King', and 'Emperor'. 'The great God tobacco' appears on stage in spectacular fashion. However, when it comes to its voice, the apparition speaks an incomprehensible language that only his sponsor 'Olfactus' (the sense of smell) is able to translate.

Following an eighteenth-century reprint of *Lingua*,[67] a gentleman styling himself 'William Whif' wrote to the editor of the *Gentleman's Magazine* asking 'can any of your correspondents make sense of Tobacco's language, or suggest why he uses it? The editor [of the reprint] takes no notice of it'.[68] Whif's query seems to epitomize the changing perspectives of the eighteenth compared to the seventeenth century. I think it unlikely a seventeenth-century audience would have seriously entertained the idea that tobacco really was speaking a potentially intelligible or translatable language, or would have accepted unquestioned the opinions of other characters, that this was an Antipodean language or the language of Arcadia whose people existed 'before the Moone'.[69] William Whif's views reflect the changes occasioned by the development of Enlightenment science and the accompanying belief that things (the language of a speaking plant, anthropomorphized as the 'King of Tobacco') need to be decoded or 'described'[70] rather than taken—as Tomkis surely intended us to do—at face value or as just a bit of amusing nonsense. No correspondents to the *Gentleman's Magazine* attempted to answer Whif's query—or if they did, the editor did not publish their responses.

A tobacco-related object does speak at least once in eighteenth-century English literature. It is a tobacco pipe which argues with a perfumed wig in a 'poetic fable' published by Christopher Smart in 1752. The wig, of the 'flaunting French' sort so popular in genteel society at the time, is critical of the tobacco pipe's 'barb'rous English! horrid Dutch!' polluting breath. The pipe retorts:

> Know, puppy, I'm an English pipe,
> Deem'd worthy of each Briton's gripe [grip],
> Who, with my cloud-compelling aid
> Help our plantations and our trade.[71]

Tobacco has shifted from an exotic, monarchical presence to an everyday product, the consumption of which was fundamental to supporting the national interest. Just as tobacco served to facilitate shifting, multinatural perspectives in its lowland South American base, the 'it-narrative' encouraged a perspectival shift for eighteenth-century English readers from one 'strictly oriented by the rank and gender of the narrator'[72] to one in which the non-human was also allowed an authoritative narrative voice. Indeed, Hudson goes on to suggest the 'it'-narrative was a necessary precursor to the frequently omniscient, 'non-focalized' narrative style of the nineteenth-century novel. From the perspective of historical, comparative anthropology, it seems likely that tobacco itself might have played a part in effecting such changes.

Conclusion

I have argued for the need to add tobacco, a hitherto unrecognized or taken for granted substance, to the list of elements we should consider in seeking to explain why the Enlightenment emerged where and when it did. The 'eloquence and oracle' offered by tobacco, I suggest, helped to promote some key intellectual and literary changes in the seventeenth and eighteenth centuries. Commonalities can be discerned between the long-term entanglement of tobacco and people in lowland South America and characteristics of certain new genres of English literature in the eighteenth century. Tobacco constitutes persons, changes perspectives, and generates dualisms. All these features have echoes in the literary outputs I have identified. They seem to me as much the result of the imbroglio with tobacco as of the other religious, political, and social changes that were taking place in England at the time. All were changes that helped attune people to the concepts and principles of the Enlightenment. The psychosocial trends and developments that I have suggested relate, at least in part, to the increased consumption of tobacco include an increased awareness of self/non-self, a sense of enhanced attentiveness, and the embodied cognitive ability to shift perspective and reflect on one's own positionality. These, I have argued, can be clearly discerned in two eighteenth-century English literary genres, the 'it'-narratives and the 'poetry of attention', both of them resonant with the 'tobacco ideology' associated with a much longer cultural history of tobacco use in lowland South America.

Notes

1. F. Van De Poel-Knottnerus and J. D. Knottnerus, 'Social Life Through Literature: A Suggested Strategy for Conducting a Literary Ethnography', *Sociological Focus* 27/1 (1994), 67–80, 67.
2. Amiria Salmond, 'Uncommon Things', *Anthropologica* 59/2 (2017), 251–66, 252.
3. Christina Lupton, Sean Silver, and Adam Sneed, 'Introduction: Latour and Eighteenth-Century Literary Studies', *The Eighteenth Century* 57/2 (2016), 165–79, 174.
4. Richard Craze, *The Voice of Tobacco: A Dedicated Smoker's Diary of Not Smoking* (Great Ambrook: White Ladder Press, 2003).
5. Haidy Geismar, '"Material Culture Studies" and Other Ways to Theorize Objects: A Primer to a Regional Debate', *Comparative Studies in Society and History* 53/1 (2011), 210–18, 213.

6. Margaret Koehler, *Poetry of Attention in the Eighteenth Century* (London: Palgrave, 2012).
7. Carlos David Londoño-Sulkin, *People of Substance: An Ethnography of Morality in the Colombian Amazon* (Toronto: University of Toronto Press, 2012), 102–3.
8. Elizabeth Rahman, 'Tobacco and Water: Everyday Blessings', in *The Master Plant: Tobacco in Lowland South America*, ed. by Andrew Russell and Elizabeth Rahman (London: Bloomsbury, 2015), 131–52.
9. Note the double-meaning in Londoño-Sulkin's title, *People of Substance*.
10. Aparecida Vilaça, 'Chronically Unstable Bodies: Reflections on Amazonian Corporalities', *Journal of the Royal Anthropological Institute* 11/3 (2005), 445–64.
11. Istvan Praet, 'Shamanism and Ritual in South America: An Inquiry into Amerindian Shape-Shifting', *Journal of the Royal Anthropological Institute* 15/4 (2009), 737–54.
12. Audrey Butt Colson, 'The Akawaio Shaman', in *Carib-Speaking Indians: Culture, Society and Language*, ed. by Ellen B. Basso (Tucson: University of Arizona Press, 1977), 43–65.
13. Johannes Wilbert, *Tobacco and Shamanism in South America* (New Haven, CT: Yale University Press, 1987).
14. Vilaça, 'Chronically Unstable Bodies', 458.
15. Vilaça, 'Chronically Unstable Bodies', 451.
16. Eduardo Viveiros de Castro, 'Cosmological Deixis and Amerindian Perspectivism', *Journal of the Royal Anthropological Institute* 4/3 (1998), 469–88, 469.
17. Tânia Stolze Lima, 'The Two and Its Many: Reflections on Perspectivism in a Tupi Cosmology', *Ethnos* 64/1 (1999), 107–31, 116.
18. Ibid.
19. Wilbert, *Tobacco and Shamanism*, 202.
20. Terry S. Turner, 'The Crisis of Late Structuralism. Perspectivism and Animism: Rethinking Culture, Nature, Spirit, and Bodiliness', *Tipití: Journal of the Society for the Anthropology of Lowland South America* 7/1 (2009), 3–42, 11.
21. Claude Lévi-Strauss, *Introduction to a Science of Mythology, Vol. 2: From Honey to Ashes*, trans. by J. and D. Weightman (London: Jonathan Cape, [1966] 1973), 29.
22. Wilbert, *Tobacco and Shamanism*, 202.
23. K. G. Davies, *The North Atlantic World in the Seventeenth Century* (Minneapolis: University of Minnesota Press, 1974), 146.
24. Jordan Goodman, *Tobacco in History: The Cultures of Dependence* (London: Routledge, 1993), 60.
25. Roy Porter, *Enlightenment: Britain and the Creation of the Modern World* (London: Penguin, 2000), 22.

26. Porter, *Enlightenment*, 15, his italics.
27. Sidney W. Mintz, *Sweetness and Power: The Place of Sugar in Modern History* (New York: Penguin, 1985), 185.
28. Anonymous, *Wine, Beere, Ale, and Tobacco. Contending for Superiority. A Dialogue* (London, 1630).
29. Marshall Sahlins, *The Western Illusion of Human Nature* (Chicago: Prickly Paradigm Press, 2008), 84.
30. James Grehan, 'Smoking and "Early Modern" Sociability: The Great Tobacco Debate in the Ottoman Middle East (seventeenth to eighteenth centuries)', *The American Historical Review* 111/5 (2006), 1352–77, 1377.
31. Jonathan Israel, *Democratic Enlightenment: Philosophy, Revolution, and Human Rights 1750–1790* (Oxford: Oxford University Press, 2011), 2 fn4.
32. Anonymous, 'Hail Indian Plant', *The London Medley* (London, 1731), 8–9.
33. Anonymous, 'The Convert to Tobacco, A Tale', in *A Collection of Merry Poems* (London, 1736), 111–14.
34. Anonymous, 'The Triumph of Tobacco over Sack and Ale', in *A Collection of Old Ballads*, 3 vols (London: J. Ambrose Philips, 1725), vol. 3, 154–56, 154.
35. Wolfgang Schivelbusch, *Tastes of Paradise: A Social History of Spices, Stimulants, and Intoxicants* (New York: Pantheon Books, [1980] 1992), 131.
36. Timothy Brook, *Vermeer's Hat: The Seventeenth Century and the Dawn of the Global World* (London: Bloomsbury, 2008), 126.
37. Edward Ward, *A Compleat and Humorous Account of All the Remarkable Clubs and Societies in the Cities of London and Westminster* (London, 1756), 276.
38. Simon Schama, *The Embarrassment of Riches: An Interpretation of Dutch Culture in the Golden Age* (New York: Alfred A. Knopf, 1987), 212.
39. Brian Cowan, *The Social Life of Coffee: The Emergence of the British Coffeehouse* (New Haven: Yale University Press, 2005), 86.
40. Andrew Russell, *Anthropology of Tobacco: Ethnographic Adventures in Non-human Worlds* (New York: Routledge, 2019), 124.
41. Philips, 'An Ode to Henry Saint John, Esq', in *Poems on Several Occasions* (London, 1728), 41–46.
42. Anon., 'The Convert to Tobacco', 111.
43. Ralph Erskine, 'Smoking Spiritualized', in *Gospel Sonnets; Or, Spiritual Songs* (Glasgow: John Pryde, [1739] 1870), 339–40.
44. Anon., 'Hail Indian Plant', 8–9.
45. Opie, for example, talks of moving from a 'God-intoxicated' to a 'Man-intoxicated' era. John Opie, 'Renaissance Origins of the Environmental Crisis', *Environmental Review* 11/1 (1987), 2–17.

46. David Riesman, with Nathan Glazer and Reuel Denney, *The Lonely Crowd: A Study of the Changing American Character* (New Haven: Yale University Press, 1950).
47. John Jeffries Martin, 'The Myth of Renaissance Individualism', in *A Companion to the Worlds of the Renaissance*, ed. by Guido Ruggiero (Oxford: Blackwell, 2002), 208–24, 209.
48. Jacob Burckhardt, *The Civilization of the Renaissance in Italy*, vol. 1, trans. S. G. C. Middlemore (New York: Harper, [1860] 1958), 142–3.
49. Tim Ingold, 'Dreaming of Dragons: On the Imagination of Real Life', *Journal of the Royal Anthropological Institute* 19/4 (2013), 734–52, 734.
50. Phil Withington, *Society in Early Modern England: The Vernacular Origins of some Powerful Ideas* (Cambridge: Polity Press, 2010), 239.
51. Martin Holbraad, 'Can the Thing Speak?', *Working Papers Series #7* (Open Anthropology Cooperative Press, 2011).
52. Scott A. Nowka, *Character Matters: Enlightenment Materialism and the Novel* (PhD thesis, University of Iowa, 2006), 7.
53. Aileen Douglas, 'Britannia's Rule and the It-Narrator', in *The Secret Life of Things: Animals, Objects, and It-Narratives in Eighteenth-Century England*, ed. by Mark Blackwell (Lewisburg: Bucknell University Press, 2007), 147–61, 147.
54. James Arbuckle, *Snuff: A Poem* (Edinburgh: James McEuen, 1719), 28.
55. John Hill, *Cautions Against the Immoderate Use of Snuff, Founded on the Known Qualities of the Tobacco Plant* (London: R. Baldwin & J. Ridley, 1761).
56. Nowka, *Character Matters*, 59.
57. Richard Bentley, *Matter and Motion cannot Think, or A Confutation of Atheism from the Faculties of the Soul* (London: Thomas Parkhurst and Henry Mortlock, 1692), 13.
58. Liz Bellamy, 'It-Narrators and Circulation: Defining a Subgenre', in *The Secret Life of Things*, 117–46.
59. Nowka, *Character Matters*, 7.
60. Bellamy, 'It-Narrators and Circulation', 118.
61. Helenus Scott, *The Adventures of a Rupee* (London: J. Murray, 1782).
62. Deidre Lynch, *The Economy of Character: Novels, Market Culture and the Business of Inner Meaning* (Chicago: University of Chicago Press, 1998), 96.
63. Nicholas Hudson, 'It-Narratives: Fictional Point of View and Constructing the Middle Class', in *The Secret Life of Things*, 292–306.
64. Christopher Flint, 'Speaking Objects: The Circulation of Stories in Eighteenth-Century Prose Fiction', in *The Secret Life of Things*, 162–86.
65. Koehler, *Poetry of Attention*, 2.
66. Ibid.

67. *Lingua* was reprinted in vol. 5 of Robert Dodsley and Isaac Reed, eds, *A Select Collection of Old Plays: In Twelve Volumes* (London: J. Nichols, 1780).
68. W. Whif, Correspondence, *Gentleman's Magazine* 58/2 (1788), 122.
69. The character Memoria, commenting in *Lingua*.
70. Ingold, 'Dreaming of Dragons', 743.
71. Christopher Smart, 'The Bag-Wig and the Tobacco-Pipe: A Fable', *Poems on Several Occasions* (London: W. Strahan, 1752), 211–13.
72. Hudson, 'It-Narratives', 294.

Selected Bibliography

Blackwell, Mark, ed. 2007. *The Secret Life of Things: Animals, Objects, and It-Narratives in Eighteenth-Century England*. Lewisburg: Bucknell University Press.

Goodman, Jordan. 1993. *Tobacco in History: The Cultures of Dependence*. London: Routledge.

Holbraad, Martin. 2011. Can the Thing Speak? *Working Papers Series #7*, Open Anthropology Cooperative Press.

Koehler, Margaret. 2012. *Poetry of Attention in the Eighteenth Century*. London: Palgrave.

Latour, Bruno. 1993 (1991). *We Have Never Been Modern*. Trans. by Catherine Porter. Cambridge, MA: Harvard University Press.

Lynch, Deidre. 1998. *The Economy of Character: Novels, Market Culture and the Business of Inner Meaning*. Chicago: University of Chicago Press.

Russell, Andrew, and Elizabeth Rahman, eds 2015. *The Master Plant: Tobacco in Lowland South America*. London: Bloomsbury.

Schivelbusch, Wolfgang. 1992 (1980). *Tastes of Paradise: A Social History of Spices, Stimulants, and Intoxicants*. New York: Pantheon Books.

Wilbert, Johannes. 1987. *Tobacco and Shamanism in South America*. New Haven, CT: Yale University Press.

Withington, Phil. 2010. *Society in Early Modern England: The Vernacular Origins of some Powerful Ideas*. Cambridge: Polity Press.

PART V

The Romantic and Victorian Periods

CHAPTER 14

Romantic Consumption: The Paradox of Fashionable Breath

Clark Lawlor

> 'I look pale. I should like to die of a consumption'. 'Why?' asked his [Byron's] guest. 'Because the ladies would all say, Look at that poor Byron, how interesting he looks in dying.'[1]

Lord Byron, poetic superstar of the Romantic period, makes the notorious claim—only half-joking—to desire the killer disease of consumption of the lungs. This assertion presents several problems, or perhaps opportunities, for a volume dedicated to the subject of breath. Consumption was the main disease of the lungs for people in the Romantic period and for the nineteenth century in general.[2] There were others, but they received nowhere near as much attention either in the medical literature or in wider representation and narrative. For the wildly popular Bryon, and for many of his contemporaries, consumption was a disease that conferred a certain social—even sexual—cachet, and the same was true for 'the ladies', albeit framed according to conventions of gender.

C. Lawlor (✉)
Department of Humanities, University of Northumbria,
Newcastle upon Tyne, UK

D. Fuller et al. (eds.), *The Life of Breath in Literature, Culture and Medicine*, Palgrave Studies in Literature, Science and Medicine,
https://doi.org/10.1007/978-3-030-74443-4_14

Consumption was fashionable, even amongst some of the lower classes, or 'orders' as they were known before the Industrial Revolution at the end of the eighteenth century. The status of breath within the construction of this disease, however variously manifested in different social groups (striated according to gender, class, ethnicity, religion, and race), is key to explaining how such a paradoxical phenomenon came to be. Breath, as most of the essays in this volume show, is not just a biological entity, not merely a physical process. The full force of literary, artistic, philosophical, economic, and social representations and narratives comes into play when breath is experienced by individual human subjects. If we want to understand our experience of breathing we had better consider our own templates for the experience of disease and health, and some of those templates include *belles lettres*, creative literature, and works of art (both high and low).[3] As this essay will show, for some conditions and entities, breath being prime among them, we need a long historical view of the sort taken in this volume in order to comprehend the function of breath in our lives and—potentially—how to deal with it better in both the clinic and the wider context of the social worlds in which we, 'all the breathers on this star', move.[4]

We must not forget that we are dealing with a disease of the breath before we turn to the apparently counter-intuitive phenomenon of consumption as a Romantic fashion. An explanation of some basic features of what was understood to be a consumption of the lungs is in order. The collection of symptoms associated with consumption remained mostly static even after the scientific revolutions that displaced the humours as the core medical theory. We see some of the spiritual connotations of breath as life, as *ruach*, as divine inspiration and communication in A. A. Long's essay, and as conspiration and co-breathing, as expounded by Phillip Horky, but here we must first focus on breath as death, as disease, as the destruction of the lungs. This is, in the physical world, conspiration gone wrong, failing, and—as often, in the case of this richly symbolic process—leading to religious assumptions about moral failure.

We can take as our starting point a horrific passage from the novel *Liberty Hall, Oxon.* (1860) by the atheist and iconoclast William W. Reade. Reade rejected the idea of the Christian good death, prevalent in his age, stemming from the medieval concept of the *ars moriendi*, or art of dying well. Reade takes the symptoms of consumption of the lungs, or pulmonary tuberculosis, as the dramatic basis of a distinctly bad death:

> Her body was bent forward on her knees; the joints of this body so thin, that it was almost deformed, were swelled and red and painful. She laboured and coughed for her breath; each time that she breathed she coughed up blood ...
>
> But while they were still gazing at her, they saw the colour fade from her cheeks, the smile from her lips. Her face, with wonderful rapidity, became sallow and dusky; her lips and her finger-nails tinged with blue. Her breathing became rapid, and was expelled in sharp and hurried gasps. Long, deep streams of perspiration ran down her face; her features assumed an expression of anxiety, her eye-balls protruded, and her heart throbbed savagely and loudly.[5]

The beatific 'smile' of the willing Christian giving up the soul to God, the gentle and measured breathing, the calm acceptance of death and the afterlife in the set-piece good death of the Victorian era, are replaced here with shocking speed by 'anxiety' and its attendant physical symptoms, such as 'rapid' breathing (which is also part of the consumptive's difficulty in breathing), blood mingled with the very breath itself, sweat, and tachycardia. Reade's prose reflects this shift in the paragraph, with the even tripartite rhythm of the first sentence descending into the disrupted central sentence that mimics the staccato gasping for breath of the heroine. Here breath is not an isolated physical phenomenon: it interacts with the other physical and mental systems to manifest an holistic experience rather than a monological process. Breath here signifies the death of spiritual hope, the failure of Christian faith, and the subject's realisation of this fact, as well as the body's disintegration via the lungs.

The facts of pulmonary tuberculosis as we presently understand them give us necessary information on how to read the disease process itself, although we should not consider this tuberculosis bacterium separately from its social contexts, but simultaneously. Before Robert Koch discovered the tubercle bacillus in 1882, 'consumption' was a term that covered a range of wasting conditions that (the clue is in the name) consumed the lungs and the body's flesh. Koch found that consumption was due to an infectious germ, a bacterium that travelled through the air and the breath and attacked the lungs when activated. Lesions or tubercles were manufactured in the lungs and destroyed the lung tissue, either slowly in a 'decline', or quickly in a 'galloping' consumption. Before the arrival of streptomycin for clinical use in 1946, a confirmed consumption was regarded as a death sentence, most famously in the case of the Romantic poet, John Keats.

The 'narrative' that consumption follows partly determines the different ways it has been constructed in human experience and representation: at first it can be mild, and the cough may seem like that of a cold or minor irritation. Then the sufferer can lose weight, appetite, be pale or flushed with a 'hectic fever' and night sweats. As things become worse, the patient can become skeletal, tortured with a never-ending cough, and bring up blood and fetid, stinking matter from the lungs. One end to this narrative is the patient choking to death on disintegrated lungs, a fate possibly made worse by the fact that the sufferer is entirely *compos mentis* throughout.[6]

This was one 'reality' of consumption, and one that was modelled in the humoral theory that underpinned representations of consumption from the classical period to the 'long' eighteenth century, the point at which the theory of the humours was discarded in favour of mechanical-hydraulic notions of the body and ideas giving an enhanced role to the nervous system, although not the nervous system we have today. According to humoral theory, the lungs were corroded by pus from the nodes or ulcers created by bad blood stagnating in the lungs. The putrid humour—often thought to be caused by the dregs of food from malfunctioning digestion—was understood to rot the lungs and, consequently, waste the body.[7] As the Welsh cleric Rowland Watkyns (*ca.* 1614–64) put it in his *Flamma sine Fumo: or, poems without fictions* (1662):

> The Consumption.
> Foul humors do descend: thin and sharp rume
> Fall from the head, and doth the Lungs consume.
> Short cough, short breath, and faintness, never cease
> To be companions of this sad disease.[8]

Hence the core imagery associated with the classical and early modern consumption of the lungs, drawn from the medical model, was disgusting decay caused by 'foul humours'.

Christopher Bennet's popular *Theatrum tabidorum* (1656) described the wasted lungs of the consumptive in terms of their excremental quality,[9] while the famed eighteenth-century Dutch physician Gerard van Swieten could hardly stand the vicious stench of the fetid product of a consumptive's cough.[10] The equation of the stinking breath of the consumptive with sin was an easy one for early modern authors to make, and the influential Enlightenment anatomist Gianbattista Morgagni's description of a consumptive prostitute did nothing to dispel the symbolism: 'This woman had thrown up pus by expectoration, and had been long macerated with a slight fever, from a venereal cause; so that in the body after death, no traces remained of her breasts, but the nipples only were discerned'.[11] So repulsive were the potentially contagious dregs and vapours emitted by the decomposed lungs of consumptives that many physicians were afraid to perform anatomies on those who had perished.

Humoral theory would die out during the course of the seventeenth and eighteenth centuries, but the moral imagery of sin and filth continued on into the nineteenth. Dr Burney Yeo observed that 'Sir William Gull has said that he can smell syphilis. I think I can smell phthisis. There is a peculiar nauseating odour in the breath'.[12] Phthisis was the Greek term for consumption, and commonly used from classical times onwards. The smell of the breath was an invisible marker for physical and moral corruption, just as marks on the skin, in diseases like smallpox or the bubonic plague, could be understood as visible signs of immorality.

The morality of consumptive breath is made abundantly clear in the eighteenth-century novel too, most notably in the bad death of the libertine Thomas Belton in Samuel Richardson's compendious novel, *Clarissa*. Belton's consumptive death is in stark contrast to Clarissa's, and is a powerful instance of the ways in which one disease (or actually set of diseases in the eighteenth century, as there were different forms of consumption) can have divergent narratives, symptomatologically and morally. Belton's horrific death presages the pains of hell that await his rapist friend and the villain of the novel, Lovelace:

> He is now at the last gasp—rattles in the throat—has a new convulsion every minute almost! What horror he is in! His eyes look like breath-stained glass! They roll ghastly no more; are quite set: his face distorted and drawn out by his sinking jaws, and erected, staring eyebrows, with his lengthened furrowed forehead, to double its usual length, as it seems. It is not, it cannot be, the face of Belton, thy Belton and my Belton.[13]

If the good death must be in the classical bodily mode, to deploy Mikhail Bakhtin's term, this bad one is most certainly in the grotesque, in which the body's boundaries are well and truly traversed and 'distorted'.[14] Belton's breath has become a death-rattle, less a living breath than a 'convulsion', a spasm, a form of torture. The eyes are windows to the soul, and here that soul is 'breath-stained glass', which depicts both the departure of the soul to domains less-than-holy, and the moral confusion and fog that has characterised Belton's sinful and rakish life.

These depictions of consumptive breath as malfunctioning, rotting, stinking, arhythmical, failing, and dying are bound up with various discourses that combine to create the different forms of representation that we have seen thus far. However, these were not the only possibilities for the representation and experience—both are mutually informing—of consumptive breath.

Fashionable Consumptive Breath

As we saw in the opening quotation of Byron's words, consumption and its concomitant breath could be viewed in almost the opposite way. Dr Thomas Beddoes (1760–1808) commented on the popular literary representation of consumption that had become prevalent by the end of the eighteenth century in his *Essay on the Causes, Early Signs, and Prevention of Pulmonary Consumption* (1799):

> Writers of romance (whether from ignorance or because it suits the tone of their narrative) exhibit the slow decline of the consumptive, as a state on which the fancy may agreeably repose and in which not much more misery is felt, than is expressed by a blossom, nipped by untimely frosts. Those who only see the sufferers in passing, are misled by the representation.
>
> And I have heard many persons thus prepossessed, after closely attending a sick friend, declare their surprise not less than their horror, at the unexpected scenes of varied and protracted misery which they have been condemned to witness.[15]

The second paragraph chimes better with the kind of consumption in which 'horror' is a key term. The first paragraph cites the imagery of wilting and dying flowers so frequent in literary descriptions of consumption, a comparison that reduces the breathless gaspings of dying consumptives and replaces them with the narrative of a 'slow decline' (and 'decline' was commonly substituted for the word consumption itself in both literature and life). In Beddoes' disapproving professional view, the demands of literary genre appear to overrule the need to be truthful about the effects of this deadly disease.

Beddoes was a political radical who supported the French Revolution, a materialist (this ungodly attitude probably prevented him from gaining a Chair in Chemistry at Oxford University), and was not amenable to Romantic and/or Christian interpretations of consumption as a blessed or glamorous disease. For Beddoes, the mind was not separate from the body, and his efforts to determine the life-giving role of breath were confined to his experiments at his Pneumatic Institution in Bristol, a city that, like Bath, was a fashionable resort for consumptives. The new-found field of pneumatic medicine, itself stimulated by the discovery and isolation of major gases such as carbon dioxide, hydrogen, and oxygen between 1754 and 1774, gave hope that cures for diseases such as consumption (a major killer at the time) might be found. If the interaction of chemistry and physiology could be understood, via experiments in combustion in particular, then the way might be clear for the process of respiration to be demystified.

For Beddoes and the materialists, consumptive breath was a physical phenomenon to be eliminated by empirical science, and the matter of a transcendent soul, spirit, and its relationship to God or some form of pantheistic 'One' was beyond their remit, notwithstanding Beddoes' friendship with writers like Samuel Taylor Coleridge, who certainly were interested in metaphysical and religious issues as well as in experimenting with Beddoes at his Pneumatic Institute.[16]

We need to understand how it can be that there were two such divergent discourses of consumptive breath. To do this we need to take a long-historical view. Previous essays in this volume discuss the Galenic notion of the lungs as a kind of bellows in which the 'natural spirits', the innate form of the vivifying *pneuma,* transported by the blood from the liver and then heart to the lungs, is further refined (via further *pneuma* drawn from the air by the lungs) into 'vital spirits', a higher form of *pneuma*.[17] Symbolically and literally, breath was life. These classical ideas of a vital spirit that animated the body translated easily in Christian contexts as the divine spark of the soul.[18]

If we fast-forward to nearer the time of the Romantics, we can see how the idea of consumptive breath plays out in a Protestant Christian context inflected by gender and genre. *Clarissa* represents consumptive breath in surprisingly positive ways, given the moral example of Belton. The eponymous heroine is raped by the rake Lovelace and develops a 'nervous consumption' as a result:

> I don't think I am dying yet—not *actually* dying, Mrs. Lovick—for I have no bodily pain—no numbnesses; no signs of immediate death, I think.— And my breath, which used of late to be so short, is now tolerable, my head clear, my intellects free—I think I cannot be dying yet ... yet how merciful is the Almighty, to give His poor creature such a sweet serenity! 'Tis what I have prayed for![19]

Clarissa embodies the tradition of the Christian 'good death' from consumption, which is based on the idea of the disease as so symptomless that one scarcely knows one is dying, or even has contracted the disease. Clarissa embraces death in a Neo-Platonic displacement of thwarted secular and sexual love, redirecting her libidinal energies towards Christ.

Consumption was thought to be a 'Golden Disease' (John Bunyan's term) in this context: it left one's mind clear to make one's peace with God and man; was usually chronic, so one had time to accomplish one's repentance and arrange one's estate; and the shedding of the flesh and lack of scarring or marks on the skin symbolised the notion that one was also shedding one's sins and rising to the afterlife with divine approval. It was no doubt a help that the notion of consumptives as beautiful had persisted from the Ancients: according to the scornful Aristides (fl. 160 AD), there was a myth that the phthisical were 'comely', 'and that they have many admirers ... while those who are in excellent health and whose beauty is in high esteem are inferior to them'.[20] Clarissa can truly 'pray for' such a holy disease. Her speech becomes progressively less breathless, less frequently punctuated by dashes, commas, and semicolons, until she arrives at the ecstatic and ringing final sentence that confirms her triumph over both death and breathlessness: ''Tis what I have prayed for!'.

Clarissa's, and later representations of the good death, are fed by the famous example of the poet John Donne's holy (and eccentric) death from consumption, especially as narrated in Izaak Walton's hagiographical 'Life of John Donne', in which he looked so thin that he seemed like a living *memento mori*, and was said to have been preaching his own funeral sermon.[21] The ease of the consumptive death in this discourse is crucial for Romantic ideas of breath. Thomas Browne's (1605–82) description of his consumptive friend's death in 'A Letter to a Friend, upon the occasion of the Death of his Intimate Friend' was a great influence on the attitudes of the following two centuries in Britain and America:

> his soft departure, which was scarce an expiration; and his end not unlike his beginning ... and his departure so like unto sleep, that he scarce needed the civil ceremony of closing his eyes; contrary unto the common way, wherein death draws up, sleep lets fall the eye-lids.[22]

This seductive vision of a 'soft departure' for the devout Christian was reinforced by popular and specialist medical texts. Of particular relevance to *Clarissa* is the pronouncement in William Buchan's best-selling and long-running manual of *Domestic Medicine* that 'a Nervous Consumption is a wasting or decay of the whole body, without any considerable degree of fever, cough, or difficulty of breathing. It is attended with indigestion, weakness, and want of appetite etc.' and affects 'those who are of a fretful temper'.[23] The lack of 'difficulty with breathing' is key here: it was apparently possible for some consumptions to act as what we might now think of 'silent' pneumonias, with very little evidence of their presence. Even if one did not have a 'nervous consumption', a category that chimed well with the later eighteenth-century emphasis on the nervous system, one might still not be aware that consumption was upon one, hence the classical concept of the *spes phthisica*, described by the Greek physician Aretaeus:

> haemorrhage from the lungs is particularly dangerous, although patients do not despair even when near their end. The insensibility of the lungs to pain appears to me to be the cause of this, for pain is more dreadful than precarious; whereas in the absence of it, even serious illness is unaccompanied by the fear of death and is more dangerous than dreadful.[24]

Because the lungs have fewer nerves than other parts of the body, the pain of consumption is minimised, so the trajectory of the disease as a silent assassin becomes possible.

Clarissa's death seems very like Browne's friend's 'soft death', and the breath is a key part of the process: 'her aspect was sweetly calm and serene; and though she started now and then, yet her sleep seemed easy; her breath indeed short and quick; but tolerably free, and not like that of a dying person' (Vol. 7, Letter CV, 412). This was the point: consumption could, in the Christian 'good death', and in the classical *spes phthisica*, evade the appearance of death itself, even in terms of breath.

This kind of representation of beautiful dying women was not confined to the pages of literature. We see the interaction between literary and visual works and real-life encounters between patient and physician, and the wider network of people around that, in operation in the burgeoning age of sensibility. In 1778 Dr Erasmus Darwin, physician and poet, treated the 'lovely' Lady Northesk for consumption. This incident was described by her friend, the poet Anna Seward, in her *Memoirs of Dr Darwin*, and by Sir Walter Scott in the preface to Seward's *Works*:

> Lady Northesk lay on a couch through the day, in Dr Darwin's parlour, drawing with difficulty that breath, which seemed often on the point of final evaporation. She was thin, even to transparency; her cheeks suffused at times with a flush, beautiful, though hectic. Her eyes remarkably lucid and full of intelligence. If the languor of disease frequently overshadowed them, they were always relumined by every observation to which she listened, on lettered excellence, on the power of science, or the ingenuity of art.[25]

As Clarissa's death demonstrates, Northesk would not be out of place in a sentimental novel, at least in Seward's description. Like her breath, Seward's friend's body seems so slender as to be about to achieve 'final evaporation', conspiring, or perhaps translating, into the ideal spirit of the Christian tradition, soon to be converted into the 'infinite One' of the Romantics. Not only is breath the Soul, but it is also the Body.

Consumption had been linked with male genius from classical times: Aretaeus in particular made an influential pronouncement on the nature of the consumptive in his *On the Causes and Symptoms of Chronic Diseases*. Consumptives tended to be young, thin, pale, and distinguished by 'eyes hollow, brilliant and glittering'.[26] Apart from the cough, they might be feverish, and this 'ardent fever' could burn away the humours in which 'formerly they were immersed ... as if in mud and darkness; but when the disease has drained these off, and taken away the mist from their eyes, they perceive those things which are in the air [spiritual presences of the dead], and with the naked soul become true prophets' (Aretaeus, 273). Now 'every sense is pure, the intellect acute, the gnostic powers prophetic' (272–73). Unfortunately, death was the price to pay for achieving 'such a degree of refinement in their humours and understanding', 'the vital power having been already evaporated into air' (273). Lady Northesk is a long way from the time of Aretaeus, but the notion of breath as vital power, rewritten as defining the Christian soul, persists in Anna Seward's glamorous depiction of her friend. Seward—a bluestocking—was also concerned to adapt the masculine tradition of consumptive genius for

Northesk, a tradition we see in full operation with the arrival of the consumptive Romantic author, and especially the poet.

Alastor and Romantic Death—Fashionable Breath

By the Romantic period, the paradigm of the nervous system as the dominant medical theory of the body was fully accepted, and underpinned the physical notions of 'Sensibility' and its concomitant moral concept of the 'Sentimental'. The hydraulic-mechanical thinking of the earlier part of the century had modulated into an idea of the human body that focused on the nerves, conceived broadly as vibrating strings that could transmit sensory inputs of all types to the brain, and thus affect that body, often without the direct intercession of the conscious mind. 'Sensibility' was the capacity of the body and mind to process external stimuli. The upper classes or 'orders' and intellectuals were thought to be naturally more refined, with literally thinner nerves, and women especially so.

To have too much sensibility was both a badge of honour and a problem. Blushing and fainting heroines abounded in the novel: psychological conflict would inevitably play itself out directly on the body in this model, as Gillian Skinner demonstrates here in the context of breath and breathlessness in the novel of sensibility. The old tradition of love melancholy found new energy in nerve medicine: now a heart could be broken by the relays of the nerves which transmitted emotions to the entire system, of which the heart was a part. Consumption was a common result of such nervous trauma: poets wasted away pining for their mistresses, while heroines like Clarissa shed their flesh in response to thwarted love and eventually, in her case, abduction and rape. As the influential Scottish physician of the nerves Robert Whytt (1714–66) argued, 'a *phthisis pulmonalis* may also be the consequence of nervous disorders, when the morbid matter producing them falls chiefly upon the lungs'.[27] Even before this, another famous Scot, George Cheyne, had connected consumption with hyper-sensible and sensitive poets as well as with the beautiful people identified as likely to be consumptives in the classical tradition: 'I have always foretold, that these Symptoms (especially in tender, delicat, lively young Persons) would terminat in a real sensible *Phthisis Pulmonum*'. If Cheyne could treat the nervous disease early on, consumption might be prevented and thus preserve 'some of the noblest and brightest Spirits this Age or Country produces'.[28]

Nerve medicine took another turn at what is usually seen as the beginning of the Romantic period in British literature, when the ideas of Edinburgh physician John Brown (1735–88) took hold in the medical and popular-literary imagination, persisting a long time after they had been exploded in the professional medical sphere. 'Brunonianism' argued that each person had a limited amount of 'excitability' or life-force, like a battery, and that to remain healthy this needed to be stimulated at the correct level, no more and no less. Either overstimulation (sthenia) or understimulation (aesthnia) would cause general problems that manifested themselves as localised symptoms. Cure might consist of applying stimulating substances when the person was under-stimulated, and 'lowering' methods when over-stimulated.

In the case of consumption of the lungs, Brown considered the tubercles to be a side-effect of blockages produced by disease of the whole system, the lungs being more directly susceptible to inflammation than other parts of the body because of their immediate contact with air. A 'general consumption' or 'tabes' was an overall wasting of the body. Literary and artistic types, especially poets, were liable to overwork their mental energies and their bodies at the same time: 'excess in thinking, by wasting the excitability, ends in indirect indebility'.[29] As the now-minor but contemporarily popular consumptive poet Henry Kirke White observed, 'My mind is of a very peculiar cast.—I began to think too early, and the indulgence of certain traits of thought, and too free an exercise of the imagination, have superinduced a morbid kind of sensibility, which is to the mind, what excessive irritability is to the body'.[30] All one's vital energy could be drained to the lees, and even result in death. Love melancholy, often the resort of the male poet (but certainly not confined to him), was also a prime route for consumption of the vital energy leading to a literal consumption.

The discourses of the consumptive good death, love melancholy, and poetic genius coalesce in the dying breath of Percy Shelley's poetic creation *Alastor; or, the Spirit of Solitude* (1815).[31] This poem describes the archetypal, semi-autobiographical figure of the questing consumptive poet, whose search for the unattainable love object (which can be a facet of himself in a more narcissistic and even masturbatory interpretation of the quest) ends in his death.[32] Here, breath is a complex amalgam of religious, secular, and scientific long-historical discourses as well as contemporary ideas about the (male) poet, and the function of poetry in society, and medicine.[33]

Shelley was directly influenced by Brunonian theory and, in his poem *Queen Mab,* had noted the deleterious effects of excessive sensibility in a medical case.[34] Dr Thomas Trotter, a well-known theorist of the modern nervous system and its ever-more severe effects on civilised society, warned in *A View of the Nervous Temperament* (1807) of the 'corroding sorrows' of passion sporting with the constitution of one who feels 'sensibility to excess': such a 'degree of feeling … loves and hates beyond all bounds'.[35] Trotter himself had helped promulgate the popular notion that fashionable contemporary lifestyles would lead to the fashionable nervous diseases of the upper 'orders' trickling further down to the middling ones, and perhaps even lower than that, leaving the masculine vigour of the British in a parlous state.

In *Alastor,* Shelley depicts the poet as unable to cope with the stresses and strains of the world, so full of sensibility is he, so prone to living fast and dying young in the Brunonian conception of poets burning like comets. In his climactic death scene, Alastor is

> A fragile lute, on whose harmonious strings
> The breath of heaven did wander—a bright stream
> Once fed with many-voicèd waves—a dream
> Of youth, which night and time have quenched for ever,
> Still, dark, and dry, and unremembered now. (667–71)

The 'breath' of the heavenly wind is not so much, as elsewhere in Shelley's writing, a symbol of the pressures of existence in the modern world, particularly the traumas of lost love, or love unable to be realised, or even poetic achievement in the cruel environment of a seemingly capitalised and brutal market for literature. At this point in the poem these external forces have done their work, and Alastor is at the end-stage of his poetic and glamorous consumption. Only those who lack sensibility can survive in the harsh world, so Shelley's 'fragile' poet-figure is tragically doomed—admittedly, to a beautiful death. Alastor's death arrives via consumption: he wastes away (the word 'waste' is repeatedly used in different forms) as his vital force is 'exhausted' by his incessant search for a seemingly unreachable number of goals, including his ideal woman, sexual plenitude, poetic perfection, and a blissful state of mystical oneness with the universe. This might seem a great deal to ask of the quest, but nevertheless this is Alastor's mission, and it understandably proves to be too much for his reserve of life-fuel.

The burnt-out Alastor adopts a suitably dramatic pose in this death scene, the masculine and more secular equivalent of Clarissa's beautiful good death:

> He did place
> His pale lean hand upon the rugged trunk
> Of the old pine. Upon an ivied stone
> Reclined his languid head, his limbs did rest,
> Diffused and motionless ... (632–36)

Now Alastor moves beyond earthly, physical encumbrances, in the manner of the consumptive Christian good death, especially as it had evolved in the late eighteenth and early nineteenth centuries, where an emphasis on the struggle with the Devil at the point of death had diminished as a part of the mild secularisation of death for the mainstream Christian.[36] Of course Shelley has no truck with the Christian good death as such, but he does adapt it for his secular and Romantic vision of the dying and wasted poet:

> no mortal pain or fear
> Marred his repose, the influxes of sense,
> And his own being unalloyed by pain,
> Yet feebler and more feeble, calmly fed
> The stream of thought, till he lay breathing there
> At peace, and faintly smiling:—his last sight
> Was the great moon, which o'er the western line
> Of the wide world her mighty horn suspended,
> With whose dun beams inwoven darkness seemed
> To mingle. Now upon the jaggèd hills
> It rests, and still as the divided frame
> Of the vast meteor sunk, the Poet's blood,
> That ever beat in mystic sympathy
> With nature's ebb and flow, grew feebler still. (640–53)

Alastor's breath is peaceful and 'smiling', a quiet acceptance of death and a release from the woes of the world: he is in 'mystic sympathy' with nature, and surely here we must think of co-breathing or conspiration as reworked for Shelley's particular Romantic context. If blood beats in sympathy with 'nature's ebb and flow', so too does his breath.

The following lines push this conspirational and consumptive process even further:

> And when two lessening points of light alone
> Gleamed through the darkness, the alternate gasp
> Of his faint respiration scarce did stir
> The stagnate night:—till the minutest ray
> Was quenched, the pulse yet lingered in his heart.
> It paused—it fluttered. (654–59)

The synchronicity of 'pulse', breath, and natural phenomena draws the poem towards its quietly dramatic conclusion, in which breath is life. 'The alternate gasp' hovers at the end of the line, as poetic rhythm enacts Alastor's final breaths—the 'faint respiration' fluttering like the pulse, about to cease. This is not the stinking breath of the sinful consumptive, the humoral or mechanical condemnation of fleshly and spiritual weakness. A new, Brunonian model, emerging from a vitalist conception of physical sensibility, gives Shelley the platform for a spectacular representation of the tragic poet, a figure who is a projection in part of himself, or at least an experiment in the possibilities available to a poet in this turbulent period.

Alastor finally presents us, after more description of his languorous dying minutes, with a vision of the poet-as-breath, 'fled / Like some frail exhalation' (687). Shelley's redeployment of the Christian consumptive narrative in a depiction of union with a Neo-Platonic Oneness retains the equation of breath with Soul, but overdetermines breath, giving it a quasi-orgasmic quality as well as a more radical political aspect. The identity of the over-reaching, Promethean poet is a familiar one, and the inevitability of his death due to the exhaustion of a fixed life-force is also enacted in the 'frail exhalation' at the end of his life.

Conclusions

Breath, we have found, is not merely a physiological process. In human culture, breath can be experienced and represented in startlingly different ways, even within the same disease and the same period. The stinking 'reality' of consumptive breath can be, and was, overridden, overwritten, or reframed by cultural discourses that prioritised other functions over the mere transfer of air into and out of the body. In order to understand breath in the Romantic period, it is necessary to examine vestigial religious and classical concepts of breath, new and old medical models—and, one might add, issues of gender, rank, and ethnicity.

By tracing the interaction between the diverse components that make up the fashionable Romantic representation of consumptive breath, we gain a more precise understanding of breath and its relationship to human experience in this period. We also find that breath is historically specific: how else can we explain the extreme divergence between the 'unexpected scenes of varied and protracted misery' caused by consumption, the heroine who 'laboured and coughed for her breath; each time that she breathed she coughed up blood', and the desire of Byron to attract the ladies by contracting consumption, or Alastor, 'breathing there / At peace, and faintly smiling'? Breath in our own time also deserves such attention: without a comprehension of its cultural meanings, we will find it more difficult to treat inside or outside the clinic, as Katharine Craik and Stephen Chapman also argue in this volume. Put another way, a breath is not separate from the social world of the person who breathes it.

Notes

1. *The Journal of Thomas Moore, 1826–1830*, ed. by Wilfred S. Dowden, et al., vol. 3 (Newark: University of Delaware Press, 1986), 1120. Entry dated 19 February [Tuesday] 1828.
2. See Clark Lawlor, *Consumption and Literature: The Making of the Romantic Disease* (Basingstoke: Palgrave, 2006).
3. This approach is taken up by Arthur Rose, et al., *Reading Breath in Literature* (Basingstoke: Palgrave, 2019).
4. Quotation from the Scottish working-class poet David Gray, who wrote a sonnet sequence, 'In the Shadows', about his experience of poetic consumption, fully conscious of the fate of Keats and a genealogy of other consumptive male poets: *In the Shadows, Poems by David Gray with Memoir of His Life* (Boston, MA: Roberts Brothers, 1864), 158.
5. William W. Reade, *Liberty Hall, Oxon.*, 3 vols (London: C. J. Skeet, 1860), vol. 3, 323, 327.
6. Thomas Dormandy, *The White Death: A History of Tuberculosis* (London: Hambledon Press, 1999), 22–23.
7. For digestion in the humoral model, see Ken Albala, *Eating Right in the Renaissance* (Berkeley: University of California Press, 2002), 57.
8. *Flamma sine Fumo: or, poems without fictions. Hereunto are annexed the Causes, Symptoms, or Signes of several Diseases with their Cures, and also the diversity of Urines, with their Causes in Poetical measure* (London: William Leake [etc.], 1662), ll. 163–66.
9. Christopher Bennet, *Theatrum Tabidorum* (London: Thompson, 1656), cited in Gerard van Swieten, *Commentaries Upon Boerhaave's Aphorisms*

Concerning the Knowledge and Cure of Diseases, 18 vols (Edinburgh: Charles Eliot, 1776), vol. 12, 131.
10. Gerard van Swieten, *Commentaries Upon Boerhaave's Aphorisms*, vol. 12, 130.
11. Gianbattista Morgagni, *The Seats and Causes of Diseases, Investigated by Anatomy*, 3 vols, trans. Benjamin Alexander (London: A. Millar, et al., 1769), 1, 652. See also Matthew Baillie, *The Morbid Anatomy of Some of the Most Important Parts of the Human Body* (London: J. Johnson, 1793), 48–49.
12. Burney I. Yeo, 'On the Results of Recent Researches in the Treatment of Phthisis', *British Medical Journal*, 1877, 1(841), 159–60, 195–97 (quotation on 196). See also G. L. Bayle, *Researches on Pulmonary Phthisis,* trans. by William Barrow (Liverpool: Longman, et al., 1815), 125.
13. Samuel Richardson, *Clarissa, or, the History of a Young Lady*, 3rd edn, 8 vols (London: S. Richardson, 1751), Letter LV, vol. 7, 192.
14. For a sophisticated analysis of Bakhtin's concepts of the classical and grotesque bodies, see Peter Stallybrass and Allon White's influential *The Politics and Poetics of Transgression* (London: Methuen, 1986).
15. Thomas Beddoes, *Essay on the Causes, Early Signs, and Prevention of Pulmonary Consumption for the use of Parents and Preceptors* (Bristol: Longman and Rees, 1799), 6.
16. For a useful summary of Beddoes' relationship to Romantic Medicine, see Neil Vickers, 'Coleridge, Thomas Beddoes and Brunonian Medicine', *European Romantic Review* 8/1 (1997), 47–94. Sharon Ruston has published a great deal on the relationship of Humphry Davy and his circle to pneumatic medicine and literature, and has argued that Romanticism was created by wider discourses than the literary and narrowly philosophical. See Sharon Ruston, *Creating Romanticism: Case Studies in the Literature, Science and Medicine of the 1790s* (Basingstoke: Palgrave Macmillan, 2013), 16–20.
17. See the essays in this volume by Carole Rawcliffe and Corinne Saunders, which give differently nuanced accounts of this interpretable and fragmentary material, reported partly from Galen and partly from later traditions of Galenic medicine.
18. Bruno Meinecke, *Consumption (Tuberculosis) in Classical Antiquity* (New York: Paul Hoeber, 1927). As Patrick Gray shows here, the relation of breath to soul in the early modern period was a matter of debate, as it was later when materialism became stronger as a philosophical and scientific movement.
19. Richardson, *Clarissa,* Letter XCI, vol. 7, 354.
20. Meinecke, *Consumption (Tuberculosis) in Classical Antiquity*, 389.

21. *The Lives of John Donne, Sir Henry Wotton, Richard Hooker, George Herbert and Robert Sanderson*, 4th edn, 1675, intro. by George Saintsbury (London: Oxford University Press, 1927).
22. 'A Letter to a Friend, Upon the Occasion of the Death of His Intimate Friend', in *Sir Thomas Browne: Religio Medici and other Works*, ed. by L. C. Martin (Oxford: Clarendon Press, 1964), 177–96, at 180–81; originally published posthumously in London, 1690.
23. William Buchan, *Domestic Medicine: Or, a Treatise on the Prevention and Cure of Diseases by Regimen and Simple Medicines* (Dublin: Saunders, et al., 1774), 125.
24. *The Extant Works of Aretaeus, The Cappadocian*, repr., 1856 edn, ed. and trans. by Francis Adams (Boston, MA: Longwood Press, 1978), 'On the Causes and Symptoms of Chronic Diseases', Bk. 1 Ch. 8, 'On Phthisis', 269.
25. Anna Seward, *Memoirs of Dr Darwin* (London: J. Johnson, 1804), 109–10.
26. *The Extant Works of Aretaeus*, 'On Phthisis', 309–12.
27. Robert Whytt, *Observations on the Nature, Causes and Cure of Those Disorders Which Have Been Commonly Called Nervous, Hypochondriac, or Hysteric*, 2nd edn (Edinburgh: J. Balfour, 1765), 237.
28. George Cheyne, *The Natural Method of Cureing the Diseases of the Body, and the Disorders of the Mind Depending on the Body* (London: G. Strahan, 1742), 185–87, 'A Consumption'.
29. John Brown, *Elementa Medicinae* (Edinburgh: n.p., 1770, trans. 1795), i.125.
30. To Leeson, from Nottingham, 7–4–1806, in Charles Vernon Fletcher's 'The Poems and Letters of Henry Kirke White: A Modern Edition', 3 vols (Ph.D. diss., University of Nottingham, 1980), vol. 3, 890. See also the *Remains of Henry Kirke White*, 2 vols (London: Robertson, 1824), vol. 2, 46.
31. References to *Alastor* are from *The Complete Poetry of Percy Bysshe Shelley*, vol. 3, ed. by Donald H. Reiman, et al. (Baltimore: Johns Hopkins University Press, 2012), 7–30. For a wider consideration of the role of breath in Romantic poetry, see Francis O'Gorman, 'Coleridge, Keats, and The Science of Breathing,' *Essays in Criticism: A Quarterly Journal of Literary Criticism* 61/4 (2011), 365–81; Andrew Kay, 'Conspiring with Keats: Toward a Poetics of Breathing', *European Romantic Review* 27/5 (2016), 563–81; Thomas H. Ford, *Wordsworth and the Poetics of Air* (Cambridge: Cambridge University Press, 2018).
32. Monika Lee, '"Happy Copulation": Revolutionary Sexuality in Blake and Shelley', in *'And Never Know the Joy': Sex and the Erotic in English Poetry*, ed. by C. C. Barfoot (Amsterdam: Rodopi, 2006), 189–206, at 193.

33. The poem also has a significant biographical context, as Mary Shelley pointed out in her Introductory Note to her 1839 edition of Shelley's *Poetical Works*: 'In the Spring of 1815 an eminent physician pronounced that he was dying rapidly of a consumption; abscesses were formed on his lungs, and he suffered acute spasms. Suddenly a complete change took place; and though through life he was a martyr to pain and debility, every symptom of pulmonary disease vanished. His nerves, which nature had formed sensitive to an unexampled degree, were rendered still more susceptible by the state of his health', *The Complete Poetry of Percy Bysshe Shelley*, ed. by Donald H. Reiman, et al., vol. 3 (2012), 1067.
34. *The Complete Poetry of Percy Bysshe Shelley*, ed. by Donald H. Reiman, et al., vol. 2 (2004), 308, note 17.
35. Thomas Trotter, *A View of the Nervous Temperament*, 2nd edn (London: Longman, et al., 1807), 164.
36. See Roy Porter, 'Death and the Doctors in Seventeenth-Century England', in *Death, Ritual and Bereavement*, ed. by Ralph Houlbrooke (London: Routledge, 1989), 77–94.

Selected Bibliography

Dormandy, Thomas. 1999. *The White Death: A History of Tuberculosis*. London: Hambledon Press.

Ford, Thomas H. 2018. *Wordsworth and the Poetics of Air*. Cambridge: Cambridge University Press.

Kay, Andrew. 2016. Conspiring with Keats: Toward a Poetics of Breathing. *European Romantic Review* 27/5: 563–81.

Lawlor, Clark. 2006. *Consumption and Literature: The Making of the Romantic Disease*. Basingstoke: Palgrave.

Meinecke, Bruno. 1927. *Consumption (Tuberculosis) in Classical Antiquity*. New York: Paul Hoeber.

O'Gorman, Francis. 2011. Coleridge, Keats, and the Science of Breathing. *Essays in Criticism: A Quarterly Journal of Literary Criticism* 61/4: 365–81.

Rose, Arthur, et al. 2019. *Reading Breath in Literature*. Basingstoke: Palgrave.

Ruston, Sharon. 2013. *Creating Romanticism: Case Studies in the Literature, Science and Medicine of the 1790s*. Basingstoke: Palgrave Macmillan.

Stallybrass, Peter, and Allon White. 1986. *The Politics and Poetics of Transgression*. London: Methuen.

Vickers, Neil. 1997. Coleridge, Thomas Beddoes and Brunonian Medicine. *European Romantic Review* 8/1: 47–94.

CHAPTER 15

Endless Breath? The Pipe Organ and Immortality

Francis O'Gorman

To contemplate this progress of the greatest of musical instruments with full instruction, it must be remembered that in a development necessitating artificial and consequently wind supplies of fixed pressures, the chief source of natural expression in a wind instrument as secured by the varying breath of the performer was lost to the organ. This disaster became in the end the source of its glorious development, by challenging man's ingenuity to the utmost, in the production of a vast and complicated mechanism by which natural expression was to be attained, through a multiplicity of artificial appliances.

'Brindley and Foster',
The New Monthly Magazine, 122 (February 1883), 261–67 (261).[1]

* * *

F. O'Gorman (✉)
Department of English Literature, University of Edinburgh, Edinburgh, Scotland, UK

D. Fuller et al. (eds.), *The Life of Breath in Literature, Culture and Medicine*, Palgrave Studies in Literature, Science and Medicine,
https://doi.org/10.1007/978-3-030-74443-4_15

As everyone knows, reading aloud—poetry, drama, prose—requires us to accommodate the limited capacity of our lungs. It is no good to run out of breath mid-sentence or to breathe in a place that breaks or even changes the sense. This is equally true of words to be sung. The first version in the first edition of *Hymns, Ancient and Modern* (1861) of Frances Elizabeth Cox's Easter hymn—a translation from Christian Fürchtegott Gellert—collapses if a breath is taken at the end of the first line: 'Jesus lives! no longer now | Can thy terrors, Death, appal me'.[2] There is, aptly in relation to Cox, always a contest with the grave in any voiced reading (which plainly includes singing). A long sentence tests us against the constraints of our own mortality. This is such a familiar issue that we are not always explicitly conscious of it. Judging where to breathe, and where not, is a standard feature of the oral performance of literature as much as in singing. But on poems about, for instance, the possibilities—or impossibilities—of living forever, or simply poems that represent the experience of the sorrow of loss, the question of breathing becomes more intimate with meanings. As the reader arranges his or her breathing to make best sense of lines, an expressive tension is set up with that which has, like Cordelia at the end of Shakespeare's play, no earthly breath at all. Here is Wordsworth, in a sonnet that needs careful preparation for reading aloud because of its elongated syntax and extended enjambment:

> Surprised by joy—impatient as the Wind
> I turned to share the transport—Oh! with whom
> But Thee, long buried in the silent Tomb,
> That spot which no vicissitude can find?
> Love, faithful love, recalled thee to my mind—
> But how could I forget thee?—Through what power,
> Even for the least division of an hour,
> Have I been so beguiled as to be blind
> To my most grievous loss!—That thought's return
> Was the worst pang that sorrow ever bore,
> Save one, one only, when I stood forlorn,
> Knowing my heart's best treasure was no more;
> That neither present time, nor years unborn
> Could to my sight that heavenly face restore.[3]

An impatient wind is a figure that marks as much as it diverts what Wordsworth does quietly with breath in this poem. An impatient wind, for a start, is a gust met here with an absence, one that might be eternal.

The spot where there are no vicissitudes is the grave, where no breath is ('heart's best treasure', interestingly, contains the word 'breath', scattered and broken). The briefly gusting wind is as fitful as human breath—the three-year-old Catherine's breathing that is over, and our breath as we read this poem aloud, breaking long grammatical units so that we do not run out of breath.

Breath easily marks the existence of life just as its absence works as a peculiar indicator of the grave. The substitution of breath for life, a synecdochic exchange, is part of the *lingua franca* of literature's signification both of life and its fragility. Stilled breathing is Lear's confirmation of his daughter's death; signs of breathing are part of Leontes' realisation that Hermione lives in *The Winter's Tale*. 'He breathed his last' is, far from Shakespeare, still a cliché for dying. In turn, the notion of endless breath, or rather of a capacity to breathe without ending, is both an appealing hint of immortality and an absurd one. Absurd because it has its own impossibility deep within it. It might be that part of the enormous imaginative attraction of the human mind to the whale is the cetacean's capacity to hold its breath far beyond human endurance. The whale is an evolutionary step closer, perhaps, to that which breathes but is not circumscribed by the tiny limits of human lungs. But whales must breathe eventually even if they cannot cry. Unceasing breathing, breath that is a sign of immortality not morality, haunts and bothers. When the 'Beat' poet Lawrence Ferlinghetti (1919–21) conjured the notion of endless days in his poem 'Endless Life', included in his *These Are My Rivers* (1993), he reached for the fantasy of inexhaustible breath—but then symptomatically hedged it with temporality in recognition of its real unreachability. '[E]ndless air and endless breath', Ferlinghetti wrote,

> Endless worlds without end of days
> in autumn capitals
> their avenues of leaves ablaze[.][4]

Time comes in—autumn, the Yeatsian turning of the leaves—even with the invocation of respiration that cannot cease and its implication of that which is outside time. Endless breath is too impossible a dream. It must be clocked out.

If endless breath is not for human beings, it is perhaps no wonder that some poets have attended to its figurative possibility in music. Only one inanimate object has been through history habitually described as having lungs or as being capable of a kind of mechanical 'breathing'. Sometimes, to be sure, steam locomotives have been described as if their steam were breath;[5] vast marshes or quicksands have seemed to breathe; and we are now familiar with the notion of the great forests as the 'lungs' of the world and the perils to our literal breath of their destruction by human beings. But the vocabulary of breath has for many centuries been more-than-established for describing the pipe organ: by listeners, players, and organ builders. Producing what organ builders call 'wind' through a range of different systems—bellows or blowers, essentially—the organ stores that wind at the required pressures in chests. The wind is released by the depressing of a key or keys to enter, usually via pallets (valves), the pipe or pipes that have been selected by the player. The earliest organs of which we have knowledge are not fundamentally different to the instruments played today. The earliest known are from the ancient Mediterranean—'organ' comes from the ancient Greek ὄργανον, *organon* [tool, instrument]. These devices produced a tremendous sound—W. H. Auden was not wrong in choosing that adjective for St Cecilia's instrument[6]—for public celebrations and their wind was, often enough, supplied and its pressure retained, by water.

In Europe in the Middle Ages, different types of instruments developed—table-top organs, for instance, as well as cathedral organs that were recognisably related to the Greek and Roman instruments.[7] The organ of Winchester Cathedral, for example, could apparently be heard throughout the town: it was played with fists on huge keys, blasting out a plainsong or *cantus firmus*, or playing the *organum*, a consistent interval above the sung chant.[8] Here was air made into music—of a sort. And there is no difficulty in thinking of that air as a form of breath. The association became more obviously explicit with the more refined instruments of the post-Reformation seventeenth century as documented by poets. Milton, who played the organ and drew on its vocabulary periodically, thought, for example, of the construction of Pandemonium in Book 1 of *Paradise Lost* (1667) via a metaphor from organ design. 'A third [hole] as soon had form'd within the ground', Milton writes:

> A various mould, and from the boyling cells
> By strange conveyance fill'd each hollow nook,
> As in an Organ from one blast of wind
> To many a row of Pipes the sound-board breaths.
> Anon out of the earth a Fabrick huge
> Rose like an Exhalation, with the sound

Of Dulcet Symphonies and voices sweet,
Built like a Temple, where Pilasters round
Were set, and Doric pillars overlaid
With Golden Architrave[.][9]

The organ and breath: an instrument associated for the Puritan Milton with Catholic worship (the Civil War, of course, saw the destruction of many instruments as had the Reformation earlier),[10] the organ is here made to figure the building, the breathing into being, of the capital city of Hell. It is, nevertheless, easy to see that Milton the musician cannot avoid writing also of the wonder and sweetness of the instrument that the Irish poet Nicholas Brady (1659–1726) called, in the libretto to Henry Purcell's 'Hail! Bright Cecilia' (first performed 1692), the 'Wondrous machine'.[11] And it is in Brady's libretto—derived in part from Dryden's 'Song for St Cecilia's Day' (1687)—that there is also an extended example of the trope in which I am interested. It was a trope, in relation to a fact of material history, that became of peculiar suggestiveness in the nineteenth century. Envisaging the organ, Brady sees that its breath can sometimes come from the realms of the immortal. 'With what', he asks,

sublime Celestial Lay
Can any Earthly Sounds compare?
If any Earthly Music dare,
The noble Organ may.
From Heav'n its wondrous Notes were giv'n,
(Cecilia oft convers'd with Heaven,)
Some Angel of the Sacred Choire
Did with his Breath the Pipes inspire;
And of their Notes above the just Resemblance gave,
Brisk without Lightness, without Dulness Grave.[12]

Not implicitly but directly, Brady observes that the organ's breath is that of angels. Once God breathed literal life into Adam in Genesis 2:7; so another act of respiration brings, instead, echoes of the music of the heavens to humanity through the organ.

But this connection between a musical instrument and the breath of immortal beings—breath that cannot die, breath that is a foretaste or, rather, fore-hearing of eternity—took, I think, a fresh form with the development, in the nineteenth century, of more sophisticated forms of organ blowing. A feature of the material history of a musical instrument prompted, I propose, an intensification of imaginative possibilities implicit in its earlier history. With the refining of water pressure systems, then steam and gas systems, and eventually electrically powered systems,[13] it became theoretically possible for the organ—even a very large one—to be supplied with wind indefinitely. Industrialisation in the nineteenth century dramatically enhanced the instrument's power and resources even if it did not alter its essential nature. The celebrated British organ builder Henry Willis (1821–1901) helped radically change the nature of the British organ. Together with the very different style of instruments constructed by William Hill (1789–1870), Willis created organs that remain some of the nineteenth century's finest achievements in pipe organ design.[14] Willis's instrument for the 1851 Great Exhibition was distinctively important. The Crystal Palace organ was—in the terms of the day—large, comprising seventy speaking stops over three manuals and pedals (coincidentally it now forms the basis of the organ of Winchester Cathedral).[15] With a version of pneumatic action to Great and Swell, the two principal divisions in most British organs, where wind pressure assists with the organ's internal mechanisms, the instrument could be said in several senses to rely on a figurative exhalation and inhalation. Powered by steam, Willis's even more famous instrument for St George's Hall, Liverpool (1855), required three eight horse-power engines to keep it going.[16] Not reliant on merely human force, Willis's show-case Liverpool organ could, far better than the whale, hold its breath in ways inconceivable to a man. Here the organ with its potentially endless sound could, in turn, more satisfactorily associate itself with the dreams of life that could not die. With the lungs of the modern organ—producing, said *The Musical Standard* in 1881, assuming the complete interchangeability of the nouns, 'wind or breath',[17]—it was possible to envisage intimations of eternity around us.

A small but interesting corpus of poems—and some other writing—in the second half of the nineteenth century muses on the relationship between this instrument and the divine, or at least the sublime. And that writing also muses on the eternal. Of course, this is most but not exclusively true when the organ in question is in a church, where, often enough, the invisibility of the organist and/or the loftiness of the instrument's placing (many larger organs being situated on chancel or triforium screens or in galleries) add to the solemnity of the building and enhance the sense of the organ speaking from, and of, another domain. But secular locations did not by any means exclude the possibility of the organ's other-worldliness.

The Canadian poet Arthur Weir (1864–1902), thinking of a church, published 'At the Recital' in 1890, describing, in what was not an original trope, the poet caught between earth and eternity, summoned by the organ's sound:

> Midway between the world and God we sate,
> While through the church the Spirit of Music stole,
> And in its robes harmonic wrapped us twain.
> Of thy pure soul, from evil free and hate,
> Then [it] woke my heart to hear the grand refrain,
> And yearned to reach, like thee, life's noblest goal.[18]

There is no direct mention of 'breath'. But the explicit association between the experience of listening to the organ and a desire to reach eternity catches, it might be, the reinvigorated possibility of large instruments powered by a theoretically unceasing energy source. For Edmund Gosse (1849–1928), fewer than twenty years after Willis's grand instrument for St George's Hall, the organ had become so intimately associated with the heavens, with the unchanging sphere above us, that it was no longer a mediator between humanity and the eternal but a feature of eternity itself. Gosse's Calvinist father, Edmund recalled in his memorable account of his childhood, *Father and Son: A Study of Two Temperaments* (1907), had always 'fortified his religious life by prayer as an athlete does his physical life by lung-gymnastics'.[19] In the organ, Gosse discerned the religious life fortified by nothing so strenuous but, rather, by unheard music of a different kind from Keats's. 'The golden spheres are God's sublime citole', Gosse observed in 'The Music of the Spheres' (1873),

> Whereon His spirit like a plectrum flies,
>   But those seraphic organ-harmonies
> Are silent to the hearing of man's soul;
>   Their deep symphonic rapture cannot roll
>   Across the ear of cold mortalities;[20]

This is a late nineteenth-century poem so wedded to the conception of the organ as an instrument of eternity that it cannot turn itself to the sound of the terrestrial organ at all.

But sometimes the association between the organ and the unchanging realms of eternity was much more explicitly associated with the new technologies of organ blowing—with breath itself. For the energetic poet, Lakesman, supporter of The National Trust, and clergyman, Hardwicke Drummond Rawnsley (1851–1920), the very experience of watching an organist playing—and here in a secular location—brought him to the edge of the eternal. And we can be more exact than with Gosse. For Rawnsley was listening to another instrument by Henry Willis: the four-manual and pedal instrument at Colston Hall, Bristol. This organ was built in 1870 (and completely destroyed in a fire in 1898). Demanding more wind than the Great Exhibition organ, the instrument needed supplies—produced from three hydraulic engines—which would support, for example, a 32' pedal flue and a 16' Bombarde also on the pedal, big reeds at 16', 8' and 4', and a 'Tuba Major' on the solo division.[21] It would have taken a hard-working team of foot- or hand-blowers to have created the necessary reserves and pressure to wind this instrument. 'I saw', Rawnsley writes, in his 'The Saturday Organ Recital in Colston Hall' (1877),

> thy hands ascending and descending
> The four white steps of yielding ivory;
> And, either side, a sweet-tongued company,
> That, bowing, made a welcome never ending,
> Now this, now that, by their own words commending
> The skill that taught thee so to touch the key,
> And to the palace gate conducting thee,
> Where winds and angels were in voice contending![22]

It is to the very edge of heaven, where the immortals are, that this player—his identity I cannot discover—is taken. And at that edge, on the cusp of eternity, the difference between earth and heaven is marked in Rawnsley's imagination by the blending, or rather contending, of wind and the angels. Wind, that seemingly endless supply that the new organ offered, might be in the background—in the air, so to speak—of this visionary combination of Henry Willis with Paradise, of the new modern organ and the gateway to the eternal. (It is a rather odd irony, from the perspective of modern medicine, that the donor for the Colston Hall organ was William Henry Wills, first baron Winterstoke (1830–1911), whose vast fortune came from what we now know as the breath-impeding business of selling tobacco.)

The breath of the organ and the certainty of immortality could, in the age of artificially powered wind supplies, even become part of quirky visions beyond, as Rawnsley has it, the palace gates themselves. The Manx poet and scholar T. E. Brown (1830–1897) contributed perhaps the most peculiar of all tributes to a deceased organist, his odd yet affectionate salute to Samuel Sebastian Wesley (1810–1876), who with Mendelssohn was the most influential of all organists and church composers in, and for, England in the first half of the nineteenth century. Brown does not merely envisage the organist taking one to the edge of heaven. The poem, 'The Organist in Heaven', figures Wesley *in* heaven, playing an instrument not built by hands. Brown's conceit is that when Wesley dies and is welcomed to his eternal home, the gate is temporarily left open, and the poet sneaks in. There he sees God constructing an organ out of thunders and great rivers for Wesley to play. The poet catches only the first moments of Wesley's performance on this vast instrument of and *in* eternity before he is chased out—a reversal of Catherine in *Wuthering Heights* (1847)—back to earth. 'Then certain angels', Brown imagines whimsically,

> clad the Master
> In very marvellous wise,
> Till clouds of rose and alabaster
> Concealed him from mine eyes.
> And likest to a dove soft brooding,
> The innocent figure ran;
> So breathed the breath of his preluding,
> And then the fugue began—
> Began; but, to his office turning,
> The porter swung his key;
> Wherefore, although my heart was yearning,
> I had to go; but he
> Played on; and, as I downward clomb,
> I heard the mighty bars
> Of thunder-gusts, that shook heaven's dome,
> And moved the balanced stars.[23]

However strange the poem, it makes distinctive use of the connections I am reflecting on here. The breathing of the organ, and its thunder-gusts, literally belong to the unchanging realm in this oddly unmetrical, unmusical salute to Wesley—with its half-allusion to Dante at the end—as, perhaps, the modern instrument's sources of power fortified and gave fresh life, if I may so phrase it, to what the instrument had long implied of its relationship with the heavens. We have, in this poem, reached an extreme point. The audacity of Brown's connection, the now immortal organist playing on a heavenly organ, rather—takes one's breath away.

And yet … and yet …

It is one thing to speculate that some poetry on the organ's long-established connection with the sublime was revitalised by the theoretical possibilities of an unending wind supply from the middle of the nineteenth century onwards. But it is another to consider what *in practice* the technologies, both new and old, of organ winding could actually involve in the century. The nineteenth-century organist, whether playing a hand-blown or a mechanically blown instrument, had risks and troubles. Hand blowing, to start with the technologies that would gradually be replaced, had the obvious limitation of human weariness. Breath of both literal and metaphorical kinds can fail. For the young Irish poet Edmund John Armstrong (1841–1865), such failure prompted satirical exploration of near-farcical scenes in 'Old St Patrick's Cathedral, Dublin', included in his posthumous *Poetical Works* (1877). His subject was the chaotic forms of the Cathedral's worship including an organ blower—a man not a machine—about to keel over. 'This Anthem,' one voice says in the poem, as if talking about the theatre,

> being a favourite piece,
> Obtains a two-nights' run.
>
> **Bellows-Blower I:**
> Blow, Timothy! What are you gaping at? Blow!
> The choristers form in procession below!
>
> **Bellows-Blower II:**
> This labour would bring Tubal-Cain to his death.
> I tell you, I've blown till I'm quite out of breath.[24]

The near collapse or, rather, non-starting, of the Christmas anthem is the organ running out of breath because Timothy, the poor organ blower whose name, ironically, derives from the Greek 'honouring God', has run out of breath too. Fragility and the organ, failure and the cessation of breath, belong together.

The similar weariness of the hand-blower, and thus the precariousness of the organ's breath, formed a topic, likewise, for the admired American author of *The Autocrat at the Breakfast-Table* (1858), Oliver Wendell Holmes (1809–1894). This is the subject of his evangelical poem, 'The Organ-Blower', included in his posthumous *Complete Poetical Works* (1912). Although it might be possible to read this poem ironically, or even mockingly, it is also legible as a sincere tribute to the men who, before the artificially powered systems, made the organ possible to play, and who, in turn, offered—for those willing to see it this way—a form of ministry. Where Montaigne reflected in 'On the Cannibals', that 'We grasp at everything but clasp nothing but wind',[25] wind itself, where the organ was concerned for Holmes, figured instead the summation of everything. Watching the organ-blower at work, Holmes' poet is taken up with the sense of his, the organ-blower's, vocation, and, moved by a recognition of this implicit religious calling, the poet relieves the weary man directly. Here, unlike for Armstrong, the organ's association with failing breath is transformed into an opportunity for human service in the worship of the divine. The ending and the endless are intertwined. 'Not all the preaching', Oliver Wendell Holmes writes at the close of his long poem,

O my friend,
Comes from the church's pulpit end!
Not all that bend the knee and bow
Yield service half so true as thou!
One simple task performed aright,
With slender skill, but all thy might,
Where honest labor does its best,
And leaves the player all the rest.

This many-diapasoned maze,
Through which the breath of being strays,
Whose music makes our earth divine,
Has work for mortal hands like mine.
My duty lies before me. Lo,
The lever there! Take hold and blow!
And He whose hand is on the keys
Will play the tune as He shall please.[26]

God Himself is the organist here—a more breath-taking proposition than Brown's. The 'hand [...] on the keys' is both that of the organist playing on the manuals as well as, dimly, perhaps, the Father who allows His Son to give the keys of heaven to St Peter. But failure and perfection are united here. God's music is mortal because it is dependent on the strength of men who wear themselves out. But it is also bound to the eternal because it is God's. The organ, for Holmes, becomes at once the instrument of immortality—indeed, of God-as-organist—and an emblem of the Fall, exposing man's weakness and his need for relief.

The new technologies, I have suggested, implied the theoretical possibility of endless breath. But, as every organist still knows well, those technologies do not always work. Men at the bellows weary, but machines break down. The most memorable stories, narrated with a particular relish by Laurence Elvin, the author of, to my knowledge, the only extended history in English of organ blowing, are not of endless breath but its opposite.[27] Hand-blowers suggested the organ's supply of wind could be unsteady. But so did the malfunctions of the industrial age. Here are some tales. Dr G. B. Arnold, organist of Winchester Cathedral (1865–1902) and one of the founder members of the (Royal) College of Organists, once experienced the partial failure of his hydraulic wind supply during his *sortie*, causing—as always happens—the organ to sink badly out of tune before it gave up altogether. 'After the service', Elvin says, 'an unmusical member of the congregation went up to the Doctor and said: "That voluntary you have just played was excellent, it reminded me so of the waves of the sea". One can imagine Dr Arnold's indignation!'.[28] Elvin narrates, too, in his survey of failing wind supplies, an anecdote about choirboys mischievously messing with the similarly hydraulic blowing system at Manchester Cathedral, causing the resident organist Dr Kendrick Pyne's concluding improvisation to end not with a blaze of fanfare reeds but a feeble wail as the supply packed in.[29] Famously, too, the steam engines of the already discussed Willis organ in St George's Hall, Liverpool—still one of the great Romantic instruments of the United Kingdom—once failed in a Saturday evening recital in 1913 when one of the cranks cracked.[30] The dependence of the organist's art on machinery must have felt all too real to the Liverpool audience—assuming that this story is true—as they queued to leave a prematurely silenced concert, which no one could resume without the skill of a man who might be equally, if not more, at home in the boiler rooms of a transatlantic liner. The sound of the St George's Hall organ is still remembered as being hailed an industrial triumph (rather than anything to do with the mysteries of religion: it was as much a success as, indeed, a great ship).[31]

None of this could have nurtured the idea of the organ as able to breathe indefinitely. And it is worth remembering, too, that there were other major problems with the new technologies. The early steam-power engines, for example, needed hours of notice before they produced wind at the correct pressure. However important Willis's organ was for St George's Hall, the organist had to give the engineers *twenty-four hours' notice* in order for the engines to produce sufficient steam.[32] There was no spontaneous practising here. Where the theoretical possibility of the organ's endless breath might have quietly re-energised late nineteenth-century writing about the immortals, it could well have seemed to other listeners and players, hearing wheezing, out-of-tune, or simply silent instruments, that the organ was perfectly capable of suggesting death. Its failure to breathe could remind listeners of what will in due course occur to us all. Here, then, in the semiotics of the organ blower (to use a grand phrase), is something like William Empson's ambiguity of the seventh type. That is where the expression of an idea or image has within it exact contraries. With the established notion of the organ's winding as a form of breathing we have at once a figure of the organ and endless life—and its opposite.

It is not, though, in literary writing in the second half of the nineteenth century that this seventh ambiguity—an idea given to Empson, Sir Jonathan Bate thinks, by Einstein's General Theory of Relativity[33]—was most notably explored. And it was not in writing in words either. John Cage's *ASLSP* (*As Slow as Possible*) (1985/7) is an organ piece currently in performance that exploits the potential of the instrument to continue indefinitely. The piece started being played at the church of St Burchardi, Halberstadt, in 2001, on a specially made organ with six pipes, and is planned to last for 639 years.[34] It will end, that is to say, in September 2640. But for the esteemed Estonian composer for the church, Arvo Pärt (b. 1935),[35] the organ's relationship with the seemingly perpetual and the enduring has been explored differently.

Pärt is a minimalist, absorbed by the inheritance of the Christian West, including that of plainchant. Fascinated by the associations between the contemporary and the far past, he was also, in his 1980 work for organ, *Annum per Annum*, energised by the aural experience of rising and failing breath as it could be suggested by the organ's wind chests. *Annum per Annum* was written to celebrate the 900 years in which a daily mass has been said in the Speyerer Dom, the ancient cathedral of the south-western German town of Speyer. Characteristically minimalist in musical material, with continuously repeated motifs across the five principal movements (corresponding to the Ordinary of the Mass), the piece epitomises the intensity and deep structural balance that Pärt can create through largely harmonic and simple patterns. But, ruminating on the durability of the Mass in the Speyerer Dom, Pärt was prompted to do something highly unusual in the history of the organ repertoire: to instruct the organist to switch the blower off during the course of the piece (this will not, of course, work with an electronic organ). The result, inevitably, is the rapid collapse of pitch and regularity of sound as the organ's lungs are emptied, and then its silence. György Ligeti's *Volumina* (1962), it might be added, does something related: this violently discordant work commences with the player instructed, with the instrument turned off, to depress a massive *tutti* cluster and then to switch on the wind. In its trial performance, this requirement caused the organ motor at the church at Göteborg, the planned scene of the first performance, to burst into flames.[36]

Pärt's dramatic gesture prompts the listener, perhaps, to consider music's relationship to time. Commemorating a long unbroken succession of worship in the Cathedral, the failing of the wind conjures the inevitable ending even of such durability, the last day. Mortal things will not endure. But then, with the blower switched back on, Pärt's strange, insistent music resumes as if figuring not only temporal fragility but reanimation, the God-like revival through a saving, returning breath. Another ambiguity of the seventh type, perhaps, Pärt's *Annum per Annum* deploys the organ's lungs to create in sound a reminder of what will, and will not, last.

And finally …

Is there any merit in thinking counter-historically, backwards from Pärt to previous composers for the organ? Is it possible to look backwards and speculate for a moment about earlier, more uncertain, forms of the organ's breath and how they might relate to time? J. S. Bach was, it does not need saying, a richly theological thinker in music. The polymathic Albert Schweitzer (1875–1965), encouraged by the organist and composer Charles-Marie Widor (1844–1937), began the serious modern investigation of this with *J. S. Bach: Le musicien-poète* (1905), later developed in German as *J. S. Bach* (1908). Now no Bach scholar could ignore the role of Bach's faith in his music. Take, as a single example, the chorale prelude on *Dies sind die heilgen zehn Gebot*, BWV 678: 'These are the holy Ten Commandments'. This, in the Lutheran scheme, is the hymn on the descent of the Law to Moses, first described in Exodus 20. In Judeo-Christian history, that is to say, the hymn marks the beginning of the possibility of civilised life on earth under God, the beginning of the Law, and the possibilities of righteousness. It recalls the commencement of man's capacity to please the Lord both in worship and in deed. And to the gift of the Decalogue to humankind, Bach responds with a chorale prelude shaped as a remarkably expressive double canon, where both the right-hand accompaniment and the left-hand articulation of the *cantus firmus* are in canon—the musical form where one line is imitated at a fixed interval of both time and pitch. Here is the Law of the Hebrews and the working out of a particularly complex example of counterpoint, that practice of music that is exceptionally dependent on, and expressive of, laws.

The structure of *Dies sind die heilgen zehn Gebot* is interpretable—a narration can be found that does not 'explain' the music but at least permits one to see one of the many meeting points between Bach's faith and the propositional architecture of his writing. But what about *breath* in his organ music? In the territory of the speculative, I wonder if, in other pieces, Bach had somewhere in his mind the notion of the organ's theoretical possibility to transcend human lungs' capacity—a possibility that would become even more real two centuries after his death—and thus to figure in sound a proximity to eternity. There could be many examples of this in Bach's work, supposing my point to have any merit. But one that strikes me in particular is the elongated, unbroken fugue subject of the *Prelude and Fugue in b minor*, BWV 544, a cornerstone work in the organist's repertoire from Bach's Leipzig years. (There is, I should say, no certainty that the fugue subject *should* be played unbroken: it is, of course, possible to play it in groups of two notes, though the effect to my ear is laboured, or to introduce simply one or two articulation points. It is, in my experience, most often played without them.)

No ordinary singer could keep the long lines of this fugue, assuming they *are* unbroken, going without a breath; no ordinary string player could articulate this fugue subject all in one movement of the bow. Perhaps, as a listener attends to this long-breathed fugue subject and its counter-subject—the elongated lines of the counterpoint in general—he or she might be hearing Bach's intrigued meditation on the capacity of the organ, even one blown by hand or by foot, to offer us the beginnings of a glimpse of transcendence beyond our frail lungs, an airy hint of possibilities concerning life beyond the ordinary human capacity that later generations would know better. In the organ's extended breathing, Bach, even in the first half of the eighteenth century, might have speculated about what later artificially blown instruments would more pointedly suggest: that the organ can reach towards realms where breath does not run out. Stravinsky said, with a remark sometimes attributed to Berlioz, that the organ's steady wind—to return to my epigraph—was the reason he never wrote music for it: 'the monster never breathes', he said.[37] Bach felt differently whether he was musing on eternities suggested by the organ's lungs or not. But those who played at St George's Hall, at Colston Hall, or at any of the newly powered organs of industrial modernity found, I think—so long as the wind supply worked—a more substantial confirmation of what Bach might just have been wondering. The act of switching on an artificially powered blowing system, strange though it is to say, could, perhaps, take one's poetically inclined listeners half-way to heaven.

Acknowledgments My gratitude for their comments on drafts of this essay or advice more generally to: Michael Brennan, John Butt OBE, David Fuller, Russell Goulbourne, Graham Huggan, David Pipe, and Iain Quinn. I am particularly grateful to Stephen Farr.

Notes

1. Brindley and Foster, Sheffield, was an influential provincial organ building company from 1854. They were eventually bought by Henry Willis and Co. in 1939. The view of the musicality of unstable wind for an organ—if that is what is being implied in this article—is hardly accurate (Bach, for one, complained about it). I turn at the end of the essay to the unmusical realities of poor wind supplies.
2. Frances Elizabeth Cox (1812–97); Christian Fürchtegott Gellert (1715–69). The text is the first verse of Hymn 117 in William Henry Monk, compiler and arranger, *Hymns, Ancient and Modern, for Use in the Services of the Church* (London: Novello, [1861]). The revised version, now widely used, removes the theological problem though not the need to breathe over the line end: 'Jesus lives! thy terrors now | Can, O death, no more appal us'. The original version proved particularly problematic in the index of first lines (x).
3. *The Cambridge Edition of the Poets: Wordsworth* (New York: Houghton Mifflin, 1904), 541.

4. Lawrence Ferlinghetti, *These Are my Rivers: New and Selected Poems, 1955–1993* (New York: New Directions, 1993), 244.
5. Russell Goulbourne points me to William Cosmo Monkhouse (1840–1901)'s 'The Night Express', with its lines (as spoken by the locomotive):

   'Now through the level plain,
   While, like a mighty mane,
   Stretches my endless breath in cloudy miles;'

   Arthur Quiller-Couch, ed., *The Oxford Book of Victorian Verse* (Oxford: Clarendon, 1912), 654.
6. '[...] And by ocean's margin this innocent virgin
   Constructed an organ to enlarge her prayer,
   And notes tremendous from her great engine
   Thundered out on the Roman air [...]',

   W. H. Auden, *Collected Poems*, ed. by Edward Mendelson (London: Faber, 1994), 280.
7. On the nature of the medieval organ, see, for instance, Peter Williams, *The European Organ: 1450–1850* (London: Batsford, 1966); Kimberly Marshall, *Iconographical Evidence for the Late-Medieval Organ in French, Flemish, and English Manuscripts*, 2 vols (New York: Garland, 1989); and, on England specifically, Stephen Bicknell's *The History of the English Organ* (Cambridge: Cambridge: University Press, 1998).
8. For more details, see Andrew Parker, *Winchester Cathedral Organs: One Thousand Years* (Winchester: Culverlands Press, 2002).
9. John Milton, *Paradise Lost: The Text Reproduced from the 1st Edition of 1667* (London: Basil Montagu Pickering, 1873), Bk 1, ll. 705–15.
10. The British organ builders Martin Goetze & Dominic Gwynn Ltd have constructed three Tudor instruments based on surviving evidence both written and physical: for more details, see https://www.goetzegwynn.co.uk/story-suffolk-fragments-making-tudor-organs-dominic-gwynn/, last accessed 3 September 2018.
11. Henry Purcell (music), Nicholas Brady (words), *Hail! Bright Cecilia*: see vol. 8 of the Purcell Society series: http://www.henrypurcell.org.uk/purcell-society-edition/.
12. Ibid.
13. Laurence Elvin observes that the organ-builders Wordsworth of Leeds asserted that the first organ in the UK to be blown by a device powered by electricity was that built by them for 'Rudstone' Church, Bridlington (actually All Saints', Rudston) in 1889, in *Organ Blowing: Its History and Development* (Lincoln: Elvin, 1971), 79. The National Pipe Organ Register nevertheless records this instrument as 1888 and makes no mention of its blowing system or alleged priority. That does not mean that Elvin is wrong.

14. On the history of the nineteenth-century organ, see in particular Nicholas Thistlethwaite, *The Making of the Victorian Organ* (Cambridge: Cambridge University Press, 1990) and Stephen Bicknell's *History of the English Organ*. For other accounts of the relationship between the organ and literature, see Francis O'Gorman, '"Blush, Sad Soul, What Harmonies Are These!" The Organ in Nineteenth-Century English Literature', *Yearbook of the Royal College of Organists* (2003–2004), 66–73, and Iain Quinn, *The Organist in Victorian Literature* (New York: Palgrave Macmillan, 2017).
15. Its present form is as described in the National Pipe Organ Register http://www.npor.org.uk/NPORView.html?RI=N00289, last accessed 4 September 2018.
16. Elvin, 34.
17. E. H. Turpin, 'The Organ and Organ Music, II', *The Musical Standard* 21 (16 July 1881), 34–36 (34). Turpin (1835–1907) was an early Fellow of the (Royal) College of Organists and later Honorary Secretary. He was, as it happens, interested in poetry and, among other things, edited and introduced a volume of Tennyson, *The Palace of Art, and Other Poems* (New York: Maynard, Merrill [1898]). He is commemorated in the Turpin Prize, awarded to the candidate gaining the second highest marks in the Practical Examination for the Fellowship of the Royal College of Organists (FRCO).
18. Arthur Weir, *The Romance of Sir Richard: Sonnets, and Other Poems* (Montreal: Drysdale, 1890), 117.
19. Edmund Gosse, *Father and Son: A Study of Two Temperaments* (London: Heinemann, 1907), 256.
20. Edmund Gosse, *On Viol and Flute* (London: Henry S. King, 1873), 119.
21. The National Pipe Organ Register survey is dated from 1899 (and gives the date of the fire as that year): it can be found on http://www.npor.org.uk/NPORView.html?RI=N03895, last accessed 24 August 2018.
22. Hardwicke Drummond Rawnsley, *A Book of Bristol Sonnets* (London: Hamilton, Adams, 1877), 111. Cf., in the same collection, the rather mundanely titled 'On Hearing the Organ in the Cathedral, While the Work in the Nave was Suspended', 15.
23. *The Collected Poems of T. E. Brown* (London: Macmillan, 1900), 95–96.
24. *The Poetical Works of Edmund J. Armstrong: A New Edition* (London: Longmans, 1877), 81.
25. Michel de Montaigne, 'On the Cannibals', *Les Essais*, trans. as *The Complete Essays* by M. A. Screech (London: Penguin, 1991), 229.
26. *The Complete Poetical Works of Oliver Wendell Holmes* (New York: Houghton Mifflin [1895]), 188.
27. Elvin is an important source but documentation in his study of organ blowing is almost non-existent.

28. Elvin, 72.
29. Ibid., 71. James Kendrick Pyne (1852–1938) was Organist of Manchester Cathedral, 1875–1908.
30. Ibid., 35.
31. Thanks to Michael Brennan for this observation via his father, a chorister at Liverpool before the Second World War.
32. Elvin, 34.
33. Jonathan Bate, 'Words in a Quantum World: How Cambridge Physics Led William Empson to Refuse "either/or"', *Times Literary Supplement* (25 July 1997), 14–15.
34. See https://www.aslsp.org/de/, last accessed 13 September 2018.
35. Although often pronounced as if the last name is German (<Peart>), Estonian pronunciation as I understand it means that the <ä> is like the <a> in <cat>.
36. See https://www.allmusic.com/composition/volumina-for-organ-mc0002406515, last accessed 13 September 2018.
37. See Igor Stravinsky and Robert Craft, *Dialogues and A Diary* (London: Faber, 1982), 79.

Selected Bibliography

Bicknell, Stephen. 1998. *The History of the English Organ*. Cambridge: Cambridge University Press.

d'Anchald, Philippe. 2018. The First Applications of Electricity to the Organ. *British Institute of Organ Studies Journal* 42: 143–60.

Elvin, Laurence. *Organ Blowing: Its History and Development*. Lincoln: Elvin, 1971.

Marshall, Kimberly. 1989. *Iconographical Evidence for the Late-Medieval Organ in French, Flemish, and English Manuscripts*. New York: Garland.

O'Gorman, Francis. 2003–2004. 'Blush, Sad Soul, What Harmonies Are These!' The Organ in Nineteenth-Century English Literature. *Yearbook of the Royal College of Organists*: 66–73.

Quinn, Iain. 2017. *The Organist in Victorian Literature*. New York: Palgrave Macmillan.

Thistlethwaite, Nicholas. 1990. *The Making of the Victorian Organ*. Cambridge: Cambridge University Press.

Thistlethwaite, Nicholas, and Geoffrey Webber, eds 2011. *The Cambridge Companion to the Victorian Organ*. Cambridge: Cambridge University Press.

Williams, Peter. 1966. *The European Organ: 1450–1850*. London: Batsford.

CHAPTER 16

London Fog as Food: From Pabulum to Poison

Christine L. Corton

London has always had a fog problem. The city lies in a natural basin surrounded by hills, with a major river running through it, and various waterways and marshlands all around. But these fogs are white or grey in their natural state. It was with the growth of London as the nation's industrial capital that they began to be polluted by sulphur emissions from domestic coal fires and steam-driven industries. Thick yellow fogs occurred with growing frequency in the winter months when warm air above the river basin trapped cold air in the city below in a phenomenon known to meteorologists as temperature inversion. Unable to escape, London's natural winter fogs became suffused with large sulphur particles, sometimes so intense that they turned from yellow to brown. As early as 1791, the composer Joseph Haydn, during his first triumphant musical visit to England, wrote on 5th November that 'there was a fog so thick that one might have spread it on bread. In order to write I had to light a candle as early as eleven o'clock' (Plate 16.1).[1]

C. L. Corton (✉)
Senior Member of Wolfson College, University of Cambridge, Cambridge, UK

D. Fuller et al. (eds.), *The Life of Breath in Literature, Culture and Medicine*, Palgrave Studies in Literature, Science and Medicine,
https://doi.org/10.1007/978-3-030-74443-4_16

Plate 16.1 'A Fog in the streets of London' (*Illustrated London News*, 12 January 1867. Courtesy of the Syndics of Cambridge University Library)

The fog experienced by Haydn might of course have been made worse by the bonfires and fireworks of Guy Fawkes' Night. But fogs continued to increase in frequency and intensity with the rapid growth of the city. By the mid-nineteenth century fogs occurred more often and lasted longer, often going on for several days; they reached their greatest density and frequency in the 1880s, when many literary representations of London fog are set, even if written later, as with Joseph Conrad's *The Secret Agent*, set in 1886 but published in 1907. Although they began to decline from then on, an effect even prompting an article on the decline of the Edwardian fog, they were still common in the twentieth century.[2] Occasionally a really major fog, as in 1952, would still cover the city in a thick yellow blanket for days on end. The Clean Air Act of 1956, passed in response, combined with the growing use of gas and electricity as replacements for coal in domestic heating and industrial plants to bring them finally to an end. That is not to say that air pollution disappeared entirely: a major fog lasted several days in 1962. But it was the last. As time went on, however, exhaust fumes from motor vehicles have created a much more dangerous problem as the particulates created are small—small enough to enter the bloodstream. This creates a less visible haze than the traditional sulphurous fog, although it poses no less a threat to health; but it does not capture the imagination as the traditional London fog did, and is certainly not seen as a form of nourishment.

It is difficult nowadays to convey the sheer intensity of the classic yellow London fog, so dense in places that people were unable to see their hands when they stretched out their arms in front of them. E. F. Benson described one such fog in his *The Book of Months:* 'January: Thick yellow fog, and in consequence electric light to dress by and breakfast by … The denseness … We blindly grope on the threshold of the future feeling here for a bell-handle, here for a knocker, while the door stood shut … From sick dead yellow the colour changed to grey'.[3] It was thick and viscous, it smelt of sulphur and soot, and it made it difficult to breathe. Charles Dickens's son, in his *Dictionary of London*, published in 1879, wrote that 'Nothing could be more deleterious to the lungs and the air-passages than the wholesale inhalation of the foul air and floating carbon, which, combined, form a London fog. In this connection it may be taken as an axiom that the nose is nature's respirator'.[4] Much earlier the *Illustrated London News* in 1847 quoted an unnamed Scottish physician who posed the question: 'If a person require half a gallon of pure air per minute, how many gallons of this foul atmosphere must he, as it were, have filtered by his lungs in the course of a day?'.[5]

During the winter months, fog affected every area of life and all of the five senses, including taste. In 1874 the *Illustrated London News* described walking through a dense London fog:

> around, a thick, substantial, clammy atmosphere, that weighs one down and clings and hangs to one … All that one eats or drinks has mingled with it the flavour and the odour of the dun atmosphere; it 'burrs' one's throat and clings round teeth and palate, making even more miserable that most depressing of all things, a breakfast by lamplight—coffee, toast, eggs, everything, might, as far as taste goes, be red herrings or Indian soy.[6]

London fog not only changed the taste of all foods: from early on, people even referred to it as if it were a kind of food, above all, famously of course, pea soup. The artist John Sartain recounted in 1830 how he would 'slink home through a fog as thick and as yellow as the pea-soup of the eating house' (indeed, he soon emigrated to the cleaner air of America).[7] In 1833 Thomas Carlyle complained bitterly about 'that horrid flood of Spartan black-broth one has to inhale in London'.[8] In his novel *Maxwell* (1834), Theodore Edward Hook described 'the pea-soup atmosphere of London, which chilled every living thing, while a sort of smoky, misty, foggy vapour, hovering over the ground, made "darkness visible"'.[9] Five years later, Rosina Bulwer Lytton, in her novel *Cheveley*, noted that 'London in December' presented 'a pea-soup fog, which renders the necessary and natural process of respiration almost what Dr. Johnson's idea of fine music ough't to be, impossible!'.[10] The London satirical journal *Punch* (1860) fantasised that

> a scheme has been devised for procuring the extraction of pea-soup from London fog. When the foreigner remembers that our fogs are now so frequent that the clear blue sky in England is never clearly seen, he may form a faint conception of the work which is cut out for this new Company of Soup-makers. The fog will daily furnish a lot of raw material, which English ingenuity will soon cook into soup.[11]

The American novelist Herman Melville wrote in 1860 in his *Journal of a Visit to London and the Continent*: 'Upon sallying out this morning encountered the oldfashioned pea soup London fog—of a gamboge color'.[12] Henry Vollam Morton in his book *In Search of London*, first published in 1951, recalled being caught in a 'thick pea-souper which tastes like iron filings at the back of your throat'.[13] Thomas Cook, the travel agent, used the term to promote winter cruises abroad as late as 1937, urging people to go abroad 'when London is groping its way home through a typical "pea-souper"'.[14]

Melville's description of the colour of the fog as 'gamboge' highlights its yellowness (gamboge was a deep yellow pigment derived from the resin of the gamboge tree and commonly used as a dye). We usually think of pea soup as green, not yellow like the colour of fogs; but the pea soup that lent its name to London fog was not the same as the pea soup we know today. It was indeed yellow, rather than green, made from dried split peas, and was traditionally a staple food of the urban poor, high in protein but inexpensive to purchase. In an era before household refrigeration, fresh 'garden peas' could not be kept for more than a few days; split-peas, turned yellow by the drying process, would last far longer, although they were much less palatable. The association with poverty is made clear in Thackeray's story 'A Little Dinner at Timmins's', first published in *Punch* in 1848, towards the end of the 'hungry forties': Mr. Timmins suggests to his wife that they should invite a friend to dinner, but she replies with 'a look of ineffable scorn', that 'the last time we went there, there was pea-soup for dinner!'.[15] The densest London fogs were often described, like pea soup, as a form of sustenance fit only for the poor, avoided by anyone who had the means to do so. *Punch* in 1850 observed that 'March is said to come in like a lion, and go out like a lamb: but of November, on account of its fogs, it may generally be said, that it comes in like a basin of pea soup and goes out like a plate of peas pudding'.[16]

This was another dish made of yellow split peas and consumed mainly by the poor. Known as pease-pottage, peas-pudding or peas-porridge, it was given to convicts on their way to Australia. Poor people kept a cauldron of it warming over the open fire, enriching it by throwing in all kinds of vegetables and cheap meat products day after day—as the nursery rhyme, *ca.* 1760, had it: 'Pease porridge hot, Pease porridge cold, Pease porridge in the pot, Nine days old'.[17] In 1852 Thomas Miller wrote of being in a London fog: 'It is something like being imbedded in a dilution of yellow peas-pudding, just thick enough to get through it without being wholly choked or completely suffocated. You can see through the yard of it which, at the next stride, you are doomed to swallow, and that is all'.[18] The Canadian writer Sara Jeannette Duncan emphasised these aspects in even more positive terms in her novel *An American Girl in London* (1891): 'It was no special odour … that could be distinguished—it was rather an abstract smell—and yet it gave a kind of solidity and nutriment to the air, and made you feel as if your lungs digested it. There was comfort and support and satisfaction in that smell'.[19] The American feminist Inez Haynes Irwin, who lived mainly in Boston, writing in her travel book *The Californiacs* (1916), described the white fogs found in San Francisco and the Bay Area as 'Not distilled from pea soup like the London fogs'.[20] The British detective novelist Margery Allingham published a story in 1952—the year of the great smog—entitled *The Tiger in the Smoke*, in which a serial killer commits his deadly acts within a foggy London. She used the traditional pea-soup imagery to describe it, but gave it her own singular twist: in the appropriately named Crumb Street, she wrote, 'The fog slopped over its low houses like a bucketful of cold soup over a row of dirty stoves'.[21] Here the soup is thoroughly unpalatable ('pease porridge cold'), as chilling as the fog it describes, and through the image of the chimneys pumping out dirty smoke from its 'dirty stoves', the novelist makes it clear that Crumb Street is an abode of the poorest part of London's population. Even Bob Hope, the famous London-born American comedian, could make it the basis of one of his numerous gags that Californian fog was 'fog with the vitamins removed'.[22] Fog could thus appear as a kind of food, ingested or inhaled: not air or vapour but something stronger and more solid, nourishing and filling, if not food of the grandest or most pretentious kind.

So frequent were London fogs between the 1830s and the 1960s that Londoners coined quite a variety of phrases to denote them. 'London Particular' was another term for fog, pregnant with meaning. It conveyed a familiar intimacy—the writers of *Slang and Its Analogues,* published in seven volumes between 1890 and 1904, noted that it meant 'a thick or black fog, the product of certain atmospheric conditions and carbon: formerly peculiar to London'. 'Particular', they added, however, was another name for a mistress, perhaps hinting that Londoners felt as ambivalent about fog as some married men may have felt about their extramarital affairs.[23] More importantly, a 'London particular' was a special quality of fortified brown Madeira wine produced solely for the London market, referred to, for example, in Thomas Love Peacock's comic novel *Melincourt* (1817), where the Reverend Mr Portpipe, 'a man of exquisite taste', invites his guests to enjoy a 'good ale and a few bottles of London Particular'.[24] 'London Particular', widely used as a term for the combination of fog and smoke, came during the second half of the nineteenth century to be employed with a degree of nostalgia. As early as 1855 the *New York Times*, describing a home-grown American fog, wrote that it had 'nothing of the characteristics of our old London particular except density'.[25]

The terms London Particular and pea-souper were brought together in the twentieth century by a recipe called 'London Particular' which was in fact, not surprisingly, a traditional pea-soup, brought up to date and dissociated from poverty by being described as 'low-fat and low-cholesterol'.[26] This particular recipe is illustrated with a scene of the character Krook's demise from 'Spontaneous Combustion' in Charles Dickens's novel *Bleak House*. As J. Hillis Miller writes: 'Krook is transformed into the basic elements of the world of the novel, fog and mud. The heavy odor in the air, as if bad pork chops were frying, and the "thick yellow liquor" which forms on the window sill as Krook burns into the circumambient atmosphere, are particularly horrible versions of these elements'.[27] In this

"OLD KING COAL" AND THE FOG DEMON.

Plate 16.2 '"Old King Coal" and the Fog Demon' (John Tenniel, *Punch*, 13 November 1880. Courtesy of the President and Fellows of Wolfson College, Cambridge)

central passage in the book, one of the main characters is dissolved back into the elements described in the famous opening passage: flesh turns into fog (Plate 16.2).

An attempt was made to rid London fog of its 'nourishing' qualities through a suggestion in 1904 to rename it. The scientist Henry Antoine Des Voeux, honorary treasurer of the Coal Smoke Abatement League, suggested that it should be named '"smog," to show that it consists much more of smoke than true fog'.[28] 'Smog', a portmanteau word combining the basic elements of smoke and fog, was designed to indicate that London fog was in fact a combination of the natural phenomenon of fog with the man-made smoke exuding from the many industrial and domestic chimneys in the capital, suggesting strongly that it was the latter that constituted the evil to be eliminated. But 'smog' was not popularly accepted until the 1950s, when it was used to refer to photochemical air pollution from motor vehicles, rather than the sulphurous mixtures of coal and smoke. Most people continued to prefer using 'pea-souper' or

'London particular' when yellowish fogs settled on London in the winter months.

For some, London fogs became part of the city's identity, signifiers of what George Gershwin's song called 'A foggy day in London town, had me low, had me down'.[29] For many Londoners it was an aspect of London Pride, the 'grey city.... smokily enchanted', as Noel Coward's patriotic Second World War song called it.[30] After all London is still referred to as 'The Big Smoke'. The emissions pouring out into the atmosphere from the factories of the capital city showed that the economy was thriving. Industry provided jobs, and the sight of a smoking chimney standing out tall and proud from a factory meant that there was employment, while smoke from humbler domestic chimneys meant that people could afford coal to light and heat their homes, as well as to cook their food. Writers could see the positive side of air pollution. M. H. Dziewicki wrote 'In Praise of London Fog', its beauty and mysterious charm, in *The Nineteenth Century* magazine in 1889.[31] Artists such as James McNeill Whistler, Claude Monet, and Yoshio Markino, fascinated by the shifting effects of light shining through the fog, also saw the beauty of foggy days and tried to capture them on canvas.[32]

Yet people were early on aware of the dangers that London fog posed to their health, through consuming this thick yellow mixture. A tombstone in Kensal Green Cemetery in north-west London provides an early testimony to this perception with its inscription: 'L. R. / Who died of suffocation in the Great Fog of London /1814'.[33] In the revised edition of his book, *The Climate of London*, published in 1833, the meteorologist Luke Howard reported on a fog that occurred on November 11 and 12, 1828: 'The effect was most distressing, making eyes smart and almost suffocating those who were in the street, particularly asthmatic persons'.[34] Fog, as Charles Manby Smith, author of *Curiosities of London Life*, pointed out in 1853, created 'a misty atmosphere fraught with catarrh and influenza'.[35] The growing frequency and intensity of the yellow London fogs in the 1830s paralleled the rise of miasmatic theories of disease, which reached their height in the 1850s: these ascribed infections such as cholera to harmful vapours (caused by putrefaction) rising from the ground, and people tried to protect themselves from inhaling the fog by covering their mouths with handkerchiefs.

Animals suffered as well—birds flew around unable to find a spot to land and would fall down exhausted, unable to breathe. The animals at London Zoo were noted to have 'very dirty coats ... Polar bears and birds with soft white plumage, such as Spoonbills and White Peafowl, become extremely dirty... many Felines in the Lion house die from chronic bronchitis, fibrosis and gangrene to the lungs associated with blackening of the lungs by dust deposit'.[36] The Smithfield Cattle Show, moved to the Royal Agricultural Hall in Islington in 1863, was held annually in December, a month noted for its fogs. During December 1873 one of the first fogs of the season enveloped London and lasted for five to six days, coinciding with the annual Cattle Show. Animals began to be distressed as early as the first day of the fog and 'exhibited symptoms as if they had been inhaling a noxious gas'.[37] The organisers tried to create more ventilation by leaving the doors and windows open but, of course, this meant only that more of the dirty fog came in. The American writer Mark Twain, who was in England performing some readings (which were also heavily affected by the fog), commented that 'the cattle are choking & dying in the great annual Cattle Show, & today they had to take some of the poor things out & haul them around on trucks to let them breathe the outside air & save their lives. I do wish it would let up'.[38] The *Daily News* reported on the distress of the animals: 'By daybreak two had died where they lay in the Hall, and by nine o'clock the reception yard was full of the great suffering animals, panting piteously in a state closely akin to suffocation; while in the interior of the hall sobbed and panted others nearly, if not quite, in as bad a state'.[39] By the end of the week, ninety-one cattle had been removed at the request of their owners and 'of these, 50 were slaughtered, some died in the vans in which they were taken away'.[40] Contemporaries coined the phrase 'fog fever' to describe the epidemic of respiratory ailments and deaths among the cattle. The *Daily News* had no doubt as to the cause: 'the sole ailment is suffocation ... proved by an examination of the animals that have died or been slaughtered, their lungs being found gorged with black blood'.[41] The same thing occurred to the Smithfield Cattle Show of 1952; fewer cattle died but some ingenious masks were applied including one that was soaked in whisky to act as a disinfectant.

But it was not only animals that suffered. The *Medical Times and Gazette* of that week in 1873 raised the issue under the title 'Killed by the Fog'. The title distinguished between those people who were killed by accidents caused by poor visibility in the fog and those killed because of the fog's direct impact on their health and above all their respiratory system. It cautioned people against following the age-old advice of ensuring good ventilation because this would simply let the fog into the house. Most people who could not afford to sit inside and just shut their windows to wait it out, resorted to homemade masks using a scarf or a handkerchief over the nose and mouth to protect themselves from the dirty particulates in the air.[42] Fog had an impact even on healthy people, who suffered bodily discomfort and smarting eyes, frequently accompanied by severe headaches; but for those already suffering from lung complaints, or the vulnerable, such as the very old or the very young, 'the combination of cold and smoky air' caused 'violent coughing; and whilst the patient coughs he cannot breathe, and cough materially interferes, moreover, with circulation'.[43] The publication of the weekly statistics on deaths quantified in human terms the deadly impact of the fog and sparked an ironic outburst from *The Times:* 'We are very glad indeed to hear that 780 Londoners above the average died the week before last of the fog. We do not want them to die, of course, but if they were to die, it is better that they should die of the fog, and so get rid at once of the superstition that the most disagreeable, inconvenient, dangerous, and spirit depressing visitation which falls on Londoners is somehow "good for us". It is not good for us, any more than for cattle, but bad, as the Registrar's return shows'.[44] In fact, this figure of 780 was too low—the death toll is now estimated well above 1,000: during the smog of 1952, 4,000 extra deaths were acknowledged at the time, although it is now thought that probably as many as 12,000 extra deaths occurred.

Plants also were in a way suffocated by the fog. A plant's stomata were likely to be clogged with soot, and the intensity of the light reaching the chlorophyll was also likely to be reduced. During the winter months, when fogs were most common, plants growing outdoors are usually in a dormant phase: 'Evergreens were, however, coated with a thick greasy deposit of soot … not likely to be washed off very readily with rain'.[45] Winter flower gardens at Kew were completely destroyed in the smog of 1952.[46] One answer was to find fog-resistant plants and trees. The most successful was the ubiquitous London plane tree—its leaves are shiny, which means any soot from the air can easily be washed off, while its bark very picturesquely peels off almost continuously, thus allowing it to breathe and renew itself.

Since the late eighteenth century and Lavoisier's discovery of oxygen, scientists had been investigating the constituents of the air, but it was not until a century later that they succeeded in liquefying oxygen, hydrogen, and other gases. As Alain Corbin showed in his *The Foul and the Fragrant*

(1986), the growth of the respectable middle classes stimulated the emergence of the perfume industry, replacing musky scents that enhanced natural body odours with perfumes that disguised them. This highlighted the poor even more through the sense of smell. The poor became distinguished by what the middle classes considered their repulsive smell, and, along with this, their exposure to noxious miasmas of the kind that were thought to cause the spread of cholera, explaining, it seemed, the higher death rates from the disease in the poorer districts of the city. Judges at the Old Bailey, the Central Criminal Court in London, carried nosegays of highly scented cut flowers to protect their olfactory systems from the unpleasant and supposedly disease-ridden vapours wafted towards them by the criminals who were brought up from the damp and unhealthy cells below. In addition, the prevailing westerly winds tended to concentrate London fog and smoke in the poorer districts of the East End, where much of London's industry was also located. Increasingly, however, as fog became more prevalent across London, including its wealthier West End, the distinction in air quality became less noticeable. Many who could afford to do so just moved further out of London. As *The Times* complained in 1853, industrialists 'have a vested interest in compelling us to consume their smoke'.[47] Living in the clean air of the suburbs, they come into town once a week by rail 'to see how the chimney draws and how the till fills', then go back home, leaving the inhabitants of 'modern Babylon' to suffer as 'smoke penetrates the pores of our skin and the air vessels of our lungs, converting the human larynx into an ill-swept chimney'.[48]

John Timbs, writing in 1855, noted: 'Suppose the wind to change suddenly to the east, the great body of smoke will be brought back in an accumulated mass; and as this repasses the town, augmented by the clouds of smoke from every fire therein, it causes the murky darkness'.[49] When fog spread westward during the dockers' strike of 1889, gangs of striking workers stormed into the fashionable districts of the West End, smashing shop windows and terrorising respectable passers-by. The fog and the strikers appeared combined as a threat of disorder to the placid lives of the middle classes, turning day into night in a social and political sense as well as an atmospheric one. And often fog and crime appeared in people's mind as one. London fogs might be a means of concealing crime and disorder, as Sherlock Holmes noted in 'The Bruce-Partington Plans' (1908) when he gazed from his first-floor window in Baker Street at the fog-bound streetscape below (Plate 16.3):

Plate 16.3 William Luker, 'November Mourning' (From W. J. Loftie, *London City* [1891]. Courtesy of the Syndics of Cambridge University Library)

> The London criminal is certainly a dull fellow … Look out of this window, Watson. See how the figures loom up, are dimly seen, and then blend once more into the cloud-bank. The thief or the murderer could roam London on such a day as the tiger does the jungle, unseen until he pounces, and then evident only to his victim.[50]

The image was taken up again by Margery Allingham, in whose novel *The Tiger in the Smoke* the murderer is actually known by the nickname 'the tiger', stalking the gloomy and fog-shrouded streets of the city with evil intent. But London fog also suggested an antidote to the crime and disorder seemingly veiled under a cloak of mist: a warm fire in an open hearth, perhaps especially in the poorest homes, providing security and stability for the family—after all, a smoking chimney meant that people could afford a coal fire. The domestic coal fire was sacrosanct, and politicians disturbed it at their peril, while factory owners lobbied successfully against effective smoke abatement for decade after decade, arguing that economic progress and prosperity were the best guarantees of social order.

In the last quarter of the nineteenth century, unease about the atmosphere of London and the medical consequences of these urban fogs took on a figurative form in a series of popular novellas and short stories. Depictions of fog were now beginning to enter the realm of popular fiction. One such novella published in 1880 by William Delisle Hay purported to recall events which the author imagined taking place in February 1882. *The Doom of the Great City; Being the Narrative of a Survivor, Written A.D. 1942* describes a dense smoke-fog that had lasted since Christmas and destroyed all life in the capital on the fateful day of the narrator's birthday. The story was written as a warning of an impending ecological disaster brought about by the dirty atmosphere and constituted a hyperbolically imaginative response to the medical and scientific literature of the time, which had begun to warn against the smoky atmosphere. It also reflected a widespread feeling that cities such as London were growing too rapidly at the expense of a supposedly more idyllic lifestyle in the countryside, and expressed the threat of degeneracy posed by excessive urbanisation. The narrator's highly critical description of London centres on its 'murky atmosphere, the dingy gloom'.[51] The fog is both a symbol of the corruption of the city and a veil through which the narrator sees 'the odious colours of the evil that lies hidden behind the awful pall ... Among the higher ranks of society immorality was so common as to excite but small attention ... while up from the lowest depths there constantly arose a stream of grosser, fouler moral putrescence, which it would be libel on the brutes to term merely bestiality'.[52] London is cut off by a wall of fog with the few people escaping London relaying horror stories of death. The narrator, desperate to know the fate of his mother and sister, whom he has left behind at their home, makes his way through the foggy barrier. He finds all of London dead; and death seems instantaneous. A sentry on guard outside Buckingham Palace still stands upright, as he has died at his post, and further down the social scale 'two miserable little bodies in the gutter, two poor little ragged urchins ... their meagre limbs cuddled round each other in a last embrace'.[53] No one is spared: even the narrator's mother and sister, innocent of the general corruption of the city, are seated next to each other, with a cat on the chair, but all are dead (Plate 16.4).

Thus London fog, with its sooty particles and its sulphurous constituents, could be a form of poison as well as a kind of nutrition in the

Plate 16.4 'London Sketches—a November Fog' (*Graphic*, 9 November 1872. Courtesy of Syndics of Cambridge University Library)

popular and literary imagination. Medical opinion only gradually gained a decisive influence on attitudes towards London fog, helped by the rise of epidemiology and medical statistics, modern communications and news media. In the great killer fog of 1952, daily bulletins issued over the radio charted the rapid rise in deaths from bronchitis and other pulmonary conditions, eventually totalling an excess mortality of 12,000 over the normal number of deaths during the short period in which it lasted. The founding of the National Health Service after World War II created a widespread expectation that government would act to protect people's health, and the Clean Air Acts were the result, bringing the history of London fog to a close. At the time of writing this essay, however, scientists have announced that air pollution is one of the major causes of death in our rapidly urbanising world today, having overtaken cancer and tobacco smoking.[54] So there is still much more to do.

Nowadays, we only need to see a scene on television or in a movie of a foggy street dimly illuminated by a gas lamp to know we are in London in the 1880s or 1890s, and a crime or outrage is about to be committed. Modern dramatisations of Conan Doyle's Sherlock Holmes stories or films and dramas featuring Jack the Ripper almost invariably shroud the action in a thick London fog, though the technology does not seem to be available to give it the proper yellow colour. Yet in fact only one of the Sherlock Holmes stories, 'The Bruce-Partington Plans', uses fog as a central part of the drama. And although Jack the Ripper is inevitably shown shrouded in fog from his first fictional outing in Marie Belloc Lowndes' 1911 novel *The Lodger*, via an episode of *The Avengers* in which a Ripper-like figure is called the 'gas ghoul', to an episode in the original *Star Trek* featuring a 'Jack the Ripper' figure hailing from the foggy side of a planet and butchering women, in fact all of the Ripper murders took place on clear nights.

From being the breath of life, a symbol of Victorian progress and well-being, through figuring as the breath of death, a vector of moral judgment on the evils of the big city, London fog has become a kind of free-floating signifier of Victorian crime, as artificial as the machines that produce it in the film studio or as the cans of 'London Fog' sold to tourists in the 1970s.

Notes

1. J. Cuthbert Hadden, *Haydn* (London: Dent, 1902: repr. Cambridge: Cambridge Scholars Press, 2002), 67.
2. Henry T. Bernstein, 'The Mysterious Disappearance of Edwardian London Fog', in *The London Journal* 1 (1975), 189–206.
3. E. F. Benson, *The Book of Months* (London: William Heinemann, 1903), 3–4.
4. Charles Dickens, *Dickens's Dictionary of London 1879* (London: Charles Dickens, 'All the Year Round' Office, 1879), 103.
5. *Illustrated London News*, 2 January 1847, 7.
6. *Illustrated London News*, 28 November 1874, 518.
7. https://www.country1011.com/2016/11/30/19781/.
8. Thomas Carlyle to John Stuart Mill, 12 January 1833, in *The Collected Letters of Thomas and Jane Welsh Carlyle*, ed. by Charles Richard Sanders, 34 vols (Durham, NC: Duke University Press, 1977), 6:300–1.
9. Theodore Edward Hook, *Maxwell*, 3 vols (London: Henry Colburn and Richard Bentley, 1830), 2:10.

10. Rosina Bulwer Lytton, *Cheveley: or The Man of Honour* (London: Charles H. Clarke, 1860), 280.
11. *Punch*, 'Facts for Foreigners', 18 February 1860, 71.
12. Herman Melville, *Journal of a Visit to London and the Continent, 1849–1850*, ed. by Eleanor Melville Metcalf (Cambridge, MA: Harvard University Press, 1948), 45–46.
13. Henry Vollam Morton, *In Search of London* (London: Methuen, 1951; repr. London: Methuen, 1988), 144.
14. *The Times*, 12 August 1937 and 19 August 1937. After these two appearances the advertisement was dropped.
15. William Makepeace Thackeray, *A Little Dinner at Timmins's*, in *The Works of Thackeray*, 12 vols (London: Smith, Elder, 1878), 12:142.
16. *Punch's Almanack for 1851*, included with *Punch* 19 (1850) in unnumbered pages at the beginning.
17. The origin of this rhyme is not known but the earliest reference is found *ca.* 1765 in John Newbery's *Mother Goose's Melody*, according to Iona Opie and Peter Opie, eds, in *The Oxford Dictionary of Nursery Rhymes*, 2nd edn (Oxford: Oxford University Press, 1997), 345.
18. Thomas Miller, *Picturesque Sketches of London: Past and Present* (London: Office of the National Illustrated Library, 1852), 243.
19. Sara Jeannette Duncan, *An American Girl in London* (London: Chatto and Windus, 1891), 30.
20. Inez Haynes Irwin, *The Californiacs*, 1919. See http://www.gutenberg.org/files/3311/3311-h.htm.
21. Margery Allingham, *The Tiger in the Smoke* (London: The Hogarth Press, 1987), 25.
22. Bob Hope, *I Owe Russia $1200* (London: Robert Hale, 1963), 31.
23. *Slang and Its Analogues, Past and Present: A Dictionary Historical and Comparative of the Heterodox Speech of All Classes of Society for More than Three Hundred Years*, ed. by John S. Farmer and W. E. Henley, vol. 5 of 7 (London, 1902), s.v. 'London Particular (or London Ivy)'.
24. Thomas Love Peacock, *Melincourt or Sir Oran Haut-Ton*, 2 vols (London: J. M. Dent, 1891), vol. 2, 122.
25. *New York Times*, 2 April 1855, 1.
26. *Farmhouse Cookery: Recipes from the Country Kitchen* (London: Reader's Digest, 1980), 21.
27. J. Hillis Miller, '*Bleak House* and the Moral Life', in *Dickens: Bleak House: A Casebook*, ed. by A. E. Dyson (London: Macmillan, 1969), 167.
28. *The Times*, 27 December 1904, 510.
29. George Gershwin and Ira Gershwin, 'A Foggy Day in London Town'.
30. Noel Coward, 'London Pride'. Written in 1941 at the height of the Blitz.
31. M. H. Dziewicki, 'In Praise of London Fog', in *The Nineteenth Century: A Monthly Review*, ed. by James Knowles, vol. 26, July-December 1889, 1047–55.

32. For example Whistler painted *Nocturne in Grey and Gold—Piccadilly*, now in the National Gallery of Ireland, between 1881 and 1883. Monet made a series of paintings of London Fog in the late nineteenth and early twentieth century. For more detail see Christine L. Corton, *London Fog: The Biography* (London: The Belknap Press of Harvard University Press, 2015), especially ch. 5.
33. *Magazine of the Friends of Kensal Green Cemetery*, April 1997.
34. Luke Howard, *Climate of London Deduced from Meteorological Observations, Made in the Metropolis, and at Various Places around It*, 3 vols (London: Harvey and Darton, 1833), vol. 3, 303.
35. Charles Manby Smith, *Curiosities of London Life; or, Phases, Physiological and Social, of the Great Metropolis* (London: William and Frederick, 1853), 80.
36. 'Smoke and the Zoo: How the Animals Are Affected', *Journal of the National Smoke Abatement Society* 25/7 (1936), 9.
37. Georg Hartwig, *The Aerial World: A Popular Account of the Phenomena and Life of the Atmosphere* (London: Longmans, Green, 1874), 139.
38. Mark Twain to Olivia L. Clemens, 11 December 1873, in *Mark Twain's Letters, Volume 5, 1872–1873*, ed. by Lin Salamo and Harriet Elinor Smith (Berkeley: University of California Press, 1997), 508.
39. *Daily News*, 11 December 1873, 2, col. C.
40. E. J. Powell, *History of the Smithfield Club, 1798 to 1900* (London: Smithfield Club, 1902), 54.
41. *Daily News*, 11 December 1873, 2, col. D.
42. Dissections in the historical respiratory collections of medical museums (for example, in the teaching collection of the Gordon Museum of Pathology, King's College London), commonly show nineteenth-century examples of the effects of the inhalation of atmospheric particulates (including from fog), understood in the period as contributing to morbidity. For a recent study relating fogs to death statistics in the nineteenth century see Jonathan H. Widdicombe, 'A Brief History of Bronchitis in England and Wales', *Chronic Obstructive Pulmonary Diseases: the Journal of the COPD Foundation*, 7/4 (2020), 303–14, especially the section 'Mortality … 1838 to 1920': http://doi.org/10.15326/jcopdf.7.4.2020.0135.
43. 'Killed by the Fog,' *Medical Times and Gazette: A Journey of Medical Science, Literature, Criticism and News*, 13 December 1873, 668, col. B. *Cf*. 'Fog in relation to the health of the Metropolis', *The Medical Times and Gazette*, 2 October 1880 (p. 410, col. A), the facts and conclusions of which are substantially repeated in the 'Annus Medicus 1880' report in the *Gazette*, 25 December 1880 (723–24). In a similar vein, the Medical Officer of Health for Marylebone reported an unusually high death rate from bronchitis and pneumonia for December 1873, which he attributed

partly to fog (*Gazette*, 24 January 1874, 116, col. B); and the same district officer reported high mortality rates resulting from 'diseases of the breathing organs' exacerbated by dense fog in February 1880 (*Gazette*, 3 April 1880, 378, col. A). In the issue of 14 February 1880 (183, col. A) a high death rate among *poitrinaires* (consumptives) in London is attributed to dense fog, which is compared to that of the Cattle Show week of 1873 (on which see above), the death rate for which it exceeded. A similar high death rate from pulmonary diseases in Glasgow (referred to as a 'semi-asphyxiated city') attributed to fogs is discussed in the issue of 23 January 1875 (92, col. A).

44. *The Times*, 29 December 1873, issue 27,885, 4.
45. 'Plant Life', *Smokeless Air*, issue 85 (1953), 108.
46. *Hansard*, House of Lords debate, 18 November 1953, vol. 184, cc364–74 on fog in Britain generally (Kew Gardens, cc366).
47. *The Times*, 10 August 1852, issue 21,190, 7.
48. Ibid.
49. John Timbs, *Curiosities of London: Exhibiting the Most Rare and Remarkable Objects of Interest in the Metropolis with Nearly Fifty Years' Personal Recollections* (London: D. Bogue, 1855), 310.
50. Sir Arthur Conan Doyle, 'The Adventure of the Bruce-Partington Plans,' in *His Last Bow: Some Reminiscences of Sherlock Holmes* (London: Penguin, 1997), 83–84. First published in the *Strand Magazine* in December 1908 and in *Collier's Weekly* on 12 December 1908.
51. William Delisle Hay, *The Doom of the Great City; Being the Narrative of a Survivor, Written A.D. 1942* (1880), in *British Future Fiction: The End of the World*, ed. by I. F. Clarke (London: Pickering and Chatto, 2001), 23.
52. Ibid., 30–32.
53. Ibid., 59.
54. See *The Independent*, 12 March 2020. https://www.independent.co.uk/news/health/air-pollution-smoking-deaths-compare-a8818851.html.

Selected Bibliography

Barr, Robert. 1892. The Doom of London. *The Idler* 2/2: 397–409.

Brimblecombe, Peter. 1987. *The Big Smoke: A History of Air Pollution in London since Medieval Times*. London: Methuen.

Corton, Christine L. 2015. *London Fog: The Biography*. London: The Belknap Press of Harvard University Press.

Dickens, Charles. 1974. *Bleak House*. Ed. Norman Page. London: Penguin.

Lowndes, Marie Belloc. 1996. *The Lodger*. Intro. Laura Marcus. Oxford: Oxford University Press.

Luckin, Bill. 2003. 'The Heart and Home of Horror': The Great London Fogs of the Late Nineteenth Century. *Social History* 28/1: 31–48.

CHAPTER 17

'Now—for a breath I tarry': Breath, Desire, and Queer Materialism at the *fin de siècle*

Fraser Riddell

A. E. Housman's marginalia to his copy of Lucretius's *De Rerum Natura* demonstrate a characteristically assiduous engagement with questions of textual accuracy in Book III of the philosophical poem.[1] In offering his emendations, the poet and classical scholar revisits a materialist account of the universe in which the individual soul is composed of a momentary accretion of falling atoms—a combination of heat (*calor*), air (*aer*), breath (*aura*, *ventus*), and a fourth mysterious nameless substance (*natura*).[2] Lucretius's *aura* (breeze) and *ventus* (wind) are translations into Latin of πνεῦμα (*pneuma*) from Epicurus's Ancient Greek, and both terms retain associations with the movement of breath. The motion of such an *aura* informs Lucretius's humoral model of character, in which individual temperament is determined by the dominance of one material element in the soul (47–48, ll. 288–313). Heat equates to anger (*ira*) and is aligned with the fierceness of the lion. The dominance of air equates to a character with a tranquil heart (*pectore tranquillo*), and is allied with the placidity of the cow. The dominance of breath, meanwhile, equates to timidity or

F. Riddell (✉)
Department of English Studies, Durham University, Durham, UK

D. Fuller et al. (eds.), *The Life of Breath in Literature, Culture and Medicine*, Palgrave Studies in Literature, Science and Medicine,
https://doi.org/10.1007/978-3-030-74443-4_17

fear (*pavor*), and is associated with the deer, whose limbs are pervaded by a tremulous movement (*tremulum* [...] *motum*) and who stands frozen with terror.

Housman's engagement with Lucretius occurs at a historical juncture at which Victorian writers turned with renewed interest to ancient materialist philosophers—Heraclitus, Democritus, Empedocles, Epicurus—to examine the manner in which their theories pre-figured the most significant developments in contemporary science, most notably the atomic theory of John Tyndall.[3] Lucretius's poem, in this sense, became a prism through which Victorian writers confronted profound and unsettling questions about the relationship between mind and matter, individual free will, the coherence of the self and the nature of God.[4] But it also draws to our attention something more specific: the striking queerness of breath's materiality in late-Victorian literature. In the works of some of the most important figures in the literary history of homosexuality, such as A. E. Housman, John Addington Symonds, and Walter Pater, tropes of breath and breathlessness act to articulate a queer subjectivity that is marked in its materiality by fearfulness, timidity, and withdrawal, and so often defined by a precariousness that drifts towards a breath-like insubstantiality. In tracing the boundary between inner and outer, self and other, privacy and disclosure, breath participates in the making and the unmaking of queer subjects at the *fin de siècle*.

Recent work in queer theory helps us think more carefully both about the queerness of this materialism, and about what it could mean to breathe in fear with Lucretius. Following the so-called 'negative turn' in queer studies, theorists such as Jack Halberstam and Lee Edelman have located in queer sexuality forms of resistance to normativity that arise from the materiality of those bodies which refuse coherence, subvert reproductive temporalities, and reject existing modes of sociality.[5] Heather Love has demonstrated how the stigma of social marginalization—the weight of bad feelings—finds material embodiment in queer subjects through their negotiations with shared historical pasts.[6] Sara Ahmed's work on 'queer phenomenology' has explored how non-normative subjectivities are formed through their fraught negotiation with and re-orientation of normative spaces and material objects.[7] Queer materialism, for such theorists, becomes a materialism marked by failure: a failure of matter to cohere in the right place, in the right shape, or at the right time; a failure of matter to occupy space in the right way. But it is also a materialism in which such failure might become productive—where new modes of sociality might emerge from not needing to matter. It is a way of thinking about how certain marginalized subjects come to breathe in fear, but also how they might find a space, perhaps, to breathe more freely.

Sexology, Embodiment, and Breath: *Atmungstypus Männlich*

The queer materiality of breath in *fin-de-siècle* literature is best understood in the context of late nineteenth-century sexological discourses through which the homosexual subject emerged as a specific category of material body. Recent work on the history of European sexology has built on the foundational insights of Michel Foucault to enrich our understanding of the complexity of pre-Freudian understandings of sexuality.[8] In contrast to psychoanalytic models which viewed desire as rooted in the enigmatic psychic processes of the unconscious, *fin-de-siècle* sexologists located the origins of sexuality in congenital physiological and anatomical characteristics.[9] The exhaustive empirical measurement, observation, and classification of the body was one of the central strategies adopted by sexologists such as Albert Moll, Richard von Krafft-Ebing, and Havelock Ellis in order to demarcate the pathologized 'inverted' subject from the healthy 'normal' subject. Amongst a plethora of scientific data recorded by sexologists in their case studies—ranging from body shape to muscularity to vocal quality—modes of breathing became a key diagnostic indicator of non-normative sexuality.

In the first volume of Magnus Hirschfeld's *Jahrbuch für sexuelle Zwischenstufen unter besonderer Berücksichtigung der Homosexualität* a distinction is drawn between distinctive male and female 'respiration types' ('*Atmungstypen*'):

> In civilized races, the man primarily breathes from the abdomen, i.e. with the diaphragm and stomach muscles, and the woman more from the ribs, i.e. with the chest muscles. In children, the breathing type in early years is largely abdominal. Meticulous investigations by a variety of researchers have shown that these sex differences in respiratory movements are simply the result, reinforced by heredity, of an artificial constriction by the customary women's clothing.[10]

Despite acknowledging the significance of environmental factors, such as the wearing of corsets, in shaping characteristic modes of breathing, Hirschfeld nevertheless draws upon this distinction in his sexological case studies as indicative of congenital abnormal sexuality. Subsequent issues of Hirschfeld's *Jahrbuch* repeatedly record whether 'inverted' male or female subjects or 'hermaphrodite' (intersex) subjects have, for example, a 'masculine respiration type' or a 'mixed respiration type, abdominal-thoracic'.[11] Hirschfeld's *Die Homosexualität des Mannes und des Weibes* (1914) cites the earlier work of Karl Ulrichs, which similarly suggests that homosexual subjects breathe in a manner which is atypical of their sexed bodies: 'I have seen homosexual men with heaving bosoms and homosexual women without a trace of costal breathing'.[12] Homosexual subjects, Ulrichs suggests, are also peculiarly capable of 'influencing' the type of respiration that they use (so as to change from stereotypically masculine abdominal breathing to stereotypically feminine costal breathing, or vice versa).[13]

An awareness of the intense focus on the non-normative body in these sexological texts might allow us to read with greater sensitivity the tropes of embodiment prevalent in queer literary texts of the *fin de siècle*. Posture, voice, gait, and breath—the basic gestures and movements of the body all became potentially fraught through their disclosive potential. Another reason to be particularly alert to the relationship between the presentation of breath and embodiment in sexological texts and the tropes of breath in literary works is that *fin-de-siècle* literary authors were themselves closely engaged in the consumption, production, and circulation of such sexological texts. The significance of John Addington Symonds's contribution to English sexology has been well recognized by Sean Brady and others.[14] Housman's attentiveness to contemporary developments in sexology remains less well-known.[15] He followed new work in European sexology closely. His extensive personal collection of sexological and pornographic texts includes a complete set of Hirschfeld's *Jahrbuch* (23 vols, 1899–1923) and Max Marcuse's journal *Sexual-Probleme: Zeitschrift für Sexualwissenschaft* (vols 5–10, 1909–1914), amongst the only extant copies of these texts in the United Kingdom.[16] Reading work by *fin-de-siècle* writers in the light of such an archive alerts us to the manner in which literary texts complicate and subvert such pathologizing discourses. The queer body in texts by Housman, Symonds, and Pater is often inscribed with a Lucretian timidity and fear bound up with the experience of social marginalization—yet it is also the site at which alternative materialisms first draw breath.

A. E. Housman: 'Beneath the Suffocating Night'

Slit throats, suffocation, strangling, hanging—stifled breath in A. E. Housman's Shropshire becomes the mark of a queer martyrdom sadistically enacted by the poet on the bodies of his unfortunate victims. Catching one's breath in Housman's poetry expresses the precarity of queer subjectivity and the impossibility of fulfilled queer desire. The movement of breath becomes emblematic of a pessimistic Lucretian flux, in which individuals must confront their inevitable dissolution into the material 'stuff' of the universe.[17] Such queer materialism of breath comes into sharpest focus in three consecutive texts from Housman's *A Shropshire Lad*: 'Others, I am not the first' (xxx, 31–32), 'On Wenlock Edge' (xxxi, 32–33), and 'From far, from eve and morning' (xxxii, 33–34).[18]

'Others, I am not the first' is a poem in which the immediacy of the demand for breath becomes a way of talking about the intense materiality of embodied queer desire, and how such a materiality refuses the comfort of abstract philosophical consolation:

> Others, I am not the first,
> Have willed more mischief than they durst:
> If in the breathless night I too
> Shiver now, 'tis nothing new.
>
> More than I, if truth were told,
> Have stood and sweated hot and cold,
> And through their reins in ice and fire
> Fear contended with desire.
>
> Agued once like me were they,
> But I like them shall win my way
> Lastly to the bed of mould
> Where there's neither heat nor cold.
>
> But from my grave across my brow
> Plays no wind of healing now,
> And fire and ice within me fight
> Beneath the suffocating night.

In a letter to Grant Richards in December 1920, Housman described the poem as 'contrasting the passions of youth and the unwholesome excitement of adultery with the quiet indifference of death'.[19] For Laurence Housman, this text is the most strikingly autobiographical of all those collected in *A Shropshire Lad*: it gives the 'most direct expression' of his brother's 'personal experience—and suffering'.[20] Such 'passions', 'excitement', and 'suffering' are felt viscerally here at the level of a body that 'shivers' and 'sweat[s] hot and cold'; desire is a 'fight' in the 'reins' between 'ice and fire'. 'Quiet indifference', if it can be achieved, might be reached through a Lucretian philosophy that situates the torturous nature of one's unfulfilled desires in a historical span that marks them as both continuous and inevitable: suffering is merely part of the universal order and is thus simply to be endured. However, as the final stanza makes clear, such consolation is not easily arrived at: death might offer comfort in the future, but it provides none 'now'. Rather, the weight of unacted desire is itself 'suffocating', drawing the speaker back to the intensity of the present moment felt in the body. Characteristically, the tripping lilt of Housman's regular meter—a 'contend[ing]' between trochaic and iambic tetrameter—is itself smothered by the awkward scansion demanded by this 'suffocating'.

Such 'suffocation' is a reiteration, and intensification, of the 'breathless night' through which the speaker 'shiver[s]' in the first stanza. This is another 'now' made intensely present by the shortage of breath. This shivering breathlessness presents 'mischief' left undone in terms that still evoke the doing of it; one might, of course, be pleasurably out of breath as one shivers with the 'excitement of adultery'. Housman's earlier draft, in which the speaker endures a '*smothering* night' (32), makes more immediately present the primary sense of 'breathless' here—being deprived of air, being unable to breathe freely. There are specifically queer reasons here for the manner in which 'fear contend[s] with desire' on the part of Housman's speaker, or for such desire rendering him 'agued'. As Carol Efrati has observed, sexual desire is rarely disentangled, in Housman's works, from *fin-de-siècle* legal and medical discourses that would render queer sexuality a matter of intense shame, criminality, and pathology.[21] The speaker attempts to find distance from this social reality by dissolving or diminishing the self into the historical past ('not the first', 'nothing new', 'more than I'). Yet the poem's metrical scheme works against this forcefully to assert the present moment: the trochees that open lines four ('Shiver') and eight ('Fear') place their stress to foreground the affective immediacy of embodied experience. We might also think back here to Lucretius's shivering, fearful soul, whose character is defined by a dominance of breath. Housman's poem suggests, at the last gasp, that to be intensely, even fearfully, aware of the social marginality of one's queer sexual desire feels something like being unable to breathe.

Being breathless in bed, for Housman, is also to find oneself wishing for the comforting breath of a lover and finding this to be painfully absent. In 'I lay me down and slumber' (*More Poems* xiii, 122–23), the 'night-long breathing / That keeps a man alive' (ll. 3–4) represents the unconscious, automatic respiration of the speaker himself as he sleeps. But it also indulges a wish for the benevolent care of a desired other who sleeps (and breathes) beside him: a 'kind and foolish comrade / That breathes all night for me' (ll. 19–20). To be 'breathe[d] for' by another is to be desired by them, but it is also to be sustained by them—for the body of the other to function as something akin to a phantasmatic iron lung. As Phillip Horky explores elsewhere in this volume, there is a well-established tradition in materialist ancient philosophy (with which Housman was closely familiar) in which shared breath (*conspiration*) functions as a figure for sustenance—most notably, in this context, for the reciprocal bond that sustains domestic and sexual intimacy within marriage.[22] Housman wonders aloud about the possibility of a same-sex equivalent of this breathing-together. The text reflects upon the extent to which feeling desired by another might be connected to feeling sustained; whether, that is, being *breathed* for allows one to feel breathed *for*.

Elsewhere in Housman's works, breath becomes a more immediately material vehicle for the transmission of queer desire—the Lucretian *aura* carried on the wind. In the opening stanzas of 'The winds out of the west land blow' (*ASL*, xxxviii, 38–39), the movement of breath functions to overcome the spatial distance between the speaker and the men he loves, partially restoring the lost intimacy of the speaker's far-off 'friends':

> The winds out of the west land blow,
>   My friends have breathed them there;
> Warm with the blood of lads I know
>   Comes east the sighing air.

It fanned their temples, filled their lungs,
  Scattered their forelocks free;
My friends made words of it with tongues
  That talk no more to me. (ll. 1–8)

The 'breath' of these men 'warm[s]' the air with the 'blood' of their desire, so that this desire might be transmitted by the wind through space to the speaker. The inhalation of breath into the lungs charges the neutral air with the warmth of erotic presence—a presence avowed by the inverted foot that opens line three on 'Warm'. Also made buoyant on this 'sighing air' are the displaced traces of contact with these friends' bodies. Like an attentive lover or nurse, the wind 'fan[s] their temples', and ruffles their 'forelocks free'. The repeated stressed 'f' fricatives (on 'fanned', 'filled', 'forelocks', 'free'), and the inverted foot on 'Scattered', capture something of the materiality of the animated breeze as it brushes the flesh. An earlier draft of the second stanza reinforces this underlying desire for a loving caress—here the wind 'stroke[s] their faces' (39), playing out a tactile fantasy seemingly denied to the speaker himself. There are hints of envy here too. The wind gains easy entry to the private interiors of these men's bodies, absorbed to 'fill their lungs'. The transformation of this air into speech sees a refusal of communication: when this breath becomes 'words' it does nothing to acknowledge or reciprocate the desires of the speaker. The queer desire encoded in the breath evaporates before it finds concrete verbal expression.

The 'sighing air' of 'The winds out of the west land blow' becomes the more characteristically Lucretian 'gale of life' in 'On Wenlock Edge': a blindly destructive dispersal of matter in which all individual identity must ultimately be subsumed within a constant cosmological process of flux. Here, the 'heaving hill' of Housman's Shropshire landscape seems itself to rise and fall like a diaphragm—a spectral displacement, perhaps, of the strained breathing of a long-dead Roman. The 'old wind in the old anger' functions in this poem as another form of breath that might bring queer bodies into contact.

The movement of breath in 'From far, from eve and morning' (xxxii) is similarly co-extensive with the blowing of the wind, which here enacts a blind dissolution of the self:

From far, from eve and morning
  And yon twelve-winded sky,
The stuff of life to knit me

Blew hither: here am I.

Now—for a breath I tarry
  Nor yet disperse apart—
Take my hand quick and tell me,
  What have you in your heart.

Speak now, and I will answer;
  How shall I help you, say;
Ere to the wind's twelve quarters
  I take my endless way.

Housman's text is about the connection between desire and precarity; about the difficulty of sustaining a subject position from which one can express desire. Housman's temporally and spatially displaced queer subject comes into being only long enough to experience a fleeting hope of requited love. As Archie Burnett has noted, the text evokes a specifically Lucretian weaving and un-weaving of the 'stuff of life' here: *pereat dispersa per auras* ('it perishes dispersed abroad through the air').[23] Housman's text draws upon Lucretius's associations of 'aura' with the movement of breath. This material accretion and 'dispers[al]' are notably performed at the level of sound: the fricatives of 'stuff', 'life', 'hither', and 'breath' are softly insubstantial when set in opposition to the more concrete plosives in 'knit', 'apart', 'heart'. The 'breath' for which the speaker 'tarries' functions in two ways. First, as a measurement of time—the transitory moment for which the speaker might be sustained before 'dispers[ing]', breath-like, onto the wind's 'endless way'. A quiet desperation to make the most of this moment is asserted by the inverted feet on 'Now' and 'Take my hand'. Second, as an unspoken indicator of the fleeting possibility of reciprocated desire, akin to the 'quick' touch of a hand. The speaker's repeated requests for the verbal affirmation of such desire ('tell me', 'speak now', 'say') remain conspicuously unanswered. The speaker waits, perhaps in vain, for the desiring breath of the other, just as he awaits their spoken response.

Tropes of breath and breathlessness in Housman's poems function to materialize forms of queer desire that cannot quite find expression. As 'fear contend[s] with desire' in these texts, desire often becomes encoded in breath, rendered material only at the moment of its dispersal. But these texts' engagement with the materialism of Lucretius also becomes a means for articulating a queer impulse towards disappearance and withdrawal. Here breath marks the failure and precarity of certain subjects; the inability of marginalized individuals to inhabit their bodies securely in time or space.

John Addington Symonds: 'I Seem to Feel the *Aura* in Him'

For the Victorian poet, historian and literary critic John Addington Symonds, the materiality of breath is similarly associated in his autobiographical writings, private correspondence, and essays on classical culture with the indirect transmission and fearful negotiation of queer desire. In Symonds's *Memoirs* (1893), the lung disease that came to dominate his life is presented as partially the result of his inability to 'yield to the attraction of the male'.[24] In the unfolding narrative of his 'pulmonary disease' (276), his decision to marry Catherine North—a woman for whom he feels 'the strongest admiration and the firmest friendship' but 'not the right quality of sexual passion'—is followed immediately by his being 'pronounced consumptive' (523). Only years later, at the age of twenty-nine, does he finally 'indulge his inborn homosexual instincts'. He 'rapidly recover[s]' his health, to 'some extent surmount[ing] his consumptive tendencies': 'the phthisis—which had progressed as far as profuse haemorrhage and formation of cavity—was arrested' (533). In striking opposition to the dominant Victorian pathologization of homosexuality, Symonds presents the free expression of homosexual desire by queer subjects as vital to their physical and mental well-being: he is 'quite certain that he suffers or benefits in health of mind and body according as he abstains from or indulges in moderate homosexual pleasure' (533). Symonds presents the breathlessness symptomatic of his lung disease in a manner similar to Housman's 'shiver[ing]' speaker, for whom the inability to indulge his 'mischief' feels like a form of suffocation.

After first meeting Symonds in October 1877, Henry James wrote to his brother William of his encounter with 'a mild, cultured man, with the Oxford perfume'.[25] James refers here, with characteristic restraint, to the potent mix of Hellenism, aestheticism, and homoeroticism then prevalent in certain sub-communities at the University of Oxford.[26] The turn of phrase also alerts us to the possibility of queerness as something which lingers as a material trace in the air, as something inhaled unawares—but also as something that might be caught only by those with a sharpened and subtle sensitivity. A similarly breath-like metaphor for the material encoding, transmission, and recognition of queer possibility occurs in Symonds's own writings. In his letters to close friends who share his interest in love between men, Symonds uses the word 'aura' to refer to a material presence that he senses in certain individuals or works of art. This 'aura' acts covertly to communicate and disclose the hidden homosexual subtext of an artwork or the homosexual identity of an individual. To Edmund Gosse in April 1890, Symonds recounts reading works by Christopher Marlowe, Pierre Loti, and Charles Warren Stoddard and notes that '[t]he *aura*, when I feel it, seems to me very distinct'. Openly acknowledging this 'diagnosis', he warns Gosse, remains 'difficult & dangerous'.[27] The 'aura' of an individual can also be sensed indirectly through the material referent of an artwork produced by him. In a later letter to Gosse in February 1892 he reports receiving a photograph of John Donoghue's sculpture of 'The Young Sophokles Leading the Chorus of Victory after the Battle of Salamis' and comments: 'I never saw the man. But I seem to feel the *aura* in him'.[28] In similar terms, he writes to Horatio Forbes Brown in March 1891 of a painting by Henry Strachey of boys playing football in a Somerset meadow as being 'full of the *aura*'.[29] In both cases, this 'aura' is presented in terms that draw attention to its materiality as breath; it is a quality that can be felt *in* an individual or that can *fill up* an object.

The queer 'aura' that Symonds senses is striking in two ways. First, it is a quality exuded by certain objects or persons unintentionally. Second, it can be detected only by those with a particular sensibility to 'discern' its presence. There are notable parallels here with Symonds's characterization of Lucretius's conception of sexual desire, in Book 4 of *De Rerum Natura*, as a 'a fierce and overmastering force [...] which men share in common *with the world of things*'.[30] The physical materiality of this 'force' which flows between human and non-human objects should also be understood as informing Symonds's response to contemporaneous work being undertaken by his close friends in the Society for Psychical Research (SPR), such as Frederic Myers and Henry Sidgwick.[31] In its fascination with telepathy, hypnotism, and other 'supernormal communication', the SPR was pre-occupied with the detection of 'effluences, as yet unknown to science, but perceptible by sensitive persons, [which] radiate from living human organisms'.[32] Myers's *Human Personality and its Survival of Bodily Death* (1903) offers a useful definition of an 'aura' as an 'influence environing each human being, whose limits it is not easy to define'.[33] Significantly, this means that Symonds does not posit the detection of queer 'aura' merely as the product of an inquisitive interpretative reading strategy (something akin to an early 'queer reading'). Rather, detection depends on a receptiveness to the transmission of queerness, here understood in material terms as a physical 'influence' or 'effluence' given off by bodies and things. As in Housman's poetry, Symonds's *aura* affords to breath a sensuous tactility through which queer bodies might come into closer contact.

Underlying this connection between the transmission of a breath-like *aura* and the recognition of desire between men is Symonds's interest in the Spartan pederastic tradition of the εἰσπνήλας (*eispnelas*)—the elder lover who 'breathes in' to the receptive younger 'listener'.[34] Writers of the Victorian period would have been familiar with this idea from Karl Otfried Müller's influential study of Spartan civilization, *Die Dorier* (1824), which was translated into English by Henry Tufnel and George Cornewall Lewis in 1830.[35] Symonds draws attention to this model in *Studies in the Greek Poets* (1873) in order to praise the dignity and purity of Platonic same-sex 'comradeship': 'a kind of chivalry[…], which, like the modern chivalry of love and arms, as long as it remained within due limits, gave birth to nothing but honourable deeds and noble friendships'.[36] Such 'inspiration' marks another means, for Symonds, through which the material dispersal of breath allows for a tentative affirmation of desire between men.

Walter Pater: 'Fade Out of the World like a Breath'

Queer forms of materialism can also be traced in the movement of breath in the works of Walter Pater, the most important writer in English Aestheticism. As recent work by Benjamin Morgan has suggested, Pater's status as a self-proclaimed Epicurean makes him a significant figure in the nineteenth-century re-evaluation of materialism.[37] Pater's remarks in 'The Genius of Plato' might be understood as a summary of his own literary project in materialist terms, when he speaks of 'the redemption of matter' and 'the vindication of the dignity of the body'.[38] At the same time, Pater's texts stage what Heather Love has called a 'shrinking politics' of the ephemeral 'diaphanous type'—something close, perhaps, to those fearful breath-like souls identified by Lucretius.[39] Metaphors of breath in Pater's works often function as tools for negotiating this tension between marginalized subjectivity and its precarious grounding in the experience of the material world. This is at its most striking in Pater's series of short fictional prose works known as 'Imaginary Portraits', and particularly in his treatment of consumption in 'Sebastian van Storck' (1886).[40]

Set in the Dutch Republic of the mid-seventeenth century, 'Sebastian van Storck' focusses on the fate of the only child of a wealthy merchant family. The text's queer materialism arises through its central character's 'passion for *Schwindsucht*' (153)—his 'continual effort at self-effacement' (164). Sebastian aspires towards the condition of evaporation: he is drawn to 'fade out of the world like a breath' (158). This breath-like dispersal is closely aligned in the text with both the gradually wasting symptoms of consumption and the queer refusal of a normative sexual materiality bound up with what Lee Edelman calls 'reproductive futurity'. Pater's use of the term '*Schwindsucht*' gestures in two directions. First, towards a strict medical sense of the term—it is the consumption (or *phthisis*) with which Sebastian will subsequently become infected. Second, towards Hegel's use of the phrase in *Vorlesungen über Naturrecht und Staatswissenschaft* (1818) to characterize those 'beautiful souls' that 'remain enclosed within themselves', fearing every concrete contact with the material external world lest their 'infinite self-consciousness' be forced 'to enter the sphere of limitedness'. Hegel associates this mode of subjectivity with Spinoza, whom Pater in turn uses as the explicit model for Sebastian's philosophy. Spinoza, Hegel suggests, 'died of consumption because [he] regarded pure objectivity only as something vanishing or consuming away [...] not as something actual'.[41] What is most significant here is the connection between consumption and the refusal of materiality. In one respect Pater's text invokes the common nineteenth-century trope of consumption as *the* characteristic disease of modernity; Sebastian suffers from a condition that blights 'people grown somewhat over-delicate in their nature by the effects of modern luxury' (166). But Pater inverts the other prevalent cultural association—one much in evidence in other 'Imaginary Portraits' such as 'An English Poet' (1878) and 'Duke Carl of Rosenmold' (1887)—between consumption and aesthetic creativity.[42] In Pater's densely material depiction of the Dutch Republic—full of painters, paintings, furniture, fabrics—Sebastian's consumption is the physical manifestation of his queer withdrawal from the sensuous pleasure of material objects. Instead, he holds himself apart in a realm of cold, pure, disembodied thought.

'Sebastian van Storck' has typically been read as a salutary warning of what might go wrong if idealist philosophy is not sufficiently grounded in the ethical commitments that arise from an acceptance of the embodied, material world that surrounds us.[43] In this view, Sebastian's solipsistic withdrawal into 'pure intellectual abstraction' (160)—a hyper-rational world of geometrical precision—represents a refusal of the material world per se. A different reading emerges, though, with closer attention to the text's presentation of the underlying motivations for Sebastian's refusals and rejections. The materialism of the Dutch Republic that Sebastian rejects is inextricably entangled with heteronormative values of reproduction. The text presents a society in which the dominant modes of material plenitude exclude certain subjects—an exclusion enacted through tropes of breathlessness. To Sebastian, the 'crowded and competing world of life [...] seem[s] wellnigh to suffocate' him (143). Pater thus stages a distinctively queer 'redemption of matter'—materialism redeemed from its entanglement with heteronormativity. Dominant materialisms, the text suggests, will always leave someone out.

The 'vanishing or consuming away' of the Spinozan Sebastian acts to resist the normative reproductive expectations that attach to Pater's protagonist as the only son of a rich mercantile family. Such expectations arise both from his specific family dynamics and from the imperatives of wider Dutch society. As a 'rich and distinguished youth', he is continually sized up by 'mothers of marriageable daughters' (145), who wish to make a prudent and profitable match. His 'calm, intellectual indifference' (159), however, will ultimately see him spurn the romantic attention of the hapless Mademoiselle van Westrheene. He is subject to pressure from his father to 'play his part' in an 'age [...] still fitted to evoke a generous ambition', one in which 'there was a tradition to be maintained' (146). His mother is keenly aware of the 'measure of cold in things for a woman of her age' (158)—that is, the inevitable dwindling of her family tree where she herself is no longer able to bear children, and her only son refuses to grant her grandchildren. Pater's Dutch Republic is similarly sustained by an aggressive mercantile ethos of productivity and accumulation: Sebastian's story unfolds against the backdrop of 'the nation's hard-won prosperity' (145), by 'heroic industry that had triumphed over nature' (154). The only form of 'practical career' (155) that Sebastian can countenance in this mercantile society represents a striking inversion of its capitalist values. He is drawn towards something in the spirit of exploration that leads his compatriots to the Arctic in search of a North-West passage. However, his underlying motivation is not the discovery, for instance, of shorter and swifter trading routes to Asia but rather the 'charm in the thought of that still, drowsy, spellbound world of perpetual ice' (156). The economic imperatives of activity, growth, accumulation, and expansion are here replaced by a desire for frozen stasis—a queer rejection of the unfolding of the future.

As Sara Ahmed's work on 'queer phenomenology' has described, for those who cannot fit snugly into 'straight spaces'—such as the normative family home—the weighty material presence of the stuff of bourgeois domesticity might be experienced as leaving no space to breathe.[44] Sebastian's desire to escape to 'a world of perpetual ice' forms part of the text's broader interest in scrutinizing how heteronormative values of reproduction, replication, and accumulation are entangled with material objects and environments. The text contrasts Sebastian's obsession with the 'abstract or cold in art' with the 'busy well-being', 'thriving genius', and 'delicate homeliness' of bourgeois domestic space (147). Sebastian purposefully removes himself from a house 'crowded with the furniture and the pretty little toys of many generations' (150). He sets himself in opposition to those Dutch realist painters whose detailed depictions of domestic interiors represent 'the ideal [...] of the good-fellowship of family life'; 'the ideal of home', preferring instead paintings of wide-open skies that present 'things seen from a distance' (148–49). The material world of Dutch realist domesticity is imprinted with expectations of heteronormativity: patrilinear inheritance sustained through the passing on of furniture, traditional bourgeois family structures marked through the accumulation of domestic objects. The pressures of normativity are experienced by Sebastian through the crowded profusion of these objects, leaving no space for those who exist outside of what Judith Butler calls the 'heterosexual matrix'.[45] We might recall here the passage quoted above, in which the word 'crowded' evokes Sebastian's feeling of being 'suffocated' by the 'world of life'. It is no coincidence, then, that the sympathy of this young Storck is drawn only to 'the creatures of the air' (150): 'the caged bird on the wing at last' (149) is afforded a freedom to dissolve into the 'clear breadth of atmosphere' (148) that Sebastian desires for himself.[46]

Indeed, Sebastian's 'preference in the matter of art for [...] prospects *à vol d'oiseau*' (149) is another extension of his queer rejection of the dominant materialism of the Dutch Republic: he refuses to be visually represented in the period's distinctive art of portraiture. 'All of his singularities appeared to be summed up', Pater's narrator notes, 'in his refusal to take his place in the life-sized family group [...]—painted about this time' (158). Sebastian figuratively terminates the van Storck family tree. The artists of the Dutch Republic, against whom Sebastian sets himself, become emblematic in the text of normative patterns of replication and reproduction. That this artistic community is 'exemplary in matters of domestic relationship' manifests itself most clearly in Pater's family tableau of the Hondecosters: 'three of them together, son, grandson, and grandfather [...]—Giles, Gybrecht, and Melchior' (152). Sebastian, for whom there will seemingly be no such line of descent, places himself outside this heteronormative sequence of reproduction. In absenting himself from the 'life-sized family group', he not only signals his queer opposition to the pull of 'reproductive futurity', but also acts to forestall the circulation of material objects that reinforce heteronormative discourses of family and domesticity. This queer act of disappearance represents an external manifestation of Sebastian's breath-related illness, figuratively enacting his desire to dissolve into thin air.

It is fitting, then, that Sebastian's death at the conclusion of the text—seemingly from consumption—is also a decidedly queer moment that gestures towards non-normative reproduction: 'when the body of Sebastian was found, apparently not long after death, a child lay asleep, swaddled warmly in his heavy furs' (166). In this sudden and unexpected appearance of an infant, the narrative enacts a fantasy of masculine parthenogenesis—a childbirth from within Sebastian's oddly mammalian 'heavy furs'. As Sebastian breathes his last, the text cautiously allows for a queer reformulation of materialism that might accommodate those who feel otherwise 'suffocate[d]' by the 'crowded and competing world of life' (143).

Queer writers at the *fin de siècle*, then, use images of breath as a central part of their interest in materialism, from Classical sources to contemporary science. The embodied but insubstantial quality of breath enables a language of materialism that explores and negotiates experiences of social marginalization. Examining the significance of modes of respiration in sexological texts alerts us to the centrality of the material body to emergent conceptions of the homosexual subject at the *fin de siècle*. In Housman's poetry, tropes of breathlessness and suffocation articulate negative affects that attach to queer desire, such as fear and shame. For

Symonds, the connection between breathlessness and repressed homosexual desire is marked more directly in his personal experience of lung disease. In works by Pater, the mobility of breath expresses a wish for disappearance, withdrawal, and disembodiment. An inability to breathe freely in these texts signals a precarious feeling of not occupying space in a way that aligns with heteronormative imperatives. Yet for each of the writers, the movement of breath also becomes a tool for some form of affirmation of queer desire—either as a mode of material transmission of desire between men, or in its resistance to normative materialisms. The interest of these writers in the experience of the material body represents a shared preoccupation with envisioning what Pater calls 'a really social air to breathe in'—even if their individual hopes for discovering such an atmosphere range from the avowedly pessimistic to the tentatively utopian.[47]

Notes

1. A. E. Housman's copy of *T. Lucreti Cari De Rerum Natura Libri Sex*, ed. by H. A. J. Munro (Cambridge: Deighton Bell, 1873) is now in the library of St. John's College, Oxford. See also *On the Nature of Things, Book III*, ed. by E. J. Kenney (Cambridge: Cambridge University Press, 2014), 116. Hereafter *ONT III*.
2. *ONT III*, 45–46, ll. 231–57.
3. Tyndall was President of the British Association for the Advancement of Science. His influential 'Belfast Address' of 1874 (the presidential address to the Association for that year) offered a teleological intellectual history of materialism, ranging from Democritus to Darwin and culminating in an account of the emergence of human consciousness from matter. For discussion of the address see Benjamin Morgan, *The Outward Mind: Materialist Aesthetics in Victorian Science and Literature* (Chicago: University of Chicago Press, 2017), 143–46.
4. See Martin Priestman, 'Lucretius in Romantic and Victorian Britain', in *The Cambridge Companion to Lucretius*, ed. by Stuart Gillespie (Cambridge: Cambridge University Press), 289–305.
5. Judith Halberstam, *The Queer Art of Failure* (Durham, NC: Duke University Press, 2011); Lee Edelman, *No Future: Queer Theory and the Death Drive* (Durham, NC: Duke University Press, 2004). My interest in the productive place of negativity in queer subject formation at the *fin de siècle* takes an alternative approach to recent work by Dustin Friedman, who situates this process in the intellectual lineage of Hegel's idealist aesthetics;

see *Before Queer Theory: Victorian Aestheticism and the Self* (Baltimore: Johns Hopkins University Press, 2019).

6. Heather Love, *Feeling Backward: Loss and the Politics of Queer History* (Cambridge, MA: Harvard University Press, 2009).
7. Sara Ahmed, *Queer Phenomenology: Orientations, Objects, Others* (Durham, NC: Duke University Press, 2006).
8. Michel Foucault, *The History of Sexuality, Vol. 1: The Will to Knowledge*, trans. by Robert Hurley (London: Penguin, 1978).
9. See Lucy Bland and Laura Doan, eds, *Sexology in Culture: Labelling Bodies and Desires* (Cambridge: Polity, 1998).
10. Magnus Hirschfeld, *Jahrbuch für sexuelle Zwischenstufen unter besonderer Berücksichtigung der Homosexualität*, vol. 1 (Leipzig: Spohr, 1899), 7. Translations from German by Tom Smith.
11. Hirschfeld, *Jahrbuch*, vol. 6 (1904), 260; *Jahrbuch*, vol. 10 (1908), 73.
12. Karl Ulrichs, *Memnon: Die Geschlechtsnatur des mannliebenden Urnings* (Schleiz: Hübscher, 1868), 130, cited in Magnus Hirschfeld, *Die Homosexualität des Mannes und des Weibes* (Berlin: Marcus, 1914), 140.
13. Ulrichs, *Memnon*, 130, cited in Hirschfeld, *Die Homosexualität*, 140.
14. Sean Brady, 'Introduction', in *John Addington Symonds (1840–1893) and Homosexuality: A Critical Edition of Sources*, ed. by Sean Brady (Basingstoke: Palgrave Macmillan, 2012), 1–38.
15. A notable exception is Peter Howarth, 'Housman's Dirty Postcards: Poetry, Modernism, and Masochism', *PMLA* 124/3 (2009), 764–81.
16. For further details of Housman's personal book collection, see P. N. Naiditch, 'The Extant Portion of the Library of A. E. Housman: Part IV. Non-Classical Materials', *Housman Society Journal* 31 (2005), 154–80.
17. On Housman and Lucretius, see Donald Mackenzie, 'Two Versions of Lucretius: Arnold and Housman', *Translation and Literature* 16/2 (2007), 160–77.
18. All references to Housman's poetry are to *The Poems of A. E. Housman*, ed. by Archie Burnett (Oxford: Clarendon, 1997).
19. A. E. Housman to Grant Richards, December 20, 1920, in *The Letters of A. E. Housman*, vol. 1, ed. by Archie Burnett (Oxford: Oxford University Press, 2007), 460.
20. Laurence Housman, 'A. E. Housman's "De Amicitia"', *Encounter* 29 (1967), 34.
21. Carol Efrati, *The Road of Danger, Guilt, and Shame: The Lonely Way of A. E. Housman* (Madison, NJ: Fairleigh Dickinson University Press, 2002).
22. See Phillip Horky (above, §6) on *conspiratio* or συμπνοία as 'concord' such as might obtain between a husband and wife, which (as he explains) may be given a specifically sexual colouration, as in Lucretius, *De rerum natura*, 4.1215–16: 'semina cum Veneris stimulis excita per artus / obvia conflixit conspirans mutuus ardor' (when the seeds stirred up through the

frame by the goads of Venus have been thrust together by the passion of two breathing as one [Loeb translation]).

23. See Burnett, ed., 343; *ONT III*, 55 (l. 544).
24. Symonds, *The Memoirs of John Addington Symonds: A Critical Edition*, ed. by Amber K. Regis (London: Palgrave Macmillan, 2016), 532. Hereafter referenced in the text.
25. Henry James to William James, February 28, 1877, in *The Correspondence of William James, Vol. 1: William and Henry, 1861–1884*, ed. by Ignas K. Skrupskelis and Elizabeth M. Berkeley (Charlottesville: University of Virginia Press, 1992), 280.
26. See Linda Dowling, *Hellenism and Homosexuality in Victorian Oxford* (Ithaca, NY: Cornell University Press, 1994).
27. Symonds to Gosse, April 4, 1890, in John G. Younger, 'Ten Unpublished Letters by John Addington Symonds at Duke University', *Victorian Newsletter* 95 (1999), 6.
28. Symonds to Gosse, February 23, 1892, in *The Letters of John Addington Symonds*, vol. 3, ed. by Herbert M. Schueller and Robert L. Peters (Detroit: Wayne State University Press, 1967–69), 553–56. For Donoghue's bronze, 'The Young Sophocles' (c. 1890), see a version in the Isabella Stewart Gardner Museum, Boston: https://www.gardnermuseum.org/experience/collection/13481.
29. Symonds to Forbes Brown, March 6, 1890, in *Letters*, vol. 3, 559–60.
30. Symonds, 'Lucretius,' in *Sketches and Studies in Italy and Greece* (London: Smith, Elder, 1879), 96. My italics.
31. See H. G. Cocks, 'Religion and Spirituality', in *Palgrave Advances in the Modern History of Sexuality*, ed. by H. G. Cocks and Matt Houlbrook (Basingstoke: Palgrave Macmillan, 2006), 173–74.
32. Frederic W. H. Myers, *Human Personality and its Survival of Bodily Death*, vol. 1 (London: Longmans, Green, 1903), 208.
33. Myers, *Human Personality*, vol. 1, 485.
34. See William A. Percy, *Pederasty and Pedagogy in Archaic Greece* (Chicago: University of Illinois Press, 1996), 73–89.
35. K. O. Müller, *The History and Antiquities of the Doric Race*, vol. 2, trans. by Henry Tufnell and George Cornewall Lewis (London: Murray, 1830), 306–07.
36. Symonds, *Studies in the Greek Poets* (London: Smith, Elder, 1873), 84. Walter Pater defends this tradition in similar terms in 'Lacedæmon', in *Plato and Platonism* (London: Macmillan, 1893), 233.
37. Morgan, *Outward Mind*, 133–73.
38. *Plato and Platonism*, 124–49 (at 146).
39. Love, *Feeling Backwards*, 58.
40. Pater, 'Sebastian van Storck', in *Imaginary Portraits*, ed. by Lene Østermark-Johansen (London: MHRA, 2014), 143–66. Hereafter referenced in the text.

41. G. W. F. Hegel, *Lectures on Natural Right and Political Science: The First Philosophy of Right*, ed. and trans. by J. Michael Stewart and Peter C. Hodgson (Oxford: Oxford University Press, 2012), 127–28.
42. For an overview, see Clark Lawlor and Akihito Suzuki, 'The Disease of the Self: Representing Consumption, 1700–1830', *Bulletin of the History of Medicine* 74/3 (2000), 458–94.
43. For example, Billie Andrew Inman, '"Sebastian van Storck": Pater's Exploration into Nihilism', *Nineteenth-Century Fiction* 30/4 (1976), 457–76.
44. Ahmed, *Queer Phenomenology*, 65.
45. Judith Butler, *Gender Trouble* (London: Routledge, 1999), 45–48.
46. Pater's choice of name also aligns his protagonist with an apposite tradition of St Sebastian as a figure of queer sexual martyrdom, for which see Richard A. Kaye, '"Determined Raptures": St. Sebastian and the Victorian Discourse of Decadence', *Victorian Literature and Culture* 27/1 (1999), 269–303.
47. Walter Pater, 'The Age of Athletic Prizemen: A Chapter in Greek Art', in *Greek Studies: A Series of Essays* (London: Macmillan, 1910), 276.

Selected Bibliography

Ahmed, Sara. 2006. *Queer Phenomenology: Orientations, Objects, Others*. Durham, NC: Duke University Press.

Bauer, Heike. 2009. *English Literary Sexology, Translations of Inversion, 1860–1930*. Basingstoke: Palgrave Macmillan.

Efrati, Carol. 2002. *The Road of Danger, Guilt, and Shame: The Lonely Way of A. E. Housman*. Madison, NJ: Fairleigh Dickinson University Press.

Halberstam, Judith. 2011. *The Queer Art of Failure*. Durham, NC: Duke University Press.

Howarth, Peter. 2009. Housman's Dirty Postcards: Poetry, Modernism, and Masochism. *PMLA* 124/33: 764–81.

Lawlor, Clark and Akihito Suzuki. 2000. The Disease of the Self: Representing Consumption, 1700–1830. *Bulletin of the History of Medicine* 74/3: 458–94.

Love, Heather. 2009. *Feeling Backward: Loss and the Politics of Queer History*. Cambridge, MA: Harvard University Press.

Morgan, Benjamin. 2017. *The Outward Mind: Materialist Aesthetics in Victorian Science and Literature*. Chicago: University of Chicago Press.

PART VI

The Twentieth Century

CHAPTER 18

The Forgotten Obvious: Breathing in Psychoanalysis

Oriana Walker and Arthur Rose

Vagitus-Rattles

When Samuel Beckett described his 35-second play *Breath*, he made of it a straightforward sound and light show: 'It is simply light coming up and going down on a stage littered with miscellaneous unidentifiable muck, synchronized with the sound of breath, once in and out, the whole (ha!) begun and ended by the same tiny vagitus-rattle'.[1] In its bare minimalism, *Breath* appears to reduce existence to an ontic essentialism: the passing of a single breath. More historically situated interpretations link *Breath* to transcriptions and notes Beckett made from his reading of psychology, particularly psychoanalysis, from spring 1934 into 1935 during his analysis by Wilfred Bion.[2] Read as an intertext, the 'vagitus-rattles' that frame *Breath* echo Otto Rank's controversial *The Trauma*

O. Walker
Institut Für Geschichtswissenschaften, Humboldt-Universität zu Berlin, Berlin, Germany

A. Rose (✉)
Department of English, University of Exeter, Exeter, UK

D. Fuller et al. (eds.), *The Life of Breath in Literature, Culture and Medicine*, Palgrave Studies in Literature, Science and Medicine,
https://doi.org/10.1007/978-3-030-74443-4_18

of Birth (1925), in which Rank traces the origin of anxiety to the first psycho-physical trauma: birth. Certainly Beckett's work develops a refrain that takes birth to be a foundational trauma: 'birth was the death of him', for instance, or the Beckettian neologism 'wombtomb'. More relevant to our purpose, however, are intersections between this original trauma and breath, coincidences in Rank that are granted further significance by Beckett's sporadic note-taking. Rank, according to Beckett's transcription, considers that 'all neurotic disturbances in breathing (e.g., asthma), the repeating feeling of suffocation, refer directly to physical reproductions of the birth trauma' and 'esse [to be] means to breathe'.[3] Beckett's notes return neurotic disturbances and the metaphysical problems of being to a *physical* point of intersection: the breath.

In this essay we are interested in the changing status of the breath as an object in psychoanalysis. The story of Beckett's *Breath* offers a pendant to our argument: underneath the apparently ahistorical, essential considerations of breath as a feature of hysteria, anxiety, or eroticism in psychoanalysis, there lurk histories of rivalries, misreadings, and terminal breaks. *Breath* renders breath 'uncanny' by focussing on an autonomic function that generally escapes our attention. Freud describes the uncanny as something that 'is in reality nothing new or foreign, but something familiar and old-established in the mind that has been estranged only by the process of repression'.[4] Psychoanalysis, as we will see, also turns breath into an uncanny object, precisely by estranging it. From the late nineteenth century to the 1930s, successive theoretical formulations of breath in psychoanalysis had as much to do with the personal relationships between the theorists as with the function of the breath itself: from Josef Breuer and Freud, to Freud and Wilhelm Fliess, to Jung and Rank, and finally to Otto Fenichel. Throughout these crucial decades, breath proved a surprisingly divisive topic, inciting arguments about vitalism, the status of matter and spirit, the meaning of symptoms, the hierarchies of the senses, the structures of mind, and the openings of access through the body.

On the Self-Steering Respiration

> Before I proceed ... I must apologize for revisiting the fundamental problems of the nervous system. There is always something oppressive about this kind of 'descent to the mothers', but the attempt to dig up the roots

> of a phenomenon is always bound to lead irresistibly to fundamental problems that cannot be avoided. With this in mind, I hope that the abstruse nature of the following observations will not be judged too harshly.[5]

Energy and its transformations were of fundamental interest to scientists of the nineteenth century, including the question of what drove life itself: was it mechanism alone or a 'vital force'? This was the context of the first appearance of breathing in what would become psychoanalysis.[6] Freud's mentor and collaborator, Josef Breuer, was originally called to the bedside of Anna O., the first patient of psychoanalysis, to tend her 'extremely acute' cough, which he quickly identified as *tussis nervosa*.[7] This respiratory disturbance, an involuntary contraction of the glottis, had started during a long night nursing her ailing father, 'whom she adored'. Through the window she heard the sounds of a party: dancing, laughing, and rousing music. She wanted to go and dance but could not leave her station.[8] The motor impulses that compelled her to joyful movement were transformed into pathological somatic symptoms. Later, hearing such music would trigger her neurotic respiration. This was an example of what Breuer called 'reflexes of the affect', a reflex association in the nervous system between hearing music with a beat, the stifling of a desired expression, and the closure of the glottis. The nervous system of a susceptible person was unable to discharge extreme 'intracerebral excitation' through appropriate pathways. Energy, in a scientific environment dominated by Hermann von Helmholtz's theory of its conservation, could neither be created nor destroyed; there was nowhere for excess quanta of excitation to go.[9]

The solution was conversion; extra excitation was usually transformed into a somatic symptom that bore little, if any, resemblance to whatever had originally instigated it. Through the process of conversion the original ideas or experiences could become entirely hidden, even to 'intelligent and observant patients'.[10] Though a cough was not the strangest or most troubling somatic conversion symptom displayed by his patient, Breuer paid close attention to it. Each and every somatic symptom was a 'shadow cast upon the wall by objects that are real but unknown'.[11] Symptoms as ordinary as shortness of breath, given close attention, pointed to underlying maps of association that eventually revealed their sources in complex, dynamic, and normally hidden processes.

But Breuer was attentive to his patient's respiratory complaints for another reason: as a physiologist, he was interested in the mechanisms that drove the life of the body through its breathing. Roughly a decade before he saw Anna O. in private practice, Breuer had 'dug up the roots' of respiration itself, making a major contribution (with his mentor Ewald Hering) to the physiology of respiratory control. And it was precisely involuntary closure of the glottis that he used to motivate the main questions of the 1868 'Ueber die Selbststeuerung der Atmung' (On the Self-steering of the Respiration), co-authored with Hering.[12] Their experiment was based on the clinical observation that different *mechanics* causing shortness of breath led to different *kinds* of shortness of breath. If respiration was controlled—somehow—by the absolute levels of oxygen in the blood, then *how* that oxygen arrived there, as long as it did indeed arrive, should not matter.[13] But clinical evidence made it clear that it did in fact matter. The *movements* of the body breathing, distinct from the chemistry underpinning breathing, appeared to have something to do with the mechanism governing its control.

The neologism 'self-steering' was first used in a biological context by the physician and physiologist Ernst Brücke, in whose laboratory Freud would meet Breuer.[14] Before Brücke used the term to refer to the mechanism of the beating of the heart, *selbststeuerung* was often used in mechanics to describe the design of steam engines (*dampfmachine)* and air pumps.[15] As Anson Rabinbach and others have shown, this imagining of the heart as a kind of self-regulating steam engine was part of the larger project of translating the body into an organic Helmholtzean motor. The steam of this biological motor was, in some formulations, like Wilhelm Wundt's, nothing less than the *mind* itself. Brücke had transformed Aristotle's *primum movens,* the heart, into a thing that moved itself. Breuer's use of *Selbststeuerung* clearly refers to Brücke's project and extends his teacher's idea of 'self-steering' to another bodily function closely associated with life itself: respiration. All this raised a question that suggests, in retrospect, unseen and unconscious forces at work in the body and mind: which part of the self was the breather?

In two communications of their experiments, Hering and Breuer described what we might today call a simple feedback loop. A circuit connected the vagus nerve, the surfaces of the lungs where the nerve attached, the central nervous system, and what was then a relatively unknown black box called the 'central nervous organ for breathing movements'.[16] It was the expansion and contraction of the lungs themselves that stimulated the vagus nerve, inhibiting further lung inflation, and triggering exhalation. Respiratory movements were thus produced by a self-governing automatic control system. Self-steering, a mechanism that controlled respiration before the breather was aware of it, was not a top-down or precisely localized process; rather, it was a dynamic one that occurred across various levels of the organism, allowing it to respond to and attenuate the atmospheric pressure under which it lived. Breathing worked automatically, driven by precise equilibration of the energy economy of respiration. Unlike others studying the vagus nerve and respiration, Breuer did not use electrical current to artificially stimulate a cut nerve ending, but rather studied the 'natural' source of vagal stimulus in breathing by observing the moving lungs of cats, dogs, and rabbits. Breuer minimized interference with his experimental animals, discovering the control patterns and mechanisms hidden beneath a manifest respiration pattern by allowing the body to move without inhibition, to 'talk', much as in his later Cathartic Method he would allow Anna O. simply to talk.

Passive observation did not preclude acting on the system. A feature of the Hering-Breuer reflex that may not be immediately obvious—but which was the definitive demonstration that control was centred in the reflex—was that if the breathing rhythm was indeed self-controlled (by the movement of the lungs in an anaesthetized animal), the rhythm could be driven mechanically from the 'outside' by entering the 'control circuit'. A rhythm given, say, by the bellows of an experimenter would replace the spontaneous breathing rhythm of the animal. In Breuer's experiments, a kind of mechanical 'suggestion' drove the respiratory pattern of an anaesthetized rabbit: 'During artificial ventilation of the lungs with the aid of a tracheal cannula and bellows one can therefore render the rhythm of an animal's active respiratory movements *wholly dependent* on the rhythm of inflation'[17] (italics added). The respiratory control reflex was in fact extremely sensitive and could be easily driven artificially by stimulating the vagus nerve (though in situ, Breuer triggered it via lung inflation).

Breuer's work on respiration took place before Charcot's popularity peaked with his famous public demonstrations of control over the appearance and disappearance of a patient's hysterical symptoms. Breuer, moreover, did not use hypnosis in his practice until around the time he saw Anna O. as a patient. But throughout the nineteenth century, despite its variable reputation, people were aware of the possibility of changing physical symptoms using hypnosis.[18] The image of an animal on the experimental table incorporating the breathing pattern that Breuer introduces is reminiscent of Charcot's use of hypnosis to 'look under the hood' of a patient's nervous system, gaining direct access to it, thereby producing somatic symptoms on demand (or at least so he claimed). Breuer's later account of the genesis of hysteria, rooted in the relationship between breathing, hypnosis, and the nervous system, would echo the Hering-Breuer model of the control of respiration.

Breuer believed that hysteria had 'two great pathogenic factors': being in love, and attending closely to the breathing of an infant while nursing. These were situations that might cause a susceptible person to slip into a hypnoid state, and ultimately, if they occurred with sufficient frequency or if a trauma took place during such hypnosis, cause hysteria. A mother, in a calm and perhaps dark environment, would train all her attention on her child, in particular listening for and focusing on the rhythm of her baby's breathing. As she nursed, her breathing rhythm would come to match that of her infant. Much like the entrained breathing of Breuer's experiment, this breathing could easily cause a state of auto-hypnosis in the mother.[19] Not unlike the rabbit on Breuer's experimental table, or Anna O. with her repetitive cough, a mother might find that a 'symptom'—her breathing—was driven by something 'external' that was of the right sort to enter the circuits of her nervous system. There was nothing really mysterious about this; it was simply how reflexes, psychic energy, and the very processes of life governed themselves.

Jung's Pneumograph

Given Breuer's interest in respiration, and his view of it as fundamentally 'unconscious' and automatic, it is curious that the young Carl Jung, having studied Freud's dream book and in his experimental phase at the Burghölzi Clinic following the standard psychophysics of the day, tried in 1905 to follow respiration as one of the somatic roads to the unconscious. He determined, however, that unlike the pulse or sweating, respiration was far too much under *conscious* control to be of any use, a dismissive response that paralleled Freud's. The research programme of his laboratory of 'experimental psychopathology' relied heavily on the use of word association tests.[20] Jung was interested not in the conscious but in the *unconscious* associations made by patients; the hesitation, delay, or physical reactions to a particular word were not noticed by the test subjects themselves, but were clearly visible to observers.[21] Then conventional experimental techniques (association, the pneumograph, the galvanometer, and kymograph recordings) were thus used in the service of a novel theory of mind. Out of this laboratory work came the two co-authored papers that used the galvanometer (which detected electric current) and the pneumograph (which transduced the thoracic breathing movements into wave-like tracings on a rotating cylinder) as 'indicator[s] of psychic processes'.[22] Unlike the practitioners of Wundtian psychophysics who had originated the use of such instruments, Jung was interested in revealing associations that were produced by but were not available to the study subject (patient). That is to say, he was interested in revealing an unconscious.

Writing in 1907 and entirely apprised of the not insubstantial amount of work on the subject in the German literature, Jung claimed that 'the relation of the respiratory innervation curve to psychic processes in both normal and pathological conditions has not yet been thoroughly investigated'. Nonetheless, he felt that the pneumograph was far less revealing than the galvanometer, since, as he put it starkly: 'respiration is an instrument of consciousness'. This was not merely a matter of the need to accumulate more studies, however. It was, instead, a limitation inherent in respiration itself:

> It is a matter of everyday experience that the respiration is influenced by our conscious emotions, especially when they are strong, as instanced in such expressions as 'bated breath', 'breathless astonishment', etc. Such inhibitions of breathing are noticeable in many pneumographic curves, particularly in association with expectation and tension. *But perhaps the emotions of the unconscious, roused up by questions or words that strike into the buried complexes of the soul, reveal themselves in the galvanometer curve, while the pneumographic curve is comparatively unaffected*. Respiration is an instrument of consciousness. You can control it voluntarily while you cannot control the galvanometer curve.[23]

It was impossible to connect breathing directly to 'the buried complexes of the soul', those hidden repositories that would produce visible emotions if opened with the proper keys, words, questions, or images. Far from revealing hidden contents, Jung's work with the pneumograph and galvanometer suggested to him that breath tracked *conscious* states. Breathing was a physical manifestation that the 'conscious' mind could control, while sweating was more occult. The act of continuing to think on an emotional stimulus no longer passing through the body—a memory of a meaningful event—could indeed be seen in a respiratory tracing. It was possible to see evidence of some emotional stimuli in the respiratory curve, but these were clearly more under the influence of conscious processes than what the galvanometer registered, which was 'an index or measure of *acute* feeling-tone'. The pneumographic traces could show 'traces of conscious reminiscence', but not the most valuable hidden contents of the unconscious.

Jung's decision to rule out the breath as a possible bodily pathway to or representation of the unconscious, despite Breuer's demonstration of its significance (though he did not put it in such terms), had a basis in the physiology of the day. But it also reflected an earlier antagonism on the part of his mentor, Sigmund Freud.

Against Breath: Freud and the Foreclosed Symptom

Unpacking Breuer's interest in breath in light of Jung's subsequent work opens up a discussion around Freud's dismissive attitude to respiration in *Studies on Hysteria*, which he co-authored with Breuer. In Case 4 of *Studies on Hysteria* Freud recalls a meeting with a serving woman called Katharina as he looked out from a refuge at the top of a mountain he had just climbed. Katharina complained of breathlessness, which Freud went on to diagnose as a symptom of an anxiety disorder precipitated by an abusive encounter with her 'uncle' (actually her father, as Freud later notes). 'Katharina' told Freud: 'I get so out of breath. Not always. But sometimes it catches me so that I think I'm suffocating'.[24] Freud glosses it as follows:

> Now at first this didn't sound like a nervous symptom, but soon afterwards I thought it likely that it was simply a description standing in for an anxiety attack. She was unduly singling out the one factor of restricted breathing from the whole complex of anxious feelings.[25]

Freud's gloss is telling: it betrays a refusal to accept shortness of breath as a nervous symptom, and it masks that refusal by imagining that it is Katharina who has isolated this symptom 'out of the complex' and laid 'undue stress' upon it. Katharina, however, invokes breath, or breath-like responses, when describing several moments of trauma: most notably when she catches her father *in flagrante delicto* with her cousin Franciska. In this instance, her response was to breathe abnormally, and then to vomit. For Freud, at this point, these somatic responses disclose an associative anxiety: they mark Katharina's hysteria, but they remain associative rather than significant in themselves. The consequence, as Freud would write in 'On the Grounds for detaching a particular syndrome from Neurasthenia', is an affect of anxiety rather than a neurosis of anxiety.[26] Affects of anxiety are external stimuli that come to be associated with the anxiety, not intrinsic features of the anxiety itself. So if breathlessness, as a somatoform response, becomes associated with the original affective anxiety, it will precipitate a neurotic response, not because it is intrinsically significant but because it is an associative response. In an implicit contradiction of his co-author, Breuer, Freud suggests that breathlessness *qua* breathlessness cannot tell us very much.

Why did breathlessness mean so little to Freud, if not simply as a point of divergence from Breuer? Perhaps Freud's re-envisioning of psychoanalysis as an archaeological process, wherein the analyst must always be going deeper, made breath too superficial to be interesting in its own right. If anything, marked references to breath in the *Collected Works* confirm this bias. Only when breath discloses an etymological, cultural aspect—whether it is *ruach*, *spiritus*, and *anima* in *Moses and Monotheism* or the Maori *hau* in *Totem and Taboo*—does Freud grant it any sustained attention. (We will consider Freud's particular agenda in these cases below.) At the same time, this explanation is hardly satisfactory. After all, Katharina's case study begins with Freud's own breathlessness. He has climbed a mountain, in part to take some time away from his work on neuroses. He comes across Katharina at a refugio because he needs to catch his breath after the hike. This is by no means the only time he mentions climbing stairs and mountains in his work: he returns to both as figures of anxiety production. Dora, for instance, is caught up in an associative chain in which her breathlessness while climbing mountains or stairs stands in for her father and mother having intercourse. So in a sense, though Freud observes the breathing of his patients, he remains blind to it.[27] A compelling explanation can be found in an intimate friendship that developed during the break with Breuer, as Freud found his next collaborator and confidant in the Berlin ear, nose, and throat doctor Wilhelm Fliess.

Nasus-Interruptus

> I am, of course, very much looking forward to your nose-sex. ... I am so certain that both of us have got hold of a beautiful piece of objective truth... We shall find many more things, I hope, and correct ourselves before anyone catches up with us... I hope that in this book [*Nose and Sex*] as well you will discuss some of the basic views on sexuality that we share.[28]

In his book *Sensual Relations,* the anthropologist of the senses David Howes identifies the possible sources of a curious omission in Freud's account of the erotogenic zones of the body: the nose. Attentive to olfaction, Howes asks: if, according to Freud's erotogenic zone theory, the libido attaches itself to each orifice in turn (mouth, anus, genitals), why is there no nasal stage?[29] Wilhelm Fliess was a Berlin doctor preoccupied with theorizing 'nasal reflex neurosis', the relationship between the nose and sexuality, and developing treatments for it that included cauterization and topical cocaine. In intimate letters written during the last decade of the nineteenth century, Freud and Fliess often discussed the peculiar relationship of the nasal passages to sexuality. Freud also provided periodic updates on his faltering relationship with Breuer, as well as frequent reports on the various discharges from his own nose, which had apparently been entrusted to Fliess's care. In May of 1895, for example, it 'discharged exceedingly ample amounts of pus' (throughout which Freud 'all the while felt splendid').[30]

Howes points out that the association of the libido with specific locations (orifices) of the body—an idea that became the core of Freud's erotogenic zone theory—had Fliess's nose as the paradigmatic erotically linked orifice.[31] That is to say, without Fliess's nose, and the associated inhalatory smelling that it allowed, there would have been no erotic anatomy at all. Throughout the letters, we see that Freud is thinking with Fliess, with the body, and with the nose in particular. In an 1897 letter, using a language that he surely assumes Fliess will understand, Freud works out an early version of what will become another central psychoanalytic idea, repression, in Fliess's own nasal terms: 'To put it crudely, the memory actually stinks just as in the present the object stinks; and in the same manner as we turn away our sense organ (the head and nose) in disgust, the preconscious and the sense of consciousness turn away from the memory. This is *repression*'.[32] If the nose, in this early version of repression, seems a potentially rich area of interest, Howes draws attention to Freud's later denigration of smell as a sense in his perceptual anatomy, for example, in *Civilization and its Discontents*, in which Freud explains that having evolved to walk upright, we came to *see* rather than *smell* the genitals.[33]

Freud's friendship with Fliess began to unravel dramatically around 1901, and the nose—and the significance of the odours known through inhalation—became a stinking memory for Freud. Freud's personal breaks first with Breuer and then Fliess—both of whom included breathing and olfaction in their understanding of reflex and psyche—strongly affected the image of the body that Freud would depict in his work. Freud's body would barely breathe, it would have no nose, and a poor sense of smell. With no mention of the forgotten Fliess, Otto Fenichel, known as the great systematizer of psychoanalysis, would later contribute the claim that inhalation and smelling are a single act in the unconscious.[34] But this is not breathing as it appears—or more often does not appear—in the breathers of Freud's case studies.

The Wolfman

Freud's fullest exposition of inhalation and exhalation may be found in his 1918 case study 'From a History of an Infantile Neurosis', based on the life of Sergei Pankejeff, known as 'The Wolfman'. Freud's analysis of Pankejeff focusses on a nightmare the analysand had as a young child, in which he was terrified by a pack of wolves sitting outside his bedroom window. Freud links this dream to the eighteen-month-old Sergei's witnessing of the primal scene: he awakened in his crib to see his parents having intercourse. Although he mentions it only in passing, Freud also notes a particular pattern in Pankejeff's breathing:

> At this time he used also to adhere to a peculiar ritual if he saw people who inspired pity in him, beggars, cripples, old men. He had to breathe out noisily in order not to become like one of them, and under certain conditions also had to inhale deeply.[35]

And again:

> The only way he could account for it to himself was that he did it so as not to become like them. [...] this was connected with his father [...]

> [His mother] took them to a sanatorium where they saw their father again; he looked ill and his son felt very sorry for him. His father, then, was the archetype of all those cripples, beggars and poor people, the sight of whom obliged him to breathe out, just as the father is normally the archetype of the grimaces seen in anxiety states and of the caricatures drawn to express contempt. We shall discover elsewhere that this pitying attitude goes back to a particular detail of the primal scene, which took effect at this late stage in the obsessive-compulsive neurosis. The resolution not to become like them, which was the motivation for his breathing out in front of cripples, was thus the old identification with the father transformed into a negative. And yet he was also copying his father in a positive sense, for his noisy breathing was an imitation of the sounds he had heard his father make during intercourse.[36]

Pankejeff's breathing patterns, for Freud, signalled his Oedipal anxiety, his desire to reject the father, expel him, while also mimicking him, desiring to become him and take his place. Freud then glosses this as follows: 'In my analysis of the breathing rituals prompted by the sight of cripples, beggars, etc. I was able to show that this symptom could also be traced back to the father, whom he had felt sorry for when he visited him in the clinic during his illness'.[37] In his 1926 *Techniques*, Otto Rank confirmed his break with Freud, already far advanced by the publication of *The Trauma of Birth* (1924), by devoting some attention to the question of why Pankejeff was not in fact locked in an Oedipal relation. Freud, Rank argued, had missed Pankejeff's transfer of affection, *i.e.*, transference, for the mother onto Freud himself, since Freud was essentially offering to 'rebirth' him. Freud was affronted, interpreting Rank's answer as 'calculated to cut the tablecloth between us'. It was a hypothesis that risked appearing like 'self-parody' because of its suggestion that Freud had been taken in by his patient.[38] As riling as Rank's suggestion might have been, there is another explanation for Freud's adamant rejection of any but an Oedipal diagnosis of Pankejeff, and it is found in his reasons for writing 'The History of an Infantile Neurosis'.

According to his own footnote, Freud published the study shortly after the conclusion of his treatment of Pankejeff, and 'in light of the recent attempts to re-interpret psychoanalytic material undertaken by C. G. Jung and Adolf Adler'. Against challenges to the libido theory by Jung and Adler, Freud wished to demonstrate beyond a doubt that disturbances of childhood sexuality (*Kernkomplex*) were sufficient to account for all adult neuroses.[39] But there was another line of powerful argument against

Jung at work, one that brings us back to the religious resonance of the breath. During their friendship, Freud had followed Jung into the psychology of religion, writing to him hopefully before their break: 'So you too are aware that the Oedipus complex is at the root of all religious feeling. Bravo! What evidence I have can be told in five minutes'.[40] Jung responded by saying that 'the outlook for me is very gloomy if you too get into the psychology of religion. You are a dangerous rival—if one has to speak of rivalry. … Our personal differences will make our work different'.[41] Jung was not convinced of the centrality of the Oedipus Complex.

Freud wrote *On the History of the Psychoanalytic Movement* in a defensive posture after their split, claiming that Jung could not tolerate having infantile sexuality and the Oedipus complex at the root of religion because, while it seemed for a time that Jung was 'ready to enter into a friendly relationship with me and for my sake to give up certain racial prejudices which he had previously permitted himself', he remained the Swiss son of a clergyman who would not surrender Christianity—and perhaps religiosity more generally—to Freud's theory of childhood sexuality.[42]

In Pankejeff's case, then, Freud took pains to show not only that Sergei's obsessional neurosis had a sexual origin, but also that a key symptom of his neurosis was compulsive religiosity and its major practice an obsessive breathing ritual that served as a neurotic 'diversion from sensual processes to purely spiritual ones'.[43] In the beginning, before Sergei was faced with the novel data of his father's penis disappearing into (what appeared to be) his mother's bottom, he (not coincidentally, born on Christmas Day) believed that he was Christ, and his father God; indeed, he was born of his father alone, while his sister was the child of his mother alone. Learning, through witnessing the primal scene, that his mother had been involved in his birth was deeply troubling. The ambivalence of love and hate for his father ensued, only intensifying when he saw that he had become a pitiful creature. Pankejeff's solution to his ambivalence about a father who both was castrated and could castrate was the breathing ritual: he could use it to identify with his father's capacity to castrate, 'for his noisy breathing was an imitation of the sounds he had heard his father make during intercourse', while at the same time exhaling the pathetic, castrated aspect that he had witnessed during the hospital visit. Freud summarized the solution in a serious joke intended precisely for the estranged Christian Jung: 'The Holy Spirit owed its origins to this sign of erotic excitement in a man'.[44]

Respiratory Erotism

If breath, for Freud, needed a high-stakes argument with Jung to warrant interest, Otto Fenichel's 1931 essay 'Respiratory Introjection' offered a rereading of Pankejeff's case as an introduction to respiratory erotism per se. Known as the author of what would become the standard, 'orthodox' textbook of psychoanalysis, *The Psychoanalytic Theory of Neurosis*, Fenichel explained respiration in the context of Freud's theory of infantile sexuality, the erotogenic, and object relations.[45] While Freud used breathing as a convenient weapon against Jung, Fenichel took Pankejeff's breathing much more literally. And Fenichel had another agenda: he used the breath as a lever, first taking Freud seriously on the matter of breath magic, but then gently going further to expand the functional anatomy of the psychoanalytic body. Fenichel's work indicates an alternative respiratory logic in psychoanalysis, in which the somatic begins to gain its own significance, apart from its function as a symptom of underlying neuroses.

Michael Heller has comprehensively documented the significance of Fenichel for the history of embodied approaches to psychotherapy, challenging Wilhelm Reich's placement of himself at the centre of the return of the body to psychoanalysis.[46] Heller's attention to both theorists and practitioners also allowed him to recognize and reintroduce a hidden source for both Fenichel and Reich: the Berlin teacher of 'gymnastik' Elsa Gindler. Introduced to her classes by his future wife, Fenichel found relief from terrible migraines that his analysis with Radó had not resolved. Gindler's view of breathing, as described in her one surviving essay 'Gymnastik for People Whose Lives are Full of Activity', was that 'Good undisturbed breathing happens involuntarily'. Any forcing of the breath would prevent the 'small lung vesicles' from opening and filling with air. Movements paired with forced or voluntary breathing required 'excessive and inappropriate effort'. The body would then be dominated by constriction and tension, together with the experience of anxiety.[47] Undoubtedly influenced by Gindler, in 1928 Fenichel wrote 'Organ Libidinization Accompanying the Defense against Drives', which was an attempt to integrate this idea of muscle tension with libido theory, and then in 1931 'Respiratory Introjection', which aimed to show the existence of what Fenichel dubbed the 'autonomous erotogenicity' of the respiratory tract. (As Heller points out, Fenichel does not even mention Gindler's name.)

Why respiratory *introjection*? According to Fenichel's reading of Freud and Karl Abraham,[48] at the origins of perception was the experience of gaining satisfaction by putting something in one's mouth: this is why the oral stage was the earliest focus of the libido. The libido would later be associated primarily with the anus and finally, should all go well, with the genitals: each pregenital stage was a form of 'selective incorporation'. At the very beginning, the ego of the infant was undifferentiated, with no separation between the self and the so-called external world. Freud, Fenichel explains, described the stage in which all pleasure is incorporated into the ego and all unpleasant sensations rejected and placed in the 'external world' as the *purified pleasure ego.* Incorporating something was a way to persist in being pleasurably united with whatever was taken in; taking in and identifying were a single gesture. Introjection, the executive of the 'primary identification', had been taken to be first an oral thing. But Fenichel begins with a discussion of this possibility of identification through *respiratory* incorporation:

> Freud has demonstrated that identification with an object of the external world [...] is conceived of as an oral incorporation [...] Abraham has demonstrated that there exist other ideas of incorporation besides oral ones [...] an anal one [...] an epidermal one [...] finally a respiratory one is suggested by an episode in the history of the Wolf Man: the patient, whenever he saw a cripple, had to exhale forcibly, in order not to become like him. Since he thus exhaled objects, he must have previously incorporated them by inhaling.[49]

Fenichel is cautious. He does not commit entirely to respiratory introjection—he still sees it as 'an intermediary factor in comparison to oral introjection'—but he nevertheless grants it an erotism: 'nasal introjection corresponds to respiratory erotism'.[50] If introjection was the attempt to 'make parts of the external world flow into the ego', then respiratory introjection was the pre-genital ego's attempt to use the respiratory apparatus to identify with pleasant experience and reject negative or unpleasant experience.[51] The pleasure-ego sought the 'original objectless condition': this oceanic state, and the longing to return to it, was the source of the mystic's ecstasy, exuberant patriotism, and other forms of the immature ego's need to participate in something 'unattainably high'.[52] The ubiquity of 'primitive' breathing magic in many cultures and the traditional identity of breathing with life and soul are thus explained: 'There exists no narcissistic-animistic philosophy of life in which breathing as the expression of life itself is not invested with narcissistic libido'.[53] To breathe was to be, both self and other.

Why was respiratory erotism overlooked in the past? Fenichel writes, 'The function of breathing has heretofore been treated by psychoanalysis in a rather stepmotherly way, and this is due to the fact that its erotogenicity appears in such close connection with oral and anal erotism, [in] bronchial asthma or breathing compulsion. But indications are not wanting that the respiratory tract too has autonomous erotogenicity.'[54] The personality of the asthmatic was closely aligned with that of the anal personality, and this 'anal orientation developed from interest in smelling' (here then is smelling, but relegated to the earliest stages).[55] The physiological 'model' of respiratory introjection was the act of smelling.[56]

Fenichel's reading of his own two case studies led him to suggest, against the traditional Freudian understanding that smoking stems from an oral fixation, that 'in smoking [...] respiratory erotism and introjection may play a greater role than oral erotism' because of 'a pregenital tendency to inhale'.[57] Moreover, as he goes on to show in the cases of 'sniffling', 'asphyxiation', 'coughing', and, in other texts, 'bronchial asthma', there is an autonomous respiratory erotism that has 'an archaic pregenital character', which, while minor, is nevertheless distinct from other erogenous zones. It is not hard to imagine a translation of Gindler's anxiety-producing respiratory constriction into the psychoanalytic idea of the repression of the narcissistic/pre-genital pleasures of the respiratory tract—pleasures such as breathing in and smelling, perhaps giving the infant a sense of unity with the mother.

For Fenichel, anxiety and respiration were closely linked. Failure to resolve castration anxiety could appear masked as a respiratory symptom: 'A patient imagined that the analyst might cut off his supply of air with scissors. He fantasized that his supply of air was arranged like that of a diver's and was being cut off by his analyst, thus choking him. This fantasy was a cover for the anxiety lest the scissors cut off his penis'.[58] Though he does not mention the exiled Rank as a source, Fenichel also argues, in distinction from Freud, that respiratory erotism is constitutive of anxiety: 'respiratory innervation belongs to the very essence of anxiety. They were present in the archetype of every traumatic situation, namely, the act of birth, and even later anxiety is a partial re-experience of this situation'.[59]

Having attended to the Freudian view that breathing is a symptom of more archaic conditions, Fenichel suggests that we understand breathing in its own right. This might give a better sense of (1) the analysand's own body and (2) the analysand's observation of and relation to others. Not only does this mean that we must begin to consider the breath less as a uniform, ideal patterned movement than as an idiosyncratic, dystonic rhythm liable to alter with the moods, attention, and energy levels of the breather, but it also means that the continued alteration of our breathing patterns might itself demonstrate how we respond to different people. Fenichel thus echoes Gindler's views when he writes:

> Breathing is like other muscular functions: the average person does not carry them out optimally, but exhibits remarkable phenomena of an inhibitory character [...] the continuous minor psychic changes exert their inexpedient dystonic influence on the respiratory function by changes in the rhythm of respiration, mostly in a passing stoppage of breathing, in variable, uneven participation of the parts of the thorax in breathing out.[60]

As a consequence, when we observe breathing irregularities, we can diagnose anxiety, but even more importantly, 'in inhibiting breathing at an intended new action, at a perception, at a change of attention, the ego tests out whether or not it needs to be afraid'.[61] Moreover, this testing also expresses itself in non-pathological interactions, wherein we identify or empathize with others and thus mimic or take on their expressive actions. Perhaps, Fenichel concludes his essay, we do this in large measure by following the other's breathing with ours.

We have come a long way from a similar scene in Breuer: a mother is vulnerable to hysteria and creates the conditions for it by matching her breathing to that of her nursing baby. Fenichel has also brought us some way from Freud's tendency to respond to breath as a superficial sign or symptom of more important anxious disorders. Not only might the breath and the anxiety be co-productive, but there may indeed be aspects of the breath that the analysand follows, or 'inhales', when developing anxiety responses. This may in turn be related to a mimetic faculty in which becoming the other means, in part, assuming the other's breathing pattern.

Conclusion: Respiration and Personality

Not long after Freud died in 1939, still another interpretation of breathing appeared in the newly founded *Journal of Psychosomatic Medicine*. It was by Franz Alexander, one of the journal's founders, and a 'bête-noir' of Fenichel's.[62] Alexander used respiratory traces similar to those Jung had studied in 1905 but made quite the opposite claim, arguing that an individual's breathing pattern was as personal as a fingerprint. Like the voice, no one could replicate the tracing of an individual's breathing; the breath was a unique, personal signature, identifiable whether in the panting of running for a bus or the gentle breathing of sleep.[63] A respiration tracing revealed something about who a person essentially was, even if they tried 'consciously' to hide it. Alexander promised to explore further the idea in a Part Two that never (to our knowledge) appeared.

The idea that a perfect record of the gestures of respiration would reveal individuality was, perhaps, closer to Fenichel/Gindler's earlier position than to Alexander's eventual view of bronchial asthma, a paradigmatic case study for his 'psychosomatic medicine' that sought to correlate psychoanalytic findings with laboratory ones, adding 'emotional' factors to known laboratory aetiologies.[64] In this later work, Alexander would claim that the emotional component of bronchial asthma was a 'repressed desire for the mother', a defence against the longing to return to her, with the gesture of wheezing resembling the attempt to suppress crying.[65] But this immaterial psyche did indeed exist in an environment full of allergens. A spasm of the bronchioles, where we began with Anna O., was in the psychosomatic school of the 1950s thought to be caused equally by emotional and allergenic factors. Complex aetiology was a problem, Alexander wrote, well known in the physiology lab as the 'summation of stimuli': treating either causative factor would be effective.[66] This complex and layered breathing mind—repressed and full of longing, somehow squeezing the bronchioles—would disappear as the incidence of fatal asthma rose rapidly in the 1950s and powerful inhalational bronchodilators were introduced at the end of that decade and in the early 1960s.[67]

Notes

1. 21 April 1969. Quoted in James Knowlson, *Damned to Fame: The Life of Samuel Beckett* (New York: Grove, 1996), 501.

2. See Matthew Feldman, *Beckett's Books: A Cultural History of the Interwar Notes* (London: Continuum, 2006).
3. Otto Rank, *The Trauma of Birth*, trans. unknown (London: Kegan, 1929), 51; 171.
4. Sigmund Freud, 'The "Uncanny"' [1919], in *The Complete Psychological Works*, vol. 17 (London: Hogarth Press, 1955), 217–56, 241.
5. Sigmund Freud and Josef Breuer, *Studies in Hysteria*, trans. by Nicola Luckhurst (New York: Penguin, 2004); Breuer, 'Theoretical Issues', *Studies in Hysteria*, 195.
6. See James E. Strick, *Wilhelm Reich, Biologist* (Cambridge, MA: Harvard University Press, 2015), 22.
7. Freud and Breuer, *Studies in Hysteria*, 27.
8. Freud and Breuer, *Studies in Hysteria*, 44.
9. On Helmholtz and nineteenth-century physics, see Anson Rabinbach, *The Human Motor: Energy, Fatigue, and the Origins of Modernity* (New York: Basic Books, 1990).
10. Freud and Breuer, *Studies in Hysteria*, 209. On the history of debates between localization and association and early psychoanalysis, see Katja Guenther, *Localization and Its Discontents: A Genealogy of Psychoanalysis and the Neurodisciplines* (Chicago: University of Chicago Press, 2015).
11. Freud and Breuer, *Studies in Hysteria*, 250.
12. Josef Breuer, 'Die Selbststeuerung der Atmung durch den Nervus vagus', in *Sitzungsberichte der Akademie der Wissenschaften Wien, math.-naturw.* Kl. 58/2 (1868), S. 909–37.
13. Elisabeth Ullman, 'About Hering and Breuer', in *Breathing: Hering-Breuer Centenary Symposium*, ed. by Ruth Porter, Ciba Foundation Symposium (London: Churchill, 1970), 3–15.
14. Margaret Muckenhoupt, *Sigmund Freud: Explorer of the Unconscious* (Oxford: Oxford University Press, 1997).
15. The term appears as a criticism of the anti-vitalism of Brücke's mechanistic view of the heart in Joseph Hyrtl, *Über die Selbststeuerung der Herzens: ein Beitrag zur Mechanik der Aortenklappen* (Wien: Wilhelm Braumüller, 1855). For Brücke and Hyrtl's debates about mechanism and vitalism as a context for Freud and Breuer, see Henri F. Ellenberger, *The Discovery of the Unconscious* (London: Fontana, 1970), 266.
16. Elisabeth Ullman, 'Preface. Two Original Papers by Hering and Breuer Submitted by Hering to the Akademie Der Wissenschaften Zu Wien in 1868', in Porter, *Breathing*, 357.
17. Ewald Hering, 'Self-Steering of Respiration through the *Nervus Vagus*', in Porter, *Breathing*, 359–64, 362.
18. Ellenberger, *The Discovery of the Unconscious*, 120.
19. Freud and Breuer, *Studies in Hysteria*, 220.
20. Frank McLynn, *Carl Gustav Jung* (New York: St. Martin's, 1996), 70.

21. Ibid., 70.
22. Frederick Peterson and C. G. Jung, 'Psychophysical Investigations with the Galvanometer and Pneumograph in Normal and Insane Individuals', in *The Collected Works of C. G. Jung*, ed. by Herbert Read, et al., vol. 2 (Princeton, NJ: Princeton University Press, 1973).
23. Peterson and Jung, *Collected Works of C. G. Jung*, vol. 2, 512.
24. Freud and Breuer, *Studies in Hysteria*, 129.
25. Ibid.
26. Sigmund Freud, *The Standard Edition of the Complete Psychological Works of Sigmund Freud*, vol. 3 (1893–99) (London: Hogarth Press, 1962).
27. This point is discussed both by Didier Anzieu in *The Skin Ego*, trans. by Naomi Segal (London: Karnac, 2016), and by Jean-Louis Tristani, *Le Stade du Respir* (Paris: Minuit, 1978).
28. Sigmund Freud, *The Complete Letters of Sigmund Freud to Wilhelm Fliess, 1887–1904*, trans. and ed. by Jeffrey Moussaieff Mason (Cambridge, MA: Belknap, 1985), Freud to Wilhem Fliess, February 13, 1896, 172.
29. David Howes, *Sensual Relations: Engaging the Senses in Culture and Theory* (Ann Arbor: University of Michigan Press, 2003), 180.
30. Freud to Fliess, May 25, 1895, *Letters of Freud to Fliess*, 130.
31. Howes, *Sensual Relations*, 196.
32. Freud to Fliess, November 14, 1897, *Letters of Freud to Fliess*, 280.
33. Howes, *Sensual Relations*, 197.
34. 'Respiratory eroticism is most intimately connected with the pleasure of smelling; particularly so since the function of smelling and that of breathing are not differentiated from one another in the unconscious', Otto Fenichel, *Outline of Clinical Psychoanalysis*, trans. by Bertram D. Lewin and Gregory Zilboorg (New York: The Psychoanalytic Quarterly Press and W. W. Norton, 1934), 218.
35. Sigmund Freud, *The 'Wolfman' and Other Cases*, trans. by Louise Adey Huish, intro. by Gillian Beer (London: Penguin, 2002), 214.
36. Ibid., 265–66.
37. Ibid., 284–85.
38. E. James Lieberman and Robert Kramer, eds, *The Letters of Sigmund Freud and Otto Rank: Inside Psychoanalysis*, trans. by Gregory C. Richter (Baltimore: The Johns Hopkins University Press, 2011), 260.
39. Freud, 'A Case of Obsessive-Compulsive Neurosis [The 'Ratman']', in *The 'Wolfman' and Other Cases*, 176.
40. William McGuire, ed., *The Freud/Jung Letters: The Correspondence between Sigmund Freud and C. G. Jung*, trans. by Ralph Manhein and R. F. C. Hull (Princeton, NJ: Princeton University Press, 1974), 1 September 1911, 441.
41. *Freud-Jung Letters*, Jung to Freud, 14 November 1911, 460.
42. Makari, *Revolution in Mind*, 334.

43. Freud, *The 'Wolfman' and Other Cases*, 313.
44. Ibid., 266.
45. Otto Fenichel, *The Psychoanalytic Theory of Neurosis* (orig. printing: W. W. Norton, 1945; London: Routledge, 1996).
46. Michael Heller, *Body Psychotherapy: History, Concepts, Methods* (New York: W. W. Norton, 2012), 417.
47. Elsa Gindler, 'Gymnastik for People Whose Lives Are Full of Activity', in *Breath, Bone, and Gesture: Practices of Embodiment*, ed. by Don Hanlon Johnson (Berkeley, CA: North Atlantic Books, 1995), 5.
48. Karl Abraham 'A Short Study of the Development of the Libido, Viewed in the Light of Mental Disorders', in *Selected Papers of Karl Abraham: With an Introductory Memoir by Ernest Jones,* trans. by Douglas Bryan and Alix Strachey (London: Hogarth Press: 1927), 418–501.
49. Otto Fenichel, 'Respiratory Introjection', in *The Collected Papers of Otto Fenichel,* First Series, ed. by Hanna Fenichel (New York: W. W. Norton, 1953), 221.
50. Fenichel, 'Respiratory Introjection', 222.
51. Otto Fenichel, *The Psychoanalytic Theory of Neurosis*, 39.
52. Ibid., 40.
53. Fenichel, *Outline of Clinical Psychoanalysis,* 221.
54. Fenichel, 'Respiratory Introjection', 222.
55. Fenichel, *The Psychoanalytic Theory of Neurosis*, 322.
56. Ibid., 332.
57. Fenichel, 'Respiratory Introjection', 223.
58. Fenichel, *The Psychoanalytic Theory of Neurosis,* 249.
59. Fenichel, 'Respiratory Introjection', 238.
60. Fenichel, 'Respiratory Introjection', 239.
61. Ibid.
62. Russell Jacoby, *The Repression of Psychoanalysis: Otto Fenichel and the Political Freudians* (Chicago: University of Chicago Press, 1983), 35.
63. Franz Alexander and Leon J. Saul, 'Respiration and Personality—A Preliminary Report. Part I: Description of the Curves', *Psychosomatic Medicine* 2/2 (1940), 115.
64. See Anne Harrington, *The Cure Within: A History of Mind-Body Medicine* (New York: W. W. Norton, 2008).
65. Franz Alexander, *Psychosomatic Medicine: Its Principles and Applications* (New York: W. W. Norton, 1950), 139.
66. Ibid., 140.
67. Mark Jackson, *Asthma: The Biography* (Oxford: Oxford University Press, 2009), 175.

Selected Bibliography

Anzieu, Didier. 1995. *Le Moi Peau*. Paris: Dunod. 2016. Trans. Naomi Segal. *The Skin Ego*. London: Karnac.

Heller, Michael. 2012. *Body Psychotherapy: History, Concepts, Methods*. New York: W. W. Norton.

Howes, David. 2003. *Sensual Relations: Engaging the Senses in Culture and Theory*. Ann Arbor: University of Michigan Press.

Tristani, Jean-Louis. 1978. *Le Stade du Respir*. Paris: Minuit.

CHAPTER 19

Mysterious Gear: Modernist Mountaineering, Oxygen Rigs, and the Politics of Breath

Abbie Garrington

In 'Mountains', one of the 'Ariel Poems' series of pamphlets published as Christmas cards by Faber in 1954, and the third of his series of 'Bucolics', W. H. Auden takes this moment, one year on from the close of the Everest era with the Hillary/Tenzing ascent of that mountain, as an apposite point from which to view the strange tribe of the mountaineer:[1]

> And it is curious how often in steep places
> You meet someone short who frowns,
> A type you catch beheading daisies with a stick;
> Small crooks flourish in big towns
> But perfect monsters—remember Dracula—
> Are bred on crags in castles: those unsmiling parties,
> Clumping off at dawn in the gear of their mystery
> For points up, are a bit alarming;
> They have the balance, nerve
> And habit of the Spiritual, but what God
> Does their Order serve?[2]

A. Garrington (✉)
Department of English Studies, Durham University, Durham, UK

D. Fuller et al. (eds.), *The Life of Breath in Literature, Culture and Medicine*, Palgrave Studies in Literature, Science and Medicine,
https://doi.org/10.1007/978-3-030-74443-4_19

Short, monstrous, cruel to the merest daisy, the mountaineer rises with the sun (which is in fact rather late for an Alpine start, the icepack at risk of melt and shift by that point), but he greets that dawn with an absent smile and the leaden foot of his climbing boot, weighting himself and Auden's line with a '[c]lumping' step. Such boots are part of the mysterious 'gear' of the climber, akin to the 'habit' of their monkish brothers who might also be found in the heights. '[H]abit' and 'Order' mix religious observance with the lumpen clumping of the ritualised yet most likely non-spiritual brotherhood of the rope. Their 'balance' and 'nerve' are necessities, but to what end, given the self-selected heroism of a dangerous climb, the purposelessness of the endeavour? Their 'gear'—their kit and paraphernalia—marks them as men apart, or as cranks perhaps (we note here that Basil Bunting's 'On the Flyleaf of Pound's *Cantos*' of 1949 refers to 'crags cranks climb').[3] 'Gear' also, with the oxygen rigs, snow goggles, and ever-developing ice axes of that Everest era in mind,[4] indicates the semi-technological body of the mountaineer, the near-automaton, whose incorporation of or bodily negotiation with mechanisms places him alongside the airman of the period, a second figure reliant upon systems of oxygen supply, technologies of the breath. In Cecil Day-Lewis's ballad 'Johnny Head-in-Air' of 1935, a voice asks of travellers burdened with a combination of kit and superstitions: 'Where are you going, you wan hikers, / And why this ganglion gear?'[5] In this reading, the body's uneasy appeasement with its burdens makes of them a kind of cyst, a 'ganglion'. Meanwhile, via its Middle English origins, 'gear' indicates a mechanism, suitable for ratcheting-up or raising-high—as does 'crank', of course. Whatever their balance and nerve, then, Auden's mountaineers' ascension causes 'alarm'—through sombre disposition and get-up (a get-up for getting up, we might say); through 'cranked' displacement into the rare air of high altitude; and through the opaque motivations of their high-risk, seemingly low-outcome practices.

Auden's landscape in 'Mountains' is, we gather, primarily supposed to indicate Alpine rather than Himalayan realms—if there are daisies in the foothills, and if the dawn departure of mountaineers is, it is implied, observed from a town location, then these are the 'steep places' of the Alps, and not the disorientatingly vertiginous peaks of the 'roof of the world'.[6] The monastic group hinted at in the extract's final lines may therefore be taken as a reference to the Carthusians, an enclosed Roman Catholic order founded by St Bruno of Cologne in 1084, in the valley of the Chartreuse of the French Alps.[7] Their motto, '*stat crux volvitur orbis*' [the cross is steady while the world turns], is one possible source of the surname of loyal and reliable Ian Shawcross (or sure cross), one of the climbing party featured in Auden and Christopher Isherwood's mountaineering verse drama *The Ascent of F6: A Tragedy in Two Acts*, which was first performed in 1937.[8] The Carthusians' precursors, the Desert Fathers, make an appearance in Matthew Arnold's 'Stanzas from the Grande Chartreuse' of 1855, the rhythms of which turn up in the play's oft-cited and frequently misinterpreted funeral dirge known by its opening phrase, 'Stop all the clocks'.[9] And if we are after all in an Alpine context, the 'gear of [the climbers'] mystery' might indicate only that they are supplied with sun goggles, hobnailed boots, hats, and veils for sun protection (the latter perhaps recalling a monkish 'habit'), with ice axes, and coils of rope. Yet, given the year of publication, with images of the Everest ascent fresh in the mind, and as a long-term observer of British attempts upon that mountain, Auden may also be overlapping his Alps and Himalayas, in a palimpsestic presentation of mountain vistas. Maurice Herzog, whose *Annapurna* of 1951 is often claimed to be the biggest-selling work of mountaineering literature of all time,[10] exported the Alps as a means of physical and psychological measure when confronting this unclimbed 8,000m-plus peak (the first of such height to be summitted), continually using 'Mont Blanc' as a point of comparison interposed between climber and the Himalayan landscape with which he was actually faced ('The height, which was about the same as that of Mont Blanc, was making us tired').[11] Auden's movement is in the other direction, having Everest (and its own 'Order[s]' of Buddhism, seemingly drawn from British accounts of access to the mountain via Tibet in the 1920s)[12] hover around, or loom over, a more familiar Alpine view. The Alpine/Himalayan collapse or conflation of Auden's poem is useful to us not only in suggesting that the climbers' 'gear' indicated here might reasonably include a contraption of oxygen supply (albeit that its mask or mouthpiece must be removed for us to register the climbers' 'unsmiling' faces), but also in suggesting that the shift toward the Greater Ranges, and to the use of modifications of or supplements to the human breath necessitated by climbs at higher altitudes, might cause problems for the contraption of poetry itself. If Everest had 'fallen' to the human foot (and, as Jan Morris's renowned account of the 1953 climb *Coronation Everest* hinted at its close, to the Commonwealth),[13] the poetic treatment of mountaineering struggled to transfer to these heights, weighed down by Romantic and Victorian conceptualisations of human struggle against rock, by Alpine landscapes that no longer formed the arena of greatest challenge for the contemporary mountaineer, by the remnants of mountain spirituality in an era of scientific and sporting impetus, and by the imperfect incorporation of the mysterious gear of the high-altitude mountaineer's oxygen supply.[14]

The only member of the climbing party to take part in all three British attempts upon Everest in the 1920s (that is, a reconnaissance in 1921, and bids to reach the summit in 1922 and 1924) was George Leigh Mallory. Now perhaps the most famous climber in Western mountain history, following his death alongside climbing partner Andrew 'Sandy' Irvine on that 1924 expedition,[15] Mallory is also of interest to scholars of modernist culture as a figure peripheral to the 'secret' society of the Cambridge Apostles (including biographer Lytton Strachey and economist John Maynard Keynes), and to Virginia Woolf *et al.*'s Bloomsbury Group. The latter circle's Duncan Grant photographed and painted the famously alluring Mallory from the life ('Mon dieu!—George Mallory! When that's been written, what more need be said?' writes Strachey in a self-satirisingly lascivious letter to Clive and Vanessa Bell),[16] leading the climber to declare that 'I am profoundly interested in the nude me'.[17] Yet Mallory's interest in physical form was, as we might expect of a modern mountaineer, accompanied by interests in physiology, including the intersection of the respiratory elements of the human somatic system, and the environment of high-altitude mountaineering. In an address entitled 'Purer Air than Mortals', read before the Joint Meeting of London's Alpine Club and the Royal Geographical Society on 16 October 1922, and reflecting on that year's first effort in earnest to reach the summit of Everest, Mallory describes a shift in the conception of the effort to ascend:

> When first the prospect of going to Mount Everest opened for me I used to visualize the expedition in my thoughts as a series of tremendous panting efforts up the final slopes. Later it became a symbol of adventure [...]. Now it has become a problem [...] the expedition brings to my mind's eye a view of the mountain slopes set at intervals with groups of little tents, with loads of stores and sleeping sacks, and with men.[18]

This move from armchair imaginings of noble struggle to a focus upon local geography, strategy, and expeditionary logistics shows the potentially self-aggrandising stuffing knocked out of the man in the course of two previous expeditions. Yet 'tremendous panting efforts' remain in play, not least because those loads and stores would include the oxygen rigs that were eventually to make possible the successful ascent of the mountain (albeit on Hillary and Tenzing's, not Mallory's, watch).

It is notable, however, that Mallory's first bid to explain to his audience the experience of climbing on Everest's flanks omits technological measures for the supply of supplementary oxygen, favouring instead a lung-versus-mountain model that praises his own modification of the breath, and affords him brothers-in-arms (or -in-lungs) in fellow climbers with similar respiratory abilities:

> Ultimately, the power of pushing up depended on lung capacity. Lungs governed our speed, making the pace a miserable crawl. From the Alpine point of view our lungs made us pause to admire the view oftener than is correct in the best circles. But our [Mallory; Dr Howard Somervell; Edward F. Norton] lungs were remarkably alike and went well together. Personally I contrived a looseness of the muscles by making an easy, deep-drawn breath, and by exercising deep breathing I found myself able to proceed.[19]

The repetition of the word 'lung' here (a term carrying its own weight in expired breath akin to the bell toll, and echoing into, we presume, the Great Hall of the RGS's Lowther Lodge as Mallory spoke) centres attention upon the body and its capacities, connects men of the 'right sort' in physical terms (underscoring their status as the 'right sort' in ethnic and social terms, Sherpa support on the mountain more than matching their ease at altitude, but being sidelined here),[20] and puts forward the 'deep-drawn' breath, the inspiration of the gifted mountaineer, as a means of oxygen management. This apparently retrogressive or at least rudimentary approach to retention of oxygen levels and prevention of hypoxia, or what was at this time gathered under the umbrella term 'mountain sickness', is then drawn into a curious confrontation with the scientific establishment, apparently performed to elicit 'you tell 'em's from Mallory's immediate audience:

> I imagine that a number of physiologists, especially, would be inclined to reduce these odds on the mountain [winning out against man]. I was told at Oxford last year, by Sir Walter Raleigh, that the physiologists said it was physiologically impossible to climb to the top of Mount Everest

> without oxygen—the matter had been proved by experiments in a pressure-reducing chamber. I told Sir Walter that the physiologists might explode themselves in their diabolical chamber, but we would do what we could to explode their damnable heresy—or words to that effect. I always, as a matter of course, take off my hat to scientists, as latter-day Olympians breathing a different if not purer air than common mortals. But the air of Mount Olympus (a base little lump after all) is not that of Mount Everest, and experiments made there with a pumped-out tank, interesting as they may be, are of no value in determining where precisely on that other hill of unrivalled altitude persevering man will be brought to a standstill.[21]

This statement deploys scientific terminology in such a manner as to undermine it, or at least hold it forth as a peculiarity ('the physiologists said it was *physiologically* impossible'; 'a pressure-reducing chamber', my emphasis), and shifts the experiments of scientists from laboratory to hellscape ('their diabolical chamber') or half-imagined Greek mountain ('Olympus') brought low (a 'base [...] lump' to set against Everest's unrivalled height and widely acknowledged beauty). Yet it is explosions that are central here, indicating not only the modified pressures of a barometric chamber, but the physical and psychological pressures of the mountain experiences they were aiming to replicate. Further, the 'heresies' of Science's claim to have the answer to access to the so-called 'Third Pole', the world's roof (or a heaven to stand against the heretics), will be exploded by mountaineer testimony. Mallory performs a model of exploratory masculinity here that places faith in the capacities of man in physical and mental terms. He also breaks the frame of that continent, reticent figure of the British explorer in recounting a kind of social explosion in front of Sir Walter (although 'or words to that effect' suggests this confrontation may have been embroidered after the fact, no corroboration being available, since Sir Walter had obligingly died five months before Mallory's address).

The climber's words are surprising on two further counts. First, his scepticism toward mechanisms of oxygen supply was considerably less extensive than he indicates here, and in fact he goes on to say that 'It will be remembered that Somervell and I when we went up for the third attempt this year intended to use oxygen'.[22] In 1924 Mallory returned to such use. Second, his statement that it is lived experience on the mountain's highest reaches, the testimony of the mountain elite who have actually breathed Everest's 'purer air', that takes precedence over laboratory experiments, anticipates the trajectory of respiratory physiology across the twentieth and twenty-first centuries. Mallory states that 'the idea of the man who has tried as to how much higher he might go should be of incomparably more value than any conclusion proceeding merely from a laboratory'.[23] The contemporary historian of science Vanessa Heggie states that respiration at altitude is 'an area of research where the use of laboratory-generated standard models produced deeply misleading conclusions about the embodied human experience of the natural world'.[24] As a result, Everest in the first half of the twentieth century becomes both field site and a kind of 'natural laboratory'[25]—one which provided, via experience and sacrifice, the *only* answers to the maintenance of human breath at altitude, just as Mallory had, rather theatrically, claimed in 1922. Thus an acute anticipatory observation about the location of scientific progress—'on the hill', as climbing parlance has it—is disguised in a pose of apparent scepticism toward sciences of the breath more generally.

Mallory had numerous speaking engagements upon his return to the UK following the 1922 expedition, of which the speech considered above (later published in the *Alpine Journal* under the more sober title of 'The Second Mount Everest Expedition') was perhaps the most important with a view to logistical planning for later attempts. He also undertook a lecture tour, sometimes speaking in partial repayment of sponsorship for the notoriously expensive 1922 trip, and sometimes to gather much-needed funds for the next. Yet he also troubled himself to make school visits to deliver inspiring and improving talks to young scholars, usually boys, during which he both recounted the adventures of his two Everest journeys, and gave some sense of the type of character (more properly, the type of masculinity) necessary for success in ambitious expeditions, or in the wider life for which they might be seen to provide an exemplar. One such visit was paid to Gresham's School, where Mallory gave an assembly address before an audience of boys we can reasonably presume to have included the young Wystan Auden. No records are held by the school regarding who was in attendance for the mountaineer's presentation that day, but the appearance of the poem 'Everest', written in late 1922, among the poet's juvenilia must surely confirm that Mallory had come into his ken:

> Far up into the amethystine vapours
> Towers the ridge in white immensity

Gazing at the stars that burn like tapers
  Deep in the sky that is Eternity

[…]

You see the fall of the Gods of yesterday
  And the fall of the Gods of the morrow
Yet never a *sigh* or regret you say
  For the infinite ocean of *sorrow* […][26]

The poem somewhat embarrasses the tentative Auden in cleaving too visibly to Romantic influences, and this is conspicuously verse written before the young poet finds his own voice (Auden scholars generally pinning that maturation to 'The crux left of the watershed', and therefore to the scarred landscapes of post-industrial Britain rather than to Himalayan 'immensit[ies]').[27] 'Everest' is a poem written in 1922, a high watermark year of 'high' or experimental modernism from which it is distant, but also the year marking the true opening of the Everest era with the first concerted attempt to ascend. It forms a useful point of contrast, therefore, with 1954's 'Mountains', written when Everest has recently been climbed, photographs taken and splashed across newspapers, and the heights are no longer a mystery, having been surveyed not only by distant instruments but, at last, by human eyes. At first, the earlier poem seems a more landscape-focussed work, with human struggle and point of view removed, although the 'towering ridge' does suggest the close-up view of the climber (or how else are we to see past the 'vapours'?), and 'sigh[s]' and 'sorrows' are on the page, if not attributed to a climbing figure: all assisting in the registration of effort involved in what is elsewhere in the poem referred to as the 'vain' effort to reach the 'head' of the mountain. Despite these hints of the laboured breath, the work is retrospective in its poetic model, for all that Mallory, the very model of the modern mountaineer, was its likely catalyst. In fact, it might be Mallory himself who permits Auden's at this stage customary turn toward Romantic formulations, since the climber is himself enamoured of the work of Shelley, writing to Eleanor 'Marjorie' Holmes in 1924: 'Do you know Shelley? One of the greatest spirits that have appeared on earth & a man of such moral beauty that I feel dazzled in his presence—I can't tell you how profound a feeling I have for Shelley; he has influenced my life more than any one'.[28] One possible conduit of this influence is Robert Bridges, Poet

Laureate 1913–1930, whose anthology *The Spirit of Man* was taken on the British Everest expeditions, and contains forty-three works or extracts by the poet.[29] Katherine Bucknell, annotating 'Everest' in Auden's *Juvenilia*, suggests that the poem 'apparently borrows its heightened tone and its grandiose imagery of absolutes from Shelley's "Mont Blanc"', although she does not give Mallory, or Bridges, as Auden's excuse for so conspicuous an influence.[30] By 1954, Auden is able, as an older man, and a mature poetic voice, to identify the 'crank' aspect in the breed of climbers, and to establish an ironic distance from this noble quest to the heights, a distance that seems to interpose leaden poetic feet in order to register the sensory and somatic distancing of the climber isolated behind the gear of his oxygen rig.

In the shift from 1922 to 1954, Auden in fact allows us to track the movement of mystery from the mountain itself, to the tightly knit band of the mountaineers' rope line or climbing party. One reason for this shift is that the mountain, now 'conquered' (as the newspapers of 1953 had it), had apparently relinquished its mystery. Yet we could also argue that the mountaineer had become more mysterious, less recognisable, in literal and metaphorical terms (that is, both facially obscured and less familiar), by the application of an oxygen rig and breathing mask. In the mountaineering literature, it is the obscuration of the face that proves a flashpoint for debates about the necessity or even seemliness of using additional oxygen. Everest mountaineer and author Graham Hoyland, writing in 2013, states that a 'flinching away from the use of supplementary oxygen' is discernible among climbers of Mallory's era, a refusal that he attributes to experience in the Great War: 'Poison gas during the [W]ar had been regarded with particular abhorrence as a cowardly form of warfare. Gas masks were dehumanising in appearance and were a horrific reminder to this group of men who had been so scarred by the [W]ar'.[31] This explanation complicates the more familiar reading to which Mallory's confrontation with Sir Walter seems to contribute, in which the 'natural' capacities of the male body allow for a more sportsmanlike attempt upon a peak, while also valorising embodied knowledge. In fact, it is in the matter of the *experienced* body, the notion of corporeal witnessing, that mountaineer testimony in Heggie's 'natural laboratory' of Everest and the matter of war service might be seen to overlap. While the Alpine Club and RGS's 'old guard', in charge of the plans and the purse strings, were for the most part the wrong generation to have seen active service in the Great War, those around Mallory's age and younger had known that conflict. While Hoyland is right to read the supply of oxygen as visually corresponding to the prevention of access of noxious gases, the young guns in the Everest climbing party might in fact have been more tempted to use all means necessary to assist physical survival, lest they recreate on the slopes of the mountain General Haig's slow adaptation to the machine gun on the Western Front, with ghastly results.

Wade Davis, the historian who has done most to link Great War experience and the exploits of the Everest era, concurs with aspects of Hoyland's claims, stating that to use supplementary oxygen at altitude was not so much a technical challenge as an 'ethical and aesthetic' one.[32] 'The very notion of using a gas mask was haunting', he writes. 'To climb the mountain with one's face covered in anonymity by an apparatus so powerfully evocative of the trenches seems less than heroic'.[33] The model of the 'hero' is unstable here, with the oxygen/gas mask anonymising the individual, and therefore placing him among a mass; a loss of distinctive identity that damages heroic status, even as it puts the mountaineer in the company of those whose lives were bravely sacrificed as 'cannon fodder'. Davis's focus on the notion of heroism and mountaineering's relationship with Great War service skews his reading somewhat, since broader questions from the 1920s onwards regarding the frightening physical incorporation of the machinic in cyborg figures—questions brought about by newly technologised working practices—must also be in play.[34] Yet the notion of 'facing' the mountain, making an individual, individuated, attempt to ascend, without the interposition of an apparatus between the face of human and of rock (or ice, or snow), remains strong in these years. To fail to 'face up' to the challenge, to use a get-up for getting up, could rig you up for laughter. Arthur Hinks, astronomer, academic cartographer, and joint honorary secretary to the Mount Everest Committee in the 1920s, writes in a letter to General Charles Bruce, leader of the 1922 and 1924 expeditions:

> This afternoon we go to see a gas drill. They have contrived a most wonderful apparatus, which will make you die laughing. Pray see that a picture of [George Ingle] Finch in his patent climbing outfit with the gas apparatus is taken by the official photographer. I would gladly put a little money on Mallory to go to 25,000 ft without assistance of four cylinders and a mask.[35]

Gifted chemist Finch's contribution to the development of the 'wonderful apparatus', and to high-altitude respiratory physiology more broadly, is highly significant, and what seems here to be an attempt to depict him as a 'crank' or pottering inventor might well be prejudice on grounds of nation and class, since Hinks was to go on to refuse Finch participation in the 1924 expedition, ostensibly as a punishment for his making personal gains from lecturing, but more probably to prevent an Australian being part of the otherwise British climbing party.[36] Hinks, more than others, deploys the use of supplements to the breath, or rather their rejection, to indicate the 'right sort' of man for the job, far beyond the question of lung capacity. Mallory's humorously outraged lecture performance at the RGS should be read in this light.

Even Finch's fellow expedition participants in 1922, men themselves bound as a strange 'Order' in Auden's terms, found the scientist a curiosity in his interests, for all that those interests aimed only to shore up a successful ascent. Surgeon-mountaineer Somervell was the author of a ditty, 'When George Finch Starts to Gas', whose relentless anapestic tetrameter, AABB rhyme scheme, and repetition of 'oxygen drill' both echo the disappointingly repetitive nature of the drill itself, and suggest that it might have been sung at the beleaguered Finch as a kind of ungainly ballad. '[T]o gas' here serves double duty, referring to the taking of oxygen, and to 'gassing', or relentless chatter—Finch's enthusiastic attempts to explain his methods have obviously been received as a barrage of ill-understood jargon:

> [...]
> Have you theories precise on the subject of gas?
> Respiration, and so on, and action in mass?
> The exactest of thought will appear rude and boorish
> Compared to the latest in science from Zurich.
> Do you think that you know about altitudes high
> And what kind of glass keeps the sun from your eye?
> On such questions your ignorance really is crass
> But you'll soon be made wise when George Finch starts to gas.
> So put down your books, come along learn the knack
> Of hoisting the cylinders on to your back.
> For if you'd be the victor of Everest's hill
> You must finish each morning with Oxygen Drill.[37]

The phrase 'Do you think that you know...' reiterates the attitudes discussed in this essay, where field experience is apparently superseded by ‘the latest in science’, *i.e.*, laboratory findings; a second-placing of embodied experience resisted by many an early twentieth-century mountaineer. Somervell, whose composition is signed ‘a mon ami [to my friend] G.F.’, deploys a hint of the ‘crank’ persona to convey Finch’s cherished scientific interests and his cleaving to ‘drill’, but the latter man was in fact more circumspect about what could or could not be achieved by developments in the delivery of supplemental oxygen at altitude, writing: ‘Oxygen renders available [to the climber] more of his store of energy and hastens his steps, but it does not, alas! fit the wings of Mercury on his feet’.[38] Reattributing climbing capacity to the climber himself (‘his store of energy’), Finch figures the oxygen rig as a means of access—to one’s reserves of power, and therefore to the summit—in a way that in fact recalls Mallory’s 1922 imaginings regarding the optimal placement of kit stores and camps. In both of these accounts of the oxygen apparatus, it is curious to note the presence of other substances beyond oxygen itself, with laughing gas implied by Hinks’s somewhat cruel caricature (and we might note that gas’s nitrous oxide is an oxidiser, therefore similar in function to molecular oxygen, and sharing a metallic taste with altitude—what mountain writer Nan Shepherd in 1945 called ‘the tang of height’),[39] while Mercury, primarily indicating a Roman god and thence the fleet of foot, also names a poison. Both Bruce and Finch’s statements suggest that, start breathing things in—break the continence of the manly body under the duress of ascension—and anything might happen.

There is no doubt that mountaineering at the Third Pole brought breath centre-stage, albeit that the history of supplementary oxygen was dogged not only by the imperfect nature of laboratory testing and the limited availability of the somatic testimony of high-altitude mountaineers, but also by attitudinal worries about sportsmanship and fair dealing, anonymity, and the Great War’s trenches dragged to the heights. More contemporary studies have confirmed that oxygen levels in the blood registered in living humans in Everest’s ‘death zone’ can be lower than those found in corpses.[40] Often set at 5–6,000m-plus, but actually indicating any height above which human habitation cannot be sustained, this significant zone is hinted at in Virginia Woolf’s final short story ‘The Symbol’, where Alpine climbers venture ‘too high for breathing flesh or fur-covered life’.[41] In this way, Everest mountaineers, always haunted by the ‘saints’ and ‘angels’ of past climbers (in the parlance), might also be considered, in terms of the breath, the walking (or climbing) dead.

On the Everest expeditions, whether in the 1920s or the 1950s, negotiations with the breath were necessary, maintaining the impression of 'going well' being crucial to selection as a member of the climbing team for the final push to the summit (both Mallory and Hillary recall such jockeying for position, while Auden's steady Ian Shawcross sulks about his chances).[42] Tangling with the necessity of supplemental oxygen was also important, both in terms of available strength when climbing (Hillary's oxygen rig weighed 32 pounds or around 14.5 kilograms when fitted with all four canisters),[43] and with regard to your post-climb and posthumous renown. Reputational matters shift, of course, with subsequent generations of climbers, while training regimens, kit, and logistics are ever more refined, and in 1978 it was shown, at long last, that Everest could be climbed without oxygen. In his account of the trip, successful summiteer Reinhold Messner, in a line that ties together mountaineering, poetry, and breath with beauty and brevity, states: 'In my state of spiritual abstraction, I no longer belong to myself and to my eyesight. I am nothing more than a single narrow gasping lung, floating over the mists and the summits'.[44] Yet as Auden's 'Mountains' might be read to show, the presence of an oxygen rig did not preclude such abstraction. What if (with Mallory's imagined supply lines in place) there was a breath that could never run out, and therefore your mortality in the 'death zone' could be baffled? The status of supernatural, or rather supranatural, being, in an already haunted landscape of the dead, the saints and angels, falls within your grasp, or gasp. In this way, while in the early twentieth century the oxygen rig is a symbol of human ingenuity and the cutting edge of scientific development in the mountain-laboratory, it might at the same time be seen not to escape, but rather to extend, a sense of the spiritual and the otherworldly. To breathe 'up there' is to face the mountain, albeit with an apparatus interposed between your face and that of rock and ice, but it is also perhaps to become more-than-man, in a way that ties you, via your mysterious gear, to the brotherhood, or strange 'Order', of the rope.

Notes

1. For an account of the 'Everest era' and its place within longer mountain histories, see Peter H. Hansen, *The Summits of Modern Man: Mountaineering After the Enlightenment* (Cambridge, MA: Harvard University Press, 2013).

2. W. H. Auden, *Bucolics III: Mountains*, illus. by Edward Bawden (London: Faber, 1954), u.p.
3. Basil Bunting, 'On the Flyleaf of Pound's *Cantos*', in *The Poems of Basil Bunting*, ed. by Don Share (London: Faber, 2016), 37.
4. For a complaint regarding the limited function of the newly short-shafted ice axe of the early twentieth century, see Harold Raeburn, *Mountaineering Art* (New York: Frederick A. Stokes Company, 1920), 22.
5. Cecil Day-Lewis, 'Johnny Head-in-Air', in *A Time to Dance and Other Poems* (London: Hogarth Press, 1935), 19–24. Vanessa Heggie cautions that 'there is a strong disanalogy between the pilot and the mountaineer', given the role of exhaustion and the ascent profile in the latter case. See Heggie, 'Experimental Physiology, Everest and Oxygen: From the Ghastly Kitchens to the Gasping Lung', *The British Journal for the History of Science* 46/1 (2013), 138.
6. For uses of this phrase in the contemporary Press, see *Collection of Newspaper Cuttings Relating to the First Everest Expedition* (Edinburgh: Mountaineering and Polar Archives, National Library of Scotland), GB/A.3845.
7. On this Order see Robin Bruce Lockhart, *Halfway to Heaven: The Hidden Life of the Carthusians* (Kalamazoo: Cistercian Publications, 1999). The title refers to the common notion of the Alps' height as offering relative proximity to God.
8. W. H. Auden and Christopher Isherwood, *The Ascent of F6: A Tragedy in Two Acts* (London: Faber, 1936). For further discussion of the significance of Ian Shawcross, see Abbie Garrington, 'What Does a Modernist Mountain Mean? Auden and Isherwood's *The Ascent of F6*', *Critical Quarterly* 55/2 (2013), n. 23, 43.
9. Matthew Arnold, 'Stanzas from the Grande Chartreuse', in *The Poetry of Matthew Arnold*, ed. by Kenneth Allott and Miriam Farris Allott (London: Longman, 1979), 301–05.
10. See, for example, Maurice Isserman and Stewart Weaver, *Fallen Giants: A History of Himalayan Mountaineering from the Age of Empire to the Age of Extremes* (New Haven: Yale University Press, 2008), 252–53.
11. Maurice Herzog, *Annapurna: The First Conquest of an 8,000-Metre Peak*, trans. by Nea Morin and Janet Adam Smith (London: Vintage [1952], 2011), 29.
12. For a wide selection of the newspaper reports that would have been available to Auden, see *Collection of Newspaper Cuttings Relating to the First Everest Expedition* (Edinburgh: Mountaineering and Polar Archives, National Library of Scotland), GB/A.3845.
13. Jan Morris, *Coronation Everest* (London: Faber [1958], 2003), 159.

14. The Alps are regularly associated with lung health, of course, due to the prevalence of sanatoria there. It is Thomas Mann's *The Magic Mountain*, published in German in 1924, and first available in English in 1927, that gives the most extended literary account of such a treatment centre. Mann's wife's unspecified 'lung complaint' saw her live at Dr Friedrich Jessen's *Waldsanatorium* in Davos, Switzerland, and the 'Arrival' chapter of Mann's book is generally taken to be based on this experience. If the Alps are a space of lung management and treatment, the Himalayas are their space of greatest duress, with the possible exception of high-altitude flying. See Thomas Mann, *The Magic Mountain*, trans. by H. T. Lowe-Porter (London: Vintage, [1927] 1999). For a recent account of the history of mountain sanatoria in this period, see Mariano Martini, et al., 'The History of Tuberculosis: The Social Role of Sanatoria for the Treatment of Tuberculosis in Italy Between the End of the 19th Century and the Middle of the 20th', *Journal of Preventive Medicine and Hygiene* 59/4 (2018), 323–27.
15. For a sparsely written but moving eyewitness account of the disappearance of Mallory and Irvine, 'enveloped in cloud once more', see Noel Odell, 'Mr. Odell's Story', in 'The Mount Everest Dispatches', *The Geographical Journal* 64/2 (1924), 163–65. See also Noel Odell, 'The Last Climb of Mallory and Irvine', *The Geographical Journal* 64/6 (1924), 455–61.
16. Lytton Strachey, 'Letter to Clive and Vanessa Bell, 21 May 1909', in *The Letters of Lytton Strachey*, ed. by Paul Levy (London: Viking, 2006), 178–79.
17. Peter Gillman and Leni Gillman, *The Wildest Dream: The Biography of George Mallory* (Seattle: Mountaineers Books, 2001), 98.
18. George Mallory, 'Purer Air than Mortals: Mount Everest, 1922', in *Climbing Everest: The Complete Writings of George Mallory*, ed. by Peter Gillman (London: Gibson Square, 2010), 205.
19. Mallory, 'Purer Air than Morals', 213.
20. For a summary of the claims of Everest mountaineers regarding the physical adaptation of Sherpas to high altitude environments, see Sherry B. Ortner, *Life and Death on Mount Everest: Sherpas and Himalayan Mountaineering* (Princeton: Princeton University Press, 1999), 63–67.
21. Mallory, 'Purer Air than Mortals', 217–18.
22. Mallory, 'Purer Air than Mortals', 218.
23. Mallory, 'Purer Air than Mortals', 218.
24. Heggie, 'Experimental physiology', 126.
25. Heggie, 'Experimental physiology', 123.
26. W. H. Auden, 'Everest', in *Juvenilia: Poems 1922–1928*, ed. by Katherine Bucknell (Princeton: Princeton University Press, 2003), 9–10.
27. See, for example, Edward Mendelson, *Early Auden* (London: Faber, 1981), 27.

28. George Mallory, 'Letter to Marjorie Holmes, 8 March 1924', Royal Geographical Society (with IBG), LMS/M48.
29. For more on Shelley's inclusion in *The Spirit of Man*, see Abbie Garrington, 'What does a Modernist Mountain Mean?', 39.
30. Katherine Bucknell, note to W. H. Auden, 'Everest', in *Juvenilia: Poems 1922–1928*, 10.
31. Graham Hoyland, *Last Hours on Everest: The Gripping Story of Mallory and Irvine's Fatal Ascent* (London: Collins, 2013), 61. It was Hoyland who discovered the frozen body of George Mallory, high on Everest's slopes, in 1999.
32. Wade Davis, *Into the Silence: The Great War, Mallory, and the Conquest of Everest* (London: The Bodley Head, 2011), 385.
33. Davis, *Into the Silence*, 385.
34. The best broad rehearsal of these debates is to be found in Tim Armstrong, *Modernism, Technology and the Body: A Cultural Study* (Cambridge: Cambridge University Press, 2008).
35. Cited by Davis, *Into the Silence*, 387.
36. For a summary of this clash between Finch and Hinks, see Robert Wainwright, *The Remarkable Life of George Ingle Finch: Climber, Scientist, Inventor* (London: Allen & Unwin, 2016). The oxygen rigs of the 1924 expedition were left in the care of 21-year-old Sandy Irvine, a fellow chemist but, given his youth, of necessarily limited experience. Sir Edmund Hillary discusses his own use of an updated version of Finch's oxygen rig on the successful 1953 Everest expedition. See Edmund Hillary, *View from the Summit* (London: Corgi, 2000), 61–62.
37. Cited by Wainwright, *The Remarkable Life*, 206–07.
38. Davis, *Into the Silence*, 457.
39. Nan Shepherd, *The Living Mountain: A Celebration of the Cairngorm Mountains of Scotland* (Edinburgh: Canongate, [1945] 2011), 9.
40. See 'Altitude Deterioration', in *High Altitude Medicine and Physiology*, ed. James S. Milledge, John B. West, and Robert B. Schoene (Boca Raton: CRC Press, 2007), 47.
41. Virginia Woolf, 'The Symbol' (1941?), in *A Haunted House: The Complete Shorter Fiction*, ed. by Susan Dick (London: Vintage 2003), 282–84.
42. Act II Sc. I, Auden and Isherwood, *The Ascent of F6*, 49–50.
43. Hillary, *View from the Summit*, 62. For a detailed account of 'Open-Circuit' and 'Closed-Circuit' oxygen rigs, and the management of oxygen in the 1953 expedition, see John Hunt, *The Ascent of Everest* (London: Hodder and Stoughton, 1953), 257–59.
44. Reinhold Messner, *Everest: Expedition to the Ultimate*, trans. by Audrey Salkeld (London: Kaye and Ward, 1979), 180.

Selected Bibliography

Collection of Newspaper Cuttings Relating to the First Everest Expedition. Edinburgh: Mountaineering and Polar Archives, National Library of Scotland. GB/A.3845.

Davis, Wade. 2011. *Into the Silence: The Great War, Mallory, and the Conquest of Everest*. London: The Bodley Head.

Gillman, Peter and Leni Gillman. 2001. *The Wildest Dream: The Biography of George Mallory*. Seattle: Mountaineers Books.

Hansen, Peter H. 2013. *The Summits of Modern Man: Mountaineering After the Enlightenment*. Cambridge, MA: Harvard University Press.

Heggie, Vanessa. 2013. Experimental Physiology, Everest and Oxygen: From the Ghastly Kitchens to the Gasping Lung. *The British Journal for the History of Science*, 46/1: 123–47.

Isserman, Maurice and Stewart Weaver. 2008. *Fallen Giants: A History of Himalayan Mountaineering from the Age of Empire to the Age of Extremes*. New Haven: Yale University Press.

Mallory, George. 2010. Purer Air than Mortals: Mount Everest, 1922. In *Climbing Everest: The Complete Writings of George Mallory*, ed. Peter Gillman, 205–21. London: Gibson Square.

Odell, Noel. 1924. The Last Climb of Mallory and Irvine. *The Geographical Journal* 64/6: 455–61.

Ortner, Sherry B. 1999. *Life and Death on Mount Everest: Sherpas and Himalayan Mountaineering*. Princeton: Princeton University Press.

Wainwright, Robert. 2016. *The Maverick Mountaineer: The Remarkable Life of George Ingle Finch: Climber, Scientist, Inventor*. London: Allen & Unwin.

CHAPTER 20

Hearing the Form: Breath and the Structures of Poetry in Charles Olson and Paul Celan

David Fuller

A group of mid-twentieth-century American poets, of whom Charles Olson is the most prominent theorist and William Carlos Williams the best-known poet, experimented with new forms for poetry with a declared aim of structuring poems in new ways. Olson particularly experimented with structuring by rhythms related not to the pulse (which is one way of understanding the relatively regular beat of European poetry) but to the breath—a different kind of measure, still bodily, but more fluid. One aim of Olson and his associates was a new freedom with form by which they could be both distinctively modern and distinctively American. For readers acculturated to the rhythms of European poetry there can be problems about hearing these forms, especially insofar as they are involved with American patterns of speech.[1]

To hear this poetry a reader needs to consider how these poets theorised about writing, listen to what they did in actual reading, and compare their theories of composition and practices of performance with poems

D. Fuller (✉)
University of Durham, Durham, UK

D. Fuller et al. (eds.), *The Life of Breath in Literature, Culture and Medicine*, Palgrave Studies in Literature, Science and Medicine,
https://doi.org/10.1007/978-3-030-74443-4_20

as they appear on the page. While some areas of what was new in mid-twentieth-century American poetry were taken up in Britain—Marianne Moore, Robert Lowell, John Berryman, Elizabeth Bishop—the Olson group remained on the fringes. Outside a coterie of poets (most prominently J. H. Prynne and his associates), English ears have mostly not learned to hear it.[2]

Breath is central to the issue of these new structures. The foundational manifesto is Olson's *Projective Verse* (1950).

> Verse now, 1950, if it is to go ahead, if it is to be of *essential* use, must, I take it, catch up and put into itself certain laws and possibilities of the breath, of the breathing of the man who writes as well as of his listenings.
> ...
> If I hammer, if I recall in, and keep calling in, the breath, the breathing as distinguished from the hearing, it is for cause, it is to insist upon a part that breath plays in verse which has not (due, I think, to the smothering of the power of the line by too set a concept of foot) has not been sufficiently observed or practiced, but which has to be if verse is to advance to its proper force and place in the day, now, and ahead. I take it that PROJECTIVE VERSE teaches, is, this lesson, that that verse will only do in which a poet manages to register both the acquisitions of his ear *and* the pressure of his breath. ...
> Together ... the syllable *and* the line, they make a poem, they make that thing, the—what shall we call it, the Boss of all, the 'Single Intelligence'. And the line comes (I swear it) from the breath, from the breathing of the man who writes, at the moment that he writes, and thus is, it is here that, the daily work, the WORK, gets in, for only he, the man who writes, can declare, at every moment, the line its metric and its ending—where its breathing, shall come to, termination. ...
> Let me put it baldly. The two halves are:
> the HEAD, by way of the EAR, to the SYLLABLE
> the HEART, by way of the BREATH, to the LINE. ...
> I say a projective poet will [go], down through the workings of his own throat to that place where breath comes from, where breath has its beginnings, where drama has to come from, where, the coincidence is, all act springs.[3]

Olson rejects ideas of verse rhythm recognised by millennia of European poetry—the idea of patterns of grouped stresses in the poetic line, variable, but with an underlying regularity. He replaces this with a fundamentally different and more variable structure, not grouped by stresses, but understood by the idea of each line as an out-breathing. There is no reference to pre-existing conventions of poetic structure: 'only he … who writes, can declare, at every moment, the line its metric and its ending'. Just two structural elements, the syllable and the line, make the poem's form a comprehensive address to the head and the heart, the intelligence and the feelings. This idea of forms that engage the whole person stresses the entire physiology of breathing—through the throat, down to where breathing begins, in the diaphragm, and the network of muscles and tendons that support the whole breathing apparatus.

Olson does not only describe ideas about expression: he enacts them. His prose style rejects the norms of written discourse in favour of imitating speech—repetitions for emphasis, piled up clauses, colloquial diction, extended turning aside into parenthetical clauses, slides and slithers from one syntactic construction into another. Flagrantly violating norms of 'correct' written style, Olson enjoins the use of, and he uses, phrasings by the breath. In his prose as in his poetry, the reader who realises the page as implied aural structures hears Olson breathe.

Two issues that Olson refers to but does not develop in *Projective Verse* became important for his group. Simplest is an implication of the then relatively unexplored technology of the typewriter. Olson's interest in orality and the typewriter is not a paradox: the machine's precision with visual layout could give, he thought, a more exact indication of how the structures of a poem are meant to be sounded. Gaps are pauses; spacing is a kind of musical 'scoring'. The typewriter is technology for the new orality. Though Thomas Edison had recorded poets (Browning, Tennyson, Whitman), and though some poets (including Olson's 'master', Ezra Pound) had recorded extensively before 1950, and though recording shows much more about the aural qualities of a poem than the typewriter, Olson had nothing to say about it.

Also significant is that 'breath allows *all* the speech-force of language back in' (244). This is not only about the individual: it is about reclaiming for poetry the energies of the language as spoken by the whole community, not the more cerebral discourse norms of an educated elite. In this Olson may sound in part like a succession of programmes for change in English poetry, but his meaning is more radical: by '*all* the speech-force of language' he also means that 'the conventions which logic has forced on syntax must be broken open' (244).[4]

That is Olson's programme: the breath, and through the breath the whole body; the spoken language and its syntax; new technology to notate more precisely in print how the poem is to be realised as sound. This is not a programme purely about the aesthetics of poetry. Olson's manifesto is not confined to issues of poetic form. From the beginning Olson aims to 'suggest a few ideas about what stance toward reality brings such verse into being' (239). This is not just a new aesthetic: it is about new modes of perception. As Olson puts it in the shorter and sketchier Part II of the manifesto: 'If the beginning and the end is breath, voice in its largest sense, then the material of verse shifts. It has to' (247).

In considering that shift Olson cites as exemplars Homer (oral epic) and three dramatists—poets writing not for print but for the voice. He posits a fundamental distinction between those who live in and compose for oral cultures—Homer, Euripides, Shakespeare, the Japanese playwright Zeami Motokiyo—and those who live in, think within the norms of, and write for print cultures. Olson's claim is that writers for print work from a lyric ego in which the human is experienced as separate from the rest of nature; their subject is the bounded ego, and their methods are learned from the traditions of bounded egos. Writers for the voice work from a physiology in which the human is experienced as part of nature; their subject is the whole of nature of which they are a part, and their methods are congruent with that rootedness in nature. Whatever mixture there may be in this of truth and enabling myth, understanding its possibilities requires a new approach to reading poetry—thinking about poetry in relation to the body, the breath, and the ear; reading with the eye only insofar as the eye assists and directs the ear with finding the sound; using the notation on the page as a key to realising all the sounds of a poem in the voice and to the hearing.

Olson's aim was to create new work for what he regarded as a new phase of civilisation, the world as it was being reconstructed after the cataclysm of World War II. The use of the atomic bomb at the end of the war indicated that real apocalypse was a possibility. Humanity could now destroy itself. Immediate post-war changes added to this the communist revolution in China; the huge extension of Soviet Russia's domination in Eastern Europe; and the beginning of the end of European empires in Asia and Africa: 'now, 1950' really seemed a new world. Olson also emphasised the new intellectual position of the age of relativity in physics and the intellectual and moral relativities that could be seen as following from that. Olson and his associates were searching for new modes of poetry adequate to the conditions and knowledge of this new world. While the focus of their experiments was aesthetic and small-scale—the structure of the poetic line—the ultimate contexts—social, political, and intellectual—were worldwide.

Olson enacted his theories of the breath-line literally. Regardless of the flow of sense and structure of syntax, as his recorded readings show, he usually marked the structure of the line with a break at the end. Visual recordings also show how he read using through the voice the expressivity of the whole body.

'The Kingfishers' (1949) shows the method in action.[5] The fundamental issue of the poem is that of Olson's civilization: change—changes of perspective, which mean that meanings once central are lost or transformed. This is typified by the kingfishers—their real existence observed scientifically; the different perspective of myths and legends about them; and the further perspective of a vanished culture in which their beauty was so valued that their feathers operated as currency. Shifts of perspective also mean different valuations of other phenomena—ancient civilizations of South America, seen in terms of their religion of human sacrifice or their art in gold work; the mysterious 'e'-stone of the oracle at Delphi, the range of meanings of which was already conjectural in ancient times. The issue is announced by a paradox of Heraclitus, 'change alone is unchanging', or in Olson's version:

> What does not change / is the will to change

Like the kingfishers, like other aspects of nature and culture, the human perceiver is also subject to perpetual change. Again Olson quotes Heraclitus: 'Into the same river no man steps twice' (perpetual change in nature); 'No one remains, nor is, one' (perpetual change in the human).

The mode of poetry that embodies this universal fluidity of being, culture, and perspective is the antithesis of the static, finished artwork emblematised by the New Critical image of a poem as a 'well-wrought urn'. It must embody the never settled, ever-shifting, perpetually in process. The breath-structured line is its vehicle. What Olson aimed for with this he was beginning to theorise: a mode by which the fixities of print imply the fluidities of orality. He is a performance poet for whom publication on film is a near-ideal medium. His manner of handling language on the page—the incomplete syntax suggesting speech; the breath-punctuating line structures emphasising performance in the living moment—is YouTube for the 1950s. Despite Olson's statements, it is not always the precise notation that is important so much as the mode of performance and orality suggested by his structuring of the printed layout. The breath's structuration of the line is less significant than the implied presence of the living breath in every aspect of performance.

Having begun with a paradox about change, and developed the idea through various epitomes of change, fluidity, shifting viewpoint, and perspectival interpretation in open-ended conjunction, 'The Kingfishers' ends paradoxically, with a question juxtaposed with an ambiguous affirmation.

> Shall you uncover honey / where maggots are?
>   I hunt among stones

Are the stones a source of wisdom? If they are, can they yield their wisdom to the seeker now? Or are they only a source of meanings attributed by the perspective of the interpreter rather than elicited from the object? The central stone in the poem, the stone of the Delphic oracle, epitomises these doubts. And what is the meaning of juxtaposing this ambiguous affirmation with its apparent opposite?

>   if I have any taste
> it is only because I have interested myself
> in what was slain in the sun

The poem closes with contrasting searches, of stones and maggots: mineral-immutable *and* animal-ephemeral. These searches take place in a new world the perspectives of which are undergoing profound changes. The methods of poetry's search for meanings must change with these shifts of perspective. As the poem puts it, 'When the attentions change' then the object of attention 'sounded otherwise / was differently heard' ('sounded … heard' maintains the doubt about elicited and attributed meanings). What Olson adds to Ezra Pound's Modernist re-orientations goes beyond what Pound identified as foundational: 'to break the pentameter, that was the first heave' (Canto LXXXI). Olson aims to abolish the ghost of *any* recognised meter. He proposes a new indeterminate structural fluidity of the line based on 'the breathing of the man who writes, in the moment that he writes'. Written while Olson was formulating *Projective Verse*, 'The Kingfishers' shows both the method of the breath-structured line under development and its underlying *raison d'être*.

Among the most significant contemporary responses to Olson was that of William Carlos Williams. In his *Autobiography* Williams quoted *Projective Verse* with this introduction: 'Until we have reorganized the basis of our thinking in any category we cannot understand our errors. An advance of estimable proportions is made by looking at the poem as a field rather than as an assembly of more or less ankylosed lines' ('ankylosed'—of two bones: grown together so firmly as no longer to move upon each other; referring to what Williams sees as the inflexibility of traditional metrical structures).[6] The advance to which he refers he described as 'well illustrated by Charles Olson in the following' (a quotation from *Projective Verse* running over four pages). This is often seen as straightforward endorsement of Olson, but the presentation is ambivalent. Williams counterpoints his extended quotations from Olson with an account of the house of the American painter Charles Sheeler and his Russian wife, Musya Sokolova. This house for Williams emblematised a 'poem', extending the idea to include all imaginative creations, and describing 'the reconstruction of the poem as one of the major preoccupations of the intelligence in our day' (332). Built in the American Colonial era, modelled on British and Dutch architecture, the house incorporated Sheeler's Welsh family background, the American Shakers whose furniture he collected, and Sokolova's ability 'to transfer an understanding from an aristocratic past … to this context' (334). For Williams it was an architectural 'poem' that brought into contemporary unity American and European elements. He set this beside selections from *Projective Verse* without comment, but the juxtaposition implies something other than straightforward endorsement, characteristic of Williams' manifestos, which are more tentative than Olson's polemic, more openly searching for what he professedly cannot fully articulate.[7]

And with Williams more than with Olson there is a disjunction between theory and practice. This can be considered through a reading Williams gave at Harvard 1951, where he began by offering some fundamental orientation.[8]

> All art is sensual. Listen. Never mind. Don't try to work it out. Listen to it. Let it come to you. Let it … Sit back, relax, and let the thing spray in your face. Get the feeling of it. Get the tactile sense of something … something going on. It may be that you will then perceive … have a sensation that you may later find will clarify itself as you go along. So that I say don't attempt to understand the modern poem: listen to it. And it should be

> heard. It's very difficult sometimes to get it off the page, but once you hear it, then you should be able to appraise it. In other words, if it ain't a pleasure, it ain't a poem.

Hear the poem; enjoy whatever you receive; do not puzzle about meaning—advice that can be paralleled from any number of twentieth-century poets. Williams is responding to a typical misleading way of addressing difficulty in modern poetry: the supposition that the reader needs to know something beyond the poem, whereas what the reader really needs is to dwell receptively on what the poem itself gives, and not see it but hear it. Williams acknowledges, however, that making the printed page reveal an aural structure can be difficult. His own readings suggest that the poem as printed can be positively misleading about the structure the poet imagined—as can be heard in his performance in this same reading of 'The Yachts'.[9]

'The Yachts' has features of the European poetry that in theory Williams was aiming to do without. It is written in a pattern that looks like *terza rima*—the form of the *Divina Commedia*, with some traditional patterning of verbal music and a sort of 'poetic diction' (words definitely not from American demotic). A marked feature is a change of direction in the final stanzas. The opening admires the yachts: they are beautiful; in the protected environment in which they exist, 'they appear youthful, rare // as the light of a happy eye'; though they need to negotiate difficulties, when 'the waves strike at them ... they are too / well made, they slip through'. But finally the poem prompts the reader to see the yachts as symbols of their rich owners as they operate in society: from one perspective beautiful, within limits superbly equipped for resisting natural disaster; but from another perspective—when 'the horror of the race dawns staggering the mind'—causing suffering to which they are culpably indifferent. In the Harvard reading, Williams explains that the poem was written during the Depression of the 1930s, and comments simply, 'At that time there was a good deal of misery, and that is why the poem came to be written'. Elsewhere he is more explicit about a political meaning: the last stanzas show 'the real situation (of the poor) [which] is desperate'.[10]

In giving the poem aural shape Williams is the converse of Olson: except insofar as formal breaks coincide with syntactic breaks, he ignores the printed form and phrases for the syntax. The visual form of the poem on the page could not be reconstructed from Williams' reading. In reading for the syntax Williams observes one form of the poem's breath structure, but not that notated by the poem's printed form, which follows loosely a structure from European tradition. Also notable is the change of pace and intensity with which he conveys the drama and pathos of the final stanzas. Nothing the typewriter could do with form on the page could convey this. The actual sound of the poet's voice is needed. Recording, not the typewriter, is the technology that makes audible the soundscape of the poem imagined by the poet.

This issue of some kind of structuration distinct from the forms that had been those of European poetry since Homer, modes capable of expressing profoundly new conditions of existence, has been a recurring preoccupation of modern and contemporary poetry. Olson's formulation of structures based in the breath had an immense effect on American poetry of the second half of the twentieth century.[11] How significant this was as a response to conditions after 1945 is evident in the similar experiments with the idea of a breath structure for the poetic line conducted contemporaneously, but independently, by Paul Celan in German poetry.

Like Olson and his followers, and for reasons that have some contextual similarity, Celan proposed new ways of thinking about the structures of poetry in terms of breath. To understand what Celan proposed, and why, it is necessary to consider the contexts in which he was writing, and specifically his sense of the need for new ways of writing adequate to all that is signified by the Holocaust. The extreme horror of what had happened meant that any treatment in art that presupposed recognisable aesthetic norms constituted a betrayal. Poetry could not be silent, but neither could poetry speak as it had spoken.

Celan's poetry is therefore difficult—fragmentary, oblique, enigmatic, speech on the verge of silence. It demands commitment to reading as an act requiring supreme attention: intellectual effort and imaginative engagement that respects the commitment of the poet. This demand cannot be compromised or mitigated. It is essential to the poetry's expressivity that articulation is pressing against a limit of possibility bound up with every aspect of its struggle to come into being at all. Despite this, Celan's poetry has spoken powerfully to post-War German culture, and not only to German culture but in translation to other parts of Europe and to the English-speaking world.

One way of addressing the poetry's difficulty is by reading it aloud. That is Celan's advice. 'Hör dich ein / mit dem Mund': listen your way in with the mouth ('Die Posaunestelle', 'The Trumpet's part', *Zeitgehöft*, *Homestead of Time*, 1976). In this way the reader discovers what Celan described as the breath-structure of a poem. This endorses the advice of Paul Valéry (one of the many poets Celan translated):

> Verse ... holds an admirable and very delicate balance between the sensual and intellectual forces of language. [When studying a poem] ... do not be in a hurry to reach the meaning. Approach it without forcing and, as it were, imperceptibly. Attain the tenderness and the violence only by the music and through it. ... Remain in this purely musical state until the moment the meaning, having gradually supervened, can no longer mar the musical form. You will gradually introduce it as the supreme nuance which will transfigure your piece without altering it.[12]

This is congruent with the advice of William Carlos Williams: 'don't attempt to understand the modern poem: listen to it'; and not only 'listen', but get the poem off the page into your whole auditory and acoustic being. Hear a poem by speaking it.[13]

Getting to know a poem through the voice is one method. Another, also fully appropriate to Celan, is suggested by Keats: living with a poem so that it evolves into your whole being: 'wander with it, and muse upon it, and reflect from it, and bring home to it, and prophesy upon it, and dream upon it'.[14] 'Wander with', 'muse upon'—free, undirected thinking: 'wander with', remaining in the words; 'muse upon', taking flight with the subject; 'reflect from', movement outwards, beyond the self; 'bring home to', movement inwards, to the self; 'prophesy', the response of conviction; 'dream', the response of meditation. This is not a discipline of criticism: it is a gamut of how to live with a poem and make it your own.

Reading in this way, accepting the freedom of the reader to wander, to muse, letting the mind move outwards into the world, engaging the idiosyncrasies of experience—this accepts that meaning is unfixed. Meaning comes from the poem; meaning comes also from the reader. Celan was interested in images with meanings that exist 'beyond our wakeful thinking', images the light of which is not daylight.[15] Even in his early poetry, where articulation is less oblique, he encourages reading open to the reader's imaginative inflection.

Mit wechselndem Schlüssel schließt du das Haus auf, darin der Schnee des Verschwiegenen treibt. Je nach dem Blut, das dir quillt aus Aug oder Mund oder Ohr, wechselt dein Schlüssel.	With a variable key you unlock the house in which drifts the snow of that left unspoken. Always what key you choose depends on the blood that spurts from your eye or your mouth or your ear.
Wechselt dein Schlüssel, wechselt das Wort, das treiben darf mit den Flocken. Je nach dem Wind, der dich fortstößt, ballt um das Wort sich der Schnee.	You vary the key, you vary the word that is free to drift with the flakes. What snowball will form round the word depends on the wind that rebuffs you.[16]

The house to be unlocked is the poem; the variable key is the reader—with his or her own embodied mindscape (eye, mouth, ear), his or her own context (wind that rebuffs). In Celan's later work the reader's freedom crucially includes freedom in eliciting from the poem, and finding through his or her own discovery of the poem, its expressive structuration by the breath.

With poetry that presses language to the extreme boundaries of expressibility, there are special difficulties for the non-native speaker, especially since Celan so consistently exploits nuances of etymology, semantic implication, and verbal sound, ultimately contributing to what he came to regard as the all-important breath-structure implied by all these elements taken together. No translation can preserve all the elements that contribute to this. Only a reader with a native speaker's fluency can grasp all the implications of Celan's diction and syntax, phrasing, inflection, and intonation. Nevertheless, as a copious translator himself Celan actively endorsed reading poetry in translation, and he has been served by dedicated and brilliant translators.[17] Though some meanings of music and structure will always be lost in translation, major aspects of a poem can be carried over into the new language; and Celan suggests principles of listening to the structures of poetry, hearing the music that is meaning, that are applicable to poetry in any language.

Celan was a Jewish German-speaking Romanian. His parents were deported from Romania in 1942 to camps in German-occupied Ukraine. Shortly after their deportation he became part of a Jewish forced labour battalion. His father died in late 1942, and in late 1942 or early 1943 his mother was shot as unfit for work. There would evidently be problems for any writer of finding modes of articulation adequate to these experiences. For a Jewish writer, writing in German, the problems were acute. German was for Celan the language of murderers.[18]

The most famous of Celan's early articulations about Nazism is 'Todesfuge' (Deathfugue). The poem is uncharacteristic in its amplitude and in its direct treatment of the subject of the camps and the Holocaust. Its repeated lines and parallel phrases, which suggest both the horror of terrors and the obsessions caused by fear and suffering, are quite unlike the pared-down implicatory understatement of Celan's later work. The poem is spoken by a prisoner who reports the actions and words of a guard addressing prisoners. The guard has blue eyes and writes to a woman with blonde hair called Margarete: they are archetypal Aryans. The other woman addressed in the poem has a Hebrew name, Shulamith, the name of the beloved in the Old Testament Song of Songs. Even for somebody who does not read German, hearing Celan's commitment to the poem in his reading is a striking experience.[19] Unlike William Carlos Williams, who supplies in performance the punctuation he eschews on the printed page, Celan reads with the same absence of punctuation as he writes, which is a notable aspect of how the text signifies. The breath structure is not that of normal speech but of somebody too terrified to breathe freely. Read in this way, the poem presents not clearly differentiated voices but a single mind echoing with alien voices so burned into it as to have become inescapable horrifying presences.

The poem is not typical of Celan in its manner, and it has been criticised as too conventionally 'poetic' for its terrible subject—a critique congruent with a famous statement about poetry and the Holocaust by the cultural theorist Theodor Adorno: 'To write poetry after Auschwitz is barbaric. And this corrodes even the knowledge of why it has become impossible to write poetry today'.[20] This has attracted a library of comment. It is a formulation that achieved wide currency more for the glamour of extremity than the humility of truth. It would make sense in a more modest form: 'Poetry must change radically after …', after the Holocaust, as also (one focus of Olson and his associates) after Hiroshima and Nagasaki—because of the knowledge of what has been done, and the knowledge of what can now be done. But 'poetry must change' is a completely different proposition from the contention that to write poetry is barbaric—by which Adorno indicted poets such as Celan who continued to write, and to write specifically in response to the Holocaust. Celan, who wrote partly as a way of keeping some sense of humane activity alive, offered a bitterly ironic comment on this in a letter about a politically correct journal which followed Adorno's line.[21] Nevertheless, under his own internal pressure to find adequate modes of utterance for the all but unutterable, Celan did look for radically new modes of writing. It was in his search for these that he became newly conscious of the role of breath in poetic structure.

He expressed a preliminary idea of the need for fundamental change in response to a Paris bookshop questionnaire in 1958—the need for German poetry to have a new sound, because the old music of poetry could not match up to the knowledge of the present, that is, after the Holocaust.

> No matter how alive its traditions [the traditions of German poetry], with most sinister events in its memory, most questionable developments around it, it can no longer speak the language which many willing ears seem to expect. Its language has become more sober, more factual. It distrusts 'beauty'. It tries to be truthful. If I may search for a visual analogy while keeping in mind the polychrome of apparent actuality: it is a 'greyer' language, a language which wants to locate even its 'musicality' in such a way that it has nothing in common with the 'euphony' which more or less blithely continued to sound alongside the greatest horrors.[22]

How did Celan evolve this new mode of utterance? Unlike many twentieth-century poets who wrote criticism copiously as a way of proselytising for the principles by which their own work should be understood, Celan wrote little critical prose. His only extended statement of poetics is his 1960 speech of acceptance of the George Büchner prize, *Der Meridian*.[23]

This is a difficult text, in part inherently, in part because of the way in which Celan articulates his poetics by reference to Büchner. As with Celan's poetry, so with his prose: his modes of utterance embody the difficulty of speaking at all. His statements are the reverse of polemical, manifesto-like, announcing the programme for a new movement, in the manner of Olson. On the contrary, they are tentative, fragmentary, gnomic, working out ideas that remain all but resistant to articulation. He incorporates a defence of the mode from Pascal: '*Ne nous reprochez pas le manque de clarté puisque nous en faisons profession*' (Do not reproach us for a lack of clarity because we profess that for ourselves; §27). The context of this in Pascal is a discussion of how the Old Testament is interpreted in the New Testament, following the interpretations of Christ himself (Luke, 24.44–45)—that is, in terms that require considerable reading in. The commentator on Pascal from whom Celan took Pascal's remark introduces it with injunctions similarly in accord with Celan's thinking: 'flee from brightly lit places because bright light makes you see a lie; love the half-lights, the shadows' (227).

Characteristically, Celan can only approach his central idea of the 'breathturn' essential to the really new in poetry obliquely. What is really new does not admit of formulation without experiment and indirection groping towards some element of what is to be discovered. One form of indirection is extreme syntactic convolution.

> Gewiss, das Gedicht—das Gedicht heute—zeigt, und das hat, glaube ich, den noch mittelbar mit den—nicht zu unterschätzenden—Schwierigkeiten der Wortwahl, dem rapideren Gefälle der Syntax oder dem wacheren Sinn für die Ellipse zu tun,—das Gedicht zeigt, das ist unverkennbar, eine starke Neigung zum Verstummen.
> Certainly, the poem—the poem today—shows, and this, I believe, has to do only indirectly with the—not to be underestimated—difficulties of word choice, the faster fall of syntax or the more lucid sense of ellipsis—the poem shows, unmistakably, a strong tendency to fall silent. (§32a)

The expressive aurality of that is written into the syntax and punctuation, though fully to realise those the reader must use not the eyes but the mouth. But what would one not give to hear the breath structures (intonation, inflection, pause) by which Celan delivered what written notations can only imply within limits.

Hesitation, interruption, circuitous syntactic involution, is combined here with overt expression of difficulty: a 'lucid sense of ellipsis', that is, perceived *lacunae*—things not expressed that must be supplied, whether by the implications of the poem or the imagination of the reader we are not told; and a 'faster fall of syntax', that is, syntax that omits connectives, that does not include all the elements that might fix the syntactic relations of words. Both of these aspects of expression, present in this prose as in the poetry it searches to envisage, problematise meaning. Insofar as articulation can, they foreshadow a 'tendency to fall silent'. They are to be heard with '*Hasenöhrchen*' (hare's ears)—with a special sensitivity to implication, listening 'beyond …the words' (§48c). With poetry this means the implications of image, syntax, and formal structure—and, as Celan goes on to say, finding within the syntax, the form, and the inflections and intonations implied by meaning the structure of the living breath.

Occasionally in all this there are syntactically simple sentences, but then another kind of difficulty supervenes: '*Das Gedicht ist einsam. Es ist einsam und unterwegs. Wer es schreibt, bleibt ihm mitgegeben*' (The poem is lonely. It is lonely and *en route*. Its author remains added to it; §34). The wisest expositor may leave that to the reader's contemplation.

Obliquity, indirection, qualification, the gnomic are built into the mode of expression. The struggle to achieve articulation is also built into the document itself, more evidently than is represented by print. The typed reading script from which Celan delivered the speech is covered in re-phrasings, deletions, and additions. Everything about *Der Meridian*, from its style to its material form, is an expression of living speech, of unconcluded search, the continued presence in the language, and the material record of the occasion, of the unrepeatable moment of its spoken delivery, incorporating varied refusals of finality.

Another problem of the speech is that Celan articulates his poetics by continuous reference to Büchner, with whom he signifies a sense of deep kinship. The fundamental starting point of this is Büchner's modernity. Born in 1813, Büchner was a political revolutionary who lived in the reactionary aftermath in Germany of the Napoleonic Wars. His writings are so unusual that they were ignored in his lifetime and for the half-century following. He was rediscovered in the early twentieth century as a precursor of Modernism. What he has in common with Celan is an aesthetic related to new political and social realities that require new ways of writing and reading. What he signifies for Celan is letting life back into art by a radical overthrow of aesthetic conventions. While Celan regularly shies away from affirmations, he asks of Büchner's work whether it does not 'propose … a truly radical calling-into-question of art' (§19). His implied answer is that it does.

Celan extracts central ideas about his own aesthetics from two of Büchner's works, the play *Danton's Death* (about the execution of the French revolutionary leader) and the novella *Lenz*. The central figures in these for Celan are Lucile Desmoulins, wife of an executed revolutionary, in *Danton's Death*, and Jakob Lenz, the 'hero' of the novella, both alienated characters who do not fit into their surroundings. Though 'blind to art'—that is, valuably ignorant of its conventions—Lucile is a proto-poet in seeing language as 'person-like and tangible' (§6c). The evidence of her alienation comes in the final moment of the drama, when, next to the revolutionary scaffold, she shouts '*Vive le roi!*', thus bringing about her own execution. For Celan this has nothing to do with the meaning of the words—tradition, the past, support for the *ancien régime*: it is an act of pure freedom (§7b). Lenz's alienation is evident throughout the novella but becomes more intense towards the end when he is unable to bear what he hears as the voice of silence all around him. Celan picks one detail of his alienation to epitomise the whole, his momentary wish to walk on his head with the sky an abyss above him: this epitomises the 'breathturn' towards a new way of being which increasingly drives him away from people who see the world in conventional terms. Both figures are analogies for Celan as poet.

Celan's association with Büchner suggests that the problems he addresses are trans-historical, the problems of the revolutionary artist in any age. Other invocations in the speech carry the same implication, especially the invocation of Mallarmé (§19). Mallarmé signifies for Celan the previous fundamental revolution in poetics, at the end of the nineteenth century, which generated endless experiment and interpretation, and which, like Celan, emphasised the centrality of sound and form in poetry to meaning. 'Should we before all … think … Mallarmé through to the end?', Celan asks. The Mallarmé revolution has implications yet to be realised—new and more intense explorations of aurality, more entirely by the ear, but with no 'pure music' separation from semantic content and history. More simply, the invocations of Mallarmé, Pascal, and others imply that the problems Celan addresses are transhistorical.

Celan also uses Büchner, however, to suggest that the permanent problem of the really new in art also has a precise historical locatedness for the difficulties of poetry in his historical moment. *Lenz* begins on 20th January. In the novella this is a minor detail, but in the speech Celan mentions it four times, twice (§44e) affirming its importance in relation to his own poetry. 20th January 1942 was the date of the Wannsee Conference, held to arrange implementation of the so-called 'final solution' of the Jewish question—the extermination of Jews in all German-held territories. Celan certainly knew this date: the minutes of the meeting (the so-called 'Wannsee Protocol') figured prominently in the Nuremberg war-crimes trials. Celan's stress on poetry being mindful of its dates must have this in view.

> Perhaps one can say that each poem has its own '20th of January' inscribed in it? Perhaps what's new in the poems written today is exactly this: theirs is the clearest attempt to remain mindful of such dates? (§30)

The date signifies the historical context of the Holocaust as prompting post-war poetry's necessary search for new modes of expression.

Celan leads through Büchner to these new modes, to the necessary radical 'breathturn'.

> Lenz—or rather Büchner—has here gone a step further than Lucile. His 'Long live the king' is no longer a word, it is a terrifying falling silent. It takes away his—and our—breath and words.
>
> Poetry: that can mean an *Atemwende*, a breathturn. Who knows, perhaps poetry travels this route—also the route of art—for the sake of such a breathturn? ... Perhaps here, with the I—with the estranged I set free *here* and *in this manner*—perhaps here a further Other is set free? (§29a/b)

Identifying the 'breathturn' with two moments in Büchner that he interprets as evidence of full human presence, Celan speculates, or postulates, that this may be what a poem can give. With his usual tentativeness, Celan does not affirm: he questions, and with a reiterated 'perhaps'. This 'breathturn', this moment of full human presence, is found not by any means that could be thought of as connected with skill in manipulating words. The primary issue is living a life of spiritual openness and discovery. But the postulated route is 'also the route of art'. To reject the musicality, craft, skill with words associated with styles of poetry that the new conditions of the post-Holocaust world have rendered archaic is not to reject art: there is still a new art to be found.

The *Meridian* drafts, and the poems written in the last decade of Celan's life, associate that with the sense of living presence located in a poem's breath structure.

> Twice, with Lucile's 'Long live the king', and when the sky opened as an abyss beneath Lenz, the *Atemwende*, the breathturn seemed to happen. Perhaps also when I tried to set course toward that inhabitable distance which finally becomes visible only in the figure of Lucile. (§43)

Again the same tentativeness: it seemed to happen; and again 'perhaps', as preface to his own 'breathturn', possibility of fully human being. 'When

I tried to set course towards …': when he attempted to understand and to write without evasion the experiences resulting from the war.

In the final paragraphs of *Der Meridian* (§45c) Celan quotes from his own sequence, *Stimmen* (Voices).

Stimmen vom Nesselweg her:	*Voices* from the nettle path:
Komm auf den Händen zu uns.	Come on your hands to us.
Wer mit der Lampe allein ist,	Whoever is alone with the lamp
hat nur die Hand, draus zu lesen.	has only his hand to read from.[24]

Voices from the path of pain invite him to join them, through whatever difficulty and suffering; they tell him of a moment of illumination which has to be approached alone, and understood in terms of one's own experience. Of this poem he says, 'I had … encountered myself' (§45f), a comment he elaborates with a paradox about poems reaching into the future to encounter the past: poems are 'a sending oneself ahead towards oneself, in search of oneself … A kind of homecoming' (§46).

Two later poems in the same sequence are especially significant:

Stimmen, vor denen dein Herz	*Voices* from which your heart
ins Herz deiner Mutter zurückweicht.	shrinks back into your mother's heart.
Stimmen vom Galgenbaum her,	Voices from the gallows tree,
wo Spätholz und Frühholz die Ringe	where latewood and springwood
tauschen und tauschen.	change and exchange their rings.

This encounter echoes with the past: his love for his mother, and her murder; and with the same sense of the future being an encounter with the past, 'latewood and springwood': the existent changing through its encounter with the new discovered by the poem. Similarly the final lines of a section that begins from the archetypal conflict in Judaic history between the brothers Esau and Jacob:

Wir wohnen darin.	We dwell inside.
Atme, dass	Breathe, that
sie sich löse.	it come loose.

In the sequence as a whole people are trapped in what confines their full humanity by histories which are immediate and personal but also archaic and cultural. The freed breath epitomises everything that can lead through those confinements to a more full humanity.

Fragments from drafts for *The Meridian* that did not find their way into the speech itself treat the issue of breath somewhat differently. One group is less metaphysical, less mystical, more like the ideas of Olson and his associates. They imply '*hör dich ein / mit dem Mund*': understand a poem through familiarity with its shape—but not only its shape on the page; also shapes discovered by finding its structures of phrasing through performance.

> 'What's on the lung, put on the tongue', my mother used to say. Which has to do with breath. One should finally learn how to also read this breath, this breath-unit in the poem: in the cola meaning is often more truthfully joined and fugued than in the rhyme; shape of the poem: that is presence of the single, breathing one.— (108)

Three fragmentary notes help to explain this:

> 'Breath-units' (Buber); cola (109)
> Mora, cola (109)
> In the mora meaning clarifies itself— (108)

The great historical *Oxford English Dictionary* does not recognise 'cola' as having a meaning in relation to prosody; but it recognises Celan's synonym, 'mora': *OED*, *n.*[1], 3.a. *Prosody*. A unit of metrical time equal to the duration of a short syllable. Celan understood 'cola' as equivalent to a term he apparently found in the theologian, Martin Buber, 'breath-units' (*Atemeinheiten*). Whether the term indicates the unit, or the space marking the end of the unit, the claim is that hearing the breath structure of a poem is fundamental to feeling its meaning. But hearing this structure is much more an act of interpretation with Celan than with Olson, because Celan does not aim to convey this by the lineation of the poem as printed. This is apparently Celan's meaning when he describes a poem as 'the trace of our breath in language' (115).

There are also more metaphysical fragments not worked into the speech related to the idea of the 'breathturn'.

> I have survived some things,—but survival / Überstehn / hopefully isn't 'everything'—, I had a bad conscience; I was searching for—maybe I can call it that?—my breathturn. (123)

The bad conscience that prompted Celan to seek the profound redirection for which he again uses his invented word, breathturn, can be, but need not be, interpreted in personal terms. What is clear is that survival is not enough. There must be a search beyond trauma—personal trauma and cultural trauma—to recover a more full humanity.

In another fragment, poetry takes away the false breath by being a revelation of things as they really are.

> Lenz: the Medusa-likeness of poetry—to: faces you with silence, it takes your—false—breath away; you have come
> //// to the breathturn. (123)

The Medusa, whose gaze turns what it looks at to stone, indicates both a possibility and a limitation. A possibility—freezing a human moment that is utterly real, its reality not compromised by conventions of representation or the perspective introduced by a human point of view; but also a limitation: where the purpose of this freezing is to capture the reality of the living moment, the paradox that the reality of that moment depends on its having a before and after. How can art capture the living without the loss built into the very idea of capture?

This may be an irresolvable paradox: hence the tendency of the poem to fall silent. Celan's later poems have nothing like the extension and rhetorical elaboration of 'Todesfuge'. They are characteristically no more than a few lines, sometimes organised in short sequences or loosely related groups as a way of extending utterance. Everything combines to convey the fundamental difficulty of achieving articulation at all. But while the individual utterance is short the totality of utterances is extended—brief, but compelled.

'Fadensonnen' is a typical late poem from the collection *Atemwende* (Breathturn).

Fadensonnen	Threadsuns
über der grauschwarzen Ödnis.	above the greyblack wastes.
Ein baum-	A tree-
hoher Gedanke	high thought
greift sich den Lichtton: es sind	grasps the light-tone: there are
noch Lieder zu singen jenseits	still songs to sing beyond
der Menschen.	mankind.[25]

Celan drew attention to this poem by making the invented word, 'Fadensonnen', the title of his next book, suggesting that it strikes a keynote. 'Fadensonnen' can be perceived in many places: weak signs of light (threads) contrast with the context, 'the greyblack wastes' which are their antithesis. A tree-high thought may not, as some readers think, stretch up to heaven, but it is surely high, and manages to grow out of the grey-black wastes; and what it attaches itself to—another invented word,

'Lichtton'—has the power of addressing the eye and the ear. In this waste-land there is something growing; and presumably the 'light-tone' it grasps is connected to the 'thread-suns'. But what is most surprising about the tentative hope expressed in these images is the conclusion drawn from them. Poetry has to reach beyond what has so far been conceived: that is fundamental to Celan's post-Holocaust situation, and must be at least part of the idea of the 'breathturn' that is this volume's title. And what is usually conceived in terms of the barely utterable, articulation on the verge of being overwhelmed by silence, is here conceived as potentially '*Lieder zu singen*', songs to sing. The images of tentative hope prophesy a new poetry not so near the verge of silence as Celan elsewhere suggests. Its 'music' will be a new music, but that it should be conceived as song is striking.

Breath is heard in everything about the realisation of a poem in the voice, including the imagined voice. This means not only those elements of the poem which have a visual form on the page—its formal and syntactic structure—but everything implied by form and meaning. It includes not only what Ezra Pound called the 'sculpture of rhyme' (three-dimensional seeing: what the flat surface of the page implies beyond what it literally shows) but also its music—the sounds that are realisations of the visible; rhythm, rhyme, line-length, every formal shape. It also includes the sounds derived from structures implied by sense and form: pause, dynamics, inflection, intonation, all of which are variable and subject to interpretation.

Olson's ideas of the breath-line relate primarily to his poetry and the poetry of his associates and followers. Though half-endorsed by William Carlos Williams, and apparently helpful to his own different experiments and conclusions, they do not apply to how Williams actually read his own poetry. Celan's less precise but more searching ideas on breath in poetic structure apply widely to English and American poetry since Modernism. Potentially they apply to feeling meaning through structure in all poetry.

That there is a breath structure written into the poem, essential to experiencing the poem as living speech, to be elicited variously by different readers, is the central idea to which Celan's searching leads. Such a structure is not simply objectively present. Celan is clear that in their every aspect poems are read with the changing key of the individually idiosyncratic and culturally situated reader. As Celan accepts, and positively argues, the breath structure of a poem is heard differently by different readers, depending on individual sensibility and context. That does not mean that perceiving it is fanciful, or substantially make it different from all the other features of meaning in poetry that competent readers can agree are present, though they may not agree what they are or how they function. But Celan is not interested in a discipline of criticism. What he offers is a way of bringing the reader into real and living relation with a poem.

This depends in part on what T. S. Eliot calls 'the inexplicable mystery of sound' in poetry.[26] For those who have ears to hear, Celan understands those sounds as gathered together in the breathing: form, structure, syntax, inflection, intonation, tone; expressions of the whole physical being—passion, humour, intensity, volume—(Olson) reaching down through the throat of the person who writes, to the lungs, the diaphragm, and the whole physiology of breathing; (Valéry) by engaging the body, integrating the intellectual and the sensual.

Celan extends Valéry's claim that 'syntax is a faculty of the soul'.[27] In poetry every inflection of the sounds of language has meaning. For Celan Valéry's syntax is synecdoche for all the expressivities of articulation dependent on the breath: how words are uttered because of the whole framework of meaning and form within which they are situated. What this was for the poet recording tells more fully than print, but that is no more authoritative and final for the reader than a composer's recording of a musical work for a performer. Though it has a special status as an interpretation, it is finally one interpretation among many. What this might be for each reader—accepting, as Celan does, that for each reader it will be different—can be discovered by living with a poem, and living with it through the voice: '*hör dich ein mit dem Mund*'.

Notes

1. The literary contexts and the influence of Olson's work can be seen in two frequently reprinted and widely distributed anthologies: *The New American Poetry 1945–1960*, ed. by Donald M. Allen (New York: Grove, 1960); *The Poetics of the New American Poetry*, ed. by Donald Allen and Warren Tallman (New York: Grove, 1974).
2. For the influence of Olson in British poetry see *Other: British and Irish Poetry since 1970*, ed. by Richard Caddel and Peter Quartermain (Hanover, NH: Wesleyan University Press, 1999).
3. Charles Olson, *Collected Prose*, ed. by Donald Allen and Benjamin Friedlander (Berkeley: University of California Press, 1997), 239–49.

4. For a similar programme of getting the spoken language back into written prose through breath structures see Jack Kerouac, 'History of the Theory of Breath as a Separator of Statements in Spontaneous Writing' (1962, typescript with manuscript corrections and additions), reproduced in Isaac Gewirtz, *Beatific Soul: Jack Kerouac on the Road* (New York: New York Public Library, 2007), 176–77; contextualised and discussed by Stefanie Heine in Arthur Rose, et al., *Reading Breath in Literature* (Basingstoke: Palgrave Macmillan, 2019), 91–112.
5. *The Collected Poems of Charles Olson*, ed. by George F. Butterick (Berkeley: University of California Press, 1987), 86–93. For Olson's 1954 recording of 'The Kingfishers' (without section III), and an extensive archive of Olson's readings, see http://writing.upenn.edu/pennsound/x/Olson.php.
6. *The Autobiography of William Carlos Williams* (New York: New Directions, 1967), 329.
7. For Williams' views on poetic structure formulated without reference to Olson see 'The Poem as a Field of Action' (1948), and 'On Measure—Statement for Cid Corman' (1953) in Williams, *Selected Essays* (1954; New York: New Directions, 1969); cf. John C. Thirlwall, 'Ten years of a new rhythm', in Williams, *Pictures from Brueghel* (Norfolk, CT: Laughlin, 1962).
8. http://www.poetryfoundation.org/features/audioitem/2838. The comments on orality are at 2.40–3.50.
9. http://www.poetryfoundation.org/features/audioitem/2838. The reading of 'The Yachts' is at 19.50–22.10.
10. Williams, *Collected Poems*, ed. by A. Walton Litz and Christopher J. MacGowan, 2 vols. (Manchester: Carcanet, 2000), vol. 1, 541.
11. On Olson's legacy and influence, see Mark Byers, *Charles Olson and American Modernism* (Oxford: Oxford University Press, 2018).
12. Paul Valéry, 'On Speaking Verse', in *Collected Works*, ed. by Jackson Matthews, 15 vols. (London: Routledge, 1957–75); vol. 7, *On the Art of Poetry*, trans. by Denise Folliot, intro. by T. S. Eliot, 162–63.
13. For Celan's recorded readings see *'Ich hörte sagen': Gedichte und Prosa* (München: Der Hörverlag, 2004). Recordings of Celan reading are also available at www.lyrikline.org.
14. Letter of 19 February 1818 to J. H. Reynolds, in *The Letters of John Keats, 1814–1821*, ed. by Hyder Edward Rollins, 2 vols. (Cambridge: Cambridge University Press, 1958).
15. Celan, 'Edgar Jené and the Dream about the Dream', *Collected Prose*, trans. by Rosemarie Waldrop (Manchester: Carcanet, 1986), 3–10 (8).
16. 'Mit wechselndem Schlüssel', *Poems of Paul Celan* (1972), trans. by Michael Hamburger (New York: Persea, 2002), 58–59.

17. For Michael Hamburger see note 16; also John Felstiner, trans., *Selected Poems and Prose of Paul Celan* (New York: Norton, 2001); Susan H. Gillespie, trans., *Corona: Selected Poems of Paul Celan* (Barrytown, NY: Station Hill, 2013); Pierre Joris, trans., *Breathturn into Timestead: The Collected Later Poetry of Paul Celan* (New York: Farrar, Straus, Giroux, 2014). All offer parallel texts in which the translation can be read alongside the original.
18. *Collected Prose*, 34.
19. The poem can be heard at https://www.lyrikline.org/de/gedichte/todesfuge-66#.WmoOxk3cvX4 and on the CD *'Ich hörte sagen'* (note 13).
20. Theodor Adorno, 'Cultural Criticism and Society' (1949, 1951), *Prisms*, trans. by Samuel and Shierry Weber (Cambridge, MA: MIT Press, 1981), 34.
21. See John Felstiner, *Paul Celan: Poet, Survivor, Jew* (New Haven: Yale University Press, 1995), 225.
22. *Collected Prose*, 15–16.
23. *Der Meridian: Endfassung—Vorstufen—Materialien*, ed. by Bernhard Böschenstein and Heino Schmull (Frankfurt: Suhrkamp, 1999); *The Meridian: Final Version—Drafts—Materials*, ed. by Böschenstein and Schmull, trans. by Pierre Joris (Stanford: Stanford University Press, 2011). The translation is printed to parallel the layout of the German text. Section numbers (§) and page references (numerals) are therefore the same for both texts. The speech can be read as a construction in continuous prose in *Collected Prose*, 37–55.
24. 'Stimmen' (Voices), Felstiner, *Selected Poems and Prose*, 88–93.
25. *Breathturn into Timestead*, 14–15.
26. 'To Walter de la Mare', *Collected Poems* (London: Faber, 1962), 233.
27. *Collected Works*, vol. 14, trans. by Stuart Gilbert, intro. by W. H. Auden, *Analects*, 'Odds and Ends', 15.

Selected Bibliography

Aviram, Amittai F. 1994. *Telling Rhythm: Body and Meaning in Poetry*. Ann Arbor: University of Michigan Press.

Bernstein, Charles, ed. 1998. *Close Listening: Poetry and the Performed Word*. New York: Oxford University Press.

Clark, Timothy. 1997. *The Theory of Inspiration: Composition as a Crisis of Subjectivity in Romantic and Post-Romantic Writing*, Chapter 11. Manchester: Manchester University Press.

Corman, Cid, et al. Encounters: American Poets on Paul Celan. *Studies in Twentieth Century Literature* 8/1 (1983), 101–27.

Cureton, Richard D. 1991. *Rhythmic Phrasing in English Verse*. London: Longman.

Derrida, Jacques. 2005. *Sovereignties in Question: the Poetics of Paul Celan*. Trans. by Thomas Dutoit and Outi Pasanen. New York: Fordham University Press.

Finch, Annie. 1993. *The Ghost of Meter: Culture and Prosody in American Free Verse*. Ann Arbor: University of Michigan Press.

Gadamer, Hans-Georg. 1997. *Gadamer on Celan: 'Who am I and who are you?'*. Trans. by Richard Heinemann and Bruce Krajewski. New York: State University of New York Press.

Wesling, Donald. 1996. *The Scissors of Meter: Grammetrics and Reading*. Ann Arbor: University of Michigan Press.

Wright, George Thaddeus. 2001. *Hearing the Measures: Shakespearean and Other Inflections*. Madison: University of Wisconsin Press.

Sound Recordings

Olson, Charles. PennSound archive, University of Pennsylvania programmes in contemporary writing. http://writing.upenn.edu/pennsound/x/Olson.php. An extensive archive including video recordings.

Celan, Paul. *'Ich hörte sagen': Gedichte und Prosa*. Recorded 1948–67. Der Hörverlag ISBN 3-89940-450-5, 2004, compact disc.

Celan, Paul. https://www.lyrikline.org/de/gedichte/todesfuge66. Ten poems.

Film

Kasten, Ullrich, dir. *Dichter ist, wer menschlich spricht*. Video recording of television broadcast, 00:52:00. YouTube. Posted by Klaus Kloßbrühe, January 3, 2016. Stuttgart, Germany: SWR, 2015. https://www.youtube.com/watch?v=qb-dtYgrIsw. French version: *Écrire pour rester humain*. https://www.youtube.com/watch?v=oV-PR9xKmNw.

CHAPTER 21

A Panting Consciousness: Beckett, Breath, and Biocognitive Feedback

Marco Bernini

Breath and Consciousness: A Co-dependent Intimacy

Booming Western interest in mindfulness and meditation has significantly mainstreamed breath and breathing practices. Mobile apps guiding breath-focused meditation are democratising, more or less carefully, earlier Buddhist intuitions of a tight, looping relation between body and mind, breath and mental states. Cognitive sciences too are starting to take the link between breathing and consciousness seriously. Breath is treated, for example, both as a symptom of stress disorders (such as hyperventilation in anxiety or OCD)[1] and a possible attenuating cure for distressing mental states.[2] The therapeutic power of breathing is also key to the working of sensory deprivation technologies such as floatation tanks[3] or sonic cradles.[4]

The cross-methodological benefits of looking at Buddhist practices, and at the relationship between breath and consciousness, for a better

M. Bernini (✉)
Department of English Studies, University of Durham, Durham, UK

D. Fuller et al. (eds.), *The Life of Breath in Literature, Culture and Medicine*, Palgrave Studies in Literature, Science and Medicine,
https://doi.org/10.1007/978-3-030-74443-4_21

understanding of the human mind have been advocated by neurophenomenologists and philosophers of the cognitive sciences such as Francisco Varela, Evan Thompson, and Owen Flanagan.[5] Within this intercultural scientific agenda, breath is considered as a way to access an 'open awareness'[6] of pre-reflective, minimal, foundational levels of consciousness and subjectivity. More than a century ago William James reflected on the intimacy of breath and consciousness by venturing the even more radical idea that our background feeling of a unified consciousness is actually materially anchored in, if not equivalent to, our breath. This is a mysterious claim; yet one that can today be illuminated by, and integrated with, contemporary scientific frameworks.

Whether technologically enhanced in artificial spaces or modulated in natural meditative environments, focussed breathing is therefore conducive to novel psychological states. Thanks to the regulation of breathing patterns, patterns in our thinking are not just affected, but revealed, together with their entanglement with respiration. Bodily proprioception and cognitive introspection thus converge in a mutual, looping modulation that discloses a dynamic relationship between biological and cognitive domains (or what this essay will describe it as 'biocognitive feedback').[7] What makes this reciprocal feedback possible is the structural intimacy and co-dependence of body and mind, in line with contemporary post-Cartesian cognitive sciences that see cognition as deeply embodied and shaped by actions, including the enactment of different breathing patterns.[8]

This essay investigates the co-dependent intimacy between breath and cognition as represented, explored, and complicated in the narrative work of Samuel Beckett. Some of Beckett's fictional environments, notably from the 1950s onwards, might indeed feel to his readers or audience like sensory deprivation tanks. Nothing there can be seen clearly, and *therefore* something or someone in the end appears or, as in *What Where* (1983), 'reappears'.[9] Such works include, for example, *The Unnamable* (1953), where the narrating voice begins as a planet floating in a cosmic void (or so it pictures itself because 'how can one be sure in such a darkness?')[10] while hallucinating presences from Beckett's imagistic universe; or *Company* (1980), which can also be experienced as a sustained solitary training in blinded flotation. Here a character is horizontally lying 'on his back in the dark', when this sensory indeterminacy is lit by projected images from the character's past. Together with emergent autobiographical visions, however, he also gradually attunes to minimal proprioceptive knowledge, and notably to the sound of his breath ('Apart from the voice and the faint sound of his breath there is no sound. None at least that he can hear. This he can tell by the faint sound of his breath'). In line with models of biocognitive regulatory feedback between breath and cognition, he simultaneously also accesses minimal states of consciousness ('in the dark in that dim mind').[11]

Sensory deprivation is thus to Beckett's characters (and, vicariously, to his readers and audience) generative of new perceptual and attentional modes. This enhanced, regressive attention slowly makes them better acquainted with primary sensations as cognitive and physiological organisms. What they access, this essay will argue, is the restless biocognitive feedback activity between minds and bodies, including resonances or frictions between patterns of breathing and states of consciousness. In this proprioceptive-introspective new attentional mode, the relationship between breath and consciousness is revealed, in Beckett's work, as a complex, fraught intimacy. Beckett notoriously devoted an entire piece, albeit performatively brief, to the act of breathing (*Breath*, 1969).[12] This work has received extensive critical attention, including a recent full-length monograph.[13] Less explicit, more nested, incrementally distributed, and cross-medially explored references to breath and consciousness in Beckett's work have received comparatively little attention. This essay will therefore avoid an analysis of Beckett's most celebrated (but also more symbolic or metaphorical) tribute to breath, to investigate his subtler exploration of breath's intimacy with consciousness and cognition against models of their relation as formulated by classic and contemporary cognitive science.

The 'I-Breathe': Agency, Time-Consciousness, and Biocognitive Tempos

Let us begin with William James' radical view of 'consciousness-as-breath' to evaluate how it can be set in dialogue with Beckett's modelling of biocognitive feedback. James' model requires gradual unpacking and speculative interpretation to introduce aspects of consciousness and breath that are key to understanding more contemporary models. In his *Principles of Psychology* (1890) James attempted multiple metaphorical descriptions to capture the nature, essence, and phenomenology of consciousness. The most famous of these sometimes conflicting metaphors is the idea of consciousness as a 'stream'.[14] With this image

James attacked previous conceptualisations of consciousness as a jointed sequence of discrete thoughts. This static and mechanical view did not account, James argued, for the unified, dynamic flowing of consciousness, because consciousness 'does not appear to itself chopped up in bits. [...] It is nothing jointed; it flows. A "river" or a "stream" are the metaphors by which it is most naturally described. *In talking of it hereafter let us call it the stream of thought, of consciousness, or of subjective life*'.[15] The image of a stream helped James to highlight the lack of gaps in subjective experience, as well as to assign to consciousness an agency of its own ('*it* flows'); thus, accounting also for subjective experiences beyond our control (*e.g.*, mind-wandering).

The impact of James' metaphor for psychology, and even more for literary studies, can hardly be overstated. What is less acknowledged is that, only a few years later, James grew deeply dissatisfied with this model. More specifically, and quite surprisingly for the putative father of the field of consciousness study, he felt that the very concept of consciousness needed to be eliminated. In his essay 'Does Consciousness Exist?' (1904), he begins with a lapidary eradication of the concept by stating that consciousness 'is the name of a nonentity, and has no right to a place among first principles. Those who still cling to it are clinging to a mere echo, the faint rumor left behind by the disappearing "soul" upon the air of philosophy'.[16] What follows is a landmark example of James' new radical empiricist method, in which he interrogates the reasons behind the resilience of a concept defining such a diaphanous object of research. In doing so, however, and in spite of indulging in some analogical exemplifications, he refrains from devising new metaphors.

Metaphors such as the stream have previously been used by James 'to negotiate ethereal explanations for consciousness with bodily processes'.[17] This is actually what, according to embodied cognitive theories of figurative language,[18] most metaphors do: in attempting to capture or to explore a more volatile or abstract phenomenon (the target of the metaphor; here, consciousness), metaphors project into it a more concrete domain (the source; here, the physical experience of a stream of water). In his later essay, however, James is keen to specify that he actually had 'mistrusted' consciousness as a target phenomenon or entity 'for the past twenty years', and that through consciousness metaphors he only 'tried to give [his students] its pragmatic equivalent in realities of experience'. Not anymore, however, because he claims that we should dismiss altogether the existence of consciousness as a distinct target, since '[i]t seems to me that the hour is ripe for it to be openly and universally discarded'. James is aware that dismissing consciousness might sound absurd, but he says it will be less so if the usual epistemic role attributed to consciousness is transferred to thoughts themselves: what we call consciousness, James says, is a 'function in experience that thoughts perform' and 'that function is *knowing*. "Consciousness" is supposed necessary to explain that things not only are, but get reported, are known'.[19] For James, thoughts themselves can perform this epistemic function, by getting to know each other in reciprocal relations. The question is then what happens, in this rejection of the conceptual relevance of consciousness, to the phenomenological sense of a unified, sometimes agentively autonomous, flowing of conscious life that was captured by the image of the stream? Where are these qualities coming from? James' non-metaphorical answer is that they are actually generated by phenomenological biocognitive feedback coming from another unified, partly involuntary process: our breathing activity. The conclusion of his radical essay deserves full citation:

> I am as confident as I am of anything that, in myself, the stream of thinking (which I recognize emphatically as a phenomenon) is only a careless name for what, when scrutinized, reveals itself to consist chiefly of the stream of my breathing. The 'I think' which Kant said must be able to accompany all my objects, is the 'I breathe' which actually does accompany them. There are other internal facts besides breathing (intracephalic muscular adjustments, etc., of which I have said a word in my larger Psychology), and these increase the assets of 'consciousness,' so far as the latter is subject to immediate perception; but breath, which was ever the original of 'spirit,' breath moving outwards, between the glottis and the nostrils, is, I am persuaded, the essence out of which philosophers have constructed the entity known to them as consciousness. That entity is fictitious, while thoughts in the concrete are fully real.[20]

It is important not to consider this conclusion as another metaphorical descriptor for consciousness. Here James is suggesting rather that the phenomenological qualities we attribute to consciousness are instead (not also) to be found in the process of breathing. The reason we carelessly conceptualise consciousness as a separate entity, James says, is that bodily states, and chiefly breathing, generate feedback into our qualitative experience of mental states. We then erroneously consider the latter as having properties that are instead projected (phenomenologically, not metaphorically) by the former. This means that if we look more carefully at our immediate perception of breathing, we should find qualities that we usually attribute to consciousness. This might seem far-fetched, but on closer scrutiny the two processes share a number of constitutive elements.

The first can be individuated in the particularly shifty agency of both breathing and consciousness. Breath, as well as consciousness, is a diaphanous process: a process we mostly live *through*, and which goes on in the background of our experience without us having to be in control of it, or being able fully to stop by sheer will. This agentive autonomy is also at the core of the felt escorting quality of consciousness, which accompanies our thoughts and mental states without us being able to control entirely its background patterns, or to shut it down apart from when falling asleep. For James, it seems that feedback from the semi-voluntary agency of breathing feeds into conscious thinking, and informs its phenomenological properties. Related to this agentive autonomy, James' reconceptualisation of the Cartesian cogito as a misnomer for the 'I-Breathe' prompts us to look into the act of breathing as a possible contributor to the feeling of a unified temporal continuity of experience. If we turn to the phenomenology of temporality, or what phenomenologists call 'time-consciousness',[21] we can consider whether its structuring components can be found in the breathing experience too.

Since Husserl's seminal analysis of human perception of time, phenomenologists have endorsed his view that the feeling of temporal flow in experience is rooted in each perception having a 'retentive' and 'protentive' aspect. Taking musical notes as an exemplification, Husserl suggested that the reason we perceive a unified stream of notes as a melody is that each impression of a note presents itself with a 'retention' of what just passed and a 'protention' of what is likely to follow.[22] Time-consciousness therefore feels continuous because of these two crucial phenomenological edges in each conscious perception. If we are to find in breathing the Ur-model of time-consciousness, retention and protention can be seen to have phenomenological foundational roots in the process of inspiration and expiration. Each retentive inhalation ends in a protentive horizon of expiration, and vice versa. The bodily breathing cycle is therefore ordered by a pendulum of retentive-protentive moments that might feed back into our feeling of a temporal continuity of consciousness.

Together with agency and temporality, a third element that breathing shares with what we think of as our consciousness is presenting different tempos. Abnormal patterns of breathing such as hyperventilation might resonate, for instance, with accelerated, hyper-reflexive patterns of thinking.[23] In his radical eradication of consciousness, however, James would probably suggest that, even if biocognitive feedback between breath and thoughts (thoughts that are, for James, real) is occurring, it is only through our perception of an altered breathing that we perceive an analogous unified, altered tempo in our consciousness (the latter being, for James, the fictitious outcome of breathing feedback).

James' empiricist evaluation of the relationship between breath and consciousness ends in a full phenomenological overlapping of the two processes; or better, in subsuming phenomenological aspects of consciousness (agency, temporality, and tempos) into the enactive experience of breathing. For him, the intimate co-dependency turns out to be an identity. This does not rule out the importance of biocognitive feedback; if anything, it reinforces the phenomenological impact, but it makes biocognitive feedback asymmetrical and unidirectional. Its role in the 'I-Breathe' assigns to bodily respiration a phenomenological priority and the sole ontological validity. It is breath, not consciousness, whose agency, temporality, and tempos envelop our thoughts.

Beckett only had an indirect and cursory knowledge of William James.[24] And yet, James' equation of thinking with breath seems at first neatly to match Beckett's modelling of the biocognitive relationship between consciousness and breathing in relation to agency, temporality, and rhythmic tempos. Throughout his entire corpus, in fact, we find characters suffering the impossibility of agentively controlling and temporally stopping a consciousness that keeps going on with frantic tempos. As in James' model, often these qualities do not just resemble or mirror bodily patterns of breathing, but they seem experienced, in biocognitive synchronies, *as* bodily patterns. In *Texts for Nothing* (written 1950–1951), for instance, we have an unnamed character who is not just comparing, but synchronically equating thinking and breathing, 'so given am I to thinking with my breath'.[25] By the same token, full breathlessness is at times synchronically experienced as a pleasant silencing of consciousness, for example by the protagonist of the short story *The Calmative* (written in 1946) who, in his 'distant refuge', feels finally at rest with his 'head bowed, weak, *breathless*, calm, free'.[26] Another character in *Texts for Nothing* wishes his 'breath [to] fail better', to reach a state where the mind also will be silent ('to be silent'): a condition that will also shut down the epistemic function of knowing that we usually attribute to consciousness ('I'll be silent, I'll know I'm silence, no, *in the silence you can't know*').[27]

Even by an external, 'heterophenomenological' perspective',[28] the way Beckett's characters breathe is an embodied, material index indicative of the altered tempo of their mental struggles. In *Not I* (1972), for instance, a lit mouth in a darkened stage channels a cascade of thoughts through breathing patterns at the limit of human performative possibilities. The hyperventilating required of the actress marks patterns of hyper-reflexive anxiety at the level of the contents. In a letter to the director Alan Schneider about this play, Beckett wrote how he saw 'her speech [as] a purely buccal phenomenon without mental control and understanding'.[29] Here Beckett seems to be touching upon the sense of a lack of agency in consciousness, which he links, as James would have it, to a lack of control of breathing tempo: 'I hear it breathless, urgent, feverish, rhythmic, *panting along*, without undue concern with intelligibility. Addressed less to the understanding than to the nerves of the audience which should in a sense share her bewilderment'.[30] In a biocognitive synchronic view of consciousness-as-breath, the panting respiration of *Not I* is the only unifying process that grants cohesion to otherwise centripetal, disjointed thoughts. It is a panting breath, not a streaming consciousness, which pushes thoughts temporally forward, in steady acceleration. There is no separate, reflective consciousness taking these thoughts as epistemic objects, and Beckett wanted the enactive experience of a panting breathing to be the primary, material index of frantic states of mind.

Beckett had modelled a similar panting consciousness a few years before in *How It Is* (1961), this time in the novelistic medium. Assuming that literary narratives require readers' silent subvocalising of the text,[31] in *How It Is* readers are forced to a more direct, sensorimotor enactive experience of consciousness as bodily articulated. In this highly experimental novel that has only the shadow skeleton of a standard plot (in Beckett's words, 'A "man" is lying panting in the mud and dark murmuring his "life" as he hears it obscurely uttered by a voice inside him'),[32] the primary meaning seems to be found in the very experience of the articulatory efforts in subvocalising these panting words and/as thoughts. When Robert Pinget received a copy of the original French version of the text (*Comment C'est*) in 1960, he reported how this subvocalising effort was testing the limit of the reader, and daunting any prospect of translation. In his account of a conversation he had with Beckett he explains how he told him 'that for him [the third] is the weakest part. I agree, but one breathes more easily in it. The gasping in the first two parts is unbearable'.[33] In 1960 Patrick Magee also manifested worries about how to openly vocalise the text in a public performance. The only help Beckett thought he could offer was to supplement the lack of punctuation in the novel by marking rhythms, while suggesting, via his editor Barney Rosset, that '[a]ll he can do is gasp it out very short of breath, into a microphone if possible, and hope for the best. I have marked the rhythms on his copy'.[34] The rhythmic percussion of breathed words, in fact, has no diacritical signals of rest, but only typographical spaces between blocks. These are arguably the moments when, as we are told, 'the panting stops'[35] and therefore markers of longer inhaling gaps between panting gasps of thoughts. It seems that, through formal manipulations of the narrative medium, Beckett has managed to force readers, performers, and translators to a material experience of consciousness as a panting breath close to the synchronic model promoted by James. If we look at the contents that are frantically breathed into the characters, however, this model seems to be only partly explanatory of Beckett's exploration of the link between consciousness and breath.

Similar to *Not I*, in fact, the contents of these thoughts are flashes from the character's life that are pushed forward with an agency that the character disowns. More specifically, these thoughts, we are told, are coming from above: from a voice 'once without quaqua on all sides then within me when the panting stops'. The novel therefore discloses a vertical cognitive architecture of levels that complicates the idea that what we are panting through is a single, biocognitively synchronic Jamesian layer of a breathing consciousness. Early commentators have already stressed how the mud can be considered as an extended metaphor transforming consciousness into a place or space traversed by the character; and how its geography seems located at the bottom of unspecified geological layers. Ruby Cohn, for instance, claims that '[m]ud is an inspired metaphor for consciousness—warm and traversable. Above the novel's mud is life in the light, with its dreams, memories, and fantasies, but the mud itself is a performative medium: curtains can open within it, as in the theater; images and scenes can vanish through cinematic blackouts.'[36] I endorse Cohn's interpretation, which in the framework we are advancing can be reconceptualised as follows: the mud does not just stand for consciousness, but is a modelling solution that allowed Beckett to introduce an enactive performance of consciousness as biocognitively linked to breath.

However, even if the mud is the viscous material that gives panting the right phenomenological feeling of 'pushing against' some agentive imperative, it seems it is not the mud itself that generates the retentive-protentive movement. The imperative of panting forward comes from some other level, and the mud can rather be said to allow a performance of an asynchrony between the level of panting and the agentive source of the panting driving tempo. It seems therefore that James' model of consciousness-as-breath, while it has a lot to offer to Beckett's fictional modelling of the biocognitive feedback between breath and consciousness, cannot account in full for the asynchronous divorce in their primary and primitive intimacy. We need to find models accounting for different levels of interaction between breath and consciousness, for models spatialising, as Beckett does, different kinds of consciousness, and set them in biocognitive relations with breath.

Breathing Mediacy: Knowing, Experiencing, Witnessing

Knowing, Experiencing, Witnessing

Such models can be found in contemporary cognitive sciences, including Buddhist-inspired cognitive scientific frameworks. Instead of simplifying or dissolving the problem of consciousness into a single level having unidirectional biocognitive feedback (as in the 'consciousness-as-breath' model in James), contemporary cognitive sciences promote a multilevel view of different *kinds* of consciousness. Multilevel models of consciousness have taken up different names in different thinkers or approaches. All of them share, however, a distinction between more bodily, temporally local, pre-reflective stages or states, *on which* higher stages or states are building to produce reflective, autobiographical, temporally extended feelings of continuity for the organism. Once this architecture is established, biocognitive feedback between lower and higher levels may occur.

The primary, pre-reflective type of consciousness has been variously defined as a 'core'[37] or 'minimal'[38] stage and state. Here, the organism has only a 'thin' or 'experiential'[39] sense of being a bodily self, moving in time. On top of this eminently bodily, phenomenologically thin state, an 'extended' level hosting higher states of consciousness arises: this is the level of thoughts, episodic memories, mental imagery, autobiographical narratives, leading to the more conceptual continuity of an 'autobiographical self'[40] spanning larger portions of time. Core or thin levels are, in this account, necessary for more extended or autobiographical levels of consciousness to emerge. Usually, we are neither aware of this nested architecture nor of biocognitive feedback between levels.

What Buddhist-inspired cognitive models of consciousness add to this picture is the possibility of training ourselves in cultivating subtler degrees of awareness of, and an attentional navigation across, these levels. As Thompson points out, however, 'subtler or deeper aspects of consciousness aren't apparent to the ordinary untrained mind; they take a high degree of meditative awareness to discern'.[41] By training our attention to breathing we can progress, or regress, within this continuum of awareness towards a subtler sense of the nested structure of conscious levels as well as of their biocognitive intermodulation and potentially asynchronous patterns. By attuning to breathing, we can magnetise our attention away from higher levels of thoughts, images, and concepts (that we usually feel as the only constitutive elements of our life and self) towards more primordial feelings of core aliveness. We can move, in short, from extended levels of consciousness down to core bodily states of primary awareness, and this can potentially send slowing, regulatory feedback (as it were) from the bottom up.

This multilevel model of consciousness seems more capable of accounting for asynchronous biocognitive feedback within Beckett's characters. It gives a hermeneutic, spatial structure to map their attempts to move away from higher forms of consciousness towards minimal bodily feelings. Since Beckett's first novelistic efforts, in fact, we find characters not only thinking with their breath, but trying to alter thinking by modulating breathing patterns. In *Dream of Fair to Middling Women* (written in 1932, published posthumously in 1992), we see the protagonist, Belacqua, training himself to regress into a meditative darkness, away from his restless thoughts. In what reads like a wonderful slapstick rendition of a self-taught meditational practice, we assist at his active attempts to manipulate biocognitive feedback between bodily and mental states, beginning with breathing patterns:

> He *trained his little brain to hold its breath*, he made covenants of all kinds with his senses, he forced the lids of the little brain down against the flaring bric-à-brac, in every imaginable way he flogged on his coenaesthesis to enwomb him, to exclude the bric-à-brac and *expunge his consciousness*. He learned how with his knuckles to press torrents of violet from his eyeballs, he lay in his skin on his belly on the bed, his face crushed grossly into

> the pillow, pressing down towards the bearings of the earth with all the pitiful little weight of his inertia, for hours and hours, *until he would begin and all things to descend, ponderously and softly to lapse downwards through darkness, he and the bed and the room and the world.* All for nothing. He was grotesque, wanting to 'troglodyse' himself, worse than grotesque. It was impossible to switch off the inward glare, wilfully to suppress the bureaucratic mind. It was stupid to imagine that he could be organised as Limbo and wombtomb, worse than stupid.[42]

It would be possible to interpret the initial reference to the brain's breath or lids as metaphorical, 'psycho-analogical'[43] projections of bodily actions (breathing, closing eyelids) targeting the more abstract domain of conscious thoughts. Given the following extensive description of actual physical manoeuvres Belacqua undertakes to silence his higher cognitive apparatus, however, it is better to understand those references as syncretic, condensed descriptions of biocognitive practices of regulation he concretely operates to move downward levels of consciousness. Belacqua slows the pace of his brain (a noun that is a synecdoche for conscious thinking) by actually slowing down his physical breathing process. This non-metaphorical interpretation of the brain's breath is reinforced by the discovery, a few lines later, that the closing of the brain's 'lids' is also a biocognitive feedback he tries to generate by physically closing his eyelids, as in sensory deprivation practices.

Belacqua therefore attempts biocognitive covenants between mind and bodily processes, trying to affect the former through the latter. He does so in order to move away from the higher, extended level of consciousness (the kind of consciousness he wants to 'expunge') down to a more primitive, core (in his imagery, 'troglodyte') level of cognition. This meditative regression, however, is doomed to fail because of an asynchronous resistance of higher consciousness to biocognitive, upward regulation coming from bodily senses, breathing patterns included. It is rather the extended, 'bureaucratic mind' which agentively wants the organism to pant synchronically forward, in accordance with its frantic tempos. If Belacqua is not attaining any rest in his core level of consciousness we, as readers, are nonetheless getting a sense of the multilevel structure of his life as a living human organism; notably, of the asynchronous biocognitive tension between a physical breath that wants to slow down the mind, and an extended consciousness that wants the body to move in tandem panting along.

The transition that Belacqua would like to make, through biocognitive breathing modulations, is therefore one from a higher, agentive, and autobiographically charged level of consciousness to a lower, core level of minimal consciousness. This transition can be understood, in contemporary phenomenology of mind, as a downward movement *from knowing to experience*. If extended consciousness is the realm of knowledge (*e.g.*, beliefs, autobiographical information, conceptual identities, and wilful control), the core level hosts what Dan Zahavi has called an 'experiential self', which is endowed only with a 'first-personal presence of experience'.[44] Without or before having to know anything about the content of experience, the experiential self we found in the core level does not require a diachronic unity threaded with autobiographical narratives or conceptual information about the world. It only needs an 'experiential continuity':[45] a minimal givenness and presence whose thin temporality is guaranteed by strictly localised phenomenological retentions and protentions offered to his 'interoceptive sensitivity'[46] by bodily processes such as breathing patterns. For the experiential self, as Thompson puts it, the act of breathing is among the 'self-specifying processes [which] support an interoceptive sense of self—a feeling from within of one's body as that which inhabits the sensorimotor perspective from which one perceives and acts'.[47]

This experiential level of subjectivity of our core consciousness is therefore the primordial, yet always active ground of our cognitive life. Once we develop from experiential to epistemic levels of cognition, however, it mostly becomes hidden below the noisy, verbal, conceptual, imagistic surface of our extended consciousness. Given the intimate role of breath in shaping our core consciousness, as Thompson notes, if we want to regress back to our cognitive foundation 'it's therefore no surprise that so many types of meditation begin with focused attention to the sensation of the breath as a way of centering our awareness on our embodied being in the present moment'.[48] Moving away from a knowing mode towards an experiential stance, however, is not just difficult for Belacqua, but even for proper meditators. This is due to the asynchronous tensions between the extended level of conscious verbal thinking, beliefs, judgments, or mental imagery and the core, interoceptive level of bodily processes. Focused breathing can nonetheless be revealing of such tensions, as Varela, Thompson, and Rosch explain:

> Breathing is one of the most simple, basic, ever-present bodily activities. Yet beginning meditators are generally astonished at how difficult it is to be mindful of even so uncomplex an object. Meditators discover that mind and body are not coordinated. The body is sitting, but the mind is seized constantly by thoughts, feelings, inner conversations, daydreams, fantasies, sleepiness, opinions, theories, judgments about thoughts and feelings, judgments about judgments—a never ending torrent of disconnected mental events that the meditators do not even realize are occurring except at those brief instants when they remember what they are doing. Even when they attempt to return to their object of mindfulness, the breath, they may discover that they are only thinking about the breath rather than being mindful of the breath.[49]

Thinking and breathing, core and extended consciousness, are revealed to meditators as lacking coordination. Meditators, like Belacqua, attempt to attune to their core level of consciousness by expunging the seizing, agentive tempos of thoughts, opinions, judgments, theories, and much more. This proves difficult, and to Belacqua impossible, because extended consciousness keeps pulling the organism into higher levels of cognition. The organism therefore becomes the battlefield of biocognitive feedback loops coming from both levels, each trying to modulate the other towards its pace, patterns, and state.

This way, meditators become aware of how the original, ontogenetically primordial, co-dependent intimacy of breath and consciousness in core states has evolved into a divorced tension that resists synchronic convergence and remodulation. Extended and core states develop a different agency, temporality and tempos set in asynchronous negotiations. Core consciousness, on the one hand, still presents the pulsing semi-autonomous agency of an organism living in a thin present. The part of the organism in this state moves forwards at the slower pace of minimal retentions and protentions, rooted in bodily processes such as inspiration and expiration. Extended consciousness, instead, has a much more ambitious, temporally thick, and frantic kinetics. It restlessly casts narratives, beliefs, judgments, and memories retaining the distant past that it projects into the future.

These two levels coexist in the human organism, and this is why meditators can attempt to shift from a state of extended knowledge to a state of experiential mindfulness within themselves. Beckett, however, after having portrayed self-taught meditators such as Belacqua or Murphy (1938; who ties himself to a rocking chair to regress into liminal states of consciousness)[50] decides to explore the asynchronous biocognitive tensions between knowledge and experiential levels by separating them into distinct agents and agentive positions. *Company*, for example, pictures a character lying on his back in the dark, when a mysterious voice begins telling him 'of a past. With occasional allusion to a present and more rarely to a future as for example. You will end as you now are'. Within the multilevel model of consciousness, the voice can be considered as coming from the extended level, pulling the one in the dark out of his core state by the downward channelling of autobiographical memories. The one in the dark is instead located in a lower, core state ('Your mind never active at any time is now even less than ever so. This is the type of assertion he does not question'),[51] whose temporality is thinner than any narrative, only interoceptively grounded in the retentive and protentive emissions of his breath.

The short novella seems to spatialise extended and core consciousness into higher and lower positions, each with its distinct agency, temporality, and tempos. If the voice at the higher level is relentlessly pushing broader narrative arches down into the organism, the one in the dark seems mindfully attuned to the semi-voluntary agency of his breath, infusing his experience just with a presentified slowed tempo. The plot of the novella can be summed up as the story of biocognitive feedback, looping downwards and upwards, without a synchronic resolution. Other texts in Beckett's work similarly model the interplay, interpenetration, and fraught intimacy between extended and core levels as a biocognitive struggle of distinct agentive positions within the same organism. *That Time* (1976), for instance, equally stages three voices coming to an old man, listening to them with his eyes open and '*His breath audible, slow and regular*'. The slow pace of the breath of the character mirrors his minimal responsiveness to the voices that, once again, are forcing upon him memories from his past, 'back and forth without any break in general flow'.[52] This time we are not given any access to the character's mind, and the slowed tempo of his breath is the only index of the asynchronous rhythm of his core aliveness compared to the incessant pace of his voiced memories.

Both texts therefore model the assault that extended levels of consciousness are launching towards minimal stages of the organism, and the resistance opposed by the latter. We assist at biocognitive conflicts between knowledge and memories that are, through voices, agentively breathed into the characters (with obvious biblical echoes); and listening organisms that are trying to stay attuned to the different, more primordial breathing patterns of core states. These biocognitive downwards and upwards negotiations can be conceptualised as two different kinds of *breathing mediacies*.[53] On the higher position, we have the faster, agentive emissions of the breathing voices of memory and knowledge, which attempt to *mediate* lower organisms up into extended levels of cognition. On the lower level, we have organisms attempting to anchor, as meditators do, to breath as a *mediating interoceptive signal* to silence thoughts, memories, judgments, beliefs, and so on.

It could be argued that the former kind of informational breath is more of a metaphoric kind of mediacy compared to the physical breath, which mediates access to lower core states. James' idea that our extended consciousness feels the way it does because of its roots in the retentions and protentions of breathing, however, can explain why Beckett decided to render even higher levels of cognition as having breathing patterns. There is another reason, though, that can cast further light on why Beckett opted to model the mediacy of breathing thoughts as voices. Elsewhere, I have argued that this may be due to Beckett's modelling the dynamic of inner speech, and its role in verbalising knowledge into a pre-reflective, lower state of the organism.[54] As cognitive psychological research on inner speech shows,[55] our inner conversations situate us both in the position of the producer and the receiver of knowledge about ourselves and about the world. As a producer, we are located in the extended level that takes agentive charge of the information to be channelled. As a receiver, or, as Beckett's fictional models define it, a listener, we are more experientially hearing knowledge about ourselves and about the world. We become an experiential object of conversation.

The dynamic structure of inner speech is therefore *isomorphic* to respiration. In a looping circle, we listen as we inspire; and we internally talk as we expire. The inhaling moment is thus equivalent to the breathing mediacy through which an autobiographical life is channelled into a lower organism. This core organism would feel, as a meditator does, as if thoughts, images, beliefs, and judgments have been pushed on him; as a listener who then has to exhale some lesson imparted to him. This is the kind of phenomenological report we read in *The Unnamable*, in which the narrator is constantly disowning the contents of his narrative by explaining how he is only exhaling a lesson ('They gave me courses on love, on intelligence, most precious, most precious. They also taught me to count, and even to reason') that has been breathed into him ('They blow me up with their voices, like a balloon, and even as I collapse it's them I hear').[56]

In Beckett's work across different media, we therefore assist, and vicariously undertake, what it is like to be an experiential core self asynchronically battling against the agency of thoughts, beliefs, judgment, and images that core selves are forced to inhale and then exhale at accelerated tempos. They discharge the contents of an extended consciousness ('a pensum to discharge')[57] that commands them to pant forward, away from more primitive and primary patterns of breathing and consciousness. As James explains in his *Talks to Teachers* (1899), sometimes consciousness discharges in outer behaviours: 'there is no sort of consciousness whatever, be it sensation, feeling, or idea, which does not directly and of itself tend to discharge into some motor effect. […] It may be only an alteration of the heart-beats or breathing'.[58] This seems to be the kind of motoric discharge of extended states in works like *Not I, How It Is*, and *The Unnamable*. If extended and core consciousness have different agency, temporality, and tempos (hence some of the limits of James' model), the former might tyrannously force its patterns onto the latter, which then becomes a state of non-agentive exhalation of inhaled knowledge (as in *Not I*, 'just the birth cry to get her going … breathing … then no more till this';[59] or in *Ceiling*, 'Dull with breath. Endless breath. Endless ending breath.').[60] Accordingly, the breathing voices in Beckett's work are perceived by the experiential characters as masters, guardians, judges, or tyrants ('my college of tyrants')[61] with asymmetric knowledge compared to the experiential subjects.

Breath is therefore key to Beckett's exploration of asynchronous biocognitive feedback between levels of knowledge and levels of experience in conscious human organisms. Different agency, temporality, and tempos in different levels of consciousness present different breathing patterns in his work. Patterns of breath are either an index of hectically discharged knowledge, or a core, pulsing signal of experiential resistance to that knowledge by the organism. Intruding knowledge or experiential mindfulness, however, are not the only options for focussed breathing meditation, nor do they seem to be for Beckett's characters. A third, ultimate attentional shift awaits the skilful meditator: an access to a mode of consciousness behind and beyond extended and experiential states. This further state is what Thompson describes as 'open awareness', which usually begins with a focus on breathing and then 'takes the form of witnessing thoughts, emotions, and sensations as they arise from moment to moment, and observing their qualities'.[62] When a meditator reaches this stage, she is neither in the extended nor in the core state, but detaches from both to watch unjudgmentally their biocognitive feedback from an outside stance.

While in this state, the organism is aware of both levels and of their reciprocal negotiations 'without identifying with them or with the self that appears in them.'[63] Building on Vipassanā terminology, Thompson defines open awareness accordingly as 'non-reactive', 'non-clinging', or 'non-grasping'. A state where one simply 'observes or witnesses the coming and going of sensory and mental events without getting caught up in cognitive and emotional reactions'.[64] Reworking the same Vipassanā tradition in analytic philosophical terms, Miri Albahari calls this deeper level of awareness 'witness-consciousness': a further level which is 'the raw registration of any experience'.[65] Physiological breathing can therefore initiate the transition from knowledge to experience, and then itself become part of the broader, subjectless field of experience of witness-consciousness. Here the organism steps outside biocognitive synchronies and asynchronies: to some extent it steps outside chronological time altogether, watching the clashing of downwards and upwards feedback as one might watch rainstorms dampening the earth from a cosmological perspective.

The stupefied, unjudgmental, non-clinging features of witness-consciousness seem to capture some of the unique qualities of Beckett's fictional organisms when they are not discharging thoughts or interoceptively attuned to minimal states. At the beginning of *The Unnamable*, we find the protagonist unjudgmentally noticing the appearance of two other protagonists from Beckett's previous novels: 'Malone is there. Of his mortal liveliness little trace remains. He passes before me at doubtless regular intervals, unless it is I who pass before him [...]. Perhaps it is Molloy, wearing Malone's hat. But it is more reasonable to suppose it is Malone, wearing his own hat. Oh look, there is the first thing, Malone's hat.'[66] This non-reactive observation of what resemble hallucinatory perceptions is close to Thompson's report of his own attaining of open awareness, where he describes how he 'slipped into a state where for some time I watched a parade of unfamiliar faces while I remained aware of the subtle feeling of the air entering and exiting my nostrils [...]. The faces seemed imaginary—creations of my mind—but also perceptual'.[67] Here Thompson is highlighting how he experienced both mental images and bodily feelings without clinging to them.

This kind of witnessing distance can be found in several moments within Beckett's fictional minds. In one of the *Texts for Nothing*, the narrating stance of the character seems remote both from extended and core levels of consciousness, even if he is still trying to intervene when 'I say to the body. Up with you now, and I can feel it struggling [...]. I say to the head. Leave it alone, stay quiet, it stops breathing, then pants on worse than ever'.[68] In spite of still caring for how bodily breath and breathing thoughts modulate each other, the character seems positioned outside their interaction. More blissful moments of witnessing detachment can be found in *Malone Dies* (1951), whose narrating protagonist feels he has trained himself to move beyond any attachment to his bodily and mental states: 'All my senses are trained full on me, me. Dark and silent and stale, I am no prey for them. I am far from the sound of blood and breath, immured'.[69] What makes Malone's non-reactive distance a captivating reading experience is precisely his non-clinging curiosity about the interplay between his old body and his dimming consciousness: 'My body does not yet make up his mind. But I fancy it weighs heavier on the bed, flattens and spreads. My breath, when it comes back, fills the room with its din, though my chest moves no more than a sleeping child's. I open my eyes and gaze unblinkingly and long at the night sky. So a tiny tot I gaped, first at the novelties, then at the antiquities'.[70]

More often than not, however, Beckett's characters are unable to attain or fully rest in such open awareness. It is rather the audience or readers who are positioned as witnesses of the incessant, sometimes cruel biocognitive mediacies between extended and core levels within the organism. Like Buster Keaton in Beckett's *Film* (1965), we are vicariously suffering the impossibility of finding an 'angle of immunity'[71] to shelter ourselves from the breathing patterns of chasing guardians, masters, tormentors. As a result, we are panting along within Beckett's sensory-deprived storyworlds, which nonetheless turn into meditational technologies for training our attention to the intimate modulation between breathing and consciousness.

Conclusion

More could be said about Beckett's further analogical pairing between breathing and writing (see, *e.g.*, Malone's non-agentive writing where 'the air that breathes through my pages turns them without my knowing';[72] or *Texts for Nothing*, with 'breathing in and out, with words like smoke')[73] or about the guilt of breathing fictional creatures into life (since 'it is enough to breathe to qualify for asphyxiation').[74] The aim of this essay was to show, through the interpretive potential of cognitive scientific models, how Beckett's exploration of the co-dependent intimacy of breath and consciousness links to key contemporary problems in consciousness studies, notably in relation to new enactive accounts of biocognitive feedback between higher and lower levels of cognition. In broaching such a vast topic, interdisciplinarily, and in such a complex author, however, it is impossible not to feel as Beckett did when discussing with George Duthuit the relationship between art and subjectivity: 'All this is literary, over-simplified, but we breathe as best we can'.[75]

Notes

1. See, *e.g.*, J. N. Han et al., 'Influence of Breathing Therapy on Complaints, Anxiety and Breathing Pattern in Patients with Hyperventilation Syndrome and Anxiety Disorders', *Journal of Psychosomatic Research* 41/5 (1996), 481–93; Antonio E. Nardi et al., 'Panic Disorder and Control of Breathing', *Respiratory Physiology & Neurobiology* 167/1 (2009), 133–43; Luiz Dratcu, 'Panic, Hyperventilation and Perpetuation of Anxiety', *Progress in Neuro-Psychopharmacology and Biological Psychiatry* 24/7 (2000), 1069–89.
2. See, *e.g.*, Michael E. Clark and Richard Hirschman, 'Effects of Paced Respiration on Anxiety Reduction in a Clinical Population', *Biofeedback and Self-Regulation* 15/3 (1990), 273–84; Ravinder Jerath et al., 'Self-Regulation of Breathing as a Primary Treatment for Anxiety', *Applied Psychophysiology and Biofeedback* 40/2 (2015), 107–15.
3. Jared R. Lindahl et al., 'A Phenomenology of Meditation-Induced Light Experiences: Traditional Buddhist and Neurobiological Perspectives', *Frontiers in Psychology* 4 (2014), 973.
4. *Sonic Cradles* are technologically enhanced environments for meditation, where in a chamber of complete darkness users can shape soundscapes by using their respiration. See Jay Vidyarthi et al., '*Sonic Cradle*: Designing for an Immersive Experience of Meditation by Connecting Respiration to Music', *Proceedings of the Designing Interactive Systems Conference* (2012), 408–17.
5. See Francisco J. Varela, Evan Thompson, and Eleanor Rosch, *The Embodied Mind: Cognitive Science and Human Experience* (Cambridge, MA: MIT Press, 1993); Evan Thompson, *Waking, Dreaming, Being: Self and Consciousness in Neuroscience, Meditation, and Philosophy* (New York:

Columbia University Press, 2015); Owen Flanagan, *The Bodhisattva's Brain: Buddhism Naturalized* (Cambridge, MA: MIT Press, 2011).

6. Thompson, *Waking, Dreaming, Being*, xxxiii.
7. I borrow from Mario Martinez the term 'biocognitive', which he uses to define the interaction between bodily, mental states, and actions, such as the possibility of breathing patterns affecting mental states, and vice versa. See Mario E. Martinez, 'The Process of Knowing: A Biocognitive Epistemology', *The Journal of Mind and Behavior* (2001), 407–26. Biocognitive feedback should be distinguished from 'biofeedback', which refers instead to feedback artificially obtained by linking the body to external technologies such as respiration in sonic cradles or the enhanced monitoring of heartbeats; see Clark and Hirschman, 'Effects of Paced Respiration'.
8. For a critical survey see Mark Rowlands, *The New Science of Mind: From Extended to Embodied Phenomenology* (Cambridge, MA: MIT Press, 2010).
9. *Krapp's Last Tape and Other Short Plays*, preface by S. E. Gontarski (London: Faber, 2009), 149–60.
10. *Three Novels: Molloy, Malone Dies, The Unnamable*, ed. by Stanley Gontarski (New York: Groove Press, 2009), 286.
11. *Company, Ill Seen Ill Said, Worstward Ho, Stirring Still*, ed. by Dirk Van Hulle (London: Faber, 2009), 3–4.
12. *Krapp's Last Tape and Other Short Plays*, 77–79.
13. Sozita Goudouna, *Beckett's Breath: Anti-Theatricality and the Visual Arts* (Edinburgh: Edinburgh University Press, 2018).
14. On James and mind metaphors see Jill M. Kress, 'Contesting Metaphors and the Discourse of Consciousness in William James', *Journal of the History of Ideas*, 61/2 (2000): 263–83.
15. *The Principles of Psychology, Vol. I* (New York: Dover, [1890] 2009), 239.
16. *Essays in Radical Empiricism* (Whithorn: Anodos Books, [1912] 2019), 5.
17. Kress, 'Contesting Metaphors', 264.
18. For embodied theories of metaphor and figurative language see George Lakoff and Mark Johnson, *Metaphors We Live By* (Chicago: University of Chicago Press, 1980); Raymond Gibbs, *Interpreting Figurative Meaning* (New York: Cambridge University Press, 2012).
19. *Essays in Radical Empiricism*, 5.
20. *Essays in Radical Empiricism*, 15–16.
21. See Shaun Gallagher and Dan Zahavi, *The Phenomenological Mind* (New York: Routledge, 2008), 83–90.
22. See Gallagher and Zahavi, *The Phenomenological Mind*, 84.
23. On the phenomenology of hyper-reflexive states and their relation to consciousness and self see Dan Zahavi, *Subjectivity and Selfhood: Investigating the First-Person Perspective* (Cambridge, MA: MIT Press, 2005), 133–38.

24. Beckett had a mediated access to William James while reading, in his formative years, Robert Woodworth's *Contemporary Schools of Psychology* (1931); see Rubin Rabinovitz, 'Beckett and Psychology,' *Journal of Beckett Studies* 11/12 (1989), 65–77.
25. *Texts for Nothing and Other Shorter Proses 1950–1956*, ed. by Mark Nixon (London: Faber, 2010), 30.
26. *The Expelled, The Calmative, The End* with *First Love*, ed. by Christopher Ricks (London: Faber, 2009), 20 (emphasis added).
27. *Texts for Nothing*, 34.
28. 'Heterophenomenology' is how Daniel Dennett describes the study of consciousness operated from the observation of outer behaviour whereby which we can infer phenomenological inner states. See Daniel Dennett, *Consciousness Explained* (London: Penguin, 1991).
29. *The Letters of Samuel Beckett: Volume 4, 1966–1989*, ed. by Martha Dow Fehsenfeld, Lois More Overbeck et al. (Cambridge: Cambridge University Press, 2016), 311.
30. *Letters 4*, 311 (emphasis added).
31. See, *e.g.*, D. Reisberg, 'Subvocalization and Auditory Imagery: Interactions Between the Inner Ear and Inner Voice', *Auditory Imagery* (2014), 107–32. For an empirical study of inner speech, reading, and literary narratives see Ben Alderson-Day et al., 'Uncharted Features and Dynamics of Reading: Voices, Characters, and Crossing of Experiences', *Consciousness and Cognition* 49 (2017), 98–109.
32. *The Letters of Samuel Beckett: Volume 3, 1957–65*, ed. by Martha Dow Fehsenfeld et al. (Cambridge: Cambridge University Press, 2014), 326.
33. *Letters 3*, 352.
34. *Letters 3*, 327.
35. *How It Is*, ed. by Édouard Magessa O'Reilly (London: Faber, 2009), 3.
36. Ruby Cohn, *A Beckett Canon* (Ann Arbor: University of Michigan Press, 2005), 256.
37. See Antonio Damasio, *Self Comes to Mind: Constructing the Conscious Brain* (London: Heinemann, 2010).
38. See Shaun Gallagher, 'Philosophical Conceptions of the Self: Implications for Cognitive Science', *Trends in Cognitive Sciences* 4/1 (2000), 14–21.
39. The idea of a 'thin' or 'experiential' self is formulated by Dan Zahavi in 'The Experiential Self: Objections and Clarifications', in *Self, No Self?: Perspectives from Analytical, Phenomenological, and Indian Traditions*, ed. by Mark Siderits et al. (Oxford: Oxford University Press, 2011), 56–78. See also Dan Zahavi, 'Thin, Thinner, Thinnest: Defining the Minimal Self', in *Embodiment, Enaction, and Culture: Investigating the Constitution of the Shared World*, ed. by Christoph Durt et al. (Cambridge, MA: MIT Press, 2017), 193–99.

40. The multilevel distinction between 'core' and 'extended consciousness' is paired by Damasio with a 'core' and 'autobiographical self', respectively, in Damasio, *Self Comes to Mind*. For a more recent cognitive scientific account of the autobiographical self see Mark Rowlands, *Memory and the Self: Phenomenology, Science and Autobiography* (New York: Oxford University Press, 2007).
41. Thompson, *Waking, Dreaming, Being*, 7.
42. *Dream of Fair to Middling Women*, ed. by Eoin O'Brien and Edith Fournier (New York: Arcade, 1992), 123 (emphasis added).
43. The term 'psycho-analogies' has been invented by Dorrit Cohn to define analogical images that the narrator or the character use to communicate mental states, especially when these are pre-verbal or opaque. See Dorrit Cohn, *Transparent Minds: Narrative Modes for Presenting Consciousness in Fiction* (Princeton: Princeton University Press, 1978).
44. Zahavi, 'The Experiential Self', 59.
45. Zahavi, 'The Experiential Self', 75.
46. Anil Seth, 'Interoceptive Inference, Emotion, and the Embodied Self', *Trends in Cognitive Sciences* 17/11 (2013), 565.
47. Thompson, *Waking, Dreaming, Being*, 336.
48. Thompson, *Waking, Dreaming, Being*, 336.
49. Varela et al., *The Embodied Mind*, 25.
50. *Murphy*, ed. by J. C. C. Mays (London: Faber, 2009).
51. *Company*, 4.
52. *Krapp's Last Tape and Other Short Plays*, 99.
53. The concept of 'mediacy' in narrative theory defines how the access to a story and a storyworld is mediated (by a narrator; a character; an empty, inorganic, or cosmological perspective, and so on). See Monika Fludernik, 'Mediacy, Mediation, Focalization: The Squaring of Terminological Circles', in *Postclassical Narratology: Approaches and Analyses*, ed. by Jan Alber and Monika Fludernik (Columbus: Ohio State University Press, 2010), 105–33. The ontological act through which a character is breathed into life through the mediating function of a narrator can be compared with the mediating function of memories through which an organism is breathed into autobiographical consciousness; see Marco Bernini, *Beckett and the Cognitive Method: Mind, Models, and Exploratory Narratives* (New York: Oxford University Press, 2021).
54. Bernini, *Beckett and the Cognitive Method*, Chapter 2 (45–84).
55. See Ben Alderson-Day and Charles Fernyhough, 'Inner Speech: Development, Cognitive Functions, Phenomenology, and Neurobiology', *Psychological Bulletin* 141/5 (2015), 931.
56. *Three Novels*, 292, 319.
57. *Three Novels*, 304.

58. *Talks to Teachers on Psychology* (1899) (New York: Henry Holt, 1916), 170.
59. *Krapp's Last Tape and Other Short Plays*, 90.
60. Collected in *Company, Ill Seen Ill Said, Worstward Ho, Stirring Still*, 127.
61. *Three Novels*, 304.
62. Thompson, *Waking, Dreaming, Being*, 52.
63. Thompson, *Waking, Dreaming, Being*, xxxiii.
64. Thompson, *Waking, Dreaming, Being*, 53.
65. Miri Albahari, *Analytical Buddhism: The Two-Tiered Illusion of Self* (New York: Springer, 2006), 66.
66. *The Unnamable*, 286–87.
67. Thompson, *Waking, Dreaming, Being*, 121.
68. *Texts for Nothing*, 3.
69. *Three Novels*, 180.
70. *Three Novels*, 199.
71. *All That Fall and Other Plays for Radio and Screen*, ed. by Everett Frost (London: Faber, 2009), 98.
72. *Three Novels*, 228.
73. *Texts for Nothing*, 4.
74. *Three Novels*, 317.
75. *The Letters of Samuel Beckett: Volume 2: 1941–1956*, ed. by Fehsenfeld et al. (Cambridge: Cambridge University Press, 2011), 130.

Selected Bibliography

Albahari, Miri. 2006. *Analytical Buddhism: The Two-Tiered Illusion of Self*. New York: Springer.

Barry, Elizabeth, ed. 2018. Special Issue on 'Beckett, Language and the Mind'. *Journal of Beckett Studies* 17: 1–2.

Flanagan, Owen. 2011. *The Bodhisattva's Brain: Buddhism Naturalized*. Cambridge, MA: MIT Press.

Herman, David, ed. 2011. *The Emergence of Mind: Representations of Consciousness in Narrative Discourse in English*. Lincoln: University of Nebraska Press.

Kearns, Michael S. 1987. *Metaphors of Mind in Fiction and Psychology*. Lexington: University Press of Kentucky.

Lindahl, Jared R. et al. 2014. A Phenomenology of Meditation-Induced Light Experiences: Traditional Buddhist and Neurobiological Perspectives. *Frontiers in Psychology* 4: 973.

Maude, Ulrika, and Matthew Feldman, eds 2009. *Beckett and Phenomenology*. London: Continuum.

Maude, Ulrika. 2009. *Beckett, Technology and the Body*. New York: Cambridge University Press.

Sotirova, Violeta. 2013. *Consciousness in Modernist Fiction: A Stylistic Study*. Basingstoke: Palgrave Macmillan.

Thompson, Evan. 2015. *Waking, Dreaming, Being: Self and Consciousness in Neuroscience, Meditation, and Philosophy*. New York: Columbia University Press.

PART VII

The Twenty-First Century

CHAPTER 22

Syllabic Gasps: M. NourbeSe Philip and Charles Olson's Poetic Conspiration

Stefanie Heine

In response to Nathaniel Mackey's Inaugural Robert Creeley Lecture in Poetry and Poetics 2016, 'Breath and Precarity', M. NourbeSe Philip reframed her book *Zong!* in respiratory terms. The starting point of her essay 'The Ga(s)p' is Mackey's exploration of a 'poetics of breath'.[1] Mackey traces this poetics back to the 'pneumatic turn' in the 1950s and 1960s, when breath-measure became a key term in attempts to emancipate literary composition from inherited traditions and writers like Charles Olson, Allen Ginsberg, and Amiri Baraka formulated their principles for a new American literature. The breath that 'was in the air'[2] then found its most prominent and palpably 'tactile' (8) expression in black music, especially jazz. This black context of respirational poetics is inextricably linked with Mackey's main argument: 'an artistic turn to breath' is symptomatic of 'individual and collective anxiety and insecurity' (6). As he argues, artists in post-war America found themselves in a 'post-traumatic' (6) period in which 'mortal susceptibility' and 'human exposure' (7) were

S. Heine (✉)
Department of Arts and Cultural Studies, Comparative Literature, University of Copenhagen, Copenhagen, Denmark

D. Fuller et al. (eds.), *The Life of Breath in Literature, Culture and Medicine*, Palgrave Studies in Literature, Science and Medicine,
https://doi.org/10.1007/978-3-030-74443-4_22

predominant on a broader socio-demographic scale. This precarious state is compared to the ongoing black experience of being subjected to a 'long state of siege', 'a long history crystallized most recently by Eric Garner's last words in a police officer's chokehold, "I can't breathe"' (13). Mackey suggests an explanation for the turn to breath in atmospheres of heightened anxiety: it implies returning 'to the basics' in moments when everything becomes uncertain. Such a turn to the elementary physiological process that keeps us alive also means going back 'to the primal or primitive *doubts*' (6, my emphasis). Breath, especially black breath, is as such 'subject' to 'precarity' (16) and it matters 'especially if at risk' (17).

Philip, whose literary and essayistic work continually engages with racism, violence against people of colour, and colonial and postcolonial oppression, connects with Mackey's 'idea of precarity as it affects African Americans' (32). In her essay, she takes literally his claim that such precarity effects a 'return to the basics' and the 'primal'. The essay opens with an intrauterine scene:

> WE ALL BEGIN life in water
> We all begin life because someone once breathed for us
> Until we breathe for ourselves
> Someone breathes for us
> Everyone has had someone—a woman—breathe for them
> Until that first ga(s)p
> For air ('The Ga(s)p', 31)

Philip does not only give a feminist twist to Mackey's exclusively male history of a poetic 'pneumatism'.[3] The scene she sketches also reacts to a number of prominent discussions by white men of the intrauterine state and the first breath. Otto Rank claimed 'that all anxiety goes back originally to the anxiety at birth (dyspnoea)'.[4] This feeling of suffocation supposedly coincides with the 'primal trauma'[5] an infant suffers when it is forced to leave an idealized condition of '*intrauterine primal pleasure*'.[6] Another theoretical interpretation follows a passing observation taken by Merleau-Ponty from Henri Wallon: the infant's 'body is already a respiratory body'.[7] In relation to this remark, David Michael Kleinberg-Levin states that '[b]reathing is our most primordial ecstatic openness', teaching the infant 'interdependency', 'giving and receiving'.[8] Along the same lines, but drawing on Peter Sloterdijk's discussion of the 'respiratory phase' as the cultural theorist Thomas Macho reinterprets it, Leo

Bersani argues that 'inhaling and exhaling' are 'our earliest transactions with exteriority'.[9] Bersani's 'analyses of "receptive bodies"', ranging from explorations of queer to non-human receptivity, include a depiction of 'intrauterine origin'[10] that in many respects resonates with Philip's:

> With its first, exuberantly welcomed scream, the human infant announces its respiratory interdependence. Within the womb, it could rely on the mother breathing for it. Having accomplished the at once biologically and symbolically necessary severance from the mother's breathing rhythm, she is on her own, dependent on her own lungs to sustain the precarious individual life into which she has just fallen. Breathing is the tiny human's first experience of her body's inescapable receptivity: absorption and expulsion.[11]

The conception of a fundamental 'respiratory interdependence' and exposure in a particularly feminine framework, which Bersani highlights through his focus on the mother's act of 'breathing for' the infant as well as the child's gendering, is precisely what Philip develops further.

Inspired by Hannah Arendt's thoughts on natality, Philip argues concerning the 'idea of the mother breathing for the foetus' that this 'process of shared breath [...] and dependency becomes useful as a model of community and connectedness in a more female-centred, embodied, symbolic universe' ('The Ga(s)p', 36). The 'imperative to lodge the poetics of breath as identified by Mackey in the Black female imaginary' (36) immediately leads her to the question 'How do we begin to think about shared breath, circular breath, or circle breath in the context of force—historical (enslavement) and contemporary' (36)? Upon finding that the *OED* gives one definition of 'gasp' as 'the last attempt to breathe before death' (34), Philip outlines a respiratory short-circuit:

> The first ga(s)p of the newborn who has, until then, been breathed for, signals a beginning; it is the same act—that of forcefully attempting to draw air into the lungs that will mark the final moment of a life—"I can't breathe...." The last words uttered by Eric Garner as he lay dying on a New York sidewalk. (34)

In the 'Black female imaginary', a history of violence and suppression also pervades the intrauterine setting that has often been stylized as a paradisial origin. From her general observation that '[b]reathing for the other' is an 'expression of radical hospitality' (34), '[r]adical because [...] physiologically the child is also a stranger' (34–35) onwards, Philip's language is infiltrated by a vocabulary of colonial force: 'housing the stranger, which includes breathing for the *occupant*' is an 'act of acceptance of alterity' (35, my emphasis). She then turns to a scenario in which the stranger is violently imposed on the woman's body: the 'forced breeding of enslaved African women' and 'the forced impregnation by white and European masters' (35).

While the gender-political and racial concerns of Philip's delineation of shared breath are apparent, it is so far still unclear how it relates to a *poetics* of breath. The poetic implications of Philip's reflections on respiration become most specific when she turns to biochemical matters. As she points out, the mother's 'radical hospitality' of breathing for the unborn child is facilitated by the body's molecular processes: 'Ordinarily, the mother should generate activated T-cells, which would then attack the foetus's foreign antigens'. Instead, the 'mother's body turns off the functions of her T-cells, which would normally result in the rejection of the foetus' (35). Philip's excursion into the field of biology turns out to be the crucial link to the poetics of breathing and fragmentation she exemplifies with her book *Zong!*. Biochemical vocabulary is what the primal scene, the model of community, and the comments on *Zong!* share in 'The Ga(s)p'. In this context, cell-*memory* is the central focus:

> Do our cells carry this epigenetic memory of a form of sharing and exchange, a modelling of a "we" that we can take into our varied practices? Do they, our cells, remember what it was like to have someone breathe for us? And how, if at all, does the epigenetic cellular memory of the forced breathing of Black women affect this most generative and generous of acts? Are we marked by this—by having someone breathe for us? A memory that is a blueprint for community and interdependency. Despite the forced couplings. (38)

The potential 'cellular memory' that might be shared is punctured by loss and deprivation: someone once breathed for us, gave her breath away; she might have been forced to do so, after her freedom and body were taken; at the outset, our breath was not our own, and probably it never is, never will be.

What Philip outlines in 'The Ga(s)p' goes hand in hand with Jean-Luc Nancy's notion of 'inoperative community'. For Nancy, being-in-common is based on a lack and 'the common is the sharing of finitude'.[12] Thereby, he understands 'sharing' as the 'condition of being-exposed', drawing on the French implications of 'partage',[13] which also means dividing or splitting into parts. Philip imagines a very visceral *partage*: a communal sharing of particles, cell-carried memory. Her suggested 'blueprint for community', the common memory of 'what it was to have someone breathe for us', implies recalling an 'act of acceptance of alterity' ('The Ga(s)p', 35): our bodies' smallest parts remember an intimate sharing of breath with what is most disparate—the Other, the occupant. In Nancy's words: 'partage', the basis of community, always entails that 'I experience the other's alterity, or I experience [...] the alterity [...] "in me"'.[14] Nancy argues that 'community cannot be presupposed. It is only exposed'.[15] Philip calls the '*memory* that is a blueprint for community' into question: 'Do they, our cells, remember what it was like to have someone breathe for us?'. Consequently, what Philip's imagined community shares, what it rests upon, is neither given nor granted; it is precarious, left open, exposed. This is the uncertain foundation on which Philip bases her 'poetics of breath' or 'poetics of the fragment', a 'distinction without a difference' (39), as she claims. *Zong!*, which 'illuminates' this poetics, consists of 'fragments driven pneumatically' (39) and amounts to a *partage* of smallest particles, cells: 'When I perform *Zong!*, I allow the words and word clusters to breathe for the I n' I—for the we in us that we carry within the memory of our cells. When I invite the audience to read with me, we collectively engage in breathing for the Other—for those who couldn't breathe' (39).

In the context of *Zong!*, 'breathing for the Other' and sharing mortality points to a very specific scenario of racial violence. '[T]hose who couldn't breathe' refers to ~150 African Americans who were deliberately drowned on board the slave ship *Zong* in 1781, a massacre committed for financial reasons during a voyage on the Middle Passage that due to navigational errors took much longer than expected. As Philip summarizes, when water supplies were getting short, the ship's captain 'was of the belief that if the African slaves on board die a natural death, the owners of the ship will have to bear the cost, but if they "were thrown alive into the sea, it would be the loss of the underwriters"'.[16] The matter turned into a legal dispute, resulting in a trial brought by the ship's owners and an appeal by the insurers, heard in the King's Bench and recorded in the report

Gregson v. Gilbert. The fragmented language of *Zong!* consists exclusively of linguistic material from the report of the appeal. Philip pulls the legal text apart, rearranges sentences and words, '*carv[es] words out of other words*' (*Zong!*, 198) and tears them apart into isolated letters and syllables. In an interview, she describes this process of un-writing in a way that directly ties in with her speculation about the 'memory of our cells': 'It is as if in fragmenting the words, the stories locked in DNA [sic] of those words are released'.[17] The 'DNA' to which Philip alludes is unlocked most effectively when the text is ripped to the smallest pieces—the word's cells: syllables, letters.

As Philip stresses, this act of 'explod[ing] the words' (*Zong!*, 203) makes them 'break into sound' (205) and the legal text is transformed 'into a cacophony of voices—wails, cries, moans, and shouts' (203): 'o', 'oo', 'oh', 'w', 'waa', 'a', 'ah', etc. (3). In the 'Notanda' accompanying *Zong!*, we find, in precisely this context, a sentence that echoes her question 'Do they, our cells, remember what it was like to have someone breathe for us?': 'Do they, the sounds, the cries, the shouts of those thrown overboard from the Zong repeat themselves over and over until they rise from the ocean floor to resurface in *Zong*? It is a question that haunts me' (203–04). Juxtaposed thus, these questions display a transition from commemoration to haunting that lies at the heart of *Zong!* and Philip's poetics of breath: '*Zong!* is hauntological' (201). The uncertainty of a respirational cell-memory mirrors the uncertainty around the textually recovered moans and groans of the drowning slaves. At times, Philip seems to be confident that the violently silenced voices can be recovered in her poetry:

> In *Zong!*, the African, transformed into a thing by the law, is retransformed, miraculously, back into human.
> Through oath and through moan, through mutter, chant and babble, through babble and curse, through chortle and ululation. (196)

Against this wish for a magical re-humanisation of the victims, the incorporation of their dying breaths to the text turns into a 'haunting' question—a question that also unsettles the more moderate hope that they can be remembered at all. From the outset, Philip stresses that *Zong!* is composed around an irrecoverable loss. The Africans who died on the *Zong* were subject to a double silencing: their voices were suffocated in the sea and by-passed by historical documentation—the surviving report anonymizes and dehumanizes them as cargo. Addressing the erasure of the Africans in the textual basis of *Zong!*, Christina Sharpe succinctly concludes: 'Breathlessness and the archive: the archives of breathlessness'.[18] The breath taken from the slaves, their lost struggle for air in the water, is 'lessened', left unarchived, in the suffocating historical document *Gregson v. Gilbert*. As a consequence of this ultimate silencing, the murdered slaves cannot be re-called, just as their bones cannot be recovered 'exaqua' (*Zong!*, 201).[19]

Philip's poetic project to a large extent consists in exposing the impossibility of remembering the dead. The imagined transference of the dying Africans' gasps into word-cells is ruptured by the gap that was torn open when they were silenced. Such a ga(s)p in memory is poetically staged in Philip's earlier collection of poetry, *She Tries Her Tongue, Her Silence Softly Breaks*, when she tears the word 're-member' apart and thereby inscribes a blank space. In the poem 'Universal Grammar', the method of linguistic dissecting, or 'parsing', is performatively displayed with words that later become crucial for the poetics of *Zong!* as delineated in the 'Notanda' and later in 'The Ga(s)p':

> the smallest cell
> remembers
> a sound[20]

On the page before, these words, here arranged in verses, appear as follows:

> Parsing—the exercise of telling the part of speech of each sentence (Latin, pars, a part)
> The—distinguishing adjective, limiting the noun, cell.
> smallest—adjective of quantity [...], cell (unsuccessfully)
> cell—common noun [...], remembers. (Long-term memory improves cell growth in nerve cells.)
> remembers—regular verb [...] cell which remembers and so re-members.
> O—sound of exclamation [...] (*She Tries Her Tongue*, 36)

The sentence 'the smallest cell remembers a sound', or 'the smallest cell remembers O' (a sound that will conspicuously pervade *Zong!* – i.e. one of the ga(s)ping word-cells that might re-member but fails to bring back the voices of the slaves) is 'parsed': each line following the sentence's vertically arranged words mimics a dictionary entry.

On the following pages, two further explications gesture towards the ambivalence inherent in Philip's respirational poetics of the fragment: '**fragments**—common noun [...] Re-membered fragments become whole' (38). '**Parsing**—the exercise of dis-membering language into fragmentary cells that forget to re-member' (40). The first 'definition' plays with a phantasy of recovery: when scattered parts are reassembled and memory is recuperated, dismembered bodies too might become whole again. The second one alludes to the flipside of fragmentation: dissecting language—be it analytically ('parsing') or poetically (Philip's own method, employed in this poem as well as in *Zong!*)—implies that it falls to pieces which cannot be put together again seamlessly; cells might not re-member, in both senses. Or, in other words, the concept of 're-membering' as commemoration effected by scattering fragments and word-cells contains a gap, a potential loss:

> when the smallest cell
>   remembers
> lose a language (41)

The proximity of 'lose a language' and 'loose a language' is not only acoustic. The fragmented poetry of *Zong!* displays this on two levels: it both exhibits a language that has been lost and cannot be remembered, and stages 'intelligible words' turning into prelinguistic 'sound' (33), words losing what qualifies them as constituents of a system of meaningful signs.

The tension between 'intelligible words' and 'sound' brings us back to breath and to the intrauterine scene sketched by Philip in 'The Ga(s)p'. The 'Discourse on the Logic of Language', which immediately precedes 'Universal Grammar' in *She Tries Her Tongue*, contains the following mock multiple-choice exercise:

> Air is forced out of the lungs up the throat to the larynx where it causes the vocal chords to vibrate and create sound. The metamorphosis from sound to intelligible word requires
>
> a. the lip, tongue and jaw working together.
> b. a mother tongue.
> c. the overseer's whip.
> d. all of the above or none. (33)

Philip's account of respiration and language production, already shadowed by a sense of violence through the formulation '[a]ir is forced out of the lungs', is embedded in the history of slavery ('the overseer's whip'). How this relates to (a) and especially (b) is elucidated in the preceding parts of the text, which poetically negotiate the ambivalent relation from a black perspective to English as a 'mother tongue' (30). Even though it might be their first one, the 'language' inherited from the colonial masters will always be 'l/anguish /[…] a foreign anguish' (30) for those who were and are subjected and whose 'native' language was taken away. On the same page, Philip reproduces the language of slave owners to illustrate the process of linguistic expropriation as a strategic means of preventing slave rebellions. While the poem introduces 'father tongue' (30) as the term that more aptly captures the foreignness of English, the 'mother tongue' cannot be fully recuperated—it remains ambivalent, precariously close to 'the overseer's whip'.

A foreignness forcibly imposed on and incorporated by the subject of racial-colonial violence also appears in the respirational intrauterine scene depicted in 'The Ga(s)p', but in that case it is associated with the child in the mother's womb, the 'occupant' resulting from 'the forced impregnation by white and European masters'. In 'The Ga(s)p' and Philip's literary texts, the roles of mother and child are diametrically opposed. I would like to suggest that, against this background, we can define Philip's poetics of breathing and poetics of the fragment as a *poetics of conspiration*. If we read the respirational primal scene Philip proposes not only as a 'blueprint for community and interdependency' but also as a blueprint for her poetics of breathing, the '*prepositional* relationship with breath' changes ('The Ga(s)p', 31, my emphasis): breathing *for* turns into breathing *with*. The intricacies of *con*-spiration are already negotiated in 'Universal Grammar' with respect to the 'mother tongue'—the language we *cannot chose* but *have*. In *Zong!*, they become the major poetic challenge in the futile attempt of recovering a language *lost* from the only source we *have*. As Philip points out in the 'Notanda', '*The text—the reported case—is a matrix—a mother document*' (*Zong!*, 200). 'How do we "breathe with" such a smothering matrix?' 'We blow her to pieces', Philip seems to suggest, describing the composition of *Zong!* as a textual matricide:

> I murder the text, literally cut it into pieces, castrating verbs, suffocating adjectives, murdering nouns, throwing articles, prepositions, conjunctions

> overboard, jettisoning adverbs: I separate subject from verb, verb from object—create semantic mayhem, until my hands are bloodied, from so much killing and cutting. (193)

In contrast to Sara Ahmed's appeal to put up resistance by 'rebuild[ing] the master's residence' after 'shattering' its 'walls',[21] Philip's poetic engagement with the 'matrix' is neither combative nor restorative. She rather *exposes* both the violence of the text, and her own violence against it, by leaving it in scattered fragments. Thereby, the notion of residence, which Philip, akin to Ahmed, draws on in her respiratory primal scene ('hospitality', 'housing the stranger'), is abandoned. Here we can pinpoint a crucial difference between the maternal 'breathing for', giving shelter for the occupant in the enclosed space of the womb, and conspiration, 'breathing with' the violent matrix. Conspiring with a '*mother document*' articulates itself in dismembered syllabic ga(s)ps that cannot be contained, in parallel with the bodies and voices that cannot be salvaged. Conspiration also implies entering a 'zone of contamination' (*Zong!*, 199): rather than being undone, the violence of a matrix that is not enclosed unavoidably becomes contagious. As a result, neither the author nor the literary child is any longer innocent: Philip perceives herself as a killer and cutter of the text while the 'castrat[ed] verbs', 'suffocat[ed] adjectives', 'murder[ed] nouns', or 'jettison[ed] adverbs' are imbued with the matrix's oppressive air—maybe this is something the plaintive syllables too bemoan.

Through the idea of isolated syllables, or language cells, Philip's text conspires (perhaps unwittingly) with a literary predecessor through the idea of isolated syllables, or language cells: Charles Olson, whose 1951 essay 'Projective Verse' initiated the American 'poetics of breath' Mackey discusses. The passages in which Olson attends to syllables, 'the elements and minims of language' in 'Projective Verse'[22] are reproduced almost word by word from a short piece he wrote in 1946, *Mouths Biting Empty Air*. In this unpublished essay, syllables and breath are suggestively conjoined; this is not the case in 'Projective Verse', where the discussion of 'particles of sound' (17) does not feature respiratory vocabulary or images. The breath-syllable nexus as it is presented in *Mouths Biting Empty Air* strongly resonates with Philip's respiratory poetics. The words at the very end of Olson's short essay, 'I give you something *with* which to *breathe*, the syllable' (*Mouths Biting Empty Air*, my emphasis),[23] intertextually conspire with Philip's claim '"Zong!#1," [...] is in fact [...] a series of ga(s)ps with syllabic sounds attached or overlaid' ('The Ga(s)p', 39). The very different contexts of these sentences as well as of the writers' poetics of breathing in general mean their juxtaposition represents a 'forced coupling' in another sense. A particularly troubling undertone to the echoes between their poetic projects is generated by the fact that Olson's writing has, for good reasons, been criticized for its masculinist stance and colonial insensitivity.[24] However, it is precisely in these non-harmonious contacts and correspondences that the ambivalences inherent in a poetics of conspiration can be unfolded: For both Philip and Olson, syllables, the elementary particles of language as well as breath, a basic life-sustaining process, are fundamental to poetry. Olson opens 'Projective Verse' with the sentence: 'Verse now, 1950, if it is to go ahead, [...] must, I take it, catch up and put into itself certain laws and possibilities of the breath, of the breathing of the man who writes' (15). The role attributed to the 'particles of sound' in the new poetry is equal to the importance of the 'laws and possibilities of the breath':

> Let's start from the smallest particle of all, the syllable. It is the king and pin of versification, what rules and holds together the lines, the larger forms, of a poem.
>
> It would do no harm, as an act of correction to both prose and verse as now written, if both rhyme and meter, [...] both sense and sound, were less in the forefront of the mind than the syllable, if the syllable, that fine creature, were more allowed to lead the harmony on. ('Projective Verse', 17, 18)

Moreover, Philip and Olson associate syllables and breath with 'origin'. Olson hints at a poetic origin when he notes that 'the projective poet will [go], down through the workings of his own throat to that place where breath comes from, [...], where, the coincidence is, *all acts spring*' (26, my emphasis), and calls syllables 'the minimum and *source* of speech' (18, my emphasis). The notion of an *ontogenic primordiality* as it is evoked in Philip's sketch of the intrauterine respirational scene is recalled when she links syllables to an early stage in the process of language acquisition: '*The poems are about language at its most fundamental in the sense of the very basic way in which children put language together when they begin to speak, building syllable on syllable*' (195). On a broader, evolutionary, phylogenetic level, she situates the syllabic sounds pervading *Zong!* in linguistic prehistory: 'Words break into sound, return to their initial and originary phonic sound—grunts, plosives, labials—is this, perhaps, how language might have sounded at the beginning of time?' (205).

A phylogenetic notion of origin enters Olson's discussion of the syllables when he illustrates their poetic value—a capacity to enact 'figures' of 'dance' (18)—with etymological speculations that suggest the following Sanskrit roots of English words:[25]

> 'Is' comes from the Aryan root, *as*, to breathe. The English 'not' equals the Sanscrit *na*, which may come from the root *na*, to be lost, to perish. 'Be' is from *bhu*, to grow. (18)

Olson's playful demonstration of how 'the minimum and source of speech' acoustically lead 'the harmony on' (is-*as*; not-*na*; be-*bhu*) goes back to a problematic textual source. The passage is taken from Ernest Fenollosa's controversial essay 'The Chinese Written Character as a Medium for Poetry'.[26] Praised by Ezra Pound, who published it in his *Instigations*, the essay was enthusiastically received in the context of avant-garde poetics, while Fenollosa's less than perfect knowledge of Chinese[27] and, above all, his orientalist and primitivist tendencies were harshly criticized.[28] Olson's negotiation of the syllable's poetic potential therefore also draws on a mother-text tainted with colonial force. Olson, on the one hand, strongly sympathizes with the essay, calling it the 'damned best piece on language since when'.[29] On the other, however, both his theoretical and literary-creative treatment of linguistic 'roots', etymons, undermine Fenollosa's discussion of Sanskrit.

The way in which Fenollosa, who considers himself 'an enthusiastic student of beauty in Oriental culture' (359), conceives of the ancient non-Western 'natural' languages goes hand in hand with what Viktor Li conceptualizes as '[a]rcadian primitivism': 'the primitive is seen as a corrective to the malaise of Western Modernity'.[30] Thereby, 'the more affirmative' primitivism[31] utilizes a 'rhetoric of contrast' ensuring 'that the alterity or otherness of the primitive is kept in the foreground';[32] 'the primitive is conceived and valued solely as the antithesis of the modern West'.[33] Fenollosa discusses 'primitive Sanskrit' (377) and Chinese in precisely this manner, as languages closer to nature and 'primitive men' (366) that contrast with the less poetic modern Western languages, which are determined by abstraction and arbitrariness. The passages Olson cites are embedded in this line of argument. Fenollosa attempts to 'reveal' 'ancient roots' of 'our large vocabularies' that 'dealt directly with physical processes' (377): almost 'all the Sanskrit roots, which seem to underlie European languages, are primitive verbs, which express characteristic actions of visible nature' (373). This claim is enmeshed in an orientalist narrative of recovery: doing 'consciously what the primitive races did unconsciously', the 'chief work of literary men in dealing with language, and of poets especially lies in feeling back along the ancient lines of advance' (378). When Fenollosa uncovers the 'vivid verbs' (377) of 'primitive Sanskrit', it is no accident that 'breathe' serves as an illustration. By claiming that '"Is" comes from the Aryan root, *as*, to breathe' (369–70), he stages a drama of re-animation in which Western 'scholars and poets' 'feel [...] back', reaching right into the womb of the primordial mother-language:

> Only scholars and poets feel painfully back along the thread of our etymologies and piece together our diction, as best they may, from forgotten fragments. This anemia of modern speech is only too well encouraged by the feeble cohesive force of our phonetic symbols. There is little or nothing in a phonetic word to exhibit the *embryonic stages of its growth*. (379, my emphasis)

Olson's pneumatic poetic endeavour is significantly different. The way in which he reflects on etymons and integrates the 'breath-root' in his own writing resists the narrative of retrieving a submerged origin. Even though Olson flirts with the idea of a primal language to which all others can be traced back,[34] he does not conceive of such an origin as an exotic primordial mother tongue. When he claims that etymological roots, or 'resistant primes' represent an 'impetus and explosion in our alphabetic speech' that puts 'us back to the origins of their force not as history but as living oral law',[35] the origin is uprooted. The 'origin' of the roots' 'force' is not their historical situation in the beginning, but their ability to affect the language spoken and written in the present. What is for Fenollosa the 'feeble cohesive force of our phonetic symbols' is re-evaluated as a 'living oral law' that breaches the boundary between past and present: 'back to the origins' is right now. Olson counteracts a 'rhetoric of contrast' by undermining the orientalist phantasy of a bygone primordial past, a constitutive other on which Western language has to turn back in order to unfold its full poetic potential. For Olson, poetic force is not set free through restauration or reconstruction, but, on the contrary, through 'explosion'. Rather than in 'piecing' the 'forgotten fragments' together into a coherent whole, he is interested in setting them loose. In this context, the nexus of syllables and breath as it is presented in *Mouths Biting Empty Air* is particularly revealing:

> 'Is' comes from the Aryan root, *as*, to breathe. The English 'not' equals the Sanscrit *na*, which may come from the root *na*, to be lost, to perish. 'Be' is from *bhu,* to grow.
>
> From the root out the syllables come, the figures of the dance. The intricacy lies in the three extending phases from the body [...] to the arms, limbs, leaves, or fingers, and their gestures, out into that third state, movement, where the dance is, and the life proper, where they strike against and pile up one upon the other, or go and stay apart, or fuse and cease to be themselves, any and all the conditions atoms and breath can create by act and multiplication.

Olson explodes rootedness by refusing to identify the source of the etymological speculation and not even marking it as a quotation. Through this act of stealing Fenollosa's words, he gestures towards another etymological implication of breath: in French, *souffler* is associated with robbing.[36] Divorced from their context, the words are torn out of the primitivist narrative. Thereby, the colonial burden is not lost, but loosened. It is in this manner that Olson conspires with Fenollosa: rather than revealing how the abstract verb *is* goes back to *breathe*, the ultimate embodiment of a 'vivid verb', Olson gives us 'syllables to breathe with'. In his own comment on the passage, he describes a dance of the smallest particles that do not reassemble to a whole. What 'atoms and breath can create' is a *partage*: the conspiring syllables 'pile up one upon the other, or go and stay apart, or fuse and cease to be themselves'.

Olson's depiction of syllables as 'figures of the dance', bodies in motion, is echoed in Philip's comments on the fragmented language of *Zong!* when she addresses its resistance against the colonial language of her matrix:

> words arrange themselves in odd and bizarre combinations: at times the result appears the verbal equivalent of the African dance style 'crumping,' in which the body is contorted and twisted into intense positions and meanings that often appear beyond human comprehension. At times it feels as if I am getting my revenge on 'this/fuck-mother motherfuckin language' of the colonizer—the way the text forces you—me—to read differently,

> bringing chaos into the language or, perhaps more accurately, revealing the chaos that is already there. (*Zong!*, 205)

According to Philip, the resistant chaos of verbal crumping is unleashed when words are dissected into syllables. It is precisely in this context that she alludes to language in a primordial state:

> the ordering of grammar, the ordering that is the impulse of empire is subverted [...;] words are broken into and open to make non-sense or no sense at all [... and] return to their initial and originary phonic sound. (205)

In line with Olson's claim that '[t]his place of the elements and minims of language, is to engage speech where it is least careless—and least logical' ('Projective Verse', 18), Philip adds the argument that the non-sense of acoustic particles, 'phonic sound—grunts, plosives, labials', disrupts the sense and logic of the colonizer's discourse. As Fenollosa's essay shows, one trait of the 'language of the colonizer', if he is engaging in an '[a]rcadian primitivism', is to condemn the rationalism of Western languages and juxtapose it with a romanticized non-logical primitive language: all 'nations have written their strongest and most vivid literature before they invented grammar'. '[M]odern speech' suffers 'anemia' because '[a]bstract meaning gives little vividness' (376) and 'the ordinary western mind' is only 'concerned with logical categories' (376).

How does Philip's invocation of a chaotic, non-sensical language 'at the beginning of time' relate to this? The syllabic gasps she stages in *Zong!* are also deemed poetic, but in an utterly non-idyllic manner. In contrast to the 'vivid' impulses of primordial speech as Fenollosa theorizes it, Philip's 'language of grunt and groan, of moan and stutter—this language of pure sound fragmented' (*Zong!*, 205) is imbued with death and suffering. This also marks a significant difference to Olson's discussion of syllables, which, in his view, embody a 'life proper', 'lead the harmony on', and make words 'juxtapose in beauty' ('Projective Verse', 17). For Philip, the primordial asemantic sounds do not characterize a particular language; they rather mark a moment when to 'loose a language' coincides with to 'lose a language': 'The loss of language and meaning on board the *Zong* levels everyone to a place where there is, at times, no distinction between languages—everyone, European and African alike, has reverted, it appears, to a state of pre-literacy' (*Zong!*, 206). This state exposes a multilingual 'cacophony of voices—wails, cries, moans, and shouts' (203); sometimes, what might look like asemantic fragments do make sense, in w w w w a wa

w a w a t

er wa s

our wa

te r gg g g go

o oo goo d

waa wa wa

w w waa

ter o oh

on o ne w one

w o n d d d

ey d a

dey a ah ay

s one day s

wa wa

Fig. 22.1 *Zong!#1*

languages other than English: 'os: bone' (Latin) (183), 'wa àgbò: look for the ram', 'ní mi ni ran: remind me' (Yoruba) (184), etc. In *Zong!*, the language Philip associates with 'the beginning of time' thwarts the orientalist idea of a primordial Eastern/African mother tongue. By presenting dispersed languages on the verge of non-language, *Zong!* shatters the singular mother tongue, be it Western or non-Western, along with the matrix. By starting to conspire in this manner, languages are uprooted (Fig. 22.1).

Zong!#1, which, according to Philip 'is in fact an extended ga(s)p, or rather a series of ga(s)ps for air with syllabic sounds attached or overlaid', thwarts a narrative of recovery in yet another manner. It reveals that the gasp is always already a gap. In line with Philip's claim that '*the poems—need a great deal of space around them—as if* [...] *they need to breathe*' (194), *Zong!#1* marks '*breathing space*' (195) by white space. This seems to exemplify how 'breathing and the breath were built into the text of *Zong!*' ('The Ga(s)p', 38). However, the white space Philip identifies with breathing space displays an absence of respiration, pointing to the impossibility of incorporating breathing into the book. *Zong!* thus stages breath-less conspiration: a 'breathing with' that turns out to lack—be *without*—physiological breath. When Philip wonders 'whether the sounds of those murdered Africans continue to resound and echo underwater' because she heard that 'sound never ceas[es] within water' (*Zong!*, 203), she seems to imply that transferred to moans, bones might after all be recoverable 'exaqua': in syllabic gasps as they are presented in *Zong!#1*, for example, where the word 'water' itself is cut into wailing sounds. A closer look at the scattered 'syllables' Philip gives us 'to breathe with' shows that instead of the magical transformation of bones into moans we encounter the transmission of ga(s)ps. Just as Philip cannot recover the bones, her text cannot 'breathe for' the killed slaves. The *con*spiration performed is a *partage*: sharing a lack. If this conspiration functions as a 'blueprint for community and interdependency', 'acceptance of alterity' implies an acceptance of the irredeemable. *Zong!* reverses the 'metamorphosis from sound to intelligible word' based on violent premises Philip addresses in *She Tries Her Tongue*. However, blowing the violent matrix of the legal text and the colonial mother tongues to pieces, the poems do not constitute a receptacle for recovered breath. The title of the book's first section, 'os', seems to encompass the transformation from bones to moans in a most condensed manner: in Latin, the word means both bone and mouth, the place from which breath and sounds are emitted; acoustically, 'os' could be an ululation issued from an open mouth visually evoked by the letter 'o'. At the same time, 'o' figures a gaping void: what reminds us of a mouth enclosing sounds is also a mute letter that fails to contain breath and sound. It seems no coincidence that in *Zong!#1*, 'o' and 's' never fuse into the noun that—in a Western language of colonizers, and the language of the law par excellence—sparks a hope of retrieval: 'os', the potential bones-moans-metamorphosis. They do so in *Zong!#2* (5), embedded in the following words, surrounded by white space:

the loss

Notes

1. M. NourbeSe Philip, 'The Ga(s)p', in *Poetics and Precarity*, ed. by Myung Mi Kim and Cristanne Miller (Albany: State University of New York Press, 2018), 31–40, 39. All subsequences references to 'The Ga(s)p' are to this edition, cited by page number.
2. Nathaniel Mackey, 'Breath and Precarity', The Inaugural Robert Creeley Lecture in Poetry and Poetics, in *Poetics and Precarity*, 1–30, 5. All subsequences references to Mackey are to this edition, cited by page number.
3. The only woman musician he discusses is fictional: the drummer Drennette, a character of Mackey's *From a Broken Bottle Traces of Perfume Still Emanate* (cf. Mackey 10–13).
4. Otto Rank, *The Trauma of Birth* (London: Kegan Paul, 1929), 11.
5. Rank, *The Trauma of Birth*, 8.
6. Rank, *The Trauma of Birth*, 17.
7. Maurice Merleau-Ponty, 'The Child's Relation with Others', in *The Merleau-Ponty Reader*, ed. by Ted Toadvine and Leonard Lawlor (Evanston: Northwestern University Press, 2007), 143–183, at 151.
8. David Michael Kleinberg-Levin, *Before the Voice of Reason. Echoes of Responsibility in Merleau-Ponty's Ecology and Levinas's Ethics* (Albany: State University of New York Press, 2008), 79.
9. Peter Sloterdijk, *Bubbles. Spheres* I, trans. by Wieland Hoban (Pasadena, CA: Semiotext(e), 2011), 297–98. Leo Bersani, *Receptive Bodies* (Chicago: University of Chicago Press, 2018), 86.
10. Bersani, *Receptive Bodies*, viii.
11. Bersani, *Receptive Bodies*, 85.
12. Jean-Luc Nancy, *The Inoperative Community*, trans. and ed. by Peter Connor (Minneapolis: University of Minnesota Press, 1991), xxxviii.
13. Jean-Luc Nancy, *The Disavowed Community*, trans. by Philip Armstrong (New York: Fordham University Press, 2016), 1, 9.
14. Nancy, *The Inoperative Community*, 33.
15. Nancy, *The Inoperative Community*, xxxix.
16. M. NourbeSe Philip, *Zong!* (Middletown, CT: Wesleyan University Press, 2011). All subsequences references to *Zong!* are to this edition, cited by page number.
17. M. NourbeSe Philip, 'Defending the Dead, Confronting the Archive: A Conversation with M. NourbeSe Philip', *Small Axe* 26 (2008), 63–79, at 73.
18. Christina Sharpe, *In the Wake. On Blackness and Being* (Durham, NC: Duke University Press, 2016), 109.
19. Cf. Sharpe, *In the Wake*, 38, and Sarah Dowling, 'Persons and Voices: Sounding Impossible Bodies in M. NourbeSe Philip's *Zong!*', *Canadian Literature/Littérature Canadienne* 210/211 (2011), 43–58, at 45–46.

20. M. NourbeSe Philip, *She Tries Her Tongue, Her Silence Softly Breaks* (London: The Women's Press, 1993), 37. All subsequences references to *She Tries Her Tongue* are to this edition, cited by page number.
21. Sara Ahmed, *Living a Feminist Life* (Durham, NC: Duke University Press, 2017), 7, 175.
22. Charles Olson, 'Projective Verse', in *Selected Writings*, ed. by Robert Creeley (New York: New Directions, 1966), 15–26, at 18. All subsequences references to 'Projective Verse' are to this edition, cited by page number. For a detailed analysis of Olson's treatment of breath in 'Projective Verse' and beyond see Stefanie Heine, *Poetics of Breathing. Modern Literature's Syncope* (Albany, NY: State University of New York Press, 2021), 108–115.
23. Charles Olson, *Mouths Biting Empty Air* (1946), typescript, I Box 32:1630, prose no. 15, Charles Olson Research Collection, University of Connecticut. The essay, held at the Thomas J. Dodd Research Center in Connecticut, is not paginated.
24. For Olson's celebration of masculinity see Rachel Blau DuPlessis, 'Olson and His *Maximus Poems*', in *Contemporary Olson*, ed. by David Herd (Manchester: Manchester University Press, 2015), 135–48. For a discussion of his problematic treatment of Mayan glyphs see Andrew Mossin, '"In Thicket": Charles Olson, Frances Boldereff, Robert Creeley and the Crisis of Masculinity at Mid-Century', in *Olson's Prose*, ed. by Gary Grieve-Carlson (Newcastle: Cambridge Scholars Publishing, 2007), 16–46; Joel Duncan, 'Frank O'Hara Drives Charles Olson's Car', *Arizona Quarterly: A Journal of American Literature, Culture, and Theory* 72/4 (2016), 77–103; Daniel Katz, 'From Olson's Breath to Spicer's Gait: Spacing, Placing, Phonemes', in *Contemporary Olson*, 78–88.
25. These etymologies are oversimplified and partly incorrect. For a discussion of this matter with regard to Olson's poetics see Stefanie Heine, 'Fishy Etymologies', *Sprachgeschichtliche Irrwege bei Charles Olson, Colloquium Helveticum* 46 (2017), 131–43.
26. Ernest Fenollosa, 'The Chinese Written Character as a Medium for Poetry', in *Instigations*, ed. by Ezra Pound (New York: Boni and Liveright, 1920), 357–88. All subsequences references to 'The Chinese Written Character' are to this edition, cited by page number. In Fenollosa's text, the sentences do not occur directly after each other. Olson deliberately arranges them in a way that highlights a rhythmic quality and tonal continuity.
27. *E.g.*, George A. Kennedy, 'Fenollosa, Pound and the Chinese Character', *Yale Literary Magazine* 126/5 (1958): 24–36.
28. For a discussion of these tendencies, see Robert Kern, *Orientalism, Modernism, and the American Poem* (Cambridge: Cambridge University

Press, 1996), 115–45. For a very nuanced overview of Fenollosa's ambivalent reception, see Haun Saussy's introduction to Ernest Fenollosa and Ezra Pound in *The Chinese Written Character as a Medium for Poetry: A Critical Edition*, ed. by Haun Saussy, Jonathan Stalling, and Lucas Klein (New York: Fordham University Press, 2008).

29. Charles Olson, 'The Gate and the Center', in *Collected Prose*, ed. by Donald Allen and Benjamin Friedlander (Berkeley: University of California Press, 1997), 168–73, at 168.
30. Victor Li, 'Primitivism and Postcolonial Literature', in *The Cambridge History of Postcolonial Literature*, ed. by Ato Quayson (Cambridge: Cambridge University Press, 2012), 982–1005, at 986, https://doi.org/10.1017/chol9781107007031.011.
31. Li, 'Primitivism and Postcolonial Literature', 985.
32. Li, 'Primitivism and Postcolonial Literature', 984.
33. Li, 'Primitivism and Postcolonial Literature', 987.
34. Cf. 'The Gate and the Center', 168.
35. 'The Gate and the Center', 168.
36. Jacques Derrida, 'La parole soufflée', in *Writing and Difference*, trans. by Alan Bass (London: Routledge, 2001), 213–45, at 222.

Selected Bibliography

Ahmed, Sara. 2017. *Living a Feminist Life*. Durham, NC: Duke University Press.

Bersani, Leo. 2018. *Receptive Bodies*. Chicago: University of Chicago Press.

Derrida, Jacques. 2001. La parole soufflée. In *Writing and Difference*, trans. by Alan Bass, 213–45. London: Routledge.

Dowling, Sarah. 2011. Persons and Voices: Sounding Impossible Bodies in M. NourbeSe Philip's *Zong!*. *Canadian Literature/Littérature Canadienne* 210/211: 43–58.

Kim, Myung Mi, and Cristanne Miller, eds 2018. *Breath and Precarity*. The University at Buffalo Robert Creeley Lectures in Poetry and Poetics. Albany: State University of New York Press.

Li, Victor. 2012. Primitivism and Postcolonial Literature. In *The Cambridge History of Postcolonial Literature*, ed. by Ato Quayson, 982–1005. Cambridge: Cambridge University Press. https://doi.org/10.1017/chol9781107007031.011.

Merleau-Ponty, Maurice. 2007. The Child's Relation with Others. In *The Merleau-Ponty Reader*, ed. by Ted Toadvine and Leonard Lawlor, 143–83. Evanston: Northwestern University Press.

Nancy, Jean-Luc. 1991. *The Inoperative Community*. Trans. and ed. by Peter Connor. Minneapolis: University of Minnesota Press.

Nancy, Jean-Luc. 2016. *The Disavowed Community*. Trans. by Philip Armstrong. New York: Fordham University Press.
Sharpe, Christina. 2016. *In the Wake: On Blackness and Being*. Durham, NC: Duke University Press.

CHAPTER 23

Visualising the Ephemeral

Jayne Wilton

Part 1: A History in Epitomes

Our experience of breath is individual and multisensory. I shall explore what visual art can contribute, consciously or unconsciously, to our encounter with the breath. In *Feeling Beauty: the Neuroscience of Aesthetic Experience* Gabrielle Starr suggests that vision is an 'exploratory activity', and that though 'aesthetic pleasures may sometimes seem to grab us … often we must work to meet them half way'.[1] A unique, proprioceptive experience can be evoked through connecting our outer world reality to the inner world of our imagination. The artwork provides a framework through which we can create empathy and construct our personal gestalt. From my experience as a practising artist I have drawn together examples from the vast range of visual art that has engaged literally, poetically, politically, and spiritually with the breath. Through these works I hope to illustrate the potential of visual art to offer insights into this elusive animating force.

It seems pertinent that the very earliest visual art, Palaeolithic paintings in the Cueva de las Manos (Cave of Hands), Santa Cruz, Argentina, is thought to be created by the spent breath.[2] The out-breath was harnessed

J. Wilton (✉)
Berkhamsted, UK

D. Fuller et al. (eds.), *The Life of Breath in Literature, Culture and Medicine*, Palgrave Studies in Literature, Science and Medicine,
https://doi.org/10.1007/978-3-030-74443-4_23

to blow pigment against the resistance of a hand, enabling us to reach out and trace the breath of our ancestors. Throughout history artists have continued to grapple with creating vessels to capture the breath and make the ephemeral process of breathing visible. The bubble was the translucent vessel which provided material for earnest exploration in Jean-Baptiste Chardin's *Soap Bubbles* (1734) in which he aims to communicate absorption, reflection, and reverie. John Everett Millais' *Bubbles* (1886) is a more sentimental capture of the breath—of which Christiana Payne, writing on air, observes, 'the blowing of bubbles was well established in art as a metaphor for the vanity of human aspirations, along with flying kites and building card tables'.[3] Both these works encapsulate an element of wonder at the ability of this transparent substrate to capture both breath and our imagination.

From the bubble evolved the more resilient balloon. Piero Manzoni's *Fiato d'artista* (Artist's Breath, 1960) started life as an ebullient and robust red balloon tethered to a wooden block. Today this work exists as a red mass of crumpled rubber, the once cheery balloon deflated and defeated. The internal cavity which held the artist's forceful if short-lived breath has become a humble monument to the inevitable transience of human existence. Manzoni's work anticipates the genre of *arte povera* in which breath as muse and art material allowed Giuseppe Penone to create iconic sculptures and drawings including *Breath 5* and *Breath of Clay* (1978)—enigmatic works to catalyse contemplation of our own experience of the breath. The poetry of Penone's work contrasts with the harsher nature of his contemporary Wincenty Dunikowski-Duniko's *Breath* works (1976), where the literal gesture of breathing out onto glass was captured photographically and called 'Moment Art'.

The ability of visual art to be static or dynamic comes into consideration in *Pulmonary Space* (2009) by Philippe Rahm. This is a tragi-comic sculpture-installation-performance where instruments of an ensemble play into a fabric bag which responds to the interaction creating a contemporary and fluid exploration of gaseous exchange. I find the words with which Rahm describes the breath as affecting as the work itself: 'tangible mutation of air according to a specific wave length … a pressure point placed upon the air in a defined space … a bodily temperature of close to 37°… a gaseous content, measurable by the amount of oxygen, nitrogen, carbon dioxide, rare gases and water in the form of vapour'.[4]

Possibly the artist most influential on my own breath-inspired practice conceptually has been Cornelia Parker, whose *Breathless* (2001) can be viewed in the Victoria and Albert Museum. Parker's work creates many subtexts by literally taking the breath away from previously breath-activated objects. Perhaps its most poignant resonance is with what I think of as the 'breathlessness' of contemporary culture—the tendency apparent in many areas of life, encouraged by the impersonality of electronic rather than personal communication, to rush from one activity to another without pauses for meditation or considering any larger sense of shaping purpose. Parker asserts that all her work is about the in-breath and the out-breath. In *Doubtful Sound*, her 2010 exhibition at the Baltic Centre for Contemporary Art Gateshead, she presented a series of works which contrast the expansive, thrusting, independent dynamic which we relate to the expiration with that of the inspiration, a more reductive, distilled, compressed aesthetic which provides opportunities to consolidate, internalise and replenish. For Parker each instrument in *Perpetual Canon* (2004), an installation of sixty flattened brass band instruments in a circle, represents an in-breath, an 'inhalation never to be released'.[5] I see this affecting work as more relating to expiration, to having sighed in resignation and taken a final breath in the face of the sheer brute force of the 250-ton press which was needed to empty its internal cavity. I find myself imagining the force needed to re-inflate each of Parker's instruments, a process which appears impossible until we equate it with the initial inertia that the newly born lungs overcome for the first intake of breath. This effort is immense (three to four times that of a normal inspiration), and will not take place without the catalyst of external contact, whether the touch of the midwife or contact with the mother's abdomen. Whilst at the Baltic, I observed the magnetic quality of this work and its shadows. I particularly delight in the fact that viewers often connect with the work through their own breath: they literally blow into the shadow of an instrument's mouthpiece in an attempt to resuscitate the form.

Cornelia Parker's concerns about absence and presence of the breath are mirrored in Joseph Wright of Derby's *An Experiment on a Bird in the Air Pump* (1768). This work illustrates how controversial and shocking the scientific undertaking to understand the breath was in its day. The image, often misinterpreted as toying with the bird's existence, actually conveys attempts to understand the mechanics of resuscitation so as to save and improve lives. The work is rich in metaphor: for example, a *Memento mori* is suggested through the skull and candle, and the cockatoo, a rare and expensive bird, refers to the fragile and precious resource of the breath. Its setting, thought to be Erasmus Darwin's front room, is significant because it illustrates Wright's desire for collaboration to push forward the boundaries of science in a culture anchored in archaic thinking and practices.[6]

The final breath is a potent trope for our fragile hold on life. Jacques-Louis David's portrayal of *The Death of Marat* (1793) evokes the poignancy of this inescapable liminal phase. The inevitability of transition has fuelled many pieces of visual art, and in my own practice I will always be moved by revisiting the image of a patient's breath that I recorded at the Hospice of St Francis just days before his death, an image which connects me directly back to the courage and humour this man demonstrated in the face of his situation.[7]

There are many examples of imagery serving to illustrate medical practice and procedures in the respiratory field. An engraving of Francisque Crotte applying his electrical remedy for tuberculosis to a seated woman (1901) served in its time to inform and inspire about advances in medical research, and today is valuable documentation of medical practice. However, for me personally, it sits uncomfortably as art: in art I want those qualities that make me feel the empathy articulated by Gabrielle Starr. Francis Bacon's *Head VI* (1949), on the other hand, based on a portrait of Pope Innocent X by the seventeenth-century Spanish painter Diego Velazquez, creates such a response. The image appears to convey the wretchedness of breathlessness. Jan Cox suggests 'that the subject is trying to draw in air … gasping for breath',[8] and therefore perhaps refers to Bacon's own experience of asthma.[9] Bob Flanagan and Martin O'Brien, both performance artists, also channelled their abject individual experiences of living in the shadow of an incurable lung disease—cystic fibrosis—using endurance and in Flanagan's case sadomasochistic acts to communicate to audiences the protracted, excruciating reality of a life-limiting disease.

Not all breathlessness is pathological. In *Several Interruptions* (2009) Thomson & Craighead create a series of five animated triptychs depicting individuals holding their breath under water, simultaneously attempting to escape the banality of normal life whilst grounding themselves by the crude physical sensation of asphyxiation. Thomson & Craighead selected the fifteen video sketches from a resource of 61,000 possible clips posted on YouTube. Their presentation through the formal device of the triptych serves to illustrate the plurality of each individual endeavour. The holding of breath is perhaps among the most potent acts of conscious rebellion we can incite, and the resulting discomfort is manifested in the faces of those undertaking the endeavour.

Collaborative practice with the health service has given me the opportunity to explore what it is like to have lungs affected by breathing bad air, either voluntarily, through smoking, or involuntarily, through polluted air. From fracking to terrorism, pollution continues to concern us, and such concerns fuel work which reflects the implications of contemporary living. Artists such as Steven Jacobson depict the contemporary condensation trails we witness in our air space: we can admire the beauty of these chemical incisions, but are also aware of their impact on the environment. More menacing still are invisible threats: the once expansive feeling that air represented as a medium for exploration has shifted through warfare in the twentieth century towards air as a contracted zone of peril. In *Terror from the Air* (2009) Peter Sloterdijk argues that the first military use of chlorine gas in 1915 marked the beginning of a new age. The most sinister connection of air, breathing, and warfare cannot but be the use of gas chambers as instruments of civilian mass murder. Current proposals to place a Holocaust memorial in Victoria Tower Gardens, London, within sight of the British Houses of Parliament as a reminder of these atrocities, show how visual art can draw attention to the enduring nature of these threats while reminding us that we do have agency.

The breath has also been used to measure fear in a society necessarily concerned with security. In *It takes 154,000 breaths to evacuate Boston* (2007–2008) American artist kanarinka ran the entire evacuation route system in Boston and attempted to measure the distance in human breaths. The project challenges the ability of the city to respond to both terrorism and man-made disasters. Other artists have also striven to illustrate the threat to our breath of man-made issues. Kaya Hanasaki created *Portrait in a Mask* (2012) in response to the disaster at the Fukushima nuclear power plant. The work demonstrates a mistrust of government and was sponsored by Art Action UK, a collective of artists, curators, gallerists, and writers who are exploring various means to show solidarity and support for those affected by deliberate or negligent man-made disasters.

In this final section I have selected examples of visual art where breath has been used to represent literal or metaphorical spirituality. Breath holds significance in many religions. For both Judaism and Christianity the first book of the Torah and the Bible, Genesis, speaks of the 'breath of life' when God breathes life into the *adamah* (Hebrew: ground, earth). In a design in an illuminated manuscript originating in Canterbury (shown on the cover of this book), God is shown over the circle of the world with two long trumpets issuing from his mouth representing the breath of God, by which he 'breathed into [man's] nostrils the breath of life, and man became a living being' (Genesis, 2.7).[10]

The twelfth-century abbess, artist and activist Hildegard of Bingen described herself as 'a feather on the breath of God'. She is portrayed in the bottom left corner of the Cosmos, Body, and Soul mandala of the *Liber Divinorum Operum*, an image that shows the spherical earth being breathed into by ethereal creatures.[11] I enjoy the lightness of touch that this imagery evokes, and I often feel I connect with such subtlety when collaborating with people in the recording of their breath, as, for example, in *One Breath*, a monoprint I made of a child's sigh.[12] I am fascinated by the sigh, because, while it has no verbal-semantic content in an ordinary sense, it is nevertheless polysemic and so communicative. I have used many different media to capture the sigh, and I have been influenced in this quest by Bernini's sculpture, *The Ecstasy of St Teresa* (1647–1652), which depicts the nun Teresa of Avila in religious rapture after spiritual contact with the angel with whom she is portrayed. Bernini's capture of this enigmatic breathing gesture in three dimensions, although criticised for its overt sensuality after his death, arguably was a direct response to the erotic terms in which the nun herself described the encounter.

Breath has been linked with spirituality in many ancient cultures, as in Ancient Egypt, where breath was associated with the concept of the soul. The funerary texts, *The Books of Breathing* (fourth century BCE to second century CE), for example, used breathing as a metaphor for all aspects of life that the deceased hoped to experience in the afterlife. Ancient Mesoamerican religions understood breath as a vital force. An image on the sarcophagus lid of the Mayan King Pakal (603–683) has been interpreted as Pakal descending into the underworld, or preparing for a cosmic journey, or falling into the open mouth of death where he breathes in a 'black hole' and moves into creation. Eastern traditions have long associated the breath with the spiritual. In Hinduism every movement is related to cosmic breath, and breathing practices such as pranayama, illustrated in the *Bhagavad Gita* (4.29), are a central spiritual practice with international and multicultural currency today. In some yoga practices life is measured not in years but in the number of breaths, each of which some yogis assert contains an unconscious prayer. Buddhists set great store by breath meditations to free the self from fleeting external and internal distractions, which explains why, after many breaths are poured into the creation of colourful sand mandalas, they can be promptly swept away with non-attachment.

Contemporary artists such as Shirazeh Houshiary aim spiritually to embody the breath. As she expresses it, 'I set out to capture my breath, to find the essence of my own existence, transcending name, nationality, cultures'.[13] The same aim is common to *Presence* (2006–2007), in which she palimpsestically applies sacred Arabic text to the ground of a canvas, and to *Breath* (2013), which simultaneously video-projects evocative chants of Buddhist, Christian, Jewish, and Islamic prayers into a darkened space, creating an enigmatic breath-driven cacophony.

Mark Wallinger uses his breath to engage with religion through his alter ego, 'Blind Faith'. *Hymn* (1997) is part of a triptych, *Talking in Tongues*, in which breath is a central concern. It shows Wallinger standing on a soap box in a park, inhaling helium from a cylinder at his feet and holding a balloon as he sings a gloomy Victorian children's hymn in an uncanny high-pitched voice. This disconcerting vignette challenges us to consider faith, authority, and religion within contemporary society. The poignant references to childhood created visually and through the breath seem to suggest a yearning for spiritual, physical, and emotional renewal and yet acknowledge the implausibility of such an endeavour. Also a video installation, *Martyrs* (2014), by Bill Viola and Kira Perov, shows four individuals breathing stoically through assaults by the four classical elements. 'The elements represent the darkest hour of the martyr's passage through death into the light' (Bill Viola).[14] A later commission, *Mary*, deals with the cycle from birth to death of the archetypal feminine. Mary, the mother of Christ, is here represented in a symbolic Anglican framework designed by Norman Foster. Creator Kira Perov speaks of Mary as the 'embodiment of eternal sorrow… "container of the uncontainable"',[15] and as iconically holding for us the transition of breath to breathlessness.

I hope this eclectic collection of imagery and concepts gives some idea of the ways in which artists have worked with breath to capture a range of its significances, literal and metaphorical. Gabrielle Starr considers how 'art changes human experience as it reorders our perceptions and engages our emotions'.[16] Perhaps art about the breath creates an opportunity which can lead quietly back to the present—an anchor back to the experience of breathing in a breathless age. Starr argues that 'powerful aesthetic experience integrates information and sensation to redefine and revalue what we feel and know'.[17] If, therefore, accessing visual imagery of the breath allows us to understand, empathise with and articulate our embodied experience, then, to borrow from Giorgio Agamben, perhaps we can engage with the humble breath as the quotidian which generates the pearl.[18]

Part 2: My Own Work—in the Laboratory; in Hospital and Hospice

As an artist working primarily to make breath visible, my practice has embraced study of the breath in the course of a Master's degree at the Slade School of Art, through collaborations, including with physicists at Brunel University and respiratory patients at the Royal Brompton and Harefield hospitals, and recently a commission to create twelve covers for *The Lancet Respiratory Medicine* journal.[19] Over the last fifteen years, working with writers, musicians, physicists, engineers, students, and hospice and hospital patients and staff, I have produced drawings, installations, sculptures, and prints that reveal the shape and movement of the usually unseen breath.

Where does my desire to visualise breath come from? Perhaps the same sense of mystery that led Palaeolithic man to blow pigment across hands in the Cueva de las Manos. The breath is a source of fascination to me for a range of reasons. Perhaps its elusiveness and mercurial qualities most intrigue me. Breath lies literally at the centre of existence, allowing communication, movement, and thought. Perhaps it is the invisibility of this animating force which makes it so enigmatic, but I think it is also the ability of the breath to communicate experience, whether through voice or universal breathing gestures such as the sigh, that has made me long to pin down and record these moments which demonstrate the coexistence of resilience and fragility.

My interest in respiration in relation to visual art began in 2004 when I worked alongside patients in the Hospice of St Francis in Hertfordshire, UK, many of whom suffered from breathlessness. Working in the hospice as an art practitioner, I ran monthly workshops with day- and in-patients, initially recording the patients' breath moving the flame of a candle in response to the Hospice's 'Light up a Life' campaign. I was moved by the generosity of patients in donating their precious, finite breath. I felt a responsibility to create something suitably celebratory from each patient's breath—an expressive record of a poignant moment.

We often only pay attention to breathing—it only becomes *visible* to us—when it troubles us, when we are breathless in the face of some emotion, out of breath, or struggling for breath. Normally, when we can ignore it, breath is barely perceptible. When it takes a more material form, in the cold, or when the air is filled with pollutants, this is often a sign of danger, of decay, or of some limit being reached. It is probably relevant that my first exploration of all things respiratory goes back further than 2004, to my role in brand management for a pharmaceutical company where I worked on asthma management products. The commitment of respiratory physicians, GPs, asthma nurses, and respiratory health organisations to improving the lives of asthma patients inspired me, years later, to investigate breathing and breathlessness through my work. I have aimed to harness a wide and innovative range of materials to capture the breath, making the invisible visible. This led to a solo exhibition at Brunel University's Beldam Gallery in 2009, and then to a Leverhulme-funded residency, 'Inspired by Breath', in Brunel's Engineering and Design School.

The cross-disciplinary arts-science objectives of this residency reminded me of Joseph Wright of Derby's aim of using art to bring the cutting edge of science to the cultural fore. My work was less controversial—except possibly when presenting my film of cosmic ray demise, *Darkness Visible*, to a convention of particle physicists in Berlin, when I became aware of different notational priorities between the fields of arts and science. (I was delighted by the evocative indigos that resulted from inverting my film footage of the cosmic rays as they emerged within the cloud chamber. I was keen to retain the restrained monochrome palate which I felt allowed each unfolding form to speak for itself whilst also making art-historical references to traditional drawing inks used to render forms. My aesthetic decision was at odds with views of several physicists who felt colour coding each type of cosmic ray in post-production would have aligned more with scientific convention.) The primary aim of this residency was to allow me to continue my investigation into human breath as a fundamental unit of exchange between people and their environment. Our collaborative aim of developing innovative ways of making visible the normally invisible dynamics of the breath allowed me to reveal further forms that lie behind the verbal and non-verbal communication that the breath facilitates.

Professor Peter Hobson established an optical set-up in a laboratory at Brunel where we could explore the recording of breathing gestures and spoken words using the so-called 'Schlieren' system (named from the

German *Schlieren*, streaks). This is a technique for recording subtle differences in refractive indices, such as would be provided by warm breath in cold air. A laser point source of light is expanded and then refocused by a mirror onto a knife-edge that blocks the direct light. Breathing in front of the mirror disturbs the light rays so that some bypass the edge and can be recorded by a camera behind it. Words were spoken in front of the mirror while I photographically recorded the patterns produced by the turbulent air, as in Plate 23.1.

Plate 23.1 *Schlieren capture of the word 'I'.* © Jayne Wilton

Multiple recording sessions allowed us to build up a library of words and breathing gestures such as the gasp, the sigh, and breath-carried sounds of spiritual significance such as 'Om', captured as Schlieren images. We installed three of these in the Brunel Library Special Collections room where they can be seen from both inside and outside the building. An objective for the residency was to be able to make words visible, and hence, the breath that carries the voice. In order to work with words that were not merely randomly selected I drew on the research of Anna Wierzbicka on universal semantic primes, in other words, key words across cultures that are posited as critical to facilitating face-to-face communication.[20] Also using this set of sixty words, I worked on inter-semiotic translation of sound into imagery or form using fluid dynamic technology to model a series of these words. I sought volunteers at Brunel University, representing different groups of the University's community from students to academics, to speak one or more words from the list. By gathering a diverse sample of vocalised words from across the Brunel community, I hoped to allow a sampling of timbre, dialect, intonation, and articulation of the key semantic primes. This sound recording was then used to generate a three-dimensional sculpture using the 'Maya' programme, a software package which facilitates 3D animation, modelling, simulation, and rendering, and which is often used to simulate fluid dynamics. The forms were 3D-printed in the School of Engineering and Design, allowing the audience literally to 'hold' pieces of conversation. This translation of voice into form created an evocative resource through which to raise questions about how carefully or recklessly we use these tools in our daily lives. The words created in this project included 'because', 'think', 'moment', and 'happen'. One word from outside the list of semantic primes was chosen by Brunel University's chair of Creative Writing, Benjamin Zephaniah, who, reflecting on the empathetic objective of the project, wanted to record 'overstanding' (Plate 23.2), an invented word, a word that he thinks he heard in Jamaica or in a dream, which he prefers to 'understanding', which he sees as carrying potentially patronising connotations. In his view, '"overstanding" transcends "understanding" by incorporating the qualities of empathy, humility, and compassion. It implies a sense of overview, and a wider or broader comprehension of a phenomenon'.[21]

Continuing in my quest to make visible the dynamics of the human breath, I collaborated with particle physicists Peter Hobson and Ivan Reid, who both have specialist experience of in-line digital holography

Plate 23.2 Conversation piece, *Overstanding*. © Jayne Wilton

and its computer reconstruction. The creation of digital holograms of breathing patterns provides the three-dimensional illustration of the dynamics at work when a breath is expired. In-line digital holograms were recorded using a fast-pulsed green laser and a 36 × 24 mm^2 CCD camera in an optics laboratory established by Professor Hobson.

Digitalis seeds, selected for their aerodynamic properties, were expired into an acrylic chamber and their trajectory recorded. The holograms were replayed using HoloMovie software that runs on a multi-core NVIDIA GPU. Real time digital hologram replay allows the audience to move through holograms of breath-borne particles. Reid enhanced the footage using 'edge detection', which allows the seeds to be seen coming in and out of focus as the camera moves through the holographic replay.[22] The rhythm of the breath provides a fundamental measure of our functioning, so to visualise the complexity of such a universal dynamic could stimulate new insights for technological modelling in wider fields of scientific research.

Most of us will have experienced playing with the interaction of our breath with shiny surfaces such as windows and mirrors, creating temporary fields of condensation in which to doodle. In emergency situations a mirror or shiny surface is held in front of the mouth of a subject to confirm whether or not he or she is breathing. My *Breathe* series, a sequence of evocative landscapes which both employ and subvert traditional printing processes, makes use of these dynamics. In November 2012 I travelled with Peter Hobson to CERN, the European Organization for Nuclear Research, to record the breath of colleagues working on the CMS and ATLAS particle detector experiments. *Breathe* is a celebration of their role in the first observations of what may well be the long-sought Higgs boson at the Large Hadron Collider. Their breath was captured on the copper surface and etched to create a negative of the breath where it initially sat. I simultaneously recorded the breath of Brunel physicists also working on these projects, and exhibited the plates together intermingled to demonstrate the interconnectedness of the two sites and the importance of collaborative projects.

My fascination with recording human breath led me to consider notions of recording the 'breath' of the universe. Peter Hobson introduced me to the idea of cosmic rays as an indicator of universal expiration. We filmed cosmic rays using an acrylic diffusion cloud chamber in Hobson's laboratory at Brunel. The recording of cosmic rays in this way is well documented and can be viewed on social networking and sharing sites such as YouTube. The challenge was to look anew at these traces and open up new perspectives on these ephemeral forms. I wanted to create a more immersive environment where the audience could consider the decay of these supernova remnants as positive forms rather than negative trails. Inverting the film footage created such an opportunity. The negative trails become positive entities, which can be viewed undertaking their entropic demise. The viewer then experiences the dark absence of the decaying rays as ebullient present forms.

HD filming of the events under tungsten lighting allowed the yellow hues that lit the traces to become violet and deep indigo when inverted, evoking the interstellar space from which the rays have travelled. The resultant film, *Darkness Visible*, became an exploration of the use of light to make visible these normally invisible forms and generate imagery not normally associated with the monochromatic traditional capture of cosmic events.[23] (Plate 23.3)

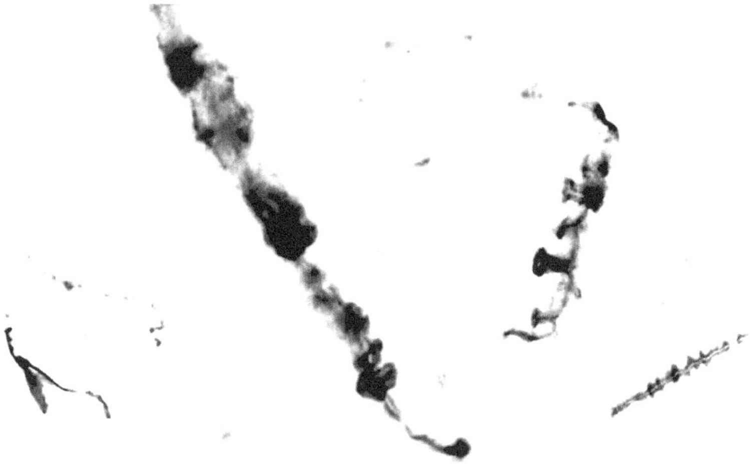

Plate 23.3 Still from *Darkness Visible* showing cosmic ray demise. © Jayne Wilton

* * *

In 2014 an Arts Council England grant enabled me to undertake a period of research and development, leading to a series of participatory workshops with patients at the Royal Brompton and Harefield hospitals. The project was an opportunity to collaborate with a hospital specialising in heart and lung disease, and resulted in a body of work including a series of copper plates onto which the breath of patients in the pulmonary rehabilitation unit was recorded. Working alongside a composer, curator, clinical research physiotherapist, and three respiratory groups, I created drawings and installations that captured the usually unseen human breath in a variety of media.[24] The outcomes of this project resulted from participation with the three patient groups: a 'Singing for Breathing' group, a 'Breathe Easy' group, and patients in the pulmonary rehabilitation unit.

'Singing for Breathing' was a singing workshop in which up to twenty-five patients met twice each week. I was intrigued by the abstract nature of breathing gestures used for the group warm-up, and the workshop leader, Joanna Foster, agreed to compose a vocal piece based on these gestures for the group to perform as a round. The outcome was an animation, *Vent*, which visualises the dynamics of interaction between the three sung parts of a round, and in which listening to the rich texture of the combined voices within each group is as important as watching the physical dynamics unfold.[25] I was struck to realise when creating this animation that the rhythm and dynamics of the piece were acting as a metronome for my own breath, an index of how deeply collaborations engage us at a personal as well as a professional level.

The 'Breathe Easy' group is a support group for patients with lung disease, supported by the British Lung Foundation (BLF). With this group I ran a series of sessions recording on photographic film the breath of individuals, which resulted in a number of installations including *Crowd* (Plate 23.4) and *Cacophony* (Plate 23.5).

With patients in the Pulmonary Rehabilitation Unit I created *Breathe*, a series of ten etched copper plates (Plate 23.6). The *Breathe* series makes use of the dynamics of breath interacting with a cold shiny surface. Capturing the condensation of a spent breath on the surface of a copper plate and then etching it into the surface creates a unique series of marks and colours, abstract in nature. Preserving an individual and nuanced trace of the individual's breath at a specific moment presents an alternative to traditional portraiture. This work employs traditional lithographic processes but also subverts and challenges the notion of etching as a specialist process by allowing compositions and drawings to be made by the humble breath. The inscription of expiration onto precious metal immortalises a discharge and makes visible an otherwise unseen exhalation.

In the case of the *Breathe* project, visual outcomes of breathing or singing onto photographic film rested on an alchemy which I associate with Marcel Duchamp's concept of 'infra-mince' (French: an invented term based on *mince*, thin: hence 'less than very little'). Though Duchamp claimed the idea could only be exemplified, not defined, one might say that this involves liminal situations that result in real and significant, albeit barely discernible, changes.[26] These occurred in the intense darkroom situation where the performance took place. There was a buzz of excitement in both workshops as patients prepared for their sessions and then discussed their experience of the darkroom 'tent' with their peers. The transaction that took place between artist, materials, and performer generated ephemeral records with the potential to open up subtle pathways and alter perceptions. Such shifts of perception can serve to invigorate the acuity of the gaze, or result in a call to compliance with clinical advice, even if only through the novelty of the encounter.

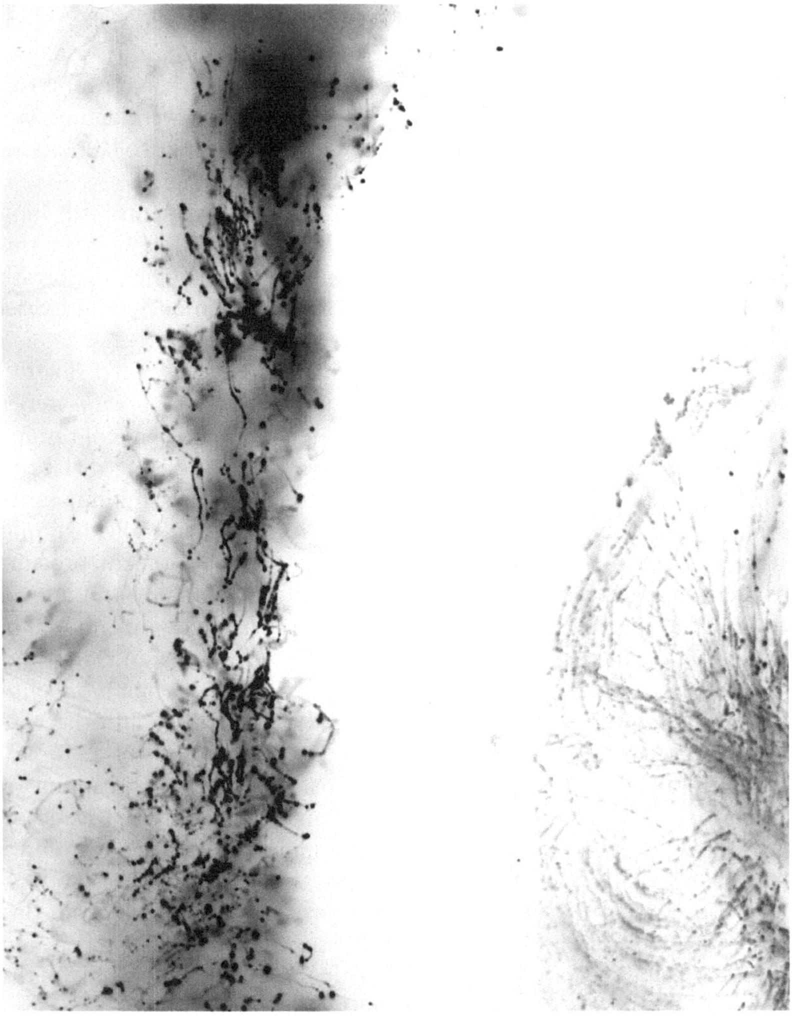

Plate 23.4 *Crowd*, layered drawings from the *Sing* series installed in Perspex blocks. © Jayne Wilton

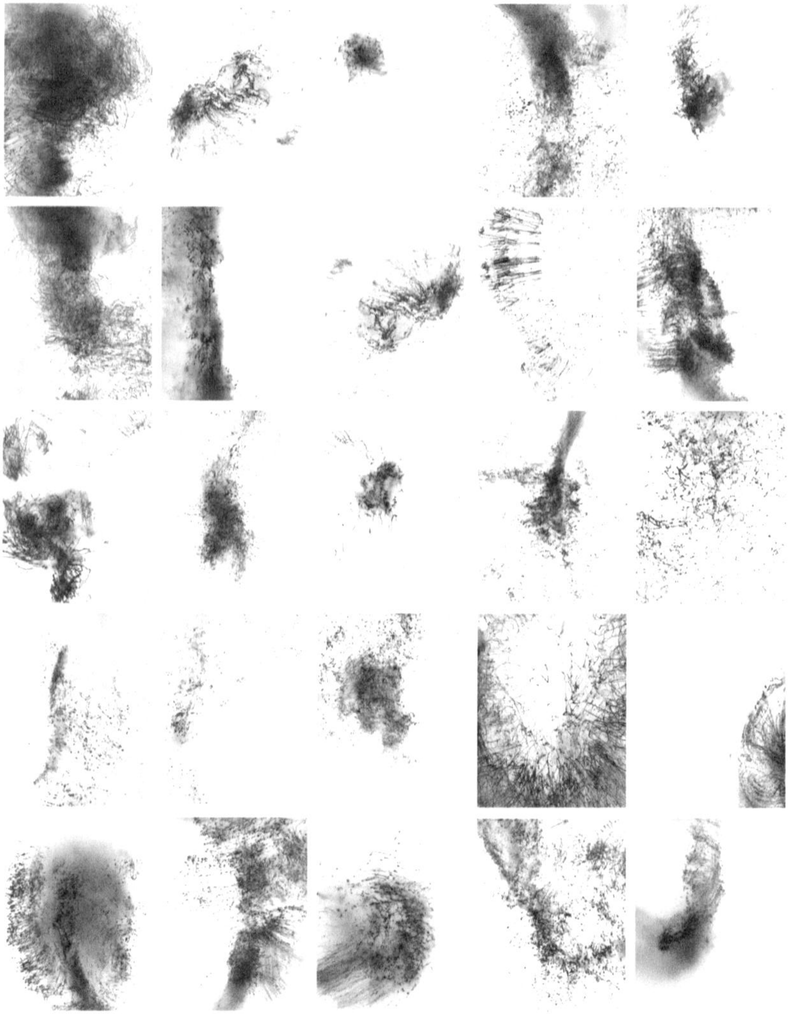

Plate 23.5 *Cacophony*, composed of twenty-five examples of the series *Sing*. © Jayne Wilton

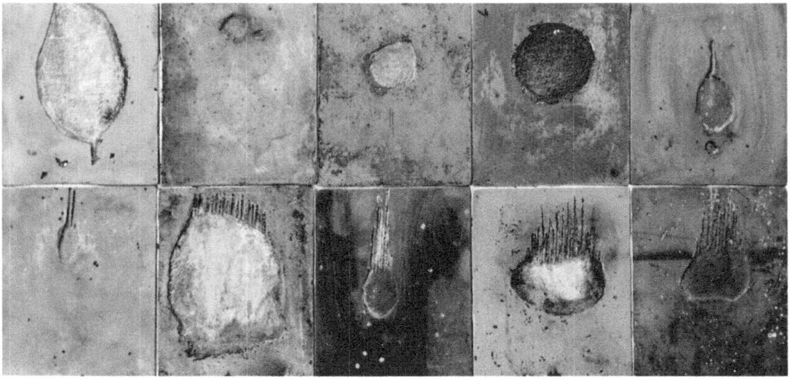

Plate 23.6 *Breathe*, plates created with patients and staff in the Royal Brompton Hospital. © Jayne Wilton

Walter Benjamin identified a modern alienation from experience initiated by the difficulties of assimilating atrocities encountered on the battlefield in World War I where 'a generation ... now stood under the open sky in a countryside in which nothing remained unchanged but the clouds, and beneath the clouds, in a field of force of destructive torrents and explosions, was the tiny, fragile human body'.[27] This 'poverty of experience' is explored further by Giorgio Agamben, who attributes it to our distancing ourselves from even day-to-day seemingly banal events, where we hold at arms' length experiences as a buffering mechanism to shield us from their intensity.[28] In this way it is possible to consider processes and media which visualise experience as serving to protect the participant from the direct gaze on his or her condition while simultaneously providing proprioceptive access to the experience. Whatever the mechanism may have been, taking part in these activities seemed to exceed patients' expectations of how much they would enjoy or benefit from them. Many reported a greater awareness of the value of the arts in terms of pleasure, wellbeing, and health.

I often wonder why collaborative projects which connect us to an expanded sensory experience of breath can command such commitment from participants. Perhaps such shared creative encounters catalyse a subtle assimilation of experience, a reconnection with the authority of felt experience. Participants from the Royal Brompton Hospital were still committed enough six months after the project had ended to travel across

London to re-perform and celebrate it. Nicolas Bourriaud's concept of relational aesthetics in contemporary art practice seems to offer a further clue: it is necessitated, he argues, by the fact that 'this is a society where human relations are no longer "directly experienced"'.[29] Perhaps the need for engagement in contemporary society is ignited by new ideas or projects making new neural pathways, reopening possibilities of sensed experience, a seeping back of connection. It may be that true conviviality, sociability, and humane exchange are in the forefront of art today because they are scarce elsewhere.

Patients are used to visualisation of the clinical indicators of their lung capacity and breathing quality through graphs and charts. However, when patients were invited to create non-clinical records of their breath as part of an art practice, there was a level of excitement and anticipation, and in some cases nervousness and reticence, when contemplating taking part. Visualisation can reinvigorate the interaction between clinician and patient, perhaps even help us to understand the emotional landscape which comes with the long-term suffering of patients with COPD, asthma, and other causes of breathlessness. Art can shift the focus to joint ownership of the problem, the novelty of the resulting image thus allowing neutral ground to mediate the gap between the physician's scientific knowledge about lung function and the sufferer's experience of it. The explicatory value of using art to reify the transitory breath may facilitate a less intimidating dialogue around clinical markers of symptomatology. An arrested moment can pierce the haze of inertia, allowing a clarity, a bridge between our experience of our breath and the knowledge of our condition.

What I also get out of collaborative work is the unexpected alchemy that unfolds. When working with the breathing of impaired lungs, while you are necessarily illustrating the limitations of the system, it does nevertheless also seem possible to see resilience alongside the fragility.

Notes

1. Gabrielle G. Starr, *Feeling Beauty: The Neuroscience of Aesthetic Experience* (Cambridge: MIT Press, 2015), 80 and 99.
2. Most of the works discussed in this chapter can be viewed by putting the name of the artist and the title of the work into an internet search engine. Here, for example, see https://en.wikipedia.org/wiki/Cueva_de_las_Manos.

3. Christiana Payne, 'The First Balloon Flights: Art and Science', in *Air: Visualising the Invisible in British Art 1768–2017*, ed. by Gemma Brace (Bristol: Sansom, 2017), 12.
4. http://www.philipperahm.com/data/projects/pulmonaryspace/index.html.
5. Cornelia Parker, quotation from *Doubtful Sound* exhibition guide, Baltic Centre for Contemporary Art, 2010.
6. I am indebted to the research of Alan Barnes (Derby University) and Stephen Leach (Keele University), delivered in the 2017 symposium, 'Breath, Flight and Atmosphere', for fresh insights into this seminal work.
7. *Courageous Breath* can be seen at https://www.jaynewilton.com/drawing-breath.
8. Jan Cox, 'Wyndham Lewis and Francis Bacon', https://www.unirioja.es/listenerartcriticism/essays/essay-Wyndham-Lewis-and-Francis-Bacon.htm.
9. Michael Peppiatt, *Francis Bacon: Anatomy of an Enigma* (London: Weidenfeld, 1996), 24.
10. British Library, Royal 1 E VII, f.1v (this leaf, third quarter of the eleventh century) https://www.bl.uk/catalogues/illuminatedmanuscripts/ILLUMIN.ASP?Size=mid&IllID=40497.
11. *Liber Divinorum Operum 1.4: Cosmos, Body, and Soul.* Biblioteca Statale di Lucca, MS 1942, fol. 38r (early 13th century) http://www.hildegard-society.org/p/liber-divinorum-operum.html.
12. *One Breath* can be seen at https://www.jaynewilton.com. It was recorded at Brunel University in 2011 during a STEM outreach project.
13. https://www.lissongallery.com/artists/shirazeh-houshiary.
14. http://billviolaatstpauls.com/martyrs/.
15. http://billviolaatstpauls.com/mary/.
16. Gabrielle Starr, *Feeling Beauty*, 28.
17. *Ibid.*, 2.
18. Giorgio Agamben, *Infancy and History: On the Destruction of Experience*, trans. by Liz Heron (London: Verso, 1993), 16.
19. All covers and cover stories created for this commission can be found at https://www.jaynewilton.com/2016-respiratory-lancet-covers-and-stories.
20. See Anna Wierzbicka, *Cross-Cultural Pragmatics: The Semantics of Human Interaction*, Trends in Linguistics: Studies and Monographs 53, 2nd edn (Berlin: de Gruyter, 2003), 8. Wierzbicka identified sixty semantic primes which form the basis of a natural semantic meta-language and are core words critical to communication in all major languages.
21. Jayne Wilton, George Mogg, and Malcolm Zammit, *Suspense* (London: Brunel University, 2013), 19. See https://itunes.apple.com/gb/book/suspense/id661082571?mt=11.

22. *The Edge of Suspense* can be seen at https://www.youtube.com/watch?v=DSVeJnlYP6k&feature=youtu.be.
23. https://www.youtube.com/watch?v=XSvw2GaMJoY&t=68s.
24. This project was delivered in partnership with the Royal Brompton and Harefield NHS Foundation Trust's charitably funded arts programme.
25. The *Vent* animation can be seen under the Drawing Breath section at http://www.jaynewilton.com.
26. For Duchamp's attempts to exemplify the idea see *Marcel Duchamp, Notes*, preface by Pontus Hulten, intro. and trans. by Paul Matisse (Paris: Centre George Pompidou, 1980).
27. Walter Benjamin, 'The Storyteller', in *Illuminations*, trans. by Harry Zohn (Pimlico: London, 1999), 83–84.
28. Agamben, *Infancy and History*, 15–17.
29. Nicolas Bourriaud, *Relational Aesthetics*, trans. by Simon Pleasance and Fronza Woods (Dijon: Presse du réel, 2002), 9.

Selected Bibliography

Agamben, Giorgio. 1993. *Infancy and History: On the Destruction of Experience*. Trans. Liz Heron. London: Verso.

Brace, Gemma (ed.). 2017. *Air: Visualising the Invisible in British Art 1768–2017*. Bristol: Sansom.

Benjamin, Walter. 1999. *Illuminations*. Trans. by Harry Zohn. Pimlico: London.

Bourriaud, Nicolas. 2002. *Relational Aesthetics*. Trans. by Simon Pleasance and Fronza Woods. Dijon: Presse du réel.

Sloterjijk, Peter. 2009. *Terror from the Air*. Trans. by Amy Patton and Steve Corcoran. Los Angeles: Semiotext(e).

Starr, Gabrielle G. 2015. *Feeling Beauty: the Neuroscience of Aesthetic Experience*. Cambridge: MIT Press.

Wierzbicka, Anna. 2003. *Cross-Cultural Pragmatics: The Semantics of Human Interaction*. Trends in Linguistics: Studies and Monographs 53, 2nd edn. Berlin: De Gruyter.

CHAPTER 24

Breath—As Subject, in Form, in Performance: An Interview with Michael Symmons Roberts

Michael Symmons Roberts with David Fuller

David Fuller: Michael Symmons Roberts is going to discuss his novel, *Breath*, and a range of earlier and later poems. One background to this is a discussion Michael and I had at the Edinburgh International Book Festival in 2017, which prompted Michael to think about how his underlying interest in writing about breath had family roots. I'd like to begin by asking Michael about those as an introduction to our discussion.

Michael Symmons Roberts: As has happened to me before, sometimes you are invited to talk about a theme in your work that you hadn't realised was there until you are invited to talk about it. I do seem to be obsessed with breath and breathlessness. Talking in Edinburgh it all started to coalesce with me. My mother's family is from Salford, near Manchester. My maternal grandfather worked in the docks. He was a clerk

M. S. Roberts
The Manchester Writing School, Department of English, Manchester Metropolitan University, Manchester, UK

D. Fuller (✉)
University of Durham, Durham, UK

D. Fuller et al. (eds.), *The Life of Breath in Literature, Culture and Medicine*, Palgrave Studies in Literature, Science and Medicine,
https://doi.org/10.1007/978-3-030-74443-4_24

for cargo ships, dealing with bills of lading, and he suffered from severe asthma. One of my childhood memories is of my grandfather struggling for breath, regularly being rushed to hospital. I had heard that in his twenties he was a good runner, and ran for the Salford Harriers, one of the great athletics clubs in Manchester. Then asthma came to dominate his life. I remembered as a child seeing family photos in which before a certain point in the 1960s he looked like a cadaver. There's no other way of saying it: he looked absolutely drawn, quite different from how I remember him in late childhood. You never know how much these things are family myth, but it was said that he met on the docks someone who had come off a merchant ship from Germany who heard him struggling for breath and said to my grandfather, 'Are you asthmatic? I am asthmatic, and my doctor in Germany has just put me on cortisone. It has turned my life around'. So my grandfather asked his doctor to put him on cortisone, and he was the first asthmatic in the country to be put on cortisone. I don't know if this is true, but this was the family's story. Cortisone shortened his life, but he lived into his early seventies. He had inhalers that he used all the time, but he was never afterwards rushed into hospital. The illness was stabilised: he lived a normal life. All the family photos after that look different. Part of that is, of course, the effect of the steroids, but also it is just that he comes to life through the record of family photos. So that is a quite dramatic thing when you are growing up, and dominant, I think, in our family. I grew up with the idea that we were a family with lung trouble, breathing troubles, and I was the kind of child who had coughs I could never get rid of. This culminated in double pneumonia. I was off school for months when I was about six. I remember being quite proud of having double pneumonia. It was a serious-sounding thing—double. I remember very little about it apart from having a cough for a long time and getting out of school. Again, I think it reinforced the sense that breath was something not to be taken for granted, something fugitive, something that our family didn't have a good record with. Later, of course, you don't think about these things. You start writing, and you realise this keeps coming up. It was a conversation with David that drew this out. That's the family background.

DF: Obviously breathing is a central issue in your novel, *Breath*.[1] I've asked Michael to talk about the relation between the novel's main elements—the aftermath of a civil war; issues of medicine and physical treatment; and a related issue of spiritual healing. One aspect of how the lung disease narrative relates to the civil war is about the sense of different kinds of air accessed by different kinds of lungs. I've also asked Michael to talk about a language issue. Several critics have said *Breath* is the novel of a poet. Sometimes on the page there are two different typefaces. In these different typefaces the novel presents the reader with different kinds of language, a language of passion and a language of bureaucracy, which find trouble meeting. All sorts of real experience, you might say, find difficulty meeting the language of bureaucracy—but I leave Michael to describe this.

MSR: It is worth saying that I don't think of myself as a novelist. I think of myself as a poet who has written two novels. There's a strong tradition of poets who write two novels!

The genesis of the novel came from an opera I was working on, *The Sacrifice*. I've worked for a long time with the composer James MacMillan. *The Sacrifice* was written for Welsh National Opera. As with a lot of new commissions in opera, the whole creative team is assembled from the start. You start talking about the work before anything is done, which is odd but also very creative. Poets just tend to get on with it, but this is a different kind of process. So we were put together, Jimmy and I and the director, Katie Mitchell. Katie's method is to immerse everybody in research from the outset—to throw lots of material at us. The story we had in mind was about the aftermath of a civil conflict in a Balkans-like state, and the difficulty of healing after that sort of neighbour on neighbour violence. Meetings were set up; we watched films; we read eye-witness accounts; we were thoroughly steeped in the subject. With the creation of an opera the librettist goes first. So we did the research, I wrote my libretto, and adjusted it, and so on. The composer comes in, and the director comes in, but I still had all this material going round in my head. I found that everything I tried to write after the opera was finished was circling back around the idea of civil conflict. I didn't know how to write about it all. Then I met somebody I worked with briefly when I was a producer at the BBC. He had given up producing at the BBC because his first love was flying: he wanted to be a pilot, and he managed it. He was flying transplant organs around Britain because that was a way of getting your flying hours up. I thought this was bizarre, so I asked more about it. 'What is it like with a heart? What is it like with a lung?' He said, 'You just strap it into the passenger seat. Sometimes you are flying all night'. I thought how strange that must be. The one that interested me most was the lung, and the lung transplant was what unlocked the idea of how to write about the aftermath of a civil conflict. In the novel a lung is transplanted from a young victim in the south to an old man in the north who has a past that he is not talking about which has something to do with the conflict. So the lung gets transferred across the border. This aspect of the novel circles around questions of what goes with the lung. Some interesting work has been done on this by the academic Fay Bound Alberti, who has written about the metaphorical and psychological implications of various parts of the body.[2] Particularly she has done interesting work with the heart, talking to medical students, who are supposed to be hardened to this kind of thing, about what it is like to handle a heart. They feel there is something different about handling a heart—that if there is a soul the soul might reside there. There is something like that for me in the breath organ, the lung, because voice is so much a part of what identifies people. The idea is that breath is the root of voice, and therefore if you carry the vessel of someone's breath in some sense you are carrying them. That all came into the novel.

The section I want to begin from (*Breath*, 31–33) concerns the protagonist of the novel, Andrews, a hospital manager in the south of this post-war-torn country. He is suddenly faced with the worst nightmare: his teenage son has been knocked off his bike and killed, and he is asked whether he is willing to allow his son's lung to be donated for a transplant. He is grief-stricken, but he is trying to be a manager as well as a father. Andrews is supposed to be an exemplary figure in healing the post-conflict divide among his staff. He has just been fine-tuning a document about this for other hospitals to use. At the same time he is about to go in to see his dead son, and it is starting to dawn on him that his son's death may be something to do with this sectarian divide. This section exploits the clash of languages David mentioned—humanely emotional from the narrator, coloured by expressions (words, turns of phrase) in which Andrews might actually think and feel; coolly bureaucratic (marked out by a different typeface) for the document he is writing. (As will become evident when I discuss a later passage [pp. 184–86], these different languages are eventually associated with different cultures of air and breathing.)

I'd also like to discuss a section about a prayer (*Breath*, 26–27). The novel takes place over a single night and flips between three settings. One is Andrews trying to come to terms with the loss of his son, and at the same time struggling with his own demons as a recovering alcoholic among other things. Another is the recipient of the lung, Baras, waiting in a hospital north of the border. He is obsessively talking to a young woman who is the hospital chaplain. He wants something from her: that dialogue becomes a key part of the novel. The third element is the lung in conversation with a pilot who is flying it across the border. The lung has a voice, though you are never sure whether or not it is just in the pilot's head. This is part of the section in which the old man north of the border, Baras, is waiting for the lung to arrive.

DF: I asked Michael to consider this passage because I was interested in the issue of a prayer memorised in childhood as like 'an implant sewn into the fabric of breath', integrated into the body by the breathing in and out. As the passage shows, Baras has had a religious education, has consciously rejected that education, but in a number of ways continues to have a religious sensibility and religious yearnings. This ultimately expresses itself in a desire for forgiveness, which is understood in religious terms as about penitence and absolution. So though Baras is consciously an atheist, his training means that his body—some preconscious level related to just continuing to breathe—retains elements of religious consciousness in a secular context.

MSR: The Jesus prayer that he recites, as you probably know, is used mainly in Orthodox traditions but has become well known in the West as well. If you look at Russian Orthodox and Greek Orthodox theology some of the main proponents of this practice of prayer felt that it became so mapped onto their breath that just being alive, just breathing, was a constant act of prayer. This intrigued me—and alarmed me.

The last section I want to draw attention to is quite different (*Breath*, 184–86). This is near the end of the novel. The leading lung transplant surgeon in the country has flown up north, across the border, to carry out Baras's operation. Baras is concerned about what this surgeon from the south is going to do to him. Everything up to this point has been exemplary. The surgeon, Ross, has his team put Baras under anaesthetic, but not fully: he is put under to a point where he can hear but cannot act. This is during that conversation. Ross is telling Baras, who cannot respond or answer back, that he knows what he has done in the war, 'And now one of my people has offered up his lung to give you breath'. His hatred of Baras is expressed in forcing him to consider his own wickedness in using poison gas with an account of the most famous form of death inflicted by extended asphyxiation: 'Do you know what crucifixion is? Do you know what kills you? Do you think they bled to death? No. It's breathlessness. … Pray for breath, Mr Baras. Pray now for the breath that you denied so many others'.

DF: Before we leave the novel, I should say it has some surprising turns in the end: the apparently southern lung turns out to be more ethnically diverse than it had appeared, and the death of Andrews' son, Jamie, is seen in a different way by the end of the novel—more directly a result of the civil war conflicts: he dies as a result of making overt through his flute-playing—a musical formalisation of the character and expressivity of breath—the northern allegiance drawn from his mixed heritage. I wonder if you could say more, Michael, about the issue of southern and northern lungs needing to breathe different air. This is a brief element of the novel, but I think a crucial one in bringing together the civil war and the medical aspects.

MSR: This has to do essentially with the difference in air between the north and the south, and how the lungs of each group differently access the air. I was trying not to map air and breathing too closely, because you are hearing about these from the perspective of one side or the other. I wanted it to be possible that everything you hear from either side about the other could be understood as propaganda. There's a sense in the novel that the north is geographically different (mountainous), that the north has a stronger poetic and musical tradition, and that breath is at the root of that. A flute from the north plays a key role in the novel, as an incendiary symbol of sectarianism but also as a form of liberation and expression. So there is something about northern breath being able to take shapes associated with art, and southern breath being more associated with the sort of bureaucratic language that Andrews finds himself able to compose for his job. I can mention just a couple of anecdotal things that have come to me recently. One is that my eldest son has moved down to London, and when he comes back to Manchester the first thing he does when he gets off the train is to take a huge breath of northern air. This is partly, I suppose, about the perceived pollution in London, but it is more than that: it is about Mancunian air meaning being home. Also I recently had a conversation with a voice coach who works with actors, partly because I thought I might be interested in doing some voice work. Like a lot of poets, I am constantly going around using my voice, and I am not sure I use it well. This voice coach comes from Liverpool. One of the first things he said to me was, 'You are from the north west. My relatives are the same. You have that slightly nasal, slightly sinus-y tone because the air is so damp'. Hence the way the breath functions; hence the way the voice functions. That's why the north west has the history it does: because the air is damp you can work with cotton. That's the story of Manchester—the damp air. So there are two anecdotal reflections on geography, culture, air, breathing, and voice.

In connection with this I'd like to discuss the poem 'The Frequency' from *Corpus*, which is a book of poems about or related to the body. 'The Frequency' was written when I was working at the BBC as a radio producer. I was interested in the relationship between voice and breath. At that point as a radio producer you spend hours editing on quarter-inch tape—a lost craft now. Breaths were a key part of editing on quarter-inch tape. Before you edited an interview you would run it through and if there were useful breaths that didn't have a hard syllable on either side of them you would cut them out and stick them with splicing tape to the side of the machine so that in future edits, where you were trying to get a blade in really fast, you could put a breath in. So you would keep your long breaths in one place and your short breaths in another. This idea of breaths being part of your kit when you were editing an interview was entirely everyday. The other thing you did when working with quarter-inch tape was move the spools to find the exact point to get the blade in when someone was speaking quickly. You would slow the tape down, and when you slow it right down a breath becomes a sound. If it's slow enough on the spooling a breath becomes a vowel sound. This poem is in that area—looking at the point where breath becomes speech. You also realise that words sound like other words when you change the frequency, and that the whole tenor and emotional content of the words changes when you change the frequency as well, so the frequency is not just about speed or pitch. There's more at stake than that.

DF: The poem is about hearing the uncensored language of the heart, hearing something that is normally inaudible. You've said elsewhere that something you go back to often is a sense of wanting or working towards an epiphany. This seems to me a poem you could describe like that, because it's about getting to something that is almost present but you can't normally hear, and that something is located in the breathing. Can you say more about that?

MSR: I think it's one of a number of poems in which I'm looking for connections and continuities between breath and voice, and where breath sits in relation to voice—the idea that the breath might be not just at the root of voice physically, but also that breath might be in some sense a pre-conscious form of voice. The other poem that gets into this territory is 'The Lung Wash', which is about getting behind the voice into breath and through into some essence or epiphany, some revelation. 'The Frequency' does that slightly differently. I was interested in looking at this idea that there could be frequencies we don't know about, frequencies we don't understand. I was asking, if you slow recorded sound almost to a standstill, what comes through? What's behind the voice? That's where breath comes in. But like all these poems, you can say certain things about them, but they are exploratory. I don't have a clear sense of what it is I'm trying to say about the relationship between breath and voice before I start the poem, and the process of writing the poem is part of how you find out and explore what you think about your voice and breath.

DF: So you're hearing vocal sounds which begin as words, but getting to a level where you've divorced them from the words but are hearing an expressive articulation which is pre-verbal: it's got a meaning, but not a meaning in the usual way that verbal articulations have.

MSR: Yes. It is that sense of some other level of meaning, some deeper level of meaning that is pre-verbal. Also I'm interested in those ideas of life being effectively bookended by one huge breath, certainly in literary understandings of life. I don't know whether these stand up medically, but I've often thought that the popular conception of the first and last acts of many lives is that you begin with a sharp intake of breath, which I've seen in my own children, and then the crying: the inhaling is the first act, and then the exhaling, famously throughout literature, is the last act, so the whole thing takes place bookended by these acts of breath, and that, if these are the bookends of life, there's something profound and perhaps related to a sense of whatever might continue after or pre-date your life—the ideas of continuity of self, of soul if you like, and that the breath is intricately connected with that.

DF: I want to offer you something of Basil Bunting's, which I am reminded of by this, about the importance of sound to meaning in poetry.

> Poetry, like music, is to be heard. It deals in sound—long sounds and short sounds, heavy beats and light beats, the tone relations of vowels, the relations of consonants to one another which are like instrumental colour in music. Poetry lies dead on the page, until some voice brings it to life, just as music, on the stave, is no more than instructions to the player. A skilled musician can imagine the sound, more or less, and a skilled reader can try to hear, mentally, what his eyes see in print; but nothing will satisfy either of them till his ears hear it as real sound in the air. Poetry must be read aloud.[3]

DF: While I fully accept what Bunting says here about reading poetry aloud in a general way, I don't want to believe what he says specifically about vowels and consonants because that seems to involve the sounds of poetry with the accent of the poet. That's an interesting idea, but I don't want to believe it, because if you think that's true then something of all poetry before Tennyson is lost, because we can't imagine at the level of vowels and consonants quite what sounds the poet was hearing. But I'm now tempted to think that this may have some truth in it—that everything about a voice is present in a reading, that is, the whole vocal mechanism, not just the vocal cords but everything that supports them, the whole breathing apparatus. Not perhaps in ways you can understand, but I think how much I like T. S. Eliot's readings, partly because

I love the sound of his voice, which seems to me to have a characteristic expressivity suited to the poetry.

MSR: I'm wary when an attempt is made to develop a programmatic step-by-step understanding of how to read the sounds of poetry, almost like a musical notation: the consonants do this, the vowels do that. It seems to me that limits the range of responses, and the range of relationships, not just the responses to poetry, but the range of relationships between the reader and the poet's voice in the poem, which differs from poet to poet. I think there is a relationship between the poet's accent, the pitch of their voice, the way they stress things, the complexity of their voice, and the way their poems are made. That seems to me self-evident. But I don't think that means you can't read Bunting unless you can hear the Northumberland accent Bunting had in mind as he was writing. I think that is needlessly precious, and to me uninteresting. The great poets of the past don't depend on being able to recreate a sense of how (say) Shakespeare might have spoken. That seems to me an academic interest not important to the reading of the poem.

DF: I extracted from Michael's jointly written book with Paul Farley, *Deaths of the Poets* (2017), an anecdote about Byron's larynx and his lungs, which are buried separately where he died at Missolonghi, and asked Michael if he would talk about that and another phrase from the book, 'poetry alive in the heart and on the tongue'—where the heart, the lungs, the larynx, the tongue, and the body as a whole fitted into his sense of where the poem comes from and how the performance of a poem works.

William Carlos Williams talks about the difficulty, but also the necessity, of getting a poem off the page. He argues that you understand a poem by getting it into your voice and constructing the poem as a soundscape for yourself. But a poem can be difficult to get off the page because the printed indicators about what it is meant to sound like as an aural structure are not always straightforward to interpret. Can you talk about those indicators and how the reader realises their implications?—the structure of the line; the structure of syntax, and how it interacts with the line; the structure of the stanza; also structures of intonation and other forms of aural inflection, some of which are the corporate possession of all the people who speak the language, some of which can be individual to particular speakers. How do you look at the printed poem and realise that as an aural structure, whether in your actual voice (lungs, larynx, patterns of breathing and intonation) or in the mind's ear?

MSR: William Carlos Williams' views and the anecdote about Byron's larynx and lungs coalesce around the issue of the poet's voice and how that works on the page, if it works on the page. Williams is interesting in relation to this. I worked for radio on his poem *Paterson*, a huge Modernist collage, which I've always very much admired. I think it's a flawed work, but also a brilliant one. We worked with fine actors who understood how to read poetry, and they struggled with parts of it. There are a number of reasons for this. One is that Williams is interested in the page as a visual field, and is therefore not necessarily interested in giving the reader a set of instructions about how to sound the poetry in your own inner voice. Also he is desperately keen to create an American poetic that is not English—bordering on anti-English. He wanted to kick against that tradition—break up and fragment what might sound like a traditional lyrical English voice in a poem. I also think on a broader level the idea of the poet's voice is a complicated and troubled one. There's an assumption that it's something innate, a post-Romantic assumption that there is something almost soul-like about the voice—going back to what I said earlier about the almost mythic properties of the voice and the breath. Actually I think most poets would say that the 'voice' of their poems—if there is a single voice, and many poets would say there are multiple voices—but if there is a recognisable single voice it's something that, if not wilfully constructed, has arisen out of everything that you've read, a negotiation with what you can and cannot achieve on the page, how interested you are in various formal structures, and so on. It also changes as you write. I certainly would not say to my students that what you have to do is find your voice, because I don't think they or I would know where to look for it. A lot of poets, when they do have a sense that people will know what a poem of theirs sounds like, resist and challenge that as well. So there are complexities around the voice, and in order to talk about poets' voice or voices you have to get beyond that myth of the authentic voice, especially because it shades into ideas of innate genius, deep-rooted inspiration that just comes through this channel of 'voice', and there is still a lot of misleading stuff assumed and written about that—the assumption, for example, that that is how Blake worked, that it all just came and he wrote it down, rather than being something intricate and crafted.

Then there is the issue about William Carlos Williams reading his own poems, and reading them with punctuation that he doesn't give the reader on the page. There are quite a few poets like that—where you know a poem, and admire it, and you hear the poet read it and think, 'they're reading it wrongly'. There was a debate a few years ago about one of the poetry prizes, when actors performed poems at the prize-giving ceremony rather than the poets reading their own work. Some—probably the majority—said that was wrong: poets know how to read poems; we don't like actorish reading of poems, etc. Then there was a minority, to which I am more sympathetic, who say that if poets are the only people who can read their poems aloud, that is not a good thing. You still read an Elizabeth Bishop poem or an Auden poem and it works—so it's about the quality of how the actors do it, rather than poets being the only possible deliverers of their poems. Just as there are actors who don't read poems well, so there are poets who don't read poems well. I think there's something again about the Romantic idea of the authenticity and power of the poet's voice that carries through into the idea that they are the only ones who can perform their poems with authenticity in a public space. Something in me wants to resist that.

DF: In a note to his recording of *Four Quartets* T. S. Eliot says that a poem of any breadth is bound to be able to be read in different ways, and he compares the issue of the composer's performance of a score, saying it's interesting to know what the composer was hearing, but there must be many other ways of performing that score, otherwise the work is too limited. That's somewhat the issue about actors and poets reading, isn't it? A poem should be able to work in different voices, and in different ways. But on what you say about actors and poets my prejudices are on the other side. Of course they are mixed: there are actors I love hearing read poetry—Richard Burton, for example. His voice is so characterful that I have the impression of hearing the whole embodied being in the sound; and I also think his Welsh background means that he doesn't mind half-singing something: that doesn't come across as affected. He can move easily between colloquial tones and quite stylised delivery. However, there's an essay by F. R. Leavis about reading poetry aloud in which he deplores actors reading poems: what he's thinking about is any assumption of an 'actorly' mode which infuses the poem with generalised feeling as a substitute for thinking what the words mean.[4] So there are two sets of judgements. With hearing poets read their own poems, I think Eliot is interesting in saying the reason for making recordings is that the listener can hear what the rhythms of the poem, the shapes of the poem that can't be notated in print, were to the poet—precisely what shapes the poet was aiming for.[5]

MSR: Probably the prevailing view is the Leavisite opinion that the authentic reading is by the poet themselves. I resist it, and think it's full of assumptions that don't hold. I don't think actors do necessarily read in an 'actorly' way. I don't think there's a single actor's mode of reading. There's a cliché that actors always ham it up, or go for easy emotion. In my experience good actors don't do that. I've had lots of poems read by actors in broadcasts and performances, and the ways of reading have differed, but not because they're actors doing an actorly thing. I also question the idea of there being a particular relationship to do with authenticity between the poet's reading voice and the poem. As I think anyone who's been to lots of poetry readings would acknowledge, some poets are terrible readers in terms of appearing to mangle their own poems, appearing not to know how to read their own poems, killing off any light of interest that the audience might have in the poems. So the hallowing of the voice of the poet as being the only way of hearing the power of the poem seems to me fraught with problems and difficulties. Usually a well-made poem—the poems you return to again and again—is susceptible to many different readings. I have often thought about the Poetry Archive opening up recordings of poets we may not have heard reading before. For example, I'd never heard Robert Graves until I heard his recordings on the Poetry Archive website. First a caveat: I am always interested to hear a poet. If a poet I admire is reading I want to go and hear them, but it's not necessarily because I need to get the authentic, or even the best performance of the poems. It's just that I'm interested to see what they do with them. With Graves you realise that these delicate, troubling, but beautiful and absolutely serious lyrics sound parodic, like Harry Enfield doing Mr Cholmondley Warner in the 1950s. It's not just a patrician public school accent. It's an accent with vowel sounds that are no longer used except in Ealing comedies: they're historically an anomaly. So when you get down to the level of the sounds behind the poem, to hear Robert Graves reading his poems is actively setting off a whole range of resonances to do with class, and with a particular period, that damage and undersell the poems now. Though it's interesting to hear him read, if students are interested in Graves, I wouldn't say, apart from just out of interest, listen to him. So I think it's complicated, and I resist the idea that the poet is the only true voice to carry the poem, or that poets are necessarily the best readers of their own work.

DF: On the issue of poets as offering particular ways of reading, Wallace Stevens is a reader I love, and the thing about Stevens that's to me a revelation—which isn't just about Stevens—is how very slowly he reads. The way I think Stevens seems so excellent is that you feel the meanings of the words are passing fully through his mind as he says them. At that pace it allows you really to think and feel the meanings, and to take pleasure in the language as a structure of sounds. The example of Stevens has persuaded me to read a lot of poetry more slowly.

MSR: I think that's why it's interesting, and that is one of the reasons I seek out poets' recordings of their own work. It's interesting to find out how they felt the pace and pitch of their poems should be set. For example, I love hearing Elizabeth Bishop read her poems, though by contemporary standards—certainly compared with Stevens—there was a lot less performative delivery and potency in her readings, but nonetheless there's something delicate and very striking about them. So I always find it interesting.

DF: I wonder if we could say more about another of your poems on the breath, 'The Lung Wash'. It took me a while to like this, because the experience the poem describes is so hideous that I found it difficult to read about.

MSR: The poem came from my family history. This is not something that happened to my asthmatic grandfather, nor to me when I had childhood pneumonia, but that is at the back of it. In early adulthood I encountered someone whose father was a chronic asthmatic, and we got talking about it because it's in my family as well. This person's father had gone to Venice to be treated by an Italian doctor who washed lungs, and he was apparently transformed by this treatment, although it's horrifying and barbaric, and sounds like drowning. I think it's a rather unorthodox treatment, and it was told to me anecdotally.[6] The idea of it then connected with those childhood things, and the sense that there's something about the washing out of the lungs which related to the way my family talked about my grandfather's surgery: that the lungs were bad, that he had asthma and emphysema. So the lungs contained things that shouldn't be there, and in some sense you did want to wash them. It related to all those childhood feelings wondering can't someone clean him out so that he can breathe again. So there was a quite visceral response to this anecdote about the doctor who washes lungs. Then in the process of exploring this in the poem I started to see (because I gathered the treatment took place over a week) that each stage of the washing got you back further and further into earlier more original, more authentic—to you—properties of voice, back behind surface levels of voice into deeper more self-revelatory elements of voice, and then right back to primal sounds, that ultimately turned to breath and potentially to silence. So there I was experimenting with how this treatment could become progressive over the week, and what that would do to voice and breath, and what that might reveal about the person. This is from my second book, *Raising Sparks*. I was already then—which I carried on being throughout all the work—preoccupied by the idea that I was interested in even as an atheist when I was first writing poems, interested in and even obsessed by, these ultimate issues of transcendence and meaning, and the possibilities of the spiritual. But I could only understand them through the body. I've never understood the category of the spiritual as being unphysical or the antithesis of physicality.

And so this becomes a sort of metaphysical poem, or a would-be metaphysical poem, partly in that it's an extended conceit, but also because it's trying to reach beyond the physical, but the only way it can go is through being more physical than ever. Your response of finding it difficult to read and unpleasant is part of what it's setting out to do—not trying to float out of and above the corporeal, but trying to find whatever transcendence might be by going deeper and deeper into that.

DF: I read it now as a poem about everybody's situation of losing touch with meaning by using words too much. I started to read it like that from the lines 'You sought him out, like countless others / who speak too much and breathe too little', and then from where you get to at the end, when the process of cleaning is completed: 'Everything you say is sudden poetry'. There's the idea that it's normal to use words without thinking acutely enough about their meanings, or what it is they're trying to search for, whereas if we could use words more chastely, or more demandingly, voice would come from deeper breathing, language would have a more vivid relation with thought and feeling. Is that too general?

MSR: No, absolutely. That's exactly what those passages are exploring.

DF: I'd like to turn to *Mancunia*, and a more recent poem about breath, 'The Cold'.

MSR: *Mancunia* is a book about Manchester but not Manchester. A lot of Mancunians call Manchester Mancunia, as a term of affection for it. I like it because it sounds recognisably Manchester. People from Manchester are Mancunians; my family are Mancunians; so there's history there. Also, it's one of those happy coincidences that during the time I was working on this book, and was thinking about Mancunia as a mythic city as a well as a real city, I was working on Thomas More's *Utopia* for a dramatisation for Radio 4. So Mancunia in this book becomes a sort of utopia and a dystopia, and it has Thomas More in the background. My premise was that any city has as many versions of itself as there are people living in it. You have your own associations and routes and maps of the city.

'The Cold', takes place in Mancunia. I was interested at the time of writing this in a poem that begins with something concrete and expands, and widens and widens. It's a shape in a poem that I've explored before, and that's what happens here. It begins with a mundane moment: it's snowing and the dog wants to be let out. There's what you could call a snowy, icy, blank whiteness, a version of the dark night of the soul here. There's an emptiness, the idea of an absent God, and the paradox that as soon as you talk about the absence of God you're making God present, and therefore the presence within the absence, one of the great paradoxes of mystical writing. I'm interested in that, and in the idea that in this apparent absence the actual presence is as close as the breath on your cheek. I wanted some expression of absolute intimacy, and breath seemed to be that—more so than the voice of God; but the breath of God, with all its biblical resonance as well. I wanted a sense of that paradox: at this moment of bleakness and coldness and emptiness, and the inability to pray, and the lack of confidence that any answer would come even if you did, there's also that absolute intimacy of breath, something so unforced that almost came back to 'The Frequency'. If you're not open to it, if you're not on the right frequency, then it feels like absence. The poem ends in a prayer that arises out of not being able to pray. So the coldness itself becomes a way of riffing into a sort of intercessory prayer for people who are only linked by the connections between them and the cold, especially at the end of their lives. It becomes a prayer for those who are caught in the cold. The image of the breath comes back in the mention of Louis MacNeice, who caught his death of cold with pneumonia in a cave recording the sound of the cave. That also was intriguing to me because what he's recording is pretty much silence, but it's a particular kind of silence in a particular place. So the whole poem is an attempt to explore emptiness, the whiteout of a blizzard, and the silence.

DF: Returning finally to the issue of reading aloud, I'm interested in the physical relation people have to a poem if they get it into their own voice. I've tried teaching about this—helping people to find their own way into the words by getting them into their own vocal mechanism, so that the meanings, forms, and inflections take on the sounds natural to their vocal tones and intonations; so that they can say them in a way that gets them into their own body, their own vocal being. It's with that that I want to go back to that issue of 'poetry alive in the heart and on the tongue'. My sense is that poetry becomes alive in the heart by a sort of mind-body interchange, partly by becoming alive 'on the tongue'. When you have made a poem belong in your own voice, and found your own way of saying it, you have interiorised it in a personal way by that act of getting it into your voice. Is there anything else to say in that area?

MSR: I think learning by heart is connected here. There are all sorts of things we could say about the phrase 'learning by heart', why we associate it with the heart, and so on. One view, which still has currency in some quarters, is that it's old fashioned. On the other hand there's the Poetry By Heart movement—getting young people to memorise and recite poems again.[7] The link between memorising and speaking has always been there in poetry. When I was a student looking around at what poets and poetry were doing, as I was starting to write myself and trying to take it seriously in my teens, the big course on poetry was taught in Boston by Derek Walcott and Joseph Brodsky. I think Seamus Heaney was involved as well. A major principle of that way of teaching was memorising by heart. I think it was driven largely by Derek Walcott's belief in the importance of that. It can become fetishistic, but I think there is a link between the way you say something and having it committed to memory that is valuable, and is to do with the sorts of metaphors you were using of ownership—taking ownership of a poem in your own voice; not just reading it off the page, having possession of it in your memory as well.

But it's also worth exploring what a poem on the page is. It's not quite a musical score, is it? If it were it would be an imprecise one. Most poets don't use punctuation in such a way that there's only one way of reading the poem. There's a pattern there which can be changed, which can be read in different ways; and that relates to poets reading when they read in a different pattern from the one you'd expect from the page. I loved hearing Les Murray read because it told you something about the man: the way he taught and the way he read were very close to the rhythm of his conversation. There was something almost offhand about his conversation, even when he was talking about profound things, and that came through in the way he read the poems, which were far from offhand but were read in an almost, or seemingly, throwaway fashion. The rub between those two things—the apparent mismatch between the tone and pace at which he read (because he often read very fast) and what seemed to be in the poem on the page—was to me interesting. But it doesn't mean that everyone has to read a Les Murray poem like that.

Seamus Heaney is perhaps one of the most effective readers of his own poetry. I find it almost impossible to read a Heaney poem without hearing his voice in my head. There's such a strong print of his voice on the poems. To some extent it's the same with Ted Hughes. This is partly a generational thing: they were the poets I grew up with and was listening to as a student. That's perhaps why I love Elizabeth Bishop's idiosyncratic readings as well, because of encountering her at the same time. But I'm wary of making a hallowed object out of the poet's delivery of their own poems, because the best poems can withstand any number of voices getting into them, and remaking them, and finding new patterns within them. That's always happened to plays, and consequently dramatists can be less precious about this. It seems part of the fabric of the play as a made object that is therefore a public object: other people can take ownership of it and interpret it in different ways, and there's a confidence underlying that that the play is strong enough to survive, and the impact be enhanced by different performances of it. There's something I resist about poetry not allowing that confidence, allowing that breadth, allowing a poem to be sung by someone if they want to sing it to a tune they've made up, or to be read at twice the pace the poet might read it. Poetry suffers from being seen primarily as a form of self-expression: hence all those metaphors about an outpouring rather than a made thing, where the expression comes through the making. If poetry is an act primarily of self-expression—I suppose we get that partly from misreading of the Romantic poets, but also from the confessional generation—then it's hard to imagine other voices reading a poem, because if it's one person's cry from the heart then you need to imagine how it sounded to them. But for me if poems are fundamentally made objects, even when they're purporting to be a cry from the heart, a made object, like a play, is a constructed thing which contains self-expression but also contains much more, and is susceptible to, and should welcome, a range of interpretations and responses, including other people's voices.

DF: What you began from there about Derek Walcott getting students to memorise poems is separate from that isn't it, because they are getting poems into their own physical being. The responsibility is thrown onto them to make the poem their own. Ted Hughes, of course, has an anthology, *By Heart*, which is about the same thing—memorising as an act of love.[8]

MSR: My own experience has borne that out as well. I don't have a vast range of poems committed to memory, but with the ones I have, I do feel you know them differently. I've got some Auden poems and some Bishop poems in my head, and some Shakespeare sonnets. You do feel there's something unique about your relationship with them because of that, and that when you speak them it's different from having them on the page in front of you and speaking. That does relate to breath, I think. You don't feel you quite take physical ownership of a poem when you're opening a page and reading it. That is more like a musician sight reading, isn't it? There is something about taking and storing it in the physical body and then being able to issue it out that means it's gone through a different process. It's gone right down into you, and is closer to breath. That is intriguing. A few years ago it seemed educationally disreputable—the idea that you had to learn poems by heart. It was associated with people of my grandparents' generation having to learn huge swathes of what they felt was incredibly dull verse, and hating doing it. But I think it is coming back, and has come back. Obviously the spoken word scene has been doing this for years. It's one of the hallmarks of it that you don't have a piece of paper or a book in front of you. But Michael Donaghy was famous for reading without a book too. He had it all by heart. Alice Oswald similarly in her *Memorial*. There's something about having poems in your memory that has a different property of voice and ownership, and again breaks that connection between the poet's voice and the authenticity of the poem.

DF: I'd like now to invite questions from the audience.

Jane Macnaughton. I'm interested by the notion that voice seems to be such an important element of how breath was realised. One of the things that we are struggling with in the *Life of Breath* project is the extent to which breath seems absent as an issue for consideration within the cultural record. You were saying that it's there, and you are not aware that it's there. People write about it, but don't reflect on the fact that they're writing about it. There's a metaphor there in the editing of the tape, isn't there, that breath is in the gaps that are cut out? It's interesting how breath needs to be expressed by a gap, by a sensation that touches your skin and walks away again. It can't be expressed by what it is itself. It doesn't have any substance that you can talk about, except in these other kind of ways.

MSR: I agree with all that. I think it's fascinating. In some of the circles I move in there is quite a lot of talking and thinking about breath. I think poets do think about and talk about it. Musicians do. Singers do. But in the wider culture perhaps what you say is true. I'm endlessly grateful that I had this early engagement working in radio, working in voice, because I think it's really affected a lot of my other work. It makes you pay minute attention to the nuances of different voices and the rhythms of them. When you are cutting them and shaping them because you have got a one-hour interview that needs to be twenty minutes, it still has to sound like them. Breaths are not just gaps. That's why you have a line of them on your machine: if you put the wrong breath in, it stands out a mile. You hear that it's wrong. It is as much a positive element of speech, in that sense, when you are editing an audio interview, as the words are. Put in the wrong breath and it's completely unconvincing. I suppose from that you become slightly obsessed with it. On radio you notice bad breath edits going through. I also get obsessed by it on television, to the annoyance of my sons.

JM: I was interested in another part of this, also in relation to the project. It's where Baras is talking about his experience of problematic breath, saying he cannot locate this in his own chest. The problem is with the air outside, and that air can change. We've raised this with some of our colleagues in British Lung Foundation 'Breathe Easy' groups, who we support with their problems of breathlessness, and they've had lightbulb moments about this. This sense of feeling no, actually it's not me: it's out there. The problem is that the air is not giving me what I want. I wonder where that idea came from for you.

MSR: It goes right back, I think. It was to do with my grandad, trying to make sense of why sometimes his condition turned critical. I suppose you were somehow trying to think, is the air somehow thinner here? Or thinking of analogies with liquids: the idea that somehow they are too thick—because of the physicality of seeing someone struggle with breathlessness. There was something about what he was getting that was wrong, that somehow because our lungs were more robust we were riding over these variations, but that he was like a finely tuned instrument, responsive to changes in the air. I suppose as a child you are trying to rationalise, why is he suddenly rushed to hospital?

This interview was conducted first as part of a public reading and discussion in Durham in March 2018. The text edited and formalised from that occasion was developed in correspondence. This was supplemented by a second interview conducted privately in Manchester in July 2019, and the whole completed through further correspondence.

Notes

1. Michael Symmons Roberts, *Breath* (London: Cape, 2008; New York: Vintage/Random House, 2009).
2. Fay Bound Alberti, *Matters of the Heart: History, Medicine, and Emotion* (Oxford: Oxford University Press, 2012).
3. 'A Statement', in *Descant on Rawthey's Madrigal: Conversations with Basil Bunting*, ed. by Jonathan Williams (Lexington, KY: Gnomon, 1968), unpaginated.
4. F. R. Leavis, 'Reading out Poetry', in *Valuation in Criticism and Other Essays*, ed. by G. Singh (Cambridge: Cambridge University Press, 1986), 253–75.
5. 'A poem … of any depth and complexity … should be capable of being read in many ways, and with a variety of emotional emphases. … The chief value of the author's record is as a guide to the rhythms. Another reader, reciting the poem, need not feel bound to reproduce these rhythms.' T. S. Eliot, sleeve note to his recording of *Four Quartets*, HMV CLP 1115.

6. 'Washing of the lung', or whole lung lavage, is still in use as a treatment for certain rare lung conditions such as Pulmonary Alveolar Proteinosis, where there is abnormal build up of surfactant, the protein which normally supports lung inflation. [JM]
7. http://www.poetrybyheart.org.uk/.
8. Ted Hughes, ed., *By Heart: 101 Poems to Remember* (London: Faber, 1997).

Selected Bibliography

Roberts, Michael Symmons. 1999. *Raising Sparks*. London: Cape.
Roberts, Michael Symmons. 2004. *Corpus*. London: Cape.
Roberts, Michael Symmons. 2008. *Breath*. London: Cape.
Roberts, Michael Symmons. 2016. *Selected Poems*. London: Cape.
Roberts, Michael Symmons. 2017. *Mancunia*. London: Cape.
Roberts, Michael Symmons. 2021. *Ransom*. London: Cape.
The Poetry Archive [sound recordings], https://poetryarchive.org/poet/michael-symmons-roberts/.

CHAPTER 25

Afterword: Breath-Taking—Ethical Impulses for Breath Studies

Peter Adey

I seem to spend most of my academic life—perhaps it is embarrassing to say or write it—trying to find inspiration. What seems a kind of nourishment from a book, conversation, or conference paper that you feel before you know it. The feeling aspirates as I notice myself inhaling. It is bound up in the feeling of the chest cavity, my body and self lift and expand, and I try holding onto the feeling for as long as possible before it is exhaled in anticipation of the idea to be worked with later. Not lost.

This contribution grew out of the *Life of Breath* conference and from a collaboration with a 'feedback' team of early career scholars who spent time behind the scenes identifying themes and organising them with me into the form of discussant comments at the end of the conference.[1] Working with and departing from those original thoughts, I want to play with the hook of this essay's title, '*breath-taking*', as a way of signalling the extraordinary scope and depth of essays working with breath across a terrain of disciplines, temporalities, and indeed terrains or spatialities of physical landscapes, cities, clinics, interiorised and imaginative geographies

P. Adey (✉)
Royal Holloway University of London, Egham, UK

D. Fuller et al. (eds.), *The Life of Breath in Literature, Culture and Medicine*, Palgrave Studies in Literature, Science and Medicine,
https://doi.org/10.1007/978-3-030-74443-4_25

of literature. These set up more than a cross-disciplinary dialogue or the loose coordinates of endeavour: they offer possibilities of a germinal field, perhaps a 'breath studies', that sees numerous approaches, disciplines, and subject matters accord in quite remarkable ways that transcend, as Tsentourou suggests, more than locations, but rather represent a 'depth of transhistorical and transliterary points of reference'. For Craik and Chapman, the book provides the possibility of drawing on a reservoir—an ocean of air—breathy meanings, experiences, and concepts, and cultural and creative experiments in writing, to 'shed light on the conceptual categories which shape medical approaches to breathless patients now'. Such illuminations work in different directions through the collection, many of them aesthetic. As Wilton surveys different artistic practices and works she identifies the 'joint ownership' of a problem, where art and other cultural forms might 'mediate the gap' between different points of view.

In this way *The Life of Breath* is breathtaking. It arrests, sometimes in moments of distinct clarity, in crystal clear prose, but I think especially in the confusions and hesitations, confluences and incongruences that work between and across the diverse chapters. *The Life of Breath* is much more than its parts.

In the time and events that have passed since the conference, I want to invoke *breathtaking* in another critical and I hope affirmative way. This builds on a political and ethical conceit or inspiration at the heart of the volume, which has since become apparent in the momentum of current events—and which the editors have noticed so acutely in their introduction in relation to COVID-19—but also in the way the chapters attune to accounts of alterity, marginality, misogyny, gendered and sexual differences, environmental degradation, classed suspicions, and bodily violence. What might be added is that, today, breathlessness, breath inhibition, breath interruption and breath-taking seem the outcome of a convergence of problems for breath: a global pandemic in the form of the respiratory virus COVID-19, the struggle to breathe in the context of air pollution and climate change, waning environmental biodiversity and species-level extinctions, and the growing movement to oppose and reject police and state violence that has suffocated people of colour by the use of force—and many acknowledge the historical (re)production of these moments of violence involving asphyxiation. The slow and fast violences and vocalisations of breath are undoubtedly among the defining moments of our times.

The overt political and contemporary taking of breath is clearly in accordance with the critique of anti-black tendencies and presences within institutions, police forces, and a wider atmosphere, or what Christina Sharpe has called the 'weather', of racism.[2] Heine, turning directly to these energies in Nathaniel Mackey and M. NourbeSe Philip, accounts for the orientation to breath as a 'returning "to the basics" in moments when everything becomes uncertain. [...] Breath, especially black breath, is as such "subject" to "precarity"'. While there is an overt absence of people of colour within this collection, and a general—although not exclusive—focus on traditions and contexts from the global north, the book and its attention to breath offer a clear and powerful, highly contextualised yet ethical impulse with which to tackle the precarity of breath more generally in uncertain times, both today and in the past, from the streets to the clinic, to the literary, spiritual, medical, and scientific imagination. I try to suggest how these breath(-)taking elaborations might therefore offer something of a critique, but not the death of hope—to paraphrase Lawlor's account of consumption in literature. What the moment demands, and what a focus on breath might nourish, are ethical sensibilities attuned, transhistorically and across disciplines, towards a generosity where breath is not taken but shared, co-produced, given. Such a generosity should be at the heart of what a Breath Studies tries to do.

Conspiration and Conspiracies

In a recent article for the *New Statesman* the journalist and academic Gary Younge, like others, has drawn a parallel between COVID-19 and the public protests and expressions of anger levelled at racism and racial inequality.[3] Younge follows the Public Health England report on *Disparities in the risk and outcomes of COVID-19* which suggests that 'People of Chinese, Indian, Pakistani, Other Asian, Caribbean and Other Black ethnicity had between 10% and 50% higher risk of death when compared to White British'.[4] Younge argues that there is some sort of 'connective tissue' between some of these conditions, between 'the most brazen forms of state violence and the more banal tribulations of the ailing pandemic patient'. As D. Asher Ghertner asks concerning the crisis of air pollution in India and premature mortalities due to air pollution,[5] 'when the source of death is diffuse and leaky, literally shifting with the winds and gathering different compositional suspensions of matter', what happens to 'social culpability'?[6] In short, a variety of commentators are noticing a manner of conspiration if not a conscious conspiracy, a tying together, a linking together through complex chains and feedback loops that define not just environmental systems, but the deep entanglements of global disease with localised air pollution levels, with global climate change, and movements for social change against systemic racism. As Bernini suggests, other feedback loops persist at more personal scales—breathing is understood as a physiological basis for consciousness. Breath feeds back into qualitative experience.

Sharpe has positioned the lack of breath, and indeed the earlier origins of the vocalisation 'I can't breathe', within what she calls the 'wake' of slavery, in the experiences of anti-blackness in the United States, signalling the suffocation of Eric Garner on Staten Island in 2014 by NYPD officers applying a 'choke hold' as a key case. Garner's death, and more recently George Floyd's, display a pattern of the suffocation of Black lives from the underbellies of ships of the Middle Passage onwards. What we are seeing is how powerful admixtures of air and breathlessness are coming together in the voice of protest and political force that Ashon T. Crawley has characterised as a 'disruption, an ethical plea'.[7]

Others have connected the vulnerability of Black, Asian, and minority ethnic communities not only to COVID-19 but also to the suffocation of the Grenfell Tower fire victims, where the tower's residents were asphyxiated and poisoned by hydrogen cyanide and other gases from the PIR insulation and ACM cladding which covered the building.[8] Grenfell has raised awareness of inadequately protected high-rise structures in Britain which are disproportionately occupied by people of colour.[9] This was precisely the relationship Professor Leslie Thomas QC observed at the resumption of the Grenfell Tower inquiry in 2020. The resounding words of dying residents in calls made to emergency 999 telephone operators, echoing with the slogans of Black Lives Matter ('We can't breathe'), were projected onto the Grenfell Tower scaffolded façade—another way to make 'visible' marginalised and invisible lives in order to make them count.[10]

Capacities

The bodies in this book reveal breath's unusual role in stabilising fields of power and control, autonomy and automation, positioning subjects in unequal and hierarchical relations to one another, by their capacities to breathe and strike speech, and their relative life chances. As Hunt argues in his essay on late antiquity, 'breath is subsumed within the discourses of mastery and property ownership by which social and political relations were understood': breath orders social relationships, whether defined by ownership, by masculinity, or by mastery over property. Gray makes a similar point through his exposition of the relationship between the Monarchy, the Church, God, and the public within a sermon of John Donne on the limits of power, where breath signals both life and spirit. In one sense the King is our breath, our breath is his. Spirit not so. For Gray breath produces and reveals particular social orders, shaping capacities to rule and be ruled, to resist, protest, and speak out.

Walker and Rose debate contentions over breath and breathing within the arguments of psychoanalysis, and reveal a view of breath within the conscious and unconscious body, and a chance to think about breathing in relation to those actions involuntary to or autonomous from the conscious self. In part, these studies reveal the body's capacity to wield breath, to speak, and to navigate social and political hierarchies—how categories of social and bodily difference position bodies in different relations to breath and to the world. Who has breath, and who is short of it? Who breathes, and who doesn't?

The breaths that dominate this book are failing sorts of breath, struggling, volatile breaths. Coughing, sneezing, choking, sighing, swooning, gasping, crying, sobbing, an excitable kind of breathing, although we see in Renevey's essay a soothing, meditating kind of breath, and a recuperative one in Knoeff's. A fragile, grotesque breath definitely turns up in the consumptive in Lawlor's essay initially—the consumptive's expiration is a horror, 'malfunctioning, rotting, stinking, arhythmical, failing and dying', as the body is turned almost inside out, before it is re-taken as a calm, a feeble and feebler, melancholic breath. Both disable the body. And yet, while there are strong parallels in literature, for Skinner literary sensibilities normatively order capacities to act within constraining fields of social propriety and moral judgement. To this extent breathlessness does not simply signal 'incapacity' but, for Skinner, 'intense and instinctive moral discernment' as a 'vital condition for action'.

We might say then that *The Life of Breath* shows us that our capacities to act, to affect and be affected, are in constant negotiation, transformation, distribution, mutually reinforcing and reverberating through micropolitical relations that breath causes us to take seriously. Thus the body gains form as breath is registered on it, just as, for Saunders, the breathing body configured in the affective thought spaces of medieval literature was a porous one into or from which the spirits in the form of breath rushed or withdrew. In moments when breath becomes vulnerable so too does the body become light and light-headed, says Renevey; deformed, concurs Lawlor, tingly, heavy, even about to evaporate on death. For Bernini, it is impossible to separate thought from feeling, where breath takes on an agency and autonomy of its own in biocognitive feedback. For Saunders in Middle English secular romances, breath and consciousness are more an interweaving. In other words, breath complicates powers to act by distributing pneumatic agencies and forces once we recognise the incorporation of the body and the world, and their transformation. Breathing bodies are trans-incorporating. Hydrocarbons previously extracted from the air and buried in the ground are unearthed, released, and re-materialise in a form of sludgy, soupy thickness, while being absorbed differently, settling in our homes, perhaps as dust, in our lungs, and in our bloodstreams, in and on plants and animals which are suffocated. Spiritual and affecting forces and elements flood through our porous selves.

Inscriptions

Certainly the breath-taking explored in this volume sees breath not only inscribed, but inscribing, through measurement, recording, and distillation in language, speech and writing, and even the techniques of reproducing breaths in the manual practice of cutting audio tape—as the interview with Symmons Roberts mentions—where breath is registered as a gap. What happens to breath when it takes to the air, and when it is taken down again in other forms, even if what is taken down might be an absence, and could appear to be superfluous? Skinner demonstrates the role of the ellipsis in literature in the way that breathlessness is written, which in itself is an index of particular emotions and of high emotional stress and intense moral problems. For Lawlor, breathlessness is similarly etched in the punctuations of 'dashes, commas and semi-colons'. The dot dot dot of the ellipsis marks out other writings through which the essays demonstrate breath has been monitored and tracked, captured in its rhythms.

Others find an excess in the taking down of breath that offers possibilities for breath to be otherwise. For Knoeff, traditions of landscape depicting rural scenes were about a capturing of air: 'Through gazing at such paintings, the beholder would follow a visually cathartic process of purification', in moving from bad air into a purer environment, to the wholesome air in which the painting moved. Knoeff's chapter shows how air was held down, stored, and preserved, surviving within the picture frame of a Dutch landscape painting for recuperative and health-giving purposes. In another way Fuller discovers breath in the 'projective' poetry and associated prose of Charles Olson and in Paul Celan's post-Holocaust poetry and criticism. In Olson, breath could structure writing in such a way that one who reads it might 'hear Olson breathe', while in Celan, the emphasis is not on reading but on speaking: 'the reader must use not the eyes but the mouth', Fuller suggests, to find the meaning of 'living breath' in the poem's structure. Breath might be a pre-conscious form of voice, a kind of pre-verbal meaning, Symmons Roberts postulates. In this sense, *The Life of Breath* encourages us to look beyond breath's taking down as a kind of capture or remote abstraction—a paradox of capture and 'loss' that Fuller and Symmons Roberts contemplate even within the body's own memory. In its very abstraction we are offered therapeutic remedies and creative impulses that animate, that give life.

Wilton reminds us of Cornelia Parker's *Breathless* (2001), a permanent sculptural installation in the V&A Museum (London), occupying the oculus space between two floors of the building. Fifty-four brass instruments have been crushed, pressed flat and arranged to radiate outwards. Parker explains that the instruments are about taking the wind out of the imperialism the institution signifies, a 'last gasp of empire': the instruments 'in my piece are permanently inhaled. They've literally had the wind taken out of them', Parker suggests. In Parker's and Wilton's wider practice, the works express a taking of breath which rejects the vestiges of colonialism through more positive and questioning inscriptions of breath that can be beautiful, such as the condensation marks from the breath of patients in Pulmonary Rehabilitation Unit I at the Royal Brompton Hospital on the surface of the copper that are etched into the material in Wilton's *Breathe*.

(DIS)PLACEMENTS

As a geographer I am particularly drawn to the different spatialisations of breath at play in the volume,[11] spaces which appear multiple and complex rather than singular. We might note that breath connects the clinic, the hospital, the vertiginous, the home, the city and its walls and gates, the church, the factory, and different surfacings and dimensions of space and breath down to, for Symmons Roberts, the meteorological conditions of a regional air that could be thicker, thinner or 'damp', shaping breath and the nasal intonation in voice. For Symmons Roberts, the body has a fleshy geography of breath, and not only in his fiction of a lung for transplant that leaves one body and crosses borders to go into another. Or for Russell, colonial circuits of trade mean breath-modulating, debilitating or stimulating substances like tobacco moving across the world, making the atmospheres of public space in the industrial city that bit denser, more social, and self-reflective, and, more widely, affecting the European arts, literature, and cognitive thought.

Despite Freud's misgivings about breath being too superficial an index, Walker and Rose see how climbing a mountain can reveal breathlessness as an expression of trauma. Corton and Rawcliffe show, with clarity and care, in the geographies of both Victorian London and the medieval city, how admixtures of pollutants and chemicals have been expelled into the skies, streets, and alleyways—Victorian atmospheric pollution; fly-tipping in the Middle Ages. Those spacings express, to some extent, class differences—depending on one's position in relation to the prevailing winds, although they too are contingent. These geographies were also imaginative: as Rawcliffe shows, the stench of medieval London meant the city could be compared with hell.

The book helps upset our assumptions about where breath might be problematic, when it becomes a problem, or comes to express multiple conflicts and violences. At the same time it flips some of our preconceptions of space, so that the home is not the refuge we might think in some of the cases of violence and misogyny shown in the essays. Such instances accord with what we are told about COVID-19's social spread in 'domestic settings', or the terrible but unsurprising rises in domestic violence reported by social care and health visitors during so-called 'lock-down'.

The spacings of breath show us where breath becomes harder, more problematic, more controlled. For Rawcliffe, medieval urban infrastructures re-routed animal carcasses and water courses, constructed new city passages and underground ways, and commenced street cleaning and disposal of other noxious waste so as not to pollute the urban breather. Other essays show how these barriers have been overcome by apparatus designed to help people occupy and breathe in different environments. Garrington's extraordinary essay explores the development and reception of apparatus for navigating and inhabiting the inhospitable spaces of the

Alps or Himalayas. Here men highlight the difficulties of ascending to climates of thinner air, and the problems that breath-supported apparatus meant for a muscular masculinity when the face was obscured. As Garrington puts it, 'masks distanced the mountaineer as a human and from the mountain', with connotations of trench warfare, cowardly and faceless contestation, and anonymity. There are of course further parallels with the face-covering debate today in the time of COVID-19, that range from a political leadership—in some quarters—seeing the mask as a signal of defeat, to, by the mask's very absence, a leadership hiding from political and global problems. The breath mask or covering seems both divisive and to distance us from each other: it becomes antagonistic, it raises suspicions, and of course elsewhere it has long been a regular feature of life living with smog, air pollution, and pandemics. For survivors of domestic abuse, face masks can re-ignite past traumas through the experience of 'hot air, not being able to breathe freely, or feeling smothered', with inhibited mobility as a consequence, a possible continuation of abusive control which has restricted women's freedoms.[12] With or without a mask then, there may be no space, no room to breathe.

Crawley has suggested that what goes 'unremarked is how breathing air is constitutive for flight, for movement, for performance',[13] noting how punishment has evoked dreadful shrieks and outpourings of breath through the violent displacement of flesh. But what of how breath constitutes movement? Crawley parallels the aestheticization of Black bodies with escape in the form of the fugitive slave, with the stigmatisation of the dancing Black body, moving apparently uncontrollably, whooping. While the examples in this book mainly stay with scenes of an intimate scale, of moving bodies at the level of the rise and fall of the chest, Heine reminds us that other movements and mobilities also matter.

Breath also expresses the social exclusions of normative spaces. Riddell's essay highlights particularly 'straight' heteronormative spaces such as the stifling setting of the bourgeois domestic home. But, he also shows, more affirmative and queer spaces are possible too. O'Gorman's essay associates the organ in the church with a kind of building by breath itself through the organ's heavenly exhalations, albeit at the Colston Hall in Bristol named after the philanthropist but also slave trader, Edward Colston. Rather than closing down, breath builds into being acoustic architectures suitable only for the heavens, that would 'take one's breath away', writes O'Gorman. In Riddell's account the precarity of straight spaces is not all-encompassing, and neither, conversely, are the organ's hands or the reliability of the machinery in O'Gorman's. For Riddell, 'breath also becomes a tool for a form of affirmation of queer desire', a more affirming 'respiratory sensuality' (Walker and Rose) perhaps, and a critique of the normative; just as for Garrington, breath became a point of connection between men as a brotherly bond, arm in arm, 'brothers in lung'. Indeed, in a moment of a sermon, in the inhalation and exhalation of a congregation, Crawley finds the possibilities of an aesthetic sociality that 'must be common and used by all, for vitality, for life'.[14]

Conspirations and Common

At a recent Black Lives Matter demonstration in London, a young music student, Geovane Silva, explained to journalists that: 'Racism is racism, and that's what we have to fight. We have to be just one, human beings, we just need to be one. We all breathe the same air, we all bleed the same color'.[15] Silva articulates a primary but intuitive sense that encompasses a much wider cultural understanding of breathing as an ethical injunction: the air is something we share via breath. In breathing together the same air through a sort of 'conspiration', explains Horky in crystal clear terms, we come to share something in common.[16] It is by breathing together that feelings of sympathy, responsibility, and care may flourish. Breath—perhaps through prayer, as Renevey and Symmons Roberts suggest—can be a kind of communication, a oneness with God in a kind of sovereign, vertical, yet intimate relation.

What is so strongly conveyed in *The Life of Breath* is how breath struggles so hard more horizontally, in multiple dimensions, to refuse the separations, enclosures, singularities, and exclusions of air that work on breath itself. Even the taking of breath becomes a loss that might be shared, as Heine shows in relation to M. NourbeSe Philip's *Zong!*: 'someone once breathed for us, gave her breath away; she might have been forced to do so, after her freedom and body were taken; at the outset, our breath was not our own, and probably it never is, never will be'. For Heine, the reformulation of Black exclusion and subjugation articulates a kind of possibility of commoning, of being in common that is founded on the sharing of a lack with others, continuing perhaps the suspicion of infinite breath O'Gorman considers. In other words, breath-taking might be a way to come together and accept difference. For Heine, Philip's suggested 'blueprint for community', the common memory of 'what it was to have someone breathe for us', implies recalling an 'act of acceptance of alterity'. Horky's 'conspiration' is for Heine a 'partage: sharing a lack'.

Throughout this book, the common and the commons treat breath as a shared resource. Breathing as one, in time, or asymmetrically, folds in other agents, maybe non-humans, and even animals. Rosalyn Diprose's notion of 'corporeal generosity' helps us to think this through.[17] Breathing might presuppose, in a giving to the air and to each other without meaning to get anything back, a manner of openness. As Long shows, this relation may be ecological: 'Our life's dependence on breath and breathing puts us into a direct and intrinsic relationship with one of the planet's great powers along with water, solar heat, and light'. Breath helps us, in the terms we started with, commune with collapsing boundaries between bodies, but also scales, from the 'microcosm or individual person to the macrocosm or the world as such'. In a different vein, Tsentourou complicates this claim even further, to see conspiration, 'not only between bodies or between bodies and the cosmos, but also between bodies and texts'. Russell suggests that the imagination of the body as 'chronically unstable' and subject to transformation in traditional societies of the Amazon further nurtures an openness between 'human and non-human, animals, plants and objects'. A substance like tobacco can stimulate a recognition of positionality and a decentering of self in Enlightenment thought and culture.

Corton's essay reminds us that these conspirations include animals, who became the unwilling sufferers of London smog, and in recent times are the subjects of species-level extinctions in an era of climate crisis. Corton's explication moves from the blackening of the lungs of a lion at London zoo, to the clogging of plant stomata by soot, and the coating of evergreens with greasy deposits. In Knoeff's essay on breathscapes, the environment, even representations of landscape—while part of a concerted effort of 'taking the air'—are surely generous in their production of air for us to breathe. In this sense, breath connotes an ethical responsibility to each other and other things. And it was—or seemed—not all bad in London's smog, for, as Long shows in classical thought, breath is often treated as a nutriment alongside food and drink, 'the prime nutriment'. As Corton concludes, London fog was not only the 'breath of death', although it could be a form of 'poison as well as a kind of nutrition in the popular and literary imagination'.

Finally, although the porosity of the body seems in one way to be sealed by the technologies shown in this book to help us breathe, or to filter air out in avoidance of pollutants or disease, we are given promise of a more than visual visage-covering to achieve openness. Walker and Rose give an account of psychoanalysis's study of a mother assuming the breathing pattern of a new-born baby. This was interpreted as a kind of hypnosis, a 'mimetic faculty' to the extent that breath and anxiety are not only co-productive, but that breathing together in respiratory response to the hearing of a breathing rhythm can be a way to generate empathy and sympathy. Craik and Chapman notice something similar in their translations and sympathies across literature and clinical medicine, suggesting that 'what it means to see, hear, and witness the breath (and breathlessness) of another' can carry what they call 'the weight of responsiveness and responsibility involved in all human relations'. Breath is a 'reciprocal or shared exchange'. Breathing together. Breath seems necessary for communion and community. Breathing in this book articulates feelings of affirmative exteriority, pushing oneself outside to become closer to others, to shared air, to blurred boundaries, from the skin's porosity to extra-bodily consciousness, as Renevey explores in relation to Holotropic Breathwork and yogic practices of breathing—an expansion of self via spiritual communion.

The Life of Breath offers the kind of affective, intellectual, critical, and ethical nutrition breath itself has been shown to promise throughout this volume. It sets up several critical incisions into the contextually specific, historico-geographical, and medico-legal, and into concerns of art and the imagination, whether for rule and governance in late antiquity, the medieval city, literature and romance, or clinical practice in the hospital today. It offers transdisciplinary and transhistorical possibilities that place apparently improbable associations, concepts, experiences, and beliefs almost side-by-side. In conspiracy, capacity, (dis)placement, inscription and conspirational commoning, across the essays are found forms of critique and diagnoses of power structures between institutions and people, and of normative conventions of sexuality or class. These modes of critique can reveal excesses offering possibilities of rupture and escape. At the heart of the book is an affirmative, commoning impulse, which conspires to see breath shared, even in its lack, across humans and non-humans.

Notes

1. This included Fredrick Nyman, Kathleen Reynolds, Maryam Golafshani, and Natalie Goodison, and we were expertly organised with imagination and good humour by the facilitator Mary Robson.
2. *In the Wake: On Blackness and Being* (Durham, NC: Duke University Press, 2016).
3. Gary Younge, 'We Can't Breathe', *New Statesman*, 5 June 2020.
4. Public Health England, *Disparities in the Risk and Outcomes of COVID-19* (London, 2020), 6.
5. On these issues see Kamal Jyoti Maji, Mohit Arora, and Anil Kumar Dikshit, 'Premature Mortality Attributable to PM2.5 Exposure and Future Policy Roadmap for 'Airpocalypse' Affected Asian Megacities', *Process Safety and Environmental Protection* 118 (2018), 371–83.
6. D. Asher Ghertner, 'Airpocalypse: Distributions of Life amidst Delhi's Polluted Airs', *Public Culture* 32/1 (2020), 133–62.
7. Ashon T. Crawley, *Blackpentecostal Breath: The Aesthetics of Possibility* (New York: Fordham University Press, 2016), 1.
8. PA Media, 'Grenfell Relative Draws Comparisons between Fire and Covid-19 Response', *The Guardian*, 13 June 2020.
9. Danny Dorling, et al., 'Counting the 21st Century Children of Britain: The Extent of Advantage and Disadvantage', *Twenty-First Century Society* 2/2 (2007), 173–89.
10. Jane Macnaughton, 'Making Breath Visible: Reflections on Relations Between Bodies, Breath and World in the Critical Medical Humanities', *Body & Society* 26/2 (2020), 30–54.
11. Sasha Engelmann, 'Toward a Poetics of Air: Sequencing and Surfacing Breath', *Transactions of the Institute of British Geographers* 40/3 (2015), 430–44.
12. See https://www.womensaid.org.uk/covid-19-domestic-abuse-survivors-experiencing-severe-distress-when-wearing-a-mask-are-exempt/.
13. *Blackpentecostal Breath*, 33.
14. *Blackpentecostal Breath*, 45.
15. Angela Dewan, '"The UK Is Just as Bad as America": Londoners Mourn George Floyd and Demand Black Justice at Home', *CNN World*, 3 June 2020.
16. Luce Irigaray, *The Forgetting of Air in Martin Heidegger*, trans. by Mary Beth Mader, Athlone Contemporary European Thinkers (London: Athlone, 1999).
17. *Corporeal Generosity: On Giving with Nietzsche, Merleau-Ponty, and Levinas*, SUNY Series in Gender Theory (New York: State University of New York Press, 2002).

Selected Bibliography

Crawley, Ashon T. 2016. *Blackpentecostal Breath: The Aesthetics of Possibility*. New York: Fordham University Press.

Diprose, Rosalyn. 2002. *Corporeal Generosity: On Giving with Nietzsche, Merleau-Ponty, and Levinas*. New York: State University of New York Press.

Engelmann, Sasha. 2015. Toward a Poetics of Air: Sequencing and Surfacing Breath. *Transactions of the Institute of British Geographers* 40/3: 430–44.

Irigaray, Luce. 1999. *The Forgetting of Air in Martin Heidegger*. Trans. by Mary Beth Mader. Athlone Contemporary European Thinkers. London: Athlone.

Macnaughton, Jane. 2020. Making Breath Visible: Reflections on Relations between Bodies, Breath and World in the Critical Medical Humanities. *Body & Society* 26/2: 30–54.

Sharpe, Christina. 2016. *In the Wake: On Blackness and Being*. Durham, NC: Duke University Press.

Index

D. Fuller et al. (eds.), *The Life of Breath in Literature, Culture and Medicine*, Palgrave Studies in Literature, Science and Medicine,
https://doi.org/10.1007/978-3-030-74443-4

The manufacturer's authorised representative in the EU is Springer Nature Customer Service Centre GmbH, Europaplatz 3, 69115 Heidelberg, Germany. If you have any concerns regarding our products, please contact ProductSafety@springernature.com

Printed and bound by CPI Group (UK) Ltd, Croydon, CR0 4YY
30/09/2026
02225696-0007